This volume, the third in a sequence that began with *The Theory of Matroids* and *Combinatorial Geometries*, concentrates on the applications of matroid theory to a variety of topics from engineering (rigidity and scene analysis), combinatorics (graphs, lattices, codes and designs), topology and operations research (the greedy algorithm). As with its predecessors, the contributors to this volume have written their articles to form a cohesive account so that the result is a volume that will be a valuable reference for research workers.

T0207181

ENCYCLOPEDIA OF MATHEMATICS AND ITS APPLICATIONS

EDITED BY G.-C. ROTA

Editorial Board

R.S. Doran, J. Goldman, T.-Y. Lam, E. Lutwak

Volume 40

Matroid Applications

ENCYCLOPEDIA OF MATHEMATICS AND ITS APPLICATIONS

ENCYCLOPEDIA OF MATHEMATICS AND ITS APPLICATIONS

Matroid Applications

Edited by

NEIL WHITE
University of Florida

CAMBRIDGE
UNIVERSITY PRESS

CAMBRIDGE UNIVERSITY PRESS
Cambridge, New York, Melbourne, Madrid, Cape Town, Singapore, São Paulo, Delhi

Cambridge University Press
The Edinburgh Building, Cambridge CB2 8RU, UK

Published in the United States of America by Cambridge University Press, New York

www.cambridge.org
Information on this title: www.cambridge.org/9780521119672

First published 1992
This digitally printed version 2009

A catalogue record for this publication is available from the British Library

Library of Congress Cataloguing in Publication data

Matroid applications / edited by Neil White.
p.   cm. — (Encyclopedia of mathematics and its applications ; v. 40)
Includes bibliographical references and index.
ISBN 0–521–38165–7
1. Matroids. I. White, Neil. II. Series.
QA166.6.M38 1992
511′.6—dc20   91–27184   CIP

ISBN 978-0-521-38165-9 hardback
ISBN 978-0-521-11967-2 paperback

CONTENTS

CONTRIBUTORS

Anders Björner
Department of Mathematics
Royal Institute of Technology
S-100 44 Stockholm
SWEDEN

Thomas Brylawski
Department of Mathematics
University of North Carolina
Chapel Hill, NC 27514 USA

M. Deza
Tokyo Institute of Technology
Department of Information Science
Ookayama, Meguro-ku, Tokyo 152
JAPAN

James Oxley
Department of Mathematics
Louisiana State University
Baton Rouge, LA 70803 USA

Ivan Rival
Department of Computer Science
34 Somerset Street E.
University of Ottawa
Ottawa, K1N 9B4 CANADA

J. M. S. Simões-Pereira
Departamento de Matemática
Faculdade de Ciéncias e Tecnologia
Universidade de Coimbra
Apartado 3008
3000 Coimbra, PORTUGAL

Miriam Stanford
Department of Mathematics
Queens University
Kingston, Ontario
K7L 3N6 CANADA

Prof. Walter Whiteley
461 Putney
St. Lambert, Quebec
J4P 3B8 CANADA

Prof. Günter M. Ziegler
Department of Mathematics
Augsburg University
Universitätsstr. 2
D-8900 Augsburg
F.R. of GERMANY

PREFACE

This is the third volume of a series that began with *Theory of Matroids* and continued with *Combinatorial Geometries*. These three volumes are the culmination of more than a decade of effort on the part of the many contributors, potential contributors, referees, the publisher, and numerous other interested parties, to all of whom I am deeply grateful. To all those who waited, please accept my apologies. I trust that this volume will be found to have been worth the wait.

This volume begins with Walter Whiteley's chapter on the applications of matroid theory to the rigidity of frameworks: matroid constructions prove to be rather useful and matroid terminology provides a helpful language for the basic results of this theory. Next we have Deza's chapter on the beautiful applications of matroid theory to a special aspect of combinatorial designs, namely perfect matroid designs. In Chapter 3, Oxley considers ways of generalizing the matroid axioms to infinite ground sets, and Simões-Pereira's chapter on matroidal families of graphs discusses other ways of defining a matroid on the edge set of a graph than the usual graphic matroid method. Next, Rival and Stanford consider two questions on partition lattices. These lattices are a special case of geometric lattices and the inclusion of this chapter will provide a lattice-theoretic perspective which has been lacking in much current matroid research (but which seems alive and well in *oriented* matroids). Then we have the comprehensive survey by Brylawski and Oxley of the Tutte polynomial and Tutte–Grothendieck invariants. These express the deletion–contraction decomposition that is so important within matroid theory and some of its important applications, namely graph theory and coding theory. Björner describes the homology and shellability properties of several simplicial complexes associated with a matroid; the complexes of independent sets, of broken circuits, and of chains in the geometric lattice. This chapter and the previous one constitute a study of the deepest known matroid invariants. We conclude with an exposition by Björner and Ziegler of greedoids, a generalization of matroids that embody the greedy algorithm and hence are very useful in operations research.

University of Florida   Neil L. White

1

Matroids and Rigid Structures

WALTER WHITELEY

Many engineering problems lead to a system of linear equations – a represented matroid – whose rank controls critical qualitative features of the example (Sugihara, 1984; 1985; White & Whiteley, 1983). We will outline a selection of such matroids, drawn from recent work on the rigidity of spatial structures, reconstruction of polyhedral pictures, and related geometric problems.

For these situations, the combinatorial pattern of the example determines a sparse matrix pattern that has both a generic rank, for general 'independent' values of the non-zero entries, and a geometric rank, for special values for the coordinates of the points, lines, and planes of the corresponding geometric model. Increasingly, the generic rank of these examples has been studied by matroid theoretic techniques. These geometric models provide nice illustrations and applications of techniques such as matroid union, truncation, and semimodular functions. The basic unsolved problems in these examples highlight certain unsolved problems in matroid theory. Their study should also lead to new results in matroid theory.

1.1. Bar Frameworks on the Line – the Graphic Matroid

We begin with the simplest example, which will introduce the vocabulary and the basic pattern. We place a series of distinct points on a line, and specify certain *bars* – pairs of joints which are to maintain their distance – defining a *bar framework on the line*. We ask whether the entire framework is 'rigid' – i.e. does any motion of the joints along the line, preserving these distances, give all joints the same velocity, acceleration, etc.? Clearly a framework has an *underlying graph* $G = (V, E)$, with a vertex v_i for each joint p_i and an undirected edge $\{i, j\}$ for each bar $\{p_i, p_j\}$. In fact, we describe the framework as $G(\mathbf{p})$, where G is a graph without multiple edges or loops, and $\mathbf{p}$ is an assignment of points p_i to the vertices v_i. If this graph is not connected,

Figure 1.1.

(a)

(b)

(c)

then each component can move separately in the framework, and the framework is not rigid (Figure 1a). Conversely, a connected graph always leads to a rigid framework (Figure 1.1b), since each bar ensures that its two joints have the same motion on the line. This gives an informal proof of the following result.

1.1.1. Proposition. *A bar framework $G(\mathbf{p})$ on the line is rigid if and only if the underlying graph G is connected.*

To extract a matrix, we make this argument a little more formal. Assume the joints p_i move along smooth paths $p_i(t)$. The length of a bar $\|p_i(t) - p_j(t)\|$, and its square, remain constant. If we differentiate, this condition becomes

$$\frac{d}{dt}[p_i(t) - p_j(t)]^2 = [p_i(t) - p_j(t)][p'_i(t) - p'_j(t)] = 0.$$

At $t = 0$, this is written $(p_i - p_j)(p'_i - p'_j) = 0$. If we have distinct joints on the line, so that $(p_i - p_j) \neq 0$, this simplifies to $(p'_i - p'_j) = 0$.

With this in mind, we define an *infinitesimal motion* of a bar framework on the line $G(\mathbf{p})$ as an assignment of a velocity u_i along the line to each joint p_i such that $u_i - u_j = 0$ for each bar $\{v_i, v_j\}$. For example, consider the framework in Figure 1.1c. The four bars lead to four equations in the unknowns $\mathbf{u} = (u_1, u_2, u_3, u_4, u_5)$:

$$\begin{bmatrix} 1 & 0 & -1 & 0 & 0 \\ 0 & 1 & -1 & 0 & 0 \\ 0 & 0 & 1 & -1 & 0 \\ 0 & 0 & 1 & 0 & -1 \end{bmatrix} \begin{bmatrix} u_1 \\ u_2 \\ u_3 \\ u_4 \\ u_5 \end{bmatrix} = \begin{bmatrix} 0 \\ 0 \\ 0 \\ 0 \end{bmatrix}.$$

In general, this system of linear equations is written $R(G, \mathbf{p}) \times \mathbf{u}^t = 0$, where the *rigidity matrix* $R(G, \mathbf{p})$ has a row for each edge of the graph and a column for each vertex, and $\mathbf{u}^t$ is the transpose of the vector of velocities. We note that $R(G, \mathbf{p})$ is the transpose of the usual matrix representation for the graph over the reals: the rows are independent in $R(G, \mathbf{p})$ if and only if the corresponding edges are a forest (an independent set of edges in the cycle matroid of the graph).

A *trivial infinitesimal motion* is the derivative of a rigid motion of the line – i.e. a translation with all velocities equal. These form a one-dimensional subspace of the solutions. *An infinitesimally rigid framework on the line* has only these trivial infinitesimal motions, so the rigidity matrix has rank $|V| - 1$. This rank corresponds to a spanning tree on the vertices, or a basis for the cycle matroid of the complete graph on $|V|$ vertices. This proves the following infinitesimal version of Proposition 1.1.1.

1.1.2. Proposition. *A bar framework $G(p)$ on the line is infinitesmially rigid if and only if the underlying graph G is connected.*

1.2. Bar Frameworks in the Plane

A *bar framework in the plane* is a graph $G = (V, E)$ and an assignment $\mathbf{p}$ of points $\mathbf{p}_i \in \mathbb{R}^2$ to the vertices v_i such that $\mathbf{p}_i \neq \mathbf{p}_j$ if $\{i, j\} \in E$. If we differentiate the condition that bars have constant length in any smooth motion, we have

$$\frac{d}{dt}[\mathbf{p}_i(t) - \mathbf{p}_j(t)]^2 = [\mathbf{p}_i(t) - \mathbf{p}_j(t)] \cdot [\mathbf{p}_i'(t) - \mathbf{p}_j'(t)] = 0.$$

Accordingly, an *infinitesimal motion* of plane bar framework is an assignment $\mathbf{u}$ of velocities $\mathbf{u}_i \in \mathbb{R}^2$ to the joint such that

$$(\mathbf{p}_i - \mathbf{p}_j) \cdot (\mathbf{u}_i - \mathbf{u}_j) = 0 \quad \text{for each } \{i, j\} \in E.$$

A plane bar framework is *infinitesimally rigid* if all infinitesimal motions are *trivial*: $\mathbf{u}_i = \mathbf{s} + \beta(\mathbf{p}_i)^{\perp}$, where $\mathbf{s}$ is a fixed translation vector, $(x, y)^{\perp} = (y, -x)$ rotates the vector 90° counterclockwise, and $\beta(\mathbf{p}_i)^{\perp}$ represents a rotation about the origin. (These infinitesimal rotations and translations are the derivatives of smooth rigid motions of the plane.)

The system of equations for an infinitesimal motion has the form $R(G, \mathbf{p}) \times \mathbf{u}^t = 0$, where the *rigidity matrix* $R(G, \mathbf{p})$ now has a row for each edge of the graph and two columns for each vertex. The row for edge $\{i, j\}$ has the form

$$[0 \quad 0 \quad \dots \quad 0 \quad 0 \quad \mathbf{p}_i - \mathbf{p}_j \quad 0 \quad 0 \quad \dots \quad 0 \quad 0 \quad \mathbf{p}_j - \mathbf{p}_i \quad 0 \quad 0 \quad \dots \quad 0 \quad 0]$$

Figure 1.2.

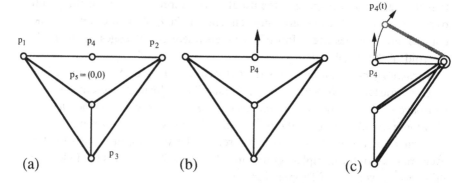

(a)   (b)   (c)

1.2.1. Example. Consider the frameworks in Figure 1.2. The framework of Figure 1.2a gives the rigidity matrix

$$
\begin{array}{c}
\{1,3\}\\
\{1,4\}\\
\{1,5\}\\
\{2,3\}\\
\{2,4\}\\
\{2,5\}\\
\{3,5\}
\end{array}
\left[
\begin{array}{cccccccccc}
x_1-x_3 & y_1-y_3 & 0 & 0 & x_3-x_1 & y_3-y_1 & 0 & 0 & 0 & 0\\
\tfrac{1}{2}(x_1-x_2) & \tfrac{1}{2}(y_1-y_2) & 0 & 0 & 0 & 0 & \tfrac{1}{2}(x_2-x_1) & \tfrac{1}{2}(y_2-y_1) & 0 & 0\\
x_1 & y_1 & 0 & 0 & 0 & 0 & 0 & 0 & -x_1 & -y_1\\
0 & 0 & x_2-x_3 & y_2-y_3 & x_3-x_2 & x_3-y_2 & 0 & 0 & 0 & 0\\
0 & 0 & \tfrac{1}{2}(x_2-x_1) & \tfrac{1}{2}(y_2-y_1) & 0 & 0 & \tfrac{1}{2}(x_1-x2) & \tfrac{1}{2}(y_1-y2) & 0 & 0\\
0 & 0 & x_2 & y_2 & 0 & 0 & 0 & 0 & -x_2 & -y_2\\
0 & 0 & 0 & 0 & x_3 & y_3 & 0 & 0 & -x_3 & -y_3
\end{array}
\right]
$$

The rows of this matrix are dependent and have rank 6. This leaves a $(10-6=4)$-dimensional space of infinitesimal motions, including the non-trivial motion shown in Figure 1.2b, which assigns zero velocity to all joints but $\mathbf{p}_4$, and gives $\mathbf{p}_4$ a velocity perpendicular to the bars at $\mathbf{p}_4$. Thus the framework is not infinitesimally rigid.

The infinitesimal motion is not the derivative of some smooth path for the vertices. The framework is *rigid* – all smooth paths, or even continuous paths, give frameworks congruent to the original framework. Figure 1.2c gives a similar framework which has the same infinitesimal motions, but is not rigid.

These examples show that there is a difference in the plane between rigid frameworks and infinitesimally rigid frameworks. A *non-rigid* plane framework will have an analytic path of positions $\mathbf{p}(t) = (\ldots, \mathbf{p}_i(t), \ldots)$, with all bar lengths of $\mathbf{p}(t)$ the same as bars in $\mathbf{p}(0)$, but $\mathbf{p}(t)$ not congruent to $\mathbf{p}(0)$, for all $0 < t < 1$ (Figure 1.2c). The first non-zero derivative of this path will be a non-trivial infinitesimal motion. However, the converse is false: many infinitesimal motions are not the derivative of an analytic path (recall Figure 1.2b). For any framework, the independence of the rows of the rigidity matrix induces a matroid on the edges of the graph. If 'rigidity' in a particular plane framework were used to define an independence structure on the edges of a graph, this need not be a matroid (see Exercise 1.6). Therefore, we will restrict ourselves, throughout this chapter, to the simpler concepts of infinitesimal motions and infinitesimal rigidity.

The space of trivial plane infinitesimal motions has dimension 3, for frameworks with at least two distinct joints. This space can be generated by two translations in distinct directions and a rotation about any fixed point. Thus an infinitesimally rigid framework with more than two joints will have an $|E|$ by $2|V|$ rigidity matrix of rank $2|V| - 3$. Our basic problem is to determine which graphs G allow this matrix to have rank $2|V| - 3$ for at least some plane frameworks $G(\mathbf{p})$.

The independence structure of the rows of the rigidity matrix defines a matroid on the edges of the complete graph on the vertices. This matroid depends on the positions of the joints. If we vary the positions there are 'generic' positions that give a maximal collection of independent sets (for example, positions where the coordinates are algebraically independent real numbers). At these positions we have the *generic rigidity matroid for $|V|$ vertices in the plane*.

1.2.2. Example. Consider the framework in Figure 1.3a. With vertices as indicated we have the rigidity matrix

$$
\begin{array}{c}
(a_1, b_1) \\
(a_1, b_2) \\
(a_1, b_3) \\
(a_2, b_1) \\
(a_2, b_2) \\
(a_2, b_3) \\
(a_2, b_1) \\
(a_3, b_2) \\
(a_3, b_3)
\end{array}
\begin{bmatrix}
1 & 0 & 0 & 0 & 0 & 0 & -1 & 0 & 0 & 0 & 0 & 0 \\
0 & 1 & 0 & 0 & 0 & 0 & 0 & 0 & 0 & -1 & 0 & 0 \\
1 & 1 & 0 & 0 & 0 & 0 & 0 & 0 & 0 & 0 & -1 & -1 \\
0 & 0 & -1 & 0 & 0 & 0 & 1 & 0 & 0 & 0 & 0 & 0 \\
0 & 0 & -2 & 1 & 0 & 0 & 0 & 0 & 2 & -1 & 0 & 0 \\
0 & 0 & -1 & 1 & 0 & 0 & 0 & 0 & 0 & 0 & 1 & 1 \\
0 & 0 & 0 & 0 & 1 & -2 & -1 & 2 & 0 & 0 & 0 & 0 \\
0 & 0 & 0 & 0 & 0 & 1 & 0 & 0 & 0 & -1 & 0 & 0 \\
0 & 0 & 0 & 0 & 1 & -1 & 0 & 0 & 0 & 0 & -1 & 1
\end{bmatrix}
$$

The graph of the framework has $|E| = 2|V| - 3$, so the framework is infinitesimally rigid if and only if the rows are independent. This independence can be checked by deleting the final three columns and seeing that the determinant of the 9×9 submatrix is non-zero. This framework is infinitesimally rigid and the graph is generically rigid, and generically independent.

Consider any realization with distinct joints $\mathbf{a}_1, \mathbf{a}_2, \mathbf{a}_3, \mathbf{b}_1, \mathbf{b}_2, \mathbf{b}_3$ on a unit circle centred at the origin (Figure 1.3b). This has a non-trivial 'in–out' infinitesimal motion (Figure 1.3c):

$$\text{for joints } \mathbf{a}_i \text{ take the velocity } \mathbf{a}_i' = \mathbf{a}_i;$$

$$\text{for joints } \mathbf{b}_j \text{ take the velocity } \mathbf{b}_j' = -\mathbf{b}_j.$$

Figure 1.3.

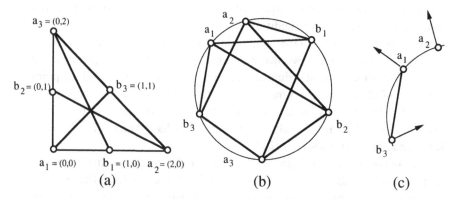

$$\begin{array}{ccc}\text{(a)} & \text{(b)} & \text{(c)}\end{array}$$

These velocities preserve the length of all bars $(\mathbf{a}_i, \mathbf{b}_j)$, since

$$(\mathbf{a}_i - \mathbf{b}_j) \cdot (\mathbf{a}'_i - \mathbf{b}'_j) = (\mathbf{a}_i - \mathbf{b}_j) \cdot (\mathbf{a}_i + \mathbf{b}_j) = (\mathbf{a}_i) \cdot (\mathbf{a}_i) - (\mathbf{b}_j) \cdot (\mathbf{b}_j) = 1 - 1 = 0.$$

This infinitesimal motion is non-trivial. Letting $\theta \neq 0$ be the angle between the unit vectors $\mathbf{a}_1$ and $\mathbf{a}_2$, we show that the distance $(\mathbf{a}_1 - \mathbf{a}_2)$ is changing instantaneously:

$$(\mathbf{a}_1 - \mathbf{a}_2) \cdot (\mathbf{a}'_1 - \mathbf{a}'_2) = (\mathbf{a}_1) \cdot (\mathbf{a}_1) - 2(\mathbf{a}_1) \cdot (\mathbf{a}_2) + (\mathbf{a}_2) \cdot (\mathbf{a}_2) = 1 + 1 - 2\cos\theta > 0.$$

Thus this special position is not generic (see Exercise 1.9).

We want to characterize the graphs of *isostatic plane frameworks* – minimal infinitesimally rigid frameworks in the sense that removing any one bar introduces a non-trivial infinitesimal motion. These graphs, of size $|E| = 2|V| - 3$, are the bases of the generic rigidity matroid 'of the complete graph' on the set of vertices.

Thus an isostatic framework corresponds to a row basis for the rigidity matrix of any infinitesimally rigid framework extending the framework. The independence of such a set of edges is determined by maximal minors of the rigidity matrix. This independence is *generic* in the sense that these minors are non-zero polynomials in the positions $\mathbf{p}_i$. If such a polynomial is non-zero for some position $G(\mathbf{p})$, then almost all $\mathbf{q} \in \mathbb{R}^{2|V|}$ give isostatic frameworks $G(\mathbf{q})$ (see Section 1.7).

More surprisingly, for points where this matrix and all its minors have the maximal rank achieved for $\mathbf{q} \in \mathbb{R}^{2|V|}$, infinitesimal rigidity and any reasonable form of local rigidity actually coincide (see, for example, Exercise 1.7).

We note that throughout this chapter the generic matroids defined on the complete graph of $|V|$ vertices are symmetric on the vertices – any permutation of the vertices does not change the independence of a set of edges. As a convention, we write the attached vertices for a subset of edges E' as V'.

1.2.3. Theorem. *For a graph G, with at least two vertices, the following are equivalent conditions:*

  (i) *G has some positions G(**p**) as an isostatic plane framework;*
  (ii) $|E| = 2|V| - 3$ *and for all proper subsets of edges E' incident with vertices V', $|E'| \leqslant 2|V'| - 3$;*
  (iii) *adding any edge to E (including doubling an edge) gives an edge set covered by two edge-disjoint spanning trees.*

Proof. (i) $\Rightarrow$ (ii): For an isostatic plane framework $G(\mathbf{p})$ on at least two vertices, the rows of the rigidity matrix have rank $|E| = 2|V| - 3$. If any proper subset of edges has $|E'| > 2|V'| - 3$, the corresponding rows are dependent. Since $G(\mathbf{p})$ is independent, we conclude that $|E'| \leqslant 2|V'| - 3$ for all proper subsets.

(ii) $\Leftrightarrow$ (iii): The count $f(E') = 2|V'| - 3$ defines a non-decreasing semimodular function on sets of edges, which is non-negative on non-empty sets (see Exercise 1.1). This semimodular function defines a matroid by the standard property:

E is independent if and only if $|E'| \leqslant f(E')$ for all proper subsets E'. (1.1)

This count has the form $f(E') = (2|V'| - 2) - 1$ which shows that the matroid for f is a Dilworth truncation of the matroid defined by the semimodular function $g(E') = 2(|V'| - 1)$. In turn, the semimodular function g represents a matroid union of two copies of the matroid given by the semimodular function $h(E') = |V| - 1$ (the cycle matroid of the graph). Thus a graph is independent in the matroid of f if and only if adding any edge (including doubling an edge) gives a graph covered by two edge-disjoint forests.

Before we prove (iii) $\Rightarrow$ (i), we need a lemma about a simpler matrix that has rank $2|V| - 2$ (matching the function g). For a graph $G = (V, E)$, including possible multiple edges, a 2-*frame* $G(\mathbf{d})$ is an assignment of directions $\mathbf{d}_e \in \mathbb{R}^2$ to the edges. An *infinitesimal motion of the 2-frame* $G(\mathbf{d})$ is an assignment of velocities $\mathbf{u}_i \in \mathbb{R}^2$ to the vertices such that

$$\mathbf{d}_e \cdot (\mathbf{u}_i - \mathbf{u}_j) = 0 \quad \text{for every edge } e \text{ joining } v_i \text{ and } v_j \ (i < j).$$

This system of equations defines the *rigidity matrix* $R(G, \mathbf{d})$ for the 2-frame.

1.2.4. Lemma. *The rows of the rigidity matrix of a generic 2-frame G(**d**) are independent if and only if G is the union of two edge-disjoint forests.*

Proof. Take the two forests F_1 and F_2. For all edges in the first forest, we assign the direction $(1, 0)$. For all edges in the second forest, we assign the direction $(0, 1)$. If we reorder the rows and columns of this rigidity matrix, placing all second columns of vertices to the right, and all rows for the second forest at the bottom, we have a pattern:

$$\begin{bmatrix} [F_1] & [0] \\ [0] & [F_2] \end{bmatrix}$$

where $[F_1]$ and $[F_2]$ are the standard matrices representing the two forests over the reals. Thus the blocks $[F_1]$ and $[F_2]$ have non-zero minors on all their rows, and the entire matrix also has a non-zero minor on all the rows. We conclude that the rows are independent.

Conversely, if the rows are independent, we can reorder the columns as above. The independence guarantees a non-zero minor on all the rows. Using a Laplace expansion on the two blocks, there are non-zero minors on complementary sets of rows. Each of these non-zero minors on the matrix representing the graphic matroid must correspond to a forest, as required.   □

Proof of Theorem 1.2.3 (continued). (iii)$\Rightarrow$(i) Assume that adding any edge to E (including doubling an edge) gives an edge set covered by two edge-disjoint forests. We show that this gives independent rows in the rigidity matrix for some (almost all) choices of the points.

Take a 2-frame $G(\mathbf{d})$ with algebraically independent directions for the edges. By our assumption, adding any edge (or doubling any edge) between any pair of vertices gives an independent 2-frame E^*. Therefore, the 2-frame on E has an infinitesimal motion $\mathbf{u}_{ij}$ that has different velocities on the two vertices of the added edge. Taking linear combinations of these $\mathbf{u}_{ij}$ there is an infinitesimal motion $\mathbf{u}$ that assigns distinct velocities $\mathbf{u}_i = (s_i, t_i)$ to each of the vertices of G.

To create the independent framework $G(\mathbf{p})$, we set $\mathbf{p}_i = (-t_i, s_i)$. Since $\mathbf{u}_i \neq \mathbf{u}_j$, we have $\mathbf{p}_i \neq \mathbf{p}_j$ for each edge $\{i, j\}$, as required in a framework. We claim that the rigidity matrix of $G(\mathbf{p})$ has rows parallel to the rows of the original 2-frame $G(\mathbf{d})$. Clearly

$$(\mathbf{p}_i - \mathbf{p}_j) \cdot (\mathbf{u}_i - \mathbf{u}_j) = (-t_i + t_j, s_i - s_j) \cdot (s_i - s_j, t_i - t_j) = 0 = \mathbf{d}_e \cdot (\mathbf{u}_i - \mathbf{u}_j).$$

Since $\mathbf{u}_i - \mathbf{u}_j \neq \mathbf{0}$, this means $(\mathbf{p}_i - \mathbf{p}_j) = \beta_e \mathbf{d}_e$ for some non-zero scalar β_e. The rigidity matrix of the framework is equivalent to the rigidity matrix of the independent 2-frame. We conclude that a set of edges satisfying condition (1.1) has been realized as an independent (therefore isostatic) bar framework. (The infinitesimal motion $\mathbf{u}$ of the 2-frame is a rotation of this framework around the origin.)

This completes the proof.   □

Figure 1.4 shows some other examples of the graphs of isostatic frameworks in the plane (Figure 1.4a) and graphs of circuits in the plane generic rigidity matroid (Figure 1.4b) with $|E| = 2|V| - 2$, and $|E'| \leqslant 2|V'| - 3$ for proper subsets.

The semimodular count of Theorem 1.2.3 (ii) converts to a criterion for graphs of infinitesimally rigid frameworks. We state the theorem without proof.

1.2.5. Corollary. (*Lovász & Yemini, 1982*) *A graph has realizations as an infinitesimally rigid plane framework if and only if for every partition of the* *edges into non-empty subsets* ($\ldots E^j \ldots$), *with vertices* V^j *incident with the edges* E^j,

$$\sum_j (2|V^j| - 3) \geqslant 2|V| - 3.$$

Figure 1.5 gives a simple example of a 5-connected graph, in a vertex sense, which is never infinitesimally rigid by this criterion: take the eight K_5 graphs as sets of the partition, and all other edges as singletons and apply Corollary 1.2.5. However, every graph which is 6-connected in a vertex sense is generically rigid (Exercise 1.16).

The dependence of rows in the rigidity matrix, or *dependence of bars in the framework*, also has a physical interpretation. A *self-stress* on a framework is an assignment of scalars ω_{ij} to the bars $\{\mathbf{p}_i, \mathbf{p}_j\}$ such that for each joint $\mathbf{p}_i$ there is an equilibrium (Figures 1.6a, b):

Figure 1.4. (a) $|E| = 2|V| - 3$, $|E'| \leqslant 2|V'| - 3$; (b) $|E| = 2|V| - 2$, $|E'| \leqslant 2|V'| - 3$.

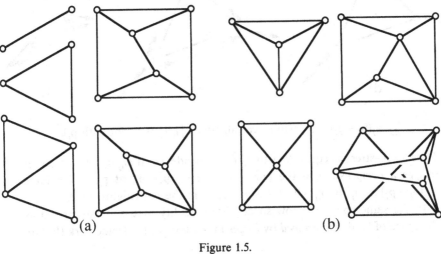

(a)   (b)

Figure 1.5.

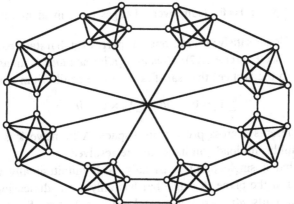

Figure 1.6.

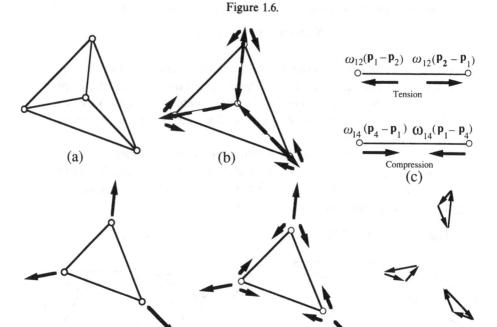

(a)   (b)

$$\omega_{12}(P_1 - P_2) \quad \omega_{12}(P_2 - P_1)$$

Tension

$$\omega_{14}(P_4 - P_1) \quad \omega_{14}(P_1 - P_4)$$

Compression

(c)

(d)   (e)   (f)

$$\sum_j \omega_{ij}(\mathbf{p}_j - \mathbf{p}_i) = 0 \quad \text{(sum over all edges } \{i, j\} \text{ incident with } \mathbf{p}_i).$$

Thus a self-stress is equivalent to a row dependence. If $\omega_{ij} < 0$, we interpret this as a *compression* in the bar – a force $\omega_{ij}(\mathbf{p}_j - \mathbf{p}_i)$ at $\mathbf{p}_i$, and a force $\omega_{ij}(\mathbf{p}_i - \mathbf{p}_j)$ at $\mathbf{p}_j$. If $\omega_{ij} > 0$, this is a *tension* in the bar (Figure 1.6c).

In the same spirit, the row space of the rigidity matrix is interpreted as the space of loads $\mathbf{L}_i$ *resolved* by forces in the bars of the framework (Figure 1.6d, e, f):

$$\mathbf{L}_i + \sum_j \omega_{ij}(\mathbf{p}_j - \mathbf{p}_i) = 0 \quad \text{(sum over all edges } \{i, j\} \text{ incident with } \mathbf{p}_i).$$

These resolved loads satisfy an additional property of global static equilibrium, defined below (see Exercise 1.17). Thus an *equilibrium load* is an assignment $\mathbf{L}_i$ of vectors to the vertices that satisfies the three equilibrium equations

$$\sum_i \mathbf{L}_i = 0 \quad \text{and} \quad \sum_i \mathbf{L}_i \times \mathbf{p}_i = 0$$

where $\times$ represents a cross product in 3-space. A framework is *statically rigid* if all equilibrium loads on its joints are resolved.

We note that a single point is trivially both infinitesimally rigid and statically rigid in the plane. A single bar has only a one-dimensional space of equilibrium loads: $\alpha(\mathbf{p}_1 - \mathbf{p}_2)$ at $\mathbf{p}_1$ and $\alpha(\mathbf{p}_2 - \mathbf{p}_1)$ at $\mathbf{p}_2$. Since these are resolved, the bar is statically rigid and infinitesimally rigid. Moreover, the equilibrium loads on any framework with at least two distinct joints form a space of dimension $2|V| - 3$, and all resolved loads are equilibrium loads (Exercise 1.17). These observations prove the following.

1.2.6. Theorem. *For a bar framework* $G(\mathbf{p})$ *in the plane, on at least two joints, the following are equivalent conditions:*
  (i) *the framework is statically rigid;*
  (ii) *the rigidity matrix has rank* $2|V| - 3$;
  (iii) *the framework is infinitesimally rigid.*

Such dual static concepts arise for the infinitesimal mechanics of most types of structure, giving dual static and infinitesimal mechanical theories for the row and column ranks of the corresponding matrix. Of course the non-zero self-stresses correspond to the dependences of the underlying matroid on the edges.

1.3. Plane Stresses and Projected Polyhedra

There is a classical theorem that interprets self-stresses in a framework with a planar graph as instructions for building a 'spatial polyhedron' with this projection. The ideas of the proof are simple, but we need a few topological definitions for our 'spherical polyhedra'.

We begin with a simple class of graphs, for which we can associate the face, edge, and vertex structures of abstract polyhedra. A *planar graph* is a graph G that can be drawn in the plane with the edges as straight lines, disjoint except at shared vertices. A graph is *vertex 3-connected* (or 3-connected for short) if deleting any two vertices leaves the graph connected. A planar drawing of a 3-connected planar graph $G = (V; E)$ divides up the plane into a set of polygonal discs, and a single infinite region with a polygonal hole. These polygonal regions oriented counterclockwise, and the infinite region, with its polygonal hole oriented counterclockwise, form the *oriented faces*, $(..., f^h, ...)$, of the associated polyhedron. (Note that a theorem of graph theory says that, up to reversing all these orientations, the face structure for a 3-connected planar graph is unique.) We call this collection of vertices, edges, and faces a *3-connected abstract spherical polyhedron* $S = (V, E, F)$.

1.3.1. Example. Figures 1.7a and 1.7b show two planar drawings of the same 3-connected graph. The reader can check that they define the same oriented faces. Figure 1.7c shows a planar graph that is not 3-connected: removing vertices a and c, or b and c, separates the graph. The results of this section actually extend to these 2-connected planar graphs – see Exercise 1.18.

Figure 1.7.

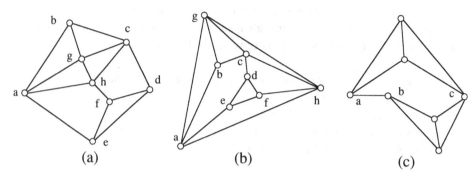

(a)   (b)   (c)

We are interested in projections of spatial polehydra. We assume that we are projecting vertically down into the xy-plane, and that no face is a vertical plane. Thus the face planes can be written $Ax + By + C = z$, or (A, B, C) for short.

A *spatial polyhedron* $S(Q, q)$ is an assignment of a plane $Q^h = (A^h, B^h, C^h)$ to each face f^h, and points $q_j = (x_j, y_j, z_j)$ to each vertex v_j, such that for each vertex v_k on a face f^i,

$$A^i x_k + B^i y_k + C^i = z_k.$$

For convenience, we also assume that for any edge $\{v_j, v_k\}$ of the polyhedron, $q_j \neq q_k$. We do not assume that the two faces at an edge have to have distinct planes, nor that the faces have to be topological discs, convex, etc. in space.

A spatial polyhedron $S(Q, q)$ *projects* to a plane framework $G(p)$ if G is the graph of edges and vertices of S, and if for each vertex with $q_j = (x_j, y_j, z_j)$, $p_j = (x_j, y_j)$.

1.3.2. Example. What geometric condition(s) on the drawings in Figure 1.8 correspond to a projected polyhedron? If a spatial polyhedron has three distinct planes for the faces $aa'b'b$, $bb'c'c$, and $cc'a'a$, these planes intersect in a single point, even in the projection. Therefore the three edges aa', bb', and cc' must have a common point. The reader can mentally reconstruct the triangular prism over Figure 1.8a. Figure 1.8b is a generic drawing, and the only spatial polyhedra are trivial, with all faces in the same plane.

Figure 1.8c corresponds to a 'degenerate' spatial polyhedron, which assigns one plane to face abc, and another plane to all other faces. This object satisfies our definition and is non-trivial.

Consider a plane framework on the graph of the polyhedron $G(p)$, with distinct vertices for each edge. A polyhedron in space that projects to this framework assigns a plane $z = A^h x + B^h y + C^h$ to each face f^h. If two faces f^h, f^i share an edge over the line through $\{p_j, p_k\}$, we have the conditions

Figure 1.8.

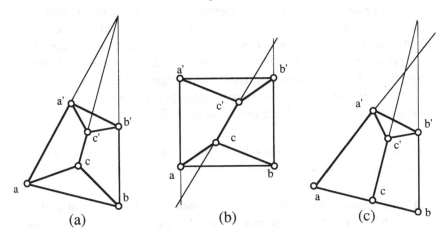

(a)   (b)   (c)

$$A^h x_j + B^h y_j + C^h = A^i x_j + B^i y_j + C^i$$

and

$$A^h x_k + B^h y_k + C^h = A^j x_k + B^j y_k + C^i$$

or equivalently, by Cramer's rule,

$$(A^h, B^h, C^h) - (A^i, B^i, C^i) = \omega_{jk}((y_j - y_k), (x_k - x_j), (x_j y_k - x_k y_j))$$

for some scalar ω_{jk}. We use the orientation of the polyhedron to associate an order of the faces (h, i) with the order of the vertices (j, k), following order of vertices in the cycle of face h. This orientation also associates (i, h) with (k, j). This generates a consistent scalar $\omega_{jk} = \omega_{kj}$ for each edge. The two planes are distinct if and only if $\omega_{jk} \neq 0$.

If we follow a cycle of faces and edges around a vertex v_j of the spatial polyhedron, the first face equals the last face. This gives the equation

$$\sum \omega_{jk}((y_j - y_k), (x_k - x_j), (x_j y_k - x_k y_j)) = (0, 0, 0) \quad \text{(sum over the cycle of edges)}.$$

These scalars look suspiciously like a self-stress on the framework $G(\mathbf{p})$.

1.3.3. Maxwell's theorem. *Given a spatial spherical polyhedron $S(\mathbf{Q}, \mathbf{q})$ projecting to the plane framework $G(\mathbf{p})$, there is a self-stress of the framework $G(\mathbf{p})$ such that an edge separates faces in distinct planes in the polyhedron if and only if the self-stress has a non-zero scalar on the edge.*

Proof. We have seen that a spatial polyhedron leads to scalars ω_{jk} satisfying

$$\sum \omega_{jk}(y_j - y_k, x_k - x_j, x_j y_k - x_k y_j) = (0, 0, 0) \quad \text{(sum over the edges at } v_j).$$

Clearly this implies that

$$\sum \omega_{jk}(x_k - x_j, y_k - y_j) = (0, 0) \quad \text{(sum over the edges at } v_j).$$

Therefore the projection of a polyhedron gives a self-stress that is non-zero on exactly the edges separating distinct planes. $\qquad\square$

1.3.2. Example (continued). By counting vertices and edges, it is found that the graph of Figure 1.8 is generically independent – hence a generic framework (Figure 1.8b) is not the projection of a spatial polyhedron. If the three edges aa', bb', and cc' have a common point, Figure 1.8a, the spatial triangular prism proves that the framework has a self-stress that is non-zero on all edges.

Figure 1.8c has a self-stress, non-zero only on edges of the collinear triangle abc. This corresponds to the degenerate spatial polyhedron that assigns one plane to face abc and another plane to all other faces.

We have a converse to Maxwell's theorem. The next lemma presents a crucial property of all self-stresses, which we need in the proof of this converse.

1.3.4. Cut lemma. *Given a framework* $G(\mathbf{p})$ *with a self-stress* ω, *and a set of bars* $(\mathbf{a}_i, \mathbf{b}_i)$ *such that they separate the graph, with all* $\mathbf{a}_i$ *in one component, and all* $\mathbf{b}_i$ *in other components, then*

$$\sum \omega_i(\mathbf{b}_i - \mathbf{a}_i) = \mathbf{0} \quad and \quad \sum \omega_i(\mathbf{b}_i \times \mathbf{a}_i) = \mathbf{0}.$$

Proof. We have used the shorthand $(\mathbf{b}_i \times \mathbf{a}_i)$ for the third component of $(\mathbf{b}_i, 0) \times (\mathbf{a}_i, 0)$. For a cut set that isolates a single vertex v_j, we have

$$(0, 0, 0) \times (x_j, y_j, 0) = \sum_k \omega_{jk}(x_k - x_j, y_k - y_j, 0) \times (x_j, y_j, 0)$$

$$= \sum_k \omega_{jk}(0, 0, x_j y_k - x_k y_j).$$

as required. Therefore:

$$\sum_k \omega_{jk}(y_j - y_k, x_k - x_j, x_j y_k - x_k y_j) = (0, 0, 0) \quad \text{(sum over the edges at } v_j\text{)}.$$

Take the component created by the cut set. If we add these equations over all vertices of the component, the terms cancel on all edges joining two ends in the component containing all the $\mathbf{a}_i$, and leave the required sums on the cut set $(\mathbf{a}_i, \mathbf{b}_i)$.   □

1.3.5. Converse of Maxwell's theorem. *Given a self-stress on the plane framework* $G(\mathbf{p})$ *built on the graph of a spherical polyhedron, there is a spatial polyhedron projecting to* $G(\mathbf{p})$ *such that an edge separates faces in distinct planes in the polyhedron if and only if the corresponding self-stress is non-zero on the edge.*

Proof. Assume that we have a self-stress $(\ldots, \omega_{jk}, \ldots)$ on the framework $G(\mathbf{p})$. We choose the arbitrary plane $z = A^0 x + B^0 y + C^0$ for an initial face f^0. For every other face f^n we take a face–edge path Π from f^0 to f^n and define the plane for f^n by

$$(A^n, B^n, C^n) = (A^0, B^0, C^0) + \sum_\Pi \omega_{jk}(y_j - y_k, x_k - x_j, x_j y_k - x_k y_j)$$

We must prove that this answer is well defined, that is, independent of the path used.

Consider two different paths Π and Π' from f^0 to f^n. Together, the path Π, $-\Pi'$ forms a face–edge cycle on the polyhedron (i.e. a cycle on the dual graph). Such a cycle is the disjoint union of simple cycles, and each simple cycle on a spherical polyhedron is a cut set satisfying Lemma 1.3.4. Summing over all these oriented cycles simultaneously:

$$\sum_\Pi \omega_{jk}(x_j - x_k, y_j - y_k, x_j y_k - x_k y_j)$$

$$- \sum_{\Pi'} \omega_{jk}(x_j - x_k. y_j - y_k, x_j y_k - x_k y_i) = (0, 0, 0).$$

It is a simple exercise to rewrite the order and the signs of coordinates in this sum, giving:

$$\sum_\Pi \omega_{jk}(y_j - y_k, x_k - x_j, x_j y_k - x_k y_j) = \sum_{\Pi'} \omega_{jk}(y_j - y_k, x_k - x_j, x_j y_k - x_k y_j).$$

This proves that the plane for face f^n is well defined and that these spatial planes have the required incidences in space over the vertices to form a spatial polyhedron projecting to $G(\mathbf{p})$.

For any edge $-jk$ of the polyhedron, we still have

$$(A^h, B^h, C^h) - (A^i, B^i, C^i) = \omega_{jk}(y_j - y_k, x_k - x_j, x_j y_k - x_k y_j).$$

Therefore the edge separates two distinct planes if and only if there is a non-zero scalar in the self-stress.   □

1.3.2. Example (completed). Consider the frameworks in Figure 1.8. We saw that Figure 1.8a was the projection of a polyhedron with distinct faces – hence it has a self-stress. Figure 1.8b is not the projection of a polyhedron – so it must be independent, by Theorem 1.3.5.

The self-stresses on a framework $G(\mathbf{p})$ actually form a vector space. The polyhedra that project to this framework, with non-vertical faces and with a face f^0 held in a fixed plane $z = 0$, also form a vector space. (Note that we include polyhedra with all faces coplanar, corresponding to the zero self-stress.) Theorems 1.3.3 and 1.3.5 actually give an isomorphism of these spaces.

1.3.6. Corollary. *The space of spatial polyhedra over a plane drawing of the graph $G(\mathbf{p})$, with the plane $z = 0$ assigned to a fixed face f^0, is isomorphic to the vector space of self-stresses of $G(\mathbf{p})$.*

1.4. Bar Frameworks in 3-space

The definitions of a framework, the rigidity matrix, and infinitesimal rigidity in the plane easily generalize to bar frameworks in $\mathbb{R}^3$. Trivial infinitesimal motions now have the form $\mathbf{u}_i = \mathbf{s} + \mathbf{r} \times \mathbf{p}_i$. These infinitesimal motions form a space of dimension 6 for any framework whose joints are not all on a line, a space generated by three non-coplanar translations and by three non-coplanar rotations about any point. Thus a spatial framework with more than two joints is infinitesimally rigid if and only if the rigidity matrix has rank $3|V| - 6$.

For statics in 3-space we have the similar modification that equilibrium loads satisfy the six equations $\sum_i \mathbf{L}_i = \mathbf{0}$ and $\sum_i \mathbf{L}_i \times \mathbf{p}_i = \mathbf{0}$. These equations are independent provided that the joints are not all collinear. The row space of the rigidity matrix still corresponds to the resolved equilibrium loads.

We also observe that a framework on the line with more than two joints cannot be statically or infinitesimally rigid even in the plane. These observations lead directly to

1.4.1. Theorem. *For bar framework $G(\mathbf{p})$ in space, with at least three joints, the following are equivalent conditions:*
  (i) *$G(\mathbf{p})$ is infinitesimally rigid;*
  (ii) *the rigidity matrix $RG(\mathbf{p})$ has rank $3|V| = 6$;*
  (iii) *$G(\mathbf{p})$ is statically rigid.*

1.4.2. Corollary. *If a framework in space with at least three joints is isostatic then $|E| = 3|V| - 6$ and $|E'| \leqslant 3|V'| - 6$ for all subsets of at least two edges.*

When we seek the converse to this corollary, we run into a fundamental problem: some graphs that satisfy the count are never realized as isostatic frameworks. Figure 1.9a gives a simple example, which clearly has a non-trivial motion based on rotations around the 'hinge' ab. If we take two copies of K_5, which are generic circuits in space (Figure 1.9b), and do a circuit exchange on the common bar ab we obtain this generic circuit. Tay (1986) surveys other generic circuits in 3-space that are not infinitesimally rigid (Figure 1.9c, Exercise 1.22).

An alternative vision of our difficulty comes from the semimodular function $f(E) = 3|V| - 6$ on sets of edges. Since $f(E)$ is negative on single edges, this function does not directly define a matroid. Instead we ask the more subtle question:

Is there a maximal matroid $M(G)$ on graphs that has rank $3|V| - 6$ on all complete graphs of at least three vertices?

Maximal means that if E'' is independent in some matroid on graphs satisfying this count, then E'' is independent in M.

Figure 1.9.

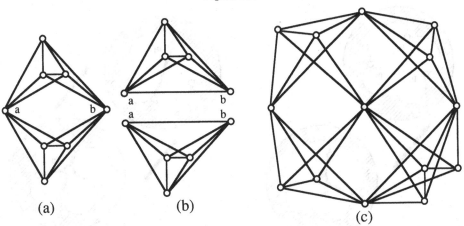

(a)   (b)   (c)

1.4.3. Graver's conjecture. *The rigidity matroid in 3-space is maximal: any independent set in another matroid on finite graphs with rank $3|V| - 6$ on each complete graph of $|V| > 2$ vertices will be independent in the rigidity matroid.*

This question can be approached through a set of inductive constructions building up the number of vertices in an isostatic set. The graph of an isostatic framework in 3-space that is larger than a triangle has a vertex of valence 3, 4, or 5. The edge set satisfies $|E| = 3|V| - 6$ and $|E'| \leqslant 3|V'| - 6$ for all subsets of at least three vertices. Since each edge is incident with two vertices, we have $2|E| = \sum_i \text{val}(v_i) < 6|V|$. There must be vertices of valence less than 6. On the other hand, if we remove vertex v_i attached to edges E_i, leaving at least three vertices V', we have

$$|E'| \leqslant 3|V'| - 6 \quad \text{and} \quad |E' \cup E_i| = |E'| + |E_i| = 3|V| - 6 = 3|V'| - 6 + 3.$$

Thus, if the graph is bigger than a triangle, each vertex must have valence of at least 3.

Accordingly, the inductive techniques describe the addition and deletion of 3-, 4-, or 5-valent vertices. Unfortunately the techniques we outline remain incomplete for 5-valent vertices in 3-space.

1.4.4. Proposition. *If an isostatic bar framework in 3-space has three non-collinear joints p_1, p_2, p_3, then inserting a 3-valent joint p_0 out of this plane with bars (p_0, p_1), (p_0, p_2), (p_0, p_3) creates an isostatic bar framework.*

Conversely, given any isostatic framework in 3-space with a 3-valent joint, then deleting this joint with its bars leaves an isostatic bar framework (Figure 1.10a).

Figure 1.10.

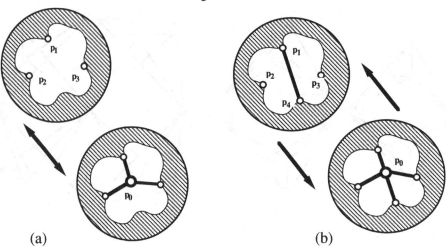

(a)   (b)

Proof. (1) For convenience, we assume that the added vertex v_0 is at the left side of the rigidity matrix, and the added bars are on the top:

$$\begin{bmatrix} \mathbf{p}_1 - \mathbf{p}_0 & \mathbf{p}_0 - \mathbf{p}_1 & 0 & 0 & \cdots & 0 \\ \mathbf{p}_2 - \mathbf{p}_0 & 0 & \mathbf{p}_0 - \mathbf{p}_2 & 0 & \cdots & 0 \\ \mathbf{p}_3 - \mathbf{p}_0 & 0 & 0 & \mathbf{p}_0 - \mathbf{p}_3 & \cdots & 0 \\ 0 & \cdots & \cdots & \cdots & \cdots & \cdots \end{bmatrix}.$$

It is clear that this vertex adds rank 3 to the lower rigidity matrix if and only if the three vectors $\mathbf{p}_1 - \mathbf{p}_0$, $\mathbf{p}_2 - \mathbf{p}_0$, and $\mathbf{p}_3 - \mathbf{p}_0$ are independent, that is, the four points are not coplanar.

(2) Conversely, removing a 3-valent joint keeps the independence of the rows of the rigidity matrix and keeps $|E'| = |E| - 3 = 3|V| - 9 = 3|V'| - 6$. □

1.4.5. Proposition. *If an isostatic bar framework in 3-space has four non-coplanar joints $\mathbf{p}_1$, $\mathbf{p}_2$, $\mathbf{p}_3$, $\mathbf{p}_4$ and a bar $(\mathbf{p}_1, \mathbf{p}_4)$ then removing this bar and inserting a joint $\mathbf{p}_0$ with bars $(\mathbf{p}_0, \mathbf{p}_1)$, $(\mathbf{p}_0, \mathbf{p}_2)$, $(\mathbf{p}_0, \mathbf{p}_3)$, $(\mathbf{p}_0, \mathbf{p}_4)$ creates an isostatic bar framework for almost all positions of $\mathbf{p}_0$.*

Conversely, given any isostatic bar framework in 3-space with a 4-valent joint $\mathbf{p}_0$ attached to four non-coplanar joints, then there is a bar $(\mathbf{p}_i, \mathbf{p}_j)$ among the vertices adjacent to $\mathbf{p}_0$ such that deleting this joint and inserting $(\mathbf{p}_i, \mathbf{p}_j)$ leaves an isostatic bar framework (Figure 1.10b).

Proof. (1) Again, we add the vertex v_0 at the left side of the rigidity matrix, with three bars to $\mathbf{p}_1$, $\mathbf{p}_2$, and $\mathbf{p}_3$, so that $\mathbf{p}_1 - \mathbf{p}_0$ is parallel to $\mathbf{p}_1 - \mathbf{p}_4$:

$$\begin{bmatrix} \mathbf{p}_1 - \mathbf{p}_0 & \mathbf{p}_0 - \mathbf{p}_1 & 0 & 0 & \cdots & 0 \\ \mathbf{p}_2 - \mathbf{p}_0 & 0 & \mathbf{p}_0 - \mathbf{p}_2 & 0 & \cdots & 0 \\ \mathbf{p}_3 - \mathbf{p}_0 & 0 & 0 & \mathbf{p}_0 - \mathbf{p}_3 & \cdots & 0 \\ 0 & \mathbf{p}_4 - \mathbf{p}_1 & 0 & 0 & \mathbf{p}_1 - \mathbf{p}_4 & 0 \\ 0 & \cdots & \cdots & \cdots & \cdots & \cdots \end{bmatrix}.$$

It is clear that this vertex adds rank 3 to the lower rigidity matrix if and only if the three vectors $\mathbf{p}_1 - \mathbf{p}_0$, $\mathbf{p}_2 - \mathbf{p}_0$, and $\mathbf{p}_3 - \mathbf{p}_0$ are independent. However, we can now row reduce with row I and row IV to get the desired independent rigidity matrix:

$$\begin{bmatrix} \mathbf{p}_1 - \mathbf{p}_0 & \mathbf{p}_0 - \mathbf{p}_1 & 0 & 0 & \cdots & 0 \\ \mathbf{p}_2 - \mathbf{p}_0 & 0 & \mathbf{p}_0 - \mathbf{p}_2 & 0 & \cdots & 0 \\ \mathbf{p}_3 - \mathbf{p}_0 & 0 & 0 & \mathbf{p}_0 - \mathbf{p}_3 & \cdots & 0 \\ \mathbf{p}_4 - \mathbf{p}_0 & 0 & 0 & 0 & \mathbf{p}_0 - \mathbf{p}_4 & 0 \\ 0 & \cdots & \cdots & \cdots & \cdots & \cdots \end{bmatrix}.$$

Since we have $|E'| = |E| + 4 - 1 = 3|V| - 3 = 3|V'| - 6$, the framework is isostatic.

(2) Conversely, deleting a 4-valent joint $\mathbf{p}_0$ leaves

$$|E'| = = |E| - 4 = 3|V| - 10 = 3|V'| - 7$$

independent bars. If some bar $(\mathbf{p}_i, \mathbf{p}_j)$, $1 \leqslant i, j \leqslant 4$, is independent, the proof is complete. If not, all six such bars are already implicit – as linear combinations of the remaining bars. These six implicit rows, plus the original four to $\mathbf{p}_0$ give a set of $|E'| = 10$ rows on five joints. This set with $|E'| > 3|V'| - 6$ rows is dependent – contradicting the assumption that the original framework was isostatic. $\square$

A 5-valent joint can be replaced by two bars in several different patterns (Figure 1.11a, Exercise 1.24). We have *conjectured* that the replacement procedures of Figure 1.11b preserve isostatic frameworks for generic positions of the joints. This would imply the basic Conjecture 1.4.3.

In the absence of a general characterization, we present another inductive technique that covers some higher valent vertices in triangulated polyhedra (Figure 1.12). Given a graph G with a vertex v_1 incident to the edges $(1, 2), (1, 3), (1, 4), \ldots, (1, k), (1, k + 1), \ldots, (1, k + m)$, then a *vertex split* of v_1 on the edges $(1, 2), (1, 3)$ is the modified graph with edges $(1, 4), \ldots, (1, k)$ removed and an added vertex p_0 incident with new edges $(0, 1)$, $(0, 2), (0, 3), (0, 4), \ldots, (0, k)$.

1.4.6. Vertex splitting theorem. *Given an independent framework $G(\mathbf{p})$ in 3-space with a joint $\mathbf{p}_1$, and all incident bars $(1, 2), (1, 3), (1, 4), \ldots, (1, k),$*

$(1, k+1), \ldots, (1, k+m)$ with $\mathbf{p}_1 - \mathbf{p}_2$ *not parallel to* $\mathbf{p}_1 - \mathbf{p}_3$, *then for any* $k + m \geqslant 2$, *the new framework on the vertex split on* $(1, 2), (1, 3)$ *is independent for almost all positions for the new joint* $\mathbf{p}_0$.

Proof. We choose, as a limiting initial case, to add $\mathbf{p}_0$ at $\mathbf{p}_1$, with the 'bar' $(\mathbf{p}_0, \mathbf{p}_1)$ assigned a direction $\mathbf{d}_{01}$ not in the plane of

$$\mathbf{d}_{02} = \mathbf{p}_2 - \mathbf{p}_0 = \mathbf{p}_2 - \mathbf{p}_1 = \mathbf{d}_{12} \quad \text{and} \quad \mathbf{d}_{03} = \mathbf{p}_3 - \mathbf{p}_0 = \mathbf{p}_3 - \mathbf{p}_1 = \mathbf{d}_{13}.$$

This is not a bar framework, but a '3-frame' with a rigidity matrix – and it is the limit of frameworks with variable $\mathbf{p}_0$ (see Exercise 1.43). This creates the following matrix:

$$\begin{bmatrix}
\mathbf{d}_{01} & -\mathbf{d}_{01} & 0 & 0 & 0 & \ldots \\
\mathbf{d}_{02} & 0 & -\mathbf{d}_{02} & 0 & 0 & \ldots \\
\mathbf{d}_{03} & 0 & 0 & -\mathbf{d}_{03} & 0 & \ldots \\
0 & \mathbf{d}_{12} & -\mathbf{d}_{12} & 0 & 0 & \ldots \\
0 & \mathbf{d}_{13} & 0 & -\mathbf{d}_{13} & 0 & \ldots \\
0 & \mathbf{p}_4 - \mathbf{p}_1 & 0 & 0 & \mathbf{p}_1 - \mathbf{p}_4 & \ldots \\
0 & \ldots & & & &
\end{bmatrix}.$$

Figure 1.11.

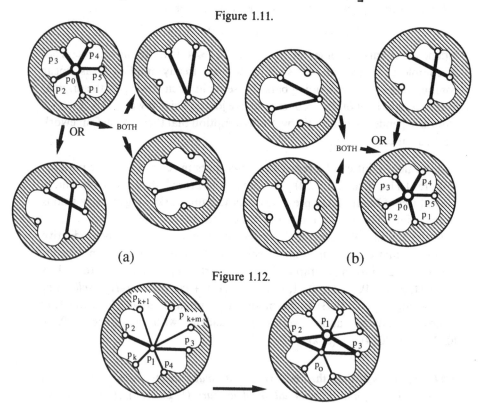

(a)   (b)

Figure 1.12.

It is clear that this vertex adds rank 3 to the lower rigidity matrix if and only if the three vectors $\mathbf{d}_{01}$, $\mathbf{d}_{02} = \mathbf{p}_2 - \mathbf{p}_0$, and $\mathbf{d}_{03} = \mathbf{p}_3 - \mathbf{p}_0$ are independent. If we add row IV to row II, as well as row V to row III, we get the row equivalent matrix

$$\begin{bmatrix} \mathbf{d}_{01} & -\mathbf{d}_{01} & 0 & 0 & 0 & \cdots \\ \mathbf{d}_{02} & -\mathbf{d}_{02} & 0 & 0 & 0 & \cdots \\ \mathbf{d}_{03} & -\mathbf{d}_{03} & 0 & 0 & 0 & \cdots \\ 0 & \mathbf{d}_{12} & -\mathbf{d}_{12} & 0 & 0 & \cdots \\ 0 & \mathbf{d}_{13} & 0 & -\mathbf{d}_{13} & 0 & \cdots \\ 0 & \mathbf{p}_4 - \mathbf{p}_1 & 0 & 0 & \mathbf{p}_1 - \mathbf{p}_4 & \cdots \\ 0 & \cdots & & & & \end{bmatrix}.$$

With these first three independent rows, and with $\mathbf{p}_0 - \mathbf{p}_4 = \mathbf{p}_1 - \mathbf{p}_4$, we can row reduce on rows such as row VI, to change the non-zero entries to:

$$[\mathbf{p}_4 - \mathbf{p}_0 \quad 0 \quad 0 \quad 0 \quad \mathbf{p}_0 - \mathbf{p}_4 \quad 0]$$

We can then return the top five rows to their original pattern, giving a matrix for the graph of the vertex split. Finally since this matrix is independent, it remains independent if we move $\mathbf{p}_0$ to any nearby position with $\mathbf{p}_1 - \mathbf{p}_0 = \mathbf{d}_{01}$. This completes the proof. $\qquad\qquad\qquad\qquad\qquad\qquad\qquad\qquad\square$

The converse is not true: if we split a vertex on two edges, we can turn a generically non-rigid graph (Figure 1.13a) into a generically rigid graph (Figure 1.13b). Figure 1.13c proves the generic rigidity of Figure 1.13b by a sequence of vertex splits from a tetrahedron which include one application of Proposition 1.4.5.

A selected vertex split on a triangulated surface will create a larger triangulated surface of the same topological type. Let v be a vertex of a triangulation of a 2-manifold (Figure 1.14a). The *star* of v is the union of the closed triangles meeting v, and the *link* of v is the simple polygon of edges in the star that do not contain v. The link can be split into two simple paths M and N, meeting at vertices x and y (Figure 1.14b). A *vertex split* of the triangulation at the vertex v consists of replacing the triangles of the star of v by two new vertices v' and v'' and the triangles determined by v' and the edges of M, v'' and the edges of N, and the two triangles $v'v''x$, $v'v''y$ (Figure 1.14c).

This vertex split in a polyhedron can be reversed by shrinking an edge $\mathbf{p}_0\mathbf{p}_1$ and identifying the vertices – provided that this edge is not part

Figure 1.13.

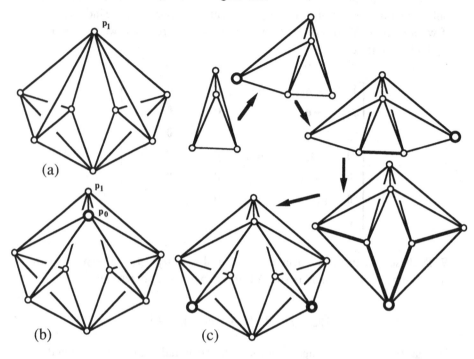

(a)

(b)

p_1
p_0

(c)

Figure 1.14.

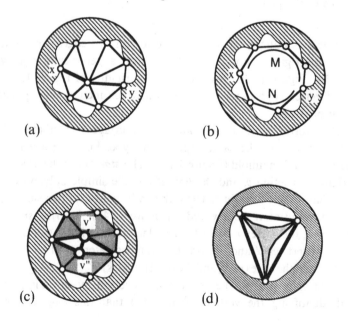

(a)   (b)

(c)   (d)

of a non-facial triangle of the surface (Figure 1.14d) and the 2-manifold is not the tetrahedron. We illustrate this process of vertex splitting within the class of triangulated surfaces with a quick proof that all triangulated spherical polyhedra are generically rigid.

1.4.7. Generic Cauchy's theorem. *Every triangulated spherical polyhedron is generically rigid in 3-space.*

Proof. We prove this by induction on $|V|$, the number of vertices in the triangulated sphere. If $|V| = 4$, we have a tetrahedron, which is generically rigid in 3-space.

Assume that generic rigidity holds for all triangulated spheres with $< |V|$ vertices and take any triangulate sphere with $|V| > 4$ vertices. We first show that there is a 'shrinkable edge'. Any *non-shrinkable edge e* is part of a non-facial triangle separating a disc of N interior vertices from the remaining triangulated sphere. We claim that there is a shrinkable edge inside each such disc. If $N = 1$, this vertex is 3-valent, and every interior edge is shrinkable.

Assume a shrinkable edge occurs for $N = k$. If a disc has $k + 1$ vertices, then there is an interior edge e. If e is shrinkable, the proof is complete. If not, e is part of a non-facial triangle – which separates the disc, with the previous boundary triangle on one side. The other component is a triangulated disc with fewer than k interior vertices, which must contain a shrinkable edge.

Take any shrinkable edge in the triangulated sphere with $|V|$ vertices. Shrinking this edge creates a smaller generically rigid triangulated sphere. Splitting back along this edge, Theorem 1.4.6 shows that the initial triangulated sphere is also generically rigid in 3-space.   □

In the exercises we offer a few other examples that can be proven by similar inductions (Figure 1.15, Exercises 1.23, 1.27). However, the central problem of characterizing the generic rigidity matroid in 3-space remains unsolved.

Figure 1.15.

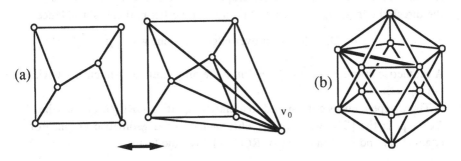

1.5. A Matroid from Splines

Consider a planar drawing of a 3-connected planar graph, with a triangular boundary. In Section 1.3 we saw that this drawing is the projection of a spatial polyhedron if and only if the drawing has a self-stress as a bar framework. If we ignore the exterior triangle (which is always a plane), such a spatial polyhedron is a globally continuous (C^0) function that is piecewise linear over the plane regions (the 'faces'). In approximation theory such a function is called a *bivariate C_1^0-spline*. Maxwell's theorem and its converse give a correspondence between C_1^0-splines and self-stresses for planar dissections of a boundary triangle.

Consider the analogous problem of finding globally C^1 functions over the planar drawing, which are piecewise second degree polynomials – *bivariate C_2^1-splines*. We shall extract a central matroid for these splines that has a deep analogy to the rigidity matroid for spatial frameworks. We outline how the techniques of sections 1.3 and 1.4 extend to this new matroid to characterize the dimension of this space of C_2^1-splines, for generic planar drawings with a triangular boundary.

A *plane geometric graph* $G(\mathbf{p})$ is a graph $G = (V; E)$, with no loops or multiple edges, and an assignment $\mathbf{p}: V \rightarrow \mathbb{R}^2$, $\mathbf{p}(v_i) = \mathbf{p}_i$, giving distinct points to the two ends of each edge. For our final application to splines these graphs will be chosen to be planar embeddings of planar graphs. However, in the general theory, including the construction of some special splines, the graph need not even be planar. A *generic* geometric graph uses algebraically independent real numbers for the coordinates of the vertices.

For each edge we have a line with equation

$$p_{ij}x + q_{ij}y + r_{ij} = (y_j - y_i)x + (x_i - x_j)y + (x_jy_i - x_iy_j) = 0.$$

The *double-line vector* for a directed edge (i, j) is the vector

$$\mathbf{D}_{ij}^2 = (p_{ij}^2, 2p_{ij}q_{ij}, q_{ij}^2) = ((y_j - y_i)^2, 2(y_j - y_i)(x_i - x_j), (x_i - x_j)^2) \quad \text{for } i < j,$$

and

$$\mathbf{D}_{ji}^2 = -\mathbf{D}_{ij}^2 \quad \text{for } i < j.$$

A *conic-dependence* of a geometric graph is an assignment of scalars ω_{ij} to the directed edges, $\omega_{ji} = \omega_{ij}$, such that at each vertex $\mathbf{p}_i$ there is a balance:

$$\sum_j \omega_{ij}\mathbf{D}_{ij}^2 = 0 \quad \text{(sum over edges incident with } \mathbf{p}_i\text{)}.$$

A geometric graph is *conic-independent* if the only conic-dependence is trivial (all $\omega_{ij} = 0$).

These dependences correspond to a matrix equation $\omega RC(G, \mathbf{p}) = \mathbf{0}$, where $RC(G, \mathbf{p})$ is the $|E| \times 3|V|$ *conic-rigidity matrix* for a geometric graph. For edges $(1, 2)$ and (i, j) the rows of $RC(G, \mathbf{p})$ have the form:

$$[\mathbf{D}_{12}^2 \quad -\mathbf{D}_{12}^2 \quad 0\,0\,0 \quad \ldots \quad 0\,0\,0 \quad \ldots \quad 0\,0\,0 \quad \ldots \quad 0\,0\,0]$$
$$[0\,0\,0 \quad 0\,0\,0 \quad 0\,0\,0 \quad \ldots \quad \mathbf{D}_{ij}^2 \quad \ldots \quad -\mathbf{D}_{ij}^2 \quad \ldots \quad 0\,0\,0]$$

1.5.1. Example. Consider a plane triangle with vertices $\mathbf{p}_1 = (0, 0)$, $\mathbf{p}_2 = (0, 1)$, $\mathbf{p}_3 = (1, 0)$. The edges of this triangle have the double line vectors

$$\mathbf{D}_{12}^2 = ((y_2 - y_1)^2,\ 2(y_2 - y_1)(x_1 - x_2),\ (x_1 - x_2)) = (1, 0, 0)$$
$$\mathbf{D}_{13}^2 = (0, 0, 1) \qquad\qquad\qquad \mathbf{D}_{23}^2 = (1, 2, 1)$$

This gives the conic-rigidity matrix

$$\begin{bmatrix} 1 & 0 & 0 & -1 & 0 & 0 & 0 & 0 & 0 \\ 0 & 0 & 1 & 0 & 0 & 0 & 0 & 0 & -1 \\ 0 & 0 & 0 & 1 & 2 & 1 & -1 & -2 & -1 \end{bmatrix}.$$

Clearly this matrix is independent with rank 3, so there is only the trivial conic-dependence.

Accordingly, the solution set to the homogeneous system $RC(G, \mathbf{p}) \times \mathbf{u}^t = \mathbf{0}$ has dimension 6. The reader can check that this solution space is spanned by the six independent vectors:

$$\mathbf{T}_1 = (1,0,0,\ 1,0,0,\ 1,0,0)$$
$$\mathbf{T}_2 = (0,1,0,\ 0,1,0,\ 0,1,0)$$
$$\mathbf{T}_3 = (0,0,1,\ 0,0,1,\ 0,0,1)$$
$$\mathbf{T}_4 = (0,0,0,\ 0,1,0,\ 2,0,0)$$
$$\mathbf{T}_5 = (0,0,0,\ 0,0,2,\ 0,1,0)$$
$$\mathbf{T}_6 = (0,0,0,\ 0,0,1,\ 1,0,0)$$

The reader can also check that a collinear triangle gives a conic-rigidity matrix of rank 2: all the $\mathbf{D}_{ij}^2$ are parallel vectors, and there is a conic-dependence.

1.5.2. Proposition. *The conic-rigidity matrix of a plane geometric graph $G(\mathbf{p})$, non-collinear, has a maximum rank $3|V| - 6$.*

Proof. We show that the homogeneous system $RC(G, \mathbf{p}) \times \mathbf{u}^t = \mathbf{0}$ has a six-dimensional space of solutions generated by:

$$\mathbf{T}_1 = (1,0,0,\ 1,0,0,\ \ldots,\ 1,0,0)$$
$$\mathbf{T}_2 = (0,1,0,\ 0,1,0,\ \ldots,\ 0,1,0)$$
$$\mathbf{T}_3 = (0,0,1,\ 0,0,1,\ \ldots,\ 0,0,1)$$
$$\mathbf{T}_4 = (2x_1,y_1,0,\ 2x_2,y_2,0,\ \ldots,\ 2x_{|V|},y_{|V|},0)$$
$$\mathbf{T}_5 = (0,x_1,2y_1,\ 0,x_2,2y_2,\ \ldots,\ 0,x_{|V|},2y_{|V|})$$
$$\mathbf{T}_6 = (x_1^2,x_1\,y_1,y_1^2,\ x_2^2,\ x_2\,y_2,y_2^2,\ \ldots,\ x_{|V|}^2,x_{|V|}\,y_{|V|},y_{|V|}^2).$$

The first three vectors, which correspond to the 'translations' in 3-space, are clearly independent solutions. We check that the last three, which correspond to 'rotations', are solutions for a typical edge joining points (x_1, y_1) and (x_2, y_2) along the corresponding line

$$(y_2 - y_1)x + (x_1 - x_2)y + (x_2 y_1 - x_1 y_2) = 0.$$

For T_4 we have

$$2x_1(y_2 - y_1)^2 + (y_1)2(y_2 - y_1)(x_1 - x_2) + 0$$
$$- [2x_2(y_1 - y_2)^2 + (y_2)2(y_1 - y_2)(x_2 - x_1) + 0]$$
$$= 2(y_2 - y_1)$$
$$\times [x_1(y_2 - y_1) + y_1(x_1 - x_2) + x_2(y_1 - y_2) + y_2(x_2 - x_1)]$$
$$= 2(y_2 - y_1)(0) = 0.$$

A similar check works for T_5.

For T_6 we have

$$x_1^2(y_2 - y_1)^2 + 2x_1 y_1(y_2 - y_1)(x_1 - x_2) + y_1^2(x_1 - x_2)^2$$
$$- [x_2^2(y_1 - y_2)^2 + 2x_2 y_2(y_1 - y_2)(x_2 - x_1) + y_2^2(x_2 - x_1)^2]$$
$$= (x_1^2 - x_2^2)(y_2 - y_1)^2 + (2x_1 y_1 - 2x_2 y_2)(y_2 - y_1)(x_1 - x_2)$$
$$+ (y_1^2 - y_2^2)(x_1 - x_2)^2$$
$$= (y_2 - y_1)(x_1 - x_2)$$
$$\times [(x_1 + x_2)(y_2 - y_1) + (2x_1 y_1 - 2x_2 y_2) - (y_1 + y_2)(x_1 - x_2)]$$
$$= (y_2 - y_1)(x_1 - x_2)(0) = 0.$$

In Example 1.5.1 we checked that these vectors were independent solutions for the three vertices of the triangle. The same check holds for any three non-collinear points.   □

By analogy with spatial frameworks, we call a plane geometric graph with at least three vertices *conic-rigid* if the conic-rigidity matrix has this maximal rank $3|V| - 6$. A single edge with distinct vertices is also conic-rigid, as is a single point with no edges. Example 1.5.1 shows that a non-collinear triangle is conic-rigid. All the inductive techniques developed for frameworks in 3-space extend to conic-rigidity (see below, and Exercises 1.35, 1.36) and some techniques only conjectured for frameworks apply to conic-rigidity. (Exercise 1.37).

With these similarities in counts and in inductive techniques, we can ask if the conic-rigidity matrix $RC(G, \mathbf{p})$ for a generic embedding in the plane, and the rigidity matrix $R(G, \mathbf{q})$ for a generic embedding in 3-space, create the same matroid on the edges.

1.5.3. Conjecture. *A set E of edges is independent for generic embeddings of the graph as a bar framework in 3-space if and only if the set is conic-independent for generic embeddings as a geometric graph in the plane.*

If the rigidity matroid is maximal, as conjectured in the previous section, Conjecture 1.5.3 claims that the generic conic-rigidity matroid for a graph is also maximal.

We have shown that a triangle is generically conic-rigid. We will show that the graph of any triangulated sphere is also generically conic-rigid, following the approach in section 1.4.

1.5.4. Vertex splitting theorem for conic-rigidity. *Given a conic independent geometric graph $G(\mathbf{p})$ in the plane with a joint $\mathbf{p}_1$, and all incident bars $(1, 2), (1, 3), (1, 4), .., (1, k), (1, k+1), ..., (1, k+m)$ with $\mathbf{p}_1 - \mathbf{p}_2$ not parallel to $\mathbf{p}_1 - \mathbf{p}_3$, then for any $k + m \geq 2$, the new geometric graph on the vertex split on $(1, 2), (1, 3)$ is conic-independent for almost all positions for the new vertex $\mathbf{p}_0$.*

Proof. We follow the proof of Theorem 1.4.6, choosing, as a limiting initial case, to add $\mathbf{p}_0$ at $\mathbf{p}_1$, with the 'line' $\mathbf{p}_0$, $\mathbf{p}_1$ assigned a direction $\mathbf{D}_{01}$ so that $\mathbf{D}_{02}^2 = \mathbf{D}_{12}^2$, $\mathbf{D}_{03}^2 = \mathbf{D}_{13}^2$, and $\mathbf{D}_{01}^2$ are independent 3-vectors. This creates the conic-rigidity matrix

$$\begin{bmatrix} \mathbf{D}_{01}^2 & \mathbf{D}_{01}^2 & 0 & 0 & 0 & \cdots \\ \mathbf{D}_{02}^2 & 0 & -\mathbf{D}_{02}^2 & 0 & 0 & \cdots \\ \mathbf{D}_{03}^2 & 0 & 0 & 0 & -\mathbf{D}_{03}^2 & \cdots \\ 0 & \mathbf{D}_{12}^2 & -\mathbf{D}_{12}^2 & 0 & 0 & \cdots \\ 0 & \mathbf{D}_{13}^2 & 0 & -\mathbf{D}_{13}^2 & 0 & \cdots \\ 0 & \mathbf{D}_{14}^2 & 0 & 0 & -\mathbf{D}_{14}^2 & \cdots \\ 0 & \cdots & & & & \end{bmatrix}.$$

It is clear that this vertex adds rank 3 to the previous conic-rigidity matrix if and only if the three vectors $\mathbf{DE}_{01}^2$, $\mathbf{D}_{02}^2$, and $\mathbf{D}_{03}^2$ are independent. The remaining steps duplicate the steps of Theorem 1.4.6, creating the pattern of a conic-rigidity matrix for the vertex split. Again this matrix remains independent if we move $\mathbf{p}_0$ to any nearby position with $\mathbf{p}_1 - \mathbf{p}_0 = \mathbf{D}_{01}$. This completes the proof. □

Theorem 1.4.7 now extends immediately to conic-rigidity, since the proof was essentially topological.

1.5.5. Proposition. *For every graph of a triangulated sphere the geometric graph* $G(\mathbf{p})$ *at a generic point* $\mathbf{p}$ *has rank* $3|V| - 6$ *and is conic-independent.*

What is the significance of these results? Consider a plane drawing of a 3-connected planar graph – with its associated faces. A C_2^1-spline assigns a quadratic function

$$z = A^h x^2 + B^h xy + C^h y^2 + D^h x + E^h y + F^h$$

for each face f^h (except the exterior unbounded region). If two faces f^h, f^i share an edge along the line $\mathbf{L}_{jk} : px + qy + r = 0$ then the C^1 requirement of a common tangent plane at all points on this edge becomes a simple algebraic condition on the functions:

$$(A^h x^2 + B^h xy + C^h y^2 + D^h x + E^h y + F^h)$$
$$- (A^j x^2 + B^i xy + C^i y^2 + D^i x + E^i y + F^i)$$
$$= \omega^{hi}(p^2 x^2 + 2pqxy + q^2 y^2 + 2prx + 2qry + r^2).$$

for some scalar ω^{hi} (see Exercise 1.38). Moreover, if we follow a cycle of faces and edges around an interior vertex of the disc, the net difference is zero, giving the equation

$$\sum \omega^{hi}(p^2 x^2 + 2pqxy + q^2 y^2 + 2prx + 2qry + r^2) = 0 \quad \text{(sum over the vertex cycle)}.$$

Thus a C_2^1-spline corresponds to a choice for a quadratic function for an initial face, and set of scalars for the edges, satisfying these vertex conditions. (The complete proof reproduces the arguments of Lemma 1.3.4 and Theorem 1.3.5 for self-stresses on a plane framework.)

These scalars look suspiciously like our conic-dependences, although these satisfy the six conditions:

$$\sum \omega^{hi}(p^2, 2pq, q^2, 2pr, 2qr, r^2) = (0, 0, 0, 0, 0, 0) \quad \text{(sum over the vertex cycle)}.$$

We can use the orientation of the polyhedron to convert from scalars for a pair of adjacent faces to scalars for a pair of adjacent vertices.

We must show that this extension from three equations to six is automatic at any vertex. Given a point $\mathbf{p}_0 = (x_0, y_0)$, and the three coordinates $(A, B, C) = (p^2, 2pq, q^2)$ for a line $px + qy + r = 0$ through $\mathbf{p}_0$, there is a unique 6-vector defined as follows:

$$(x_0, y_0, 1) \# (A, B, C)$$
$$= (A, B, C, -2Ax_0 - By_0, -Bx_0 - 2Cy_0, Ax_0^2 + Bx_0 y_0 + Cy_0^2)$$
$$= (p^2, 2pq, q^2, 2pr, 2qr, r^2) = \mathbf{L}^2.$$

Thus if $\sum_j \omega_{ij}(A_{ij}, B_{ij}, C_{ij}) = 0$ then $\sum_j \omega_{ij}(x_i, y_i, 1) \# (A_{ij}, B_{ij}, C_{ij}) = 0$. (The 6-vector

$$(A, B, C, -2Ax_0 - By_0, -Bx_0 - 2Cy_0, Ax_0^2 + Bx_0 y_0 + Cy_0^2)$$

is interpreted as the coordinates of a quadratic function that is tangent to the xy-plane at the point $\mathbf{p}_0$.)

This gives a set of scalars for all edges except the exterior boundary edges. If the boundary is a non-degenerate triangle, then this triangle has rank $3|V - 6| = 3$. This rank gives conic-rigidity – so the interior scalars extend to scalars on the boundary edges (see Exercise 1.38). The conic-dependence on the interior extends to a conic-dependence on the whole.

1.5.6. Proposition. *For a triangulated disc in the plane, with a triangular boundary and a graph $G(\mathbf{p})$, the space of C_2^1-splines with the zero quadric over a fixed triangle is isomorphic to the vector space of conic-dependences of $G(\mathbf{p})$.*

This gives, as a corollary, the basic theorem for the dimension of the space of C_2^1-splines.

1.5.7. Corollary. *A generic triangulated plane disc with a triangular boundary has only trivial C_2^1-splines (all quadrics equal).*

An analogous matroid exists for C_3^2-splines, giving a matroid similar, but not identical, to the generic rigidity matroid in 4-space (see Exercises 1.40–1.42).

1.6. Scene Analysis of Polyhedral Pictures

In section 1.3 we demonstrated a connection between polyhedral pictures and the plane rigidity matroid. We shall now introduce a matroid that characterizes pictures of 'general polyhedral scenes'. Basic similarities to rigidity, at the level of matroid theory, will again be evident.

Consider an *elementary plane picture* formed by a set P of distinct points $\mathbf{p}_j$, and a set F of unordered sets of four distinct points $f^i = \{\mathbf{p}_h, \mathbf{p}_j, \mathbf{p}_k, \mathbf{p}_m\}$. The points $\mathbf{p}_j$ are lifted to heights z_j, with the constraint that the four points of each face must remain coplanar in this *spatial scene*. For four non-collinear points, each face imposes one linear equation on the heights:

$$\begin{vmatrix} x_h & x_j & x_k & x_m \\ y_h & y_j & y_k & y_m \\ z_h & z_j & z_k & z_m \\ 1 & 1 & 1 & 1 \end{vmatrix} = 0 \quad \text{or} \quad [\bar{\mathbf{p}}_j\bar{\mathbf{p}}_k\bar{\mathbf{p}}_m]z_h - [\bar{\mathbf{p}}_h\bar{\mathbf{p}}_k\bar{\mathbf{p}}_m]z_j - [\bar{\mathbf{p}}_h\bar{\mathbf{p}}_j\bar{\mathbf{p}}_k]z_m = 0$$

where the bracket $[\bar{\mathbf{p}}_j\bar{\mathbf{p}}_k\bar{\mathbf{p}}_m]$ is the corresponding minor of the determinant, recording the oriented area of the plane triangle $\mathbf{p}_j\mathbf{p}_k\mathbf{p}_m$. (If the four points are collinear, then a vertical plane will fit any selection of heights, so there is no condition.)

We actually record a plane picture as an abstract *hypergraph* $H = (V, F)$, with all faces as unordered quadruples of vertices, and an assignment **P** of plane points to the vertices. The constraints on the spatial scenes define an $|F| \times |V|$ homogeneous system $M(H, \mathbf{P}) \times Z^t = 0$, and a corresponding matroid on the faces. Every picture whose points span the plane has a 3-space of *trivial scenes* – scenes with all points coplanar. Thus an independent set F of faces must satisfy

$$|F'| \leqslant |V'| - 3 \quad \text{for every non-empty subset } F'.$$

Figure 1.16 shows some elementary plane pictures, with the faces shown as quadrilateral regions, using heavy lines between faces sharing two vertices.

Figure 1.16.

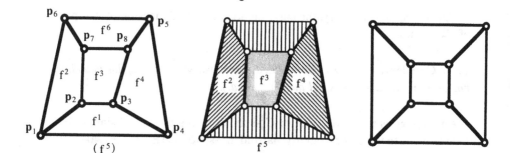

(a)   (b)   (c)

1.6.1. Example. Consider the elementary scene in Figure 1.16a, with six faces and eight vertices. The matrix for this picture has the form:

$$
\begin{bmatrix}
[\bar{p}_2\bar{p}_3\bar{p}_4] & -[\bar{p}_1\bar{p}_3\bar{p}_4] & [\bar{p}_1\bar{p}_2\bar{p}_4] & -[\bar{p}_1\bar{p}_2\bar{p}_3] & 0 & 0 & 0 & 0 \\
[\bar{p}_2\bar{p}_6\bar{p}_7] & -[\bar{p}_1\bar{p}_6\bar{p}_7] & 0 & 0 & 0 & [\bar{p}_1\bar{p}_2\bar{p}_7] & -[\bar{p}_1\bar{p}_2\bar{p}_6] & 0 \\
0 & [\bar{p}_3\bar{p}_7\bar{p}_8] & [-\bar{p}_2\bar{p}_7\bar{p}_8] & 0 & 0 & 0 & [\bar{p}_2\bar{p}_3\bar{p}_8] & -[\bar{p}_2\bar{p}_3\bar{p}_7] \\
0 & 0 & [\bar{p}_4\bar{p}_5\bar{p}_8] & -[\bar{p}_3\bar{p}_5\bar{p}_8] & [\bar{p}_3\bar{p}_4\bar{p}_8] & 0 & 0 & -[\bar{p}_3\bar{p}_4\bar{p}_5] \\
[\bar{p}_4\bar{p}_5\bar{p}_6] & 0 & 0 & -[\bar{p}_1\bar{p}_5\bar{p}_6] & [\bar{p}_1\bar{p}_4\bar{p}_6] & -[\bar{p}_1\bar{p}_4\bar{p}_5] & 0 & 0 \\
0 & 0 & 0 & 0 & [\bar{p}_6\bar{p}_7\bar{p}_8] & -[\bar{p}_5\bar{p}_7\bar{p}_8] & [\bar{p}_5\bar{p}_6\bar{p}_8] & -[\bar{p}_6\bar{p}_7\bar{p}_8]
\end{bmatrix}
$$

The matrix has rank 5 for generic pictures, giving a linear dependence and leaving only the trivial scenes with all vertices (and faces) coplanar. Even with one face removed, general pictures are independent, but have only trivial scenes. However, with two faces removed (any two), the matrix has rank 4 for a generic picture (Figure 1.16b), giving non-trivial scenes, with distinct planes for each face.

Consider a special picture, such as Figure 1.16c, with the positions:

$$\mathbf{p}_1 = (0, 0) \quad \mathbf{p}_2 = (1, 1) \quad \mathbf{p}_3 = (2, 1) \quad \mathbf{p}_4 = (3, 0)$$

$$\mathbf{p}_5 = (3, 3) \quad \mathbf{p}_6 = (0, 3) \quad \mathbf{p}_7 = (1, 2) \quad \mathbf{p}_8 = (2, 2).$$

The picture matrix for all six faces now has the form

$$
\begin{bmatrix}
-1 & 3 & -3 & 1 & 0 & 0 & 0 & 0 \\
-1 & 3 & 0 & 0 & 0 & 1 & -3 & 0 \\
0 & -1 & 1 & 0 & 0 & 0 & 1 & -1 \\
0 & 0 & 3 & -1 & 1 & 0 & 0 & -3 \\
9 & 0 & 0 & -9 & 9 & -9 & 0 & 0 \\
0 & 0 & 0 & 0 & 1 & -1 & 3 & -3
\end{bmatrix}.
$$

This matrix has rank 4, giving non-trivial scenes with distinct planes for each face.

We want to characterize the *generic elementary picture matroid*, which occurs for general (e.g. algebraically independent) points in the plane.

1.6.2. Theorem. *A hypergraph H of quadruples has realizations as an independent plane picture if and only if $|F'| \leqslant |V'| - 3$ for every non-empty subset of faces F'.*

Proof. We already have seen that an independent plane picture satisfies $|F'| \leqslant |V'| - 3$ for every non-empty subset of faces F'.

Assume that a hypergraph satisfies

$$|F'| \leqslant |V'| - 3 \quad \text{for every non-empty subset of faces } F'.$$

We begin with the simpler transversal matroid, defined by the semimodular function $g(F) = |V|$. The transversal matroid is represented over the reals by the modified incidence matrix $A(F)$:

$$a_{ij} = t_{ij} \quad \text{if } v_j \in f_i \quad \text{and} \quad a_{ij} = 0 \quad \text{if } v_j \notin f_i$$

for any set of algebraically independent reals t_{ij}.

By our assumption, F is also independent in this transversal matroid, giving independent rows in the transversal matrix. Since this matrix is $|F| \times |V|$, the system $A \times X = 0$ has a solution space of dimension at least 3. Moreover, if we add three copies of any edge of F, we get a set F'' that is still independent in the transversal matroid, so that we can choose a solution set of dimension 3 when restricted to the vertices of any edge. We write a basis for this chosen solution space as the $|F| \times 3$ matrix Y, with rows $Y_j = [x_j y_j w_j]$, satisfying $A \times Y = 0$.

We can choose the matrix Y so that no values of w_i are zero (take linear combinations of the columns). If we divide each row Y_j by the scalar w_j, creating Y', and multiply the corresponding column A^j of A by w_j, we obtain another representation A' of the same matroid, with $A' \times Y' = 0$ and $Y'_j = [x'_j y'_j 1]$. We check that this matrix A' is the picture matrix up to multiplication of rows by non-zero constants.

For a typical face of the form $f = \{v_1, v_2, v_3, v_4\}$, the equation $A' \times Y' = 0$ gives the equations

$$t_1 x_1 + t_2 x_2 + t_3 x_3 + t_4 x_4 = 0 \quad t_1 y_1 + t_2 y_2 + t_3 y_3 + t_4 y_4 = 0$$

$$t_1 1 + t_2 1 + t_3 1 + t_4 1 = 0$$

Given the matrix Y', these form three independent equations in the t_i, for which the only solutions are scalar multiples of

$$t_1 = \begin{vmatrix} x_2 & x_3 & vx_4 \\ y_2 & y_3 & y_4 \\ 1 & 1 & 1 \end{vmatrix}, \quad t_2 = \begin{vmatrix} x_1 & x_3 & x_4 \\ y_1 & y_3 & y_4 \\ 1 & 1 & 1 \end{vmatrix},$$

$$t_3 = \begin{vmatrix} x_1 & x_2 & x_4 \\ y_1 & y_2 & y_4 \\ 1 & 1 & 1 \end{vmatrix}, \quad t_4 = \begin{vmatrix} x_1 & x_2 & x_3 \\ y_1 & y_2 & y_3 \\ 1 & 1 & 1 \end{vmatrix}.$$

Therefore, the row of A' for edge f is:

$$\beta([\bar{\mathbf{p}}_2 \bar{\mathbf{p}}_3 \bar{\mathbf{p}}_4], \ -[\bar{\mathbf{p}}_1 \bar{\mathbf{p}}_3 \bar{\mathbf{p}}_4], \ [\bar{\mathbf{p}}_1 \bar{\mathbf{p}}_2 \bar{\mathbf{p}}_4], \ -[\bar{\mathbf{p}}_1 \bar{\mathbf{p}}_2 \bar{\mathbf{p}}_3]).$$

If we take the picture with $\mathbf{p}_i = (x_i, y_i)$, this row of A' is a multiple of the row of picture matrix $M(F, \mathbf{P})$ for the edge f. Thus an independent set of rows in A' corresponds to independent rows in $M(F, \mathbf{P})$.

We conclude that F has a realization as an independent elementary picture.

$\square$

This is an elementary result for elementary pictures. A general *plane picture* involves faces with arbitrary (finite) sets of points which are constrained to be coplanar in a general *spatial scene*. There is an alternate system of equations that is traditionally used for these scenes. Each face f^i in the scene is recorded as a non-vertical plane $a^i x + b^i y + z + c^i = 0$ or as a triple (a^i, b^i, c^i). If vertex $v_j \in f^i$, then we have the single linear equation in the variables z_j, a^i, b^i, c^i:

$$a^i x_j + b^i y_j + z_j + c^i = 0$$

The corresponding system of equations creates a matroid on the rows of the coefficient matrix, or a matroid on the incidences $I \subset V \times F$. A *plane picture* $S(\mathbf{P})$ is an *incidence structure* (or bipartite graph) $S = (V, F; I)$ with an assignment $\mathbf{P}$ of plane points to the vertices V. The 3-space of trivial scenes continues to exist, so a set of independent incidences satisfies

$$|I'| \leqslant |V'| + 3|F'| - 3 \quad \text{on non-empty subsets } I'.$$

Figure 1.17.

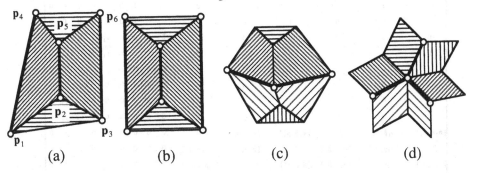

(a)   (b)   (c)   (d)

1.6.3. Example. We consider the picture in Figure 1.17a. This incidence structure of six vertices, three faces, and 12 incidences, creates the picture matrix

$$
\begin{bmatrix}
1 & 0 & 0 & 0 & 0 & 0 & x_1 & y_1 & 1 & 0 & 0 & 0 & 0 & 0 & 0 \\
0 & 1 & 0 & 0 & 0 & 0 & x_2 & y_2 & 1 & 0 & 0 & 0 & 0 & 0 & 0 \\
0 & 0 & 0 & 1 & 0 & 0 & x_4 & y_4 & 1 & 0 & 0 & 0 & 0 & 0 & 0 \\
0 & 0 & 0 & 0 & 1 & 0 & x_5 & y_5 & 1 & 0 & 0 & 0 & 0 & 0 & 0 \\
0 & 1 & 0 & 0 & 0 & 0 & 0 & 0 & 0 & x_2 & y_2 & 1 & 0 & 0 & 0 \\
0 & 0 & 1 & 0 & 0 & 0 & 0 & 0 & 0 & x_3 & y_3 & 1 & 0 & 0 & 0 \\
0 & 0 & 0 & 0 & 1 & 0 & 0 & 0 & 0 & x_5 & y_5 & 1 & 0 & 0 & 0 \\
0 & 0 & 0 & 0 & 0 & 1 & 0 & 0 & 0 & x_6 & y_6 & 1 & 0 & 0 & 0 \\
1 & 0 & 0 & 0 & 0 & 0 & 0 & 0 & 0 & 0 & 0 & 0 & x_1 & y_1 & 1 \\
0 & 0 & 1 & 0 & 0 & 0 & 0 & 0 & 0 & 0 & 0 & 0 & x_3 & y_3 & 1 \\
0 & 0 & 0 & 1 & 0 & 0 & 0 & 0 & 0 & 0 & 0 & 0 & x_4 & y_4 & 1 \\
0 & 0 & 0 & 0 & 0 & 1 & 0 & 0 & 0 & 0 & 0 & 0 & x_6 & y_6 & 1 \\
\end{bmatrix}
$$

This has rank $12 = |V| + 3|F| - 3$ for generic points – and the picture has only trivial scenes.

For a special position, such as Figure 1.17b, with vertices

$$\mathbf{p}_1 = (0, 0), \ \mathbf{p}_2 = (1, 1), \ \mathbf{p}_3 = (2, 0), \ \mathbf{p}_4 = (0, 3), \ \mathbf{p}_5 = (1, 2), \ \mathbf{p}_6 = (2, 3),$$

the matrix has rank 11 – and has non-trivial scenes, as the three quadrilateral faces of a 'triangular prism'.

Each of these three faces has exactly four vertices – and we can return to the elementary picture matroid by row reducing on the final nine columns, provided no face has all vertices collinear:

$$
\begin{bmatrix}
1 & 0 & 0 & 0 & 0 & 0 & x_1 & y_1 & 1 & 0 & 0 & 0 & 0 & 0 & 0 \\
0 & 1 & 0 & 0 & 0 & 0 & x_2 & y_2 & 1 & 0 & 0 & 0 & 0 & 0 & 0 \\
0 & 0 & 0 & 1 & 0 & 0 & x_4 & y_4 & 1 & 0 & 0 & 0 & 0 & 0 & 0 \\
0 & 1 & 0 & 0 & 0 & 0 & 0 & 0 & 0 & x_2 & y_2 & 1 & 0 & 0 & 0 \\
0 & 0 & 1 & 0 & 0 & 0 & 0 & 0 & 0 & x_3 & y_3 & 1 & 0 & 0 & 0 \\
0 & 0 & 0 & 0 & 1 & 0 & 0 & 0 & 0 & x_5 & y_5 & 1 & 0 & 0 & 0 \\
1 & 0 & 0 & 0 & 0 & 0 & 0 & 0 & 0 & 0 & 0 & 0 & x_1 & y_1 & 1 \\
0 & 0 & 1 & 0 & 0 & 0 & 0 & 0 & 0 & 0 & 0 & 0 & x_3 & y_3 & 1 \\
0 & 0 & 0 & 0 & 0 & 1 & 0 & 0 & 0 & 0 & 0 & 0 & x_4 & y_4 & 1 \\
[\bar{p}_2\bar{p}_4\bar{p}_5] & -[\bar{p}_1\bar{p}_4\bar{p}_5] & 0 & [\bar{p}_1\bar{p}_2\bar{p}_5] & -[\bar{p}_1\bar{p}_2\bar{p}_4] & 0 & 0 & 0 & 0 & 0 & 0 & 0 & 0 & 0 & 0 \\
0 & [\bar{p}_3\bar{p}_5\bar{p}_6] & -[\bar{p}_2\bar{p}_5\bar{p}_6] & 0 & [\bar{p}_2\bar{p}_3\bar{p}_6] & -[\bar{p}_2\bar{p}_3\bar{p}_5] & 0 & 0 & 0 & 0 & 0 & 0 & 0 & 0 & 0 \\
[\bar{p}_3\bar{p}_4\bar{p}_5] & 0 & -[\bar{p}_1\bar{p}_4\bar{p}_6] & [\bar{p}_1\bar{p}_3\bar{p}_6] & 0 & -[\bar{p}_1\bar{p}_3\bar{p}_4] & 0 & 0 & 0 & 0 & 0 & 0 & 0 & 0 & 0
\end{bmatrix}
$$

A similar reduction, applied to all picture matrices, is the foundation for the next theorem.

1.6.4. Theorem. *The incidences I of an independence structure $S = (V, F; I)$ are independent in a generic plane picture if and only if $|I'| \leqslant |V'| + 3|F'| - 3$ on all non-empty subsets I'.*

Proof. For a generic picture, faces with less than four vertices impose no conditions, and will not change the independence of the structure. Faces of $n > 4$ vertices can be replaced by $(n - 3)$ 4-valent faces, choosing three fixed vertices and adding the others one at a time. This converts the problem to a problem on an elementary plane picture $H^\wedge(\mathbf{P})$, and the interested reader can check that the counts $|I'| \leqslant |V'| + 3|F'| - 3$ correspond to the counts $F^\wedge \leqslant |V^\wedge| - 3$ on this elementary picture. $\qquad\square$

A more central matroid for computer science arises from one more truncation of the picture matroid. A *sharp scene* is a spatial scene in which any two faces lie in distinct planes. A sharp scene has a 4-space of trivial changes over the same picture: with one face fixed in the horizontal, we can change the vertical scale, in addition to the 3-space of scenes that change the plane of the selected initial face.

The following corollary is used in computer algorithms for recognizing pictures of sharp scenes.

1.6.5. Corollary. *For an incidence structure $S = (V, F; I)$ a generic picture has sharp scenes if and only if $|I'| \leqslant |V'| + 3|F'| - 4$ for all subsets I' with at least two faces.*

Figure 1.18 shows some generically sharp pictures (pictures with sharp scenes).

Figure 1.18. $|I| = |V| + 3|F| - 4$. (a) $|F| = 4, |V| = 8, |I| = 16$; (b) $|F| = 9, |V| = 13, |I| = 36$;
(c) $|F| = 10, |V| = 10, |I| = 36$.

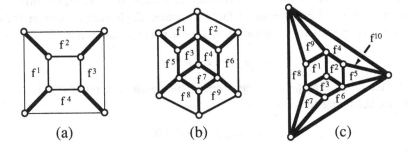

(a)   (b)   (c)

1.6.6. Example. To highlight the similarity to rigidity matroids, consider *simple pictures* in which each vertex is incident with exactly two faces. With this assumption, we have $|V'| = 2|I'|$ on relevant subsets and the count for sharp scenes $|I'| \leqslant |V'| + 3|F'| - 4$ simplifies to the count $|V'| \leqslant 3|F'| - 4$. This assumption also means that the incidence structure describes a graph, or multi-graph – with 'vertices' F and an 'edge' for each element of V connecting its two faces.

If we row reduce the picture matrix on the columns for the vertices, the matrix assumes a rigidity-like pattern. For example, the matrix from Example 1.6.3 becomes

$$
\begin{bmatrix}
1 & 0 & 0 & 0 & 0 & 0 & x_1 & y_1 & 1 & 0 & 0 & 0 & 0 & 0 & 0 \\
0 & 1 & 0 & 0 & 0 & 0 & x_2 & y_2 & 1 & 0 & 0 & 0 & 0 & 0 & 0 \\
0 & 0 & 1 & 0 & 0 & 0 & 0 & 0 & 0 & x_3 & y_3 & 1 & 0 & 0 & 0 \\
0 & 0 & 0 & 1 & 0 & 0 & x_4 & y_4 & 1 & 0 & 0 & 0 & 0 & 0 & 0 \\
0 & 0 & 0 & 0 & 1 & 0 & x_5 & y_5 & 1 & 0 & 0 & 0 & 0 & 0 & 0 \\
0 & 0 & 0 & 0 & 0 & 1 & 0 & 0 & 0 & x_6 & y_6 & 1 & 0 & 0 & 0 \\
0 & 0 & 0 & 0 & 0 & 0 & x_1 & y_1 & 1 & 0 & 0 & 0 & -x_1 & -y_1 & -1 \\
0 & 0 & 0 & 0 & 0 & 0 & x_2 & y_2 & 1 & -x_2 & -y_2 & -1 & 0 & 0 & 0 \\
0 & 0 & 0 & 0 & 0 & 0 & 0 & 0 & 0 & x_3 & y_3 & 1 & -x_3 & -y_3 & -1 \\
0 & 0 & 0 & 0 & 0 & 0 & x_4 & y_4 & 1 & 0 & 0 & 0 & -x_4 & -y_4 & -1 \\
0 & 0 & 0 & 0 & 0 & 0 & x_5 & y_5 & 1 & -x_5 & -y_5 & -1 & 0 & 0 & 0 \\
0 & 0 & 0 & 0 & 0 & 0 & 0 & 0 & 0 & x_6 & y_6 & 1 & -x_6 & -y_6 & -1
\end{bmatrix}
$$

The block in the lower right corner is a '3-frame matrix' analogous to the 2-frame used in section 1.2, and studied in the exercises. It does not directly correspond to a problem of rigid frameworks, although it does have a subtle interpretation in terms of bar and body frameworks in the plane (Exercise 1.45).

The picture matrix has an important geometric dual interpretation – describing *parallel redrawings* of faces in space. Each vertex (a, b) in the picture corresponds to a normal $(a, b, 1)$ of a non-vertical plane

$$ax + by + z + d = 0.$$

Each face containing (a, b) in the scene corresponds to a point (x_0, y_0, z_0) in space satisfying the equation

$$ax_0 + by_0 + z_0 + d = 0.$$

Thus the dual realization of the picture has a plane corresponding to each vertex and a point corresponding to each face.

Different dual realizations have corresponding faces parallel. A trivial realization has all vertices at the same point, corresponding to a solution to the picture equation with all faces at the same place. A non-trivial realization has distinct vertices, with faces parallel to the original faces (same normal). A trivial realization has a three-dimensional space of trivial parallel redrawings – the translations. A non-trivial realization has a four-dimensional space of *trivial parallel redrawings* generated by the translations and the dilations.

1.6.7. Example. Consider the dual of Example 1.6.6 (Figure 1.17c). This is, essentially, a triangular bipyramid: the top and bottom 'vertices' are implicit in the intersection of three planes. For general choices for the normals of six faces, the only realizations are trivial, with all vertices at the same point (Figure 1.17d). The 'general' bipyramid has 'general vertices' – but not general face normals – giving a lower rank for the parallel redrawing matrix, and an extra parallel redrawing: the similarity map.

Formally, this example has only three vertices, and six faces which meet in pairs along the three sides of a spatial triangle. As such, this becomes an example of the parallel redrawing polymatroid on spatial graphs (see Exercises 1.54).

1.7. Special Positions

Graphs which are isostatic (or independent) in generic realizations as a bar framework in the plane or in 3-space also have special *critical forms* that are dependent (or stressed) because of the underlying geometry of the framework.

Similar critical forms arise for conic-rigidity and for plane pictures. There is some combinatorics, and a lot of geometry, in the study of these special positions.

1.7.1. Example. Recall the frameworks of Example 1.2.2. The general rigidity matrix has the form

$$\begin{bmatrix}
x_4-x_1 & y_4-y_1 & 0 & 0 & 0 & 0 & x_1-x_4 & y_1-y_4 & 0 & 0 & 0 & 0 \\
x_5-x_1 & y_5-y_1 & 0 & 0 & 0 & 0 & 0 & 0 & x_1-x_5 & y_1-y_5 & 0 & 0 \\
x_6-x_1 & y_6-y_1 & 0 & 0 & 0 & 0 & 0 & 0 & 0 & 0 & x_1-y_6 & y_1-y_6 \\
0 & 0 & x_4-x_2 & y_4-y_2 & 0 & 0 & x_2-x_4 & y_2-y_4 & 0 & 0 & 0 & 0 \\
0 & 0 & x_5-x_2 & y_5-y_2 & 0 & 0 & 0 & 0 & x_2-x_5 & y_2-y_5 & 0 & 0 \\
0 & 0 & x_6-x_2 & y_6-y_2 & 0 & 0 & 0 & 0 & 0 & 0 & x_2-x_6 & y_2-y_6 \\
0 & 0 & 0 & 0 & x_4-x_3 & y_4-y_3 & x_3-x_4 & y_3-y_4 & 0 & 0 & 0 & 0 \\
0 & 0 & 0 & 0 & x_5-x_3 & y_5-y_3 & 0 & 0 & x_3-x_5 & y_3-y_5 & 0 & 0 \\
0 & 0 & 0 & 0 & x_6-x_3 & y_6-y_3 & 0 & 0 & 0 & 0 & x_3-x_6 & y_3-y_6
\end{bmatrix}$$

We previously saw that this has generic rank 9, but has low rank, corresponding to non-trivial infinitesimal motions, for six points on a conic section. A lower rank corresponds to all 9×9 minors being zero; thus a set of 84 conditions has to be checked. Each minor gives a polynomial equation in the coordinates of the six points:

$$p(\mathbf{p}_1, \mathbf{p}_2, \mathbf{p}_3, \mathbf{p}_4, \mathbf{p}_5, \mathbf{p}_6) = 0.$$

What do these polynomials look like? They are of degree 9 in the points. We have already seen that each polynomial is zero for six points on a conic section. This condition corresponds to an irreducible polynomial of degree 12 in the points (degree 2 in each point) that must be a factor of each of the minors. What is left in each minor after this polynomial has been factored out? Just a simple linear *factor* that turns out to be a trivial statement about distinct coordinates for the points.

For example, if we delete the first two columns and the seventh column, the remaining minor can be calculated by Laplace expansion, using the first row. This gives the trivial factor $(y_1 - y_4)$ times a minor that must be (and is) the irreducible polynomial for the conic. If we delete the first two columns and the eighth column, the linear factor is $(x_1 - x_4)$. These inessential linear by-products of the deleted columns are called *tie down factors*, and a basic theorem is stated as follows.

1.7.2. Theorem. (*White & Whiteley, 1983*) *For any graph G that is generically isostatic in the plane there is a polynomial pure condition $C_G(\mathbf{p}_1, \ldots, \mathbf{p}_{|V|})$, of degree $val_i - 1$ in each point of valence val_i in the graph, such that all $|E| \times |E|$ minors of the ridigity matrix are trivial linear factors times the pure condition.*

1.7.3. Example. We have seen that the graph of Figure 1.19a is generically isostatic in the plane (section 1.2), and, by Maxwell's Theorem of section 1.3, that this framework is dependent if and only if it is the projection of a spatial configuration with at least two distinct faces. We found three conditions which guarantee such a projection:

(i) if the three points A, B, and C form a collinear triangle;
(ii) if the three points A', B', and C' form a collinear triangle;
(iii) if the three lines AA', BB', and CC' are concurrent.

These three conditions give three irreducible polynomial factors for the pure condition of the graph which are of total degree 12 in the points. Thus they are the complete pure condition of the graph. This pure condition is also equivalent to the single geometric condition (which is algebraically reducible): the two triangles ABC and $A'B'C'$ are perspective from a line (i.e. the usual conclusion of Desargues' theorem holds).

1.7.4. Example. In 3-space, the graph of an octahedron is generically isostatic as a triangulated convex sphere (Figure 1.19b). A similar analysis of the minors shows that all $|E| \times |E|$ minors of the rigidity matrix are either simple multiples of a basic pure condition that is of degree $val_i - 2$ in each point, or are identically zero. A classical theorem shows that the pure condition is the projective condition that four opposite faces ABC, $AB'C'$, $A'BC'$, and $A'B'C$ are on planes through a common point $\mathbf{p}$ (Figure 1.19c).

Figure 1.19.

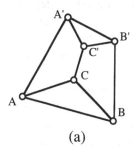 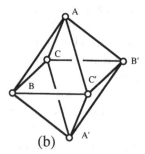 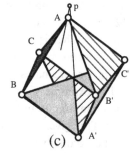

(a)   (b)   (c)

Since a rigidity matrix for these special positions represents a specialization of the variables for a generic position, the corresponding matroid is a weak-map image of the generic rigidity matroid. The new circuits created in these specializations are *geometric circuits* rather than the *generic circuits* of the generic realizations (see Exercises 1.56, 1.57, and 1.58).

As the examples have hinted, the critical forms for infinitesimal rigidity are given by *projective geometric conditions*. Engineers in the last century realized that the statics and infinitesimal mechanics of a framework are invariant under projective transformations and Exercise 1.10 asks the reader to confirm that the rank of the plane rigidity matrix is projectively invariant. There is also projective geometry buried in the special positions of the spline matroid, the scene analysis matroid, and the parallel redrawing matroid (Exercises 1.39, 1.51, and 1.53).

Example 1.7.1 and Theorem 1.7.2 indicated that single polynomial 'pure conditions' control the lowering of the rank of matrix of a generic basis in the rigidity matrix. A similar analysis leading to polynomial 'pure conditions' applies to the other matrices from splines, scene analysis, etc. There is a rich field of research in this interplay of geometry and combinatorics arising in the matroids of critical forms.

1.8. Conclusions

The families of rigidity matroids, spline matroids, and scene analysis matroids represent only the best known and more accessible examples of the large collection of matroids that arise in discrete applied geometry. A number of related matroids are presented in the exercises.

We underline some common themes of these examples from discrete applied geometry.

(1) The matroid is defined by a sparse matrix. The non-zero entries correspond to 'coordinates' of geometric objects in the real plane (3-space, etc.), and follow a combinatorial pattern based on a graph or incidence structure.

For sections 1.1–1.5, this incidence structure was the underlying graph. We then replaced each non-zero entry by a d-vector (drawn from the geometry) and each zero entry by the zero d-vector. The result was an expanded pattern which passes through d copies of graphic matroid (with variables), and then an algebraic (or geometric) specialization. Our task was to trace the effects of this specialization of the entries.

More generally, as in section 1.6, we begin with an incidence matrix, and make vector substitutions under one 'type' of vertex (e.g. the faces). In other applications, we even replace entries by rectangular matrices rather than single rows, creating a polymatroid. However, all of these examples carry properties inherited from the incidence matrix.

(2) Analysis of the matroid is based on the dual concepts of row dependences (self-stresses) and column dependences (infinitesimal motions).

(3) If variables are used for these entries, we find the 'generic' matroid for the class of geometric realizations. Independent sets, and bases, in the generic matroid can often, but not always, be characterized by simple counts (semimodular functions) on the underlying combinatorial structures.

(4) The cases where simple counts fail (rigidity in 3-space, conic-rigidity, etc.) point to a fundamental problem in defining matroids by semimodular functions that are negative on small sets. The inductive techniques being developed represent matroidal approaches to a matroid problem.

(5) There are fundamental similarities in apparently disparate geometric problems. These similarities become accessible in the patterns of the sparse matrices, and in the matroidal properties. In new examples, the patterns of the sparse matrix suggest the appropriate analog among our basic patterns, and permit an easy transfer of methods among the examples.

(6) The behaviour of non-generic examples creates polynomial conditions that give interesting interactions of projective geometry, algebraic geometry, and combinatorics in the specialized matroids.

We conclude as we began. Engineering and computer science examples create a set of represented matroids for which the combinatorics of even generic examples are currently unsolved. The existing work applies, and illustrates, a number of matroid methods, including circuit elimination, matroid union, and truncation. The examples also suggest some new approaches to represented matroids.

1.9. Historical Notes

The results of this chapter are drawn from a number of scattered papers stretching over the last 130 years. The explicit connections among the various concepts of rigidity appear in Gluck (1975) and Asimow & Roth (1978) and are presented in great detail in Connelly (1987). The explicit use of matroid methods for generic bar frameworks structures is drawn from Crapo (1985), Graver (1984), Lovász & Yemini (1982), Recski (1984, 1989), and Whiteley (1988a).

Theorem 1.2.3 is a combination of Laman's theorem (1970) and Lovasz & Yemini (1982). The proof follows Whiteley (1988a) and the 2-frame represents a matroid union of two copies of the graph. The inductive results in the last century are summarized in Henneberg (1911), and are developed in Tay & Whiteley (1985). The analysis of the semimodular functions in terms of matroid unions and truncations is related to classical matroid results of Edmonds (1970), Pym & Perfect (1970), and Brylawski (1985a, b).

Maxwell's theorem dates to the development of 'graphical statics' by engineers and geometers in the last century (Maxwell, 1864; Cremona 1890). The classical geometric approach is presented anew in Crapo & Whiteley (1988), along with the converse and its extensions. The approach to Maxwell's theorem and its converse used in section 1.2 is based on Whiteley (1982).

The infinitesimal rigidity of convex triangulated spherical polyhedra dates back implicitly to Cauchy (1831), and explicitly to Dehn (1916). The generic results were extracted by Gluck (1975) and simplified in Graver (1984), and in Tay & Whiteley (1985). The topological development of triangulated surfaces by vertex splitting is found in Barnette (1982), and greatly extended in Fogelsanger (1988). That vertex splitting preserves infinitesimal rigidity and conic-independence is found in Whiteley (1991a, 1991c).

The problem of the generic dimension of the space of C_2^1-splines was raised by Strang (1973), and studied in Schumaker (1979) and in Alfeld (1987). The basic result was proven in Billera (1986) and in Whiteley (1991b). The approach presented here is related to techniques of Chui & Wang (1983) and was developed by explicit analogy to spatial frameworks in Whiteley (1986c, 1990a, 1991b).

The explicit matroid for general scene analysis was presented in Sugihara (1984), with a conjecture for Corollary 1.6.5. Whiteley (1988b) gave the first proof of the theorem, implicitly re-proving Crapo (1985) for the result on elementary pictures. Sugihara (1986) presents this proof anew, with a full background on the motivation and related problems in computer reconstruction of spatial objects. The dual parallel redrawing matroids were developed in Whiteley (1986a, 1988b). Parallel redrawing of spatial polyhedra and convex polytopes is basic to the study of Minkowski decomposition of polytopes (Shephard, 1963; Kallay, 1982; Smilansky, 1987).

Particular examples of geometric 'critical forms' date back to the last century. The explicit study of 'pure conditions' was carried out in White & Whiteley (1983, 1987). The general behaviour of bipartite frameworks appears for statics in Bolker & Roth (1980) and for infinitesimal mechanics in Whiteley (1984a). The projective invariance of statics and infinitesimal mechanics dates back at least to Rankine (1863), and is modernized in Roth & Whiteley (1981). The projective invariance of parallel redrawing appears in Kallay (1982) and is proven for C_2^1-splines in Whiteley (1986b, 1991b).

Exercises

A difficult problem is indicated by *, an unsolved one by **.

1.1. Show that for any graph $G = (V, E)$ and any $m > 0$, the function $f(E) = m|V| + k$ defines a non-decreasing semimodular function on subsets of the edges.

1.2. If, for $k > -2m$, we define circuits $|E'|$ on the edges of a graph by $|E'| = m|V'| + k + 1$ and $|E''| \leqslant m|V'| + k$, verify directly that the circuit elimination axiom for matroids is satisfied.

1.3. Show that the semimodular function $f(E) = |V| - 1$ defines the cycle matroid of a graph or multi-graph.

1.4. Show that the semimodular function $f(E) = |V|$ defines the bicycle matroid of a graph.

1.5. (Whiteley, 1988a) Show that the semimodular function $f(E) = 2|V| - 1$ defines the matroid union of the bicycle matroid and the cycle matroid of a graph. (This is the generic matroid for bar frameworks on the surface of a cone).

1.6.   Show that the frameworks in Figures 1.20b and 1.20c are each minimal rigid plane frameworks, i.e. removing any one bar leaves a finite motion. Conclude that 'rigidity' does not induce a matroid on the plane framework in Figure 1.20a.

Figure 1.20.

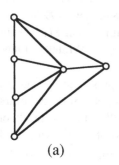 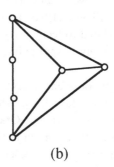 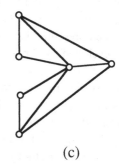

   (a)   (b)   (c)

1.7.   (a) Show that if $G(\mathbf{p})$ and $G(\mathbf{q})$ are two frameworks with the same bar lengths for all edges, then $\mathbf{p} - \mathbf{q}$ is an infinitesimal motion of the framework $G((\mathbf{p} + \mathbf{q})/2)$.

      (b) Show that $\mathbf{p} - \mathbf{q}$ is a non-trivial infinitesimal motion of $G((\mathbf{p} + \mathbf{q})/2)$ if $G(\mathbf{p})$ is not congruent to $G(\mathbf{q})$.

      (c) Give an example where $G(\mathbf{p})$ is congruent to $G(\mathbf{q})$ and $\mathbf{p} - \mathbf{q}$ is a non-trivial infinitesimal motion of $G((\mathbf{p} + \mathbf{q})/2)$.

      (d) Show that if $G(\mathbf{p}_i)$ is a sequence of frameworks all with the same bar lengths but not congruent to $\mathbf{p}$, such that $\lim \mathbf{p}_i = \mathbf{p}$, then $G(\mathbf{p})$ has a non-trivial infinitesimal motion.

      (e) (Local uniqueness) Show that if $G(\mathbf{p})$ is infinitesimally rigid, then there is an open neighborhood $N(\mathbf{p})$ in $\mathbb{R}^{2|V|}$ such that if $G(\mathbf{q})$ has the same edge lengths as $G(\mathbf{p})$ and $\mathbf{q} \in N(\mathbf{p})$, then $G(\mathbf{q})$ is congruent to $G(\mathbf{p})$.

1.8.   Show that an infinitesimal motion $\mathbf{u}$ of a non-collinear plane framework $G(\mathbf{p})$ is non-trivial if and only if there is a pair of joints $\mathbf{p}_h, \mathbf{p}_k$ (not joined by a bar) such that

$$(\mathbf{p}_h - \mathbf{p}_k) \cdot (\mathbf{p}'_h - \mathbf{p}'_k) \neq 0.$$

1.9.   (Whiteley, 1984a) Recall that a plane conic centered at the origin (an ellipse, an hyperbola, or two lines) can be written $[\mathbf{p}_1][Q][\mathbf{p}_1]^t = r$ for a symmetric matrix $[Q]$ and a constant r. Show that any bipartite graph $K_{A,B}$, a framework with all its joints on such a conic has a non-trivial infinitesimal motion

$$\mathbf{a}'_i = [\mathbf{a}_i][Q] \quad \text{and} \quad \mathbf{b}'_j = [\mathbf{b}_j][Q]$$

for $\mathbf{a}_i \in A$ and $\mathbf{b}_j \in B$. Show that this is a non-trivial infinitesimal motion unless the conic contains all lines joining $(\mathbf{a}_h, \mathbf{a}_i)$ and $(\mathbf{b}_j, \mathbf{b}_k)$.

1.10. (Projective invariance) Show that if a non-singular projective transformation in the plane, T, takes the points $\mathbf{p} = (\mathbf{p}_1, ..., \mathbf{p}_{|V|})$ to the points $\mathbf{q} = (T(\mathbf{p}_1), ..., T(\mathbf{p}_{|V|}))$ then the bar frameworks $G(\mathbf{p})$ and $G(\mathbf{q})$ define the same matroid for any graph G. (Hint: prove that T takes a dependent set to a dependent set, using self-stresses and the projective weights of the transformed points.)

1.11. (Crapo, 1988) Show that a graph $G = (V, E)$ is generically isostatic in the plane if and only if the set E of edges is the disjoint union of three trees T_i such that each vertex of G is incident with exactly two of the trees T_i, and with the further property that no two distinct subtrees of the trees T_i have the same span.

1.12. Given a graph $G = (V, E)$, the *cone graph with* v_0 is the graph $G * v_0$ formed by adding one new vertex v_0 having edges to all vertices of G (Figure 1.15a). Show, from the counts, that the cone graph $G * v_0$ is generically isostatic in the plane if and only if G is generically isostatic on the line.

1.13. (a) Show that adding a 2-valent joint to an isostatic plane framework gives an isostatic plane bar framework, if the new bars are not collinear.

    (b) Show that deleting a bar $(1, 2)$ from an isostatic plane framework, and inserting a 3-valent joint $\mathbf{p}_0$, with non-collinear bars $(0, 1), (0, 2)$, and $(0, 3)$ gives an isostatic plane bar framework for generic positions of $\mathbf{p}_0$.

    (c) (Tay & Whiteley, 1985) Show that every generically 2-isostatic graph is generated from a single edge by a *Henneberg 2-sequence* of steps (a) and (b), and that every graph with such a sequence is generically 2-isostatic.

1.14. Give a Henneberg 2-sequence for each graph in Figure 1.4a.

1.15. Show that a set of edges E' is a circuit in the plane generic rigidity matroid if and only if $|E'| = 2|V'| - 2$ and $|E''| \leqslant 2|V''| - 3$ for all proper subsets.

*1.16. (Lovász & Yemini, 1982) If a graph is 6-connected in a vertex sense, then it has realizations as an infinitesimally rigid plane bar framework.

1.17. (a) Show that all resolved loads on a plane framework are equilibrium loads.

    (b) Show that for a plane framework with at least 2 distinct joints the equilibrium loads form a space of dimensions $2|V| - 3$.

*1.18. (a) Extend the definition of polyhedra to 2-connected planar graphs.

    (b) (Crapo & Whiteley, 1988) Extend Theorems 1.3.3 and 1.3.5 to these 2-connected polyhedra.

    (c) Show that every self-stress on a plane framework is the sum of self-stresses on subframeworks with 2-connected graphs.

    (d) Conclude that a plane framework with a planar graph has a non-trivial self-stress if and only if it contains the projection of a spatial polyhedron with at least two distinct face planes.

*1.19. (a) (Crapo & Whiteley, 1988) Extend Theorem 1.3.3 to projections of arbitrary oriented polyhedra from 3-space.

    (b) Show that all self-stresses from oriented projected polyhedra satisfy the cycle condition

$$\sum_\Pi \omega_{jk}(y_j - y_k, x_k - y_j, x_j x_k - x_k y_j) = (0, 0, 0)$$

for each face–edge cycle in the oriented polyhedron.

    (c) (Crapo & Whiteley, 1988) Assume that a self-stress on the graph of an abstract oriented polyhedron satisfies the cycle condition:

$$\sum_{\Pi} \omega_{jk}(y_j - y_k,\ x_k - x_j,\ x_j y_k - x_k y_j) = (0,\ 0,\ 0)$$

for each face–edge cycle in the oriented polyhedron. Show that this framework is the projection of a spatial polyhedron, with distinct face planes over any edge with a non-zero scalar in the self-stress.

(d) Give an example of the graph of a toroidal polyhedron with a self-stress that violates the cycle condition and hence is not the projection of a toroidal polyhedron.

1.20. (a) (Whiteley, 1984a) Extend Exercise 1.9 to give an infinitesimal motion of any bipartite framework in 3-space with all joints on a quadric surface centered at the origin: $[\mathbf{p}_1][Q][\mathbf{p}_1]^t = r$.

(b) When is this infinitesimal motion non-trivial?

1.21. (a) A *simple Henneberg 3-sequence* is a sequence of graphs beginning with a triangle, and the following steps:

(1) adding a 3-valent vertex;

(2) removing one edge $\{1, 2\}$ and adding a 4-valent vertex $\mathbf{p}_0$, with edges $\{0, 1\}$, $\{0, 2\}$, $\{0, 3\}$, and $\{0, 4\}$.

Show that any graph with a simple Henneberg 3-sequence is generically 3-rigid.

(b) Show that the graph of an octahedron is generically 3-isostatic, by constructing a simple Henneberg 3-sequence.

(c) Show that the bipartite graph $K_{4,6}$ is generically 3-isostatic, by constructing a simple Henneberg 3-sequence.

1.22. (a) Show that the graph of Figure 1.9c is generically non-rigid in 3-space.

(b) Show that this graph is a generic circuit in 3-space.

1.23. Show that the cone graph $G * v_0$ is generically 3-isostatic if and only if G is generically isostatic in the plane.

1.24. (a) Show that a 5-valent vertex, in an isostatic framework in 3-space with no four joints coplanar, can be replaced by one of the two patterns in Figure 1.11a, producing smaller isostatic framework(s).

(b) Give an example of a 5-valent vertex that requires the second replacement principle.

(c) Give an example of a graph, with all vertices 5-valent, which is generically isostatic in 3-space.

1.25. (a) Give an example of a graph that is 11-connected in a vertex sense, but for which no spatial realizations are infinitesimally rigid in 3-space.

** (b) (Lovász & Yemini, 1982) Show that if a graph is 12-connected in a vertex sense, then it has realizations as an infinitesimally rigid bar framework in 3-space.

**1.26. (a) Show that the graph of any triangulated sphere has realizations as an isostatic framework with all joints on the unit sphere.

(b) Characterize the graphs that give isostatic frameworks for some realization with joints on the unit sphere.

*1.27. (Whiteley, 1987) Show that the graph of a triangulated sphere that is 4-connected in a vertex sense, plus one edge, is a circuit in the generic rigidity matroid for 3-space (Figure 1.15b). (Hint: use vertex splitting.)

*1.28. (Fogelsanger, 1988) Show that the graph of a triangulated surface in 3-space spans the generic rigidity matroid for its vertices in 3-space.

1.29. Give the plane analog of the vertex splitting theorem 1.4.6.

1.30. (a) Give the definition of infinitesimal rigidity for a bar framework in d-space. Show that an isostatic bar framework $G(\mathbf{p})$ in d-space with $|V| \geq d$ satisfies

$$|E| = d|V| - \frac{d(d+1)}{2}$$

and

$$|E'| \leq d|V'| - \frac{d(d+1)}{2}$$

for all subsets $|V'| \geq d$.

(b) Give the definition of static rigidity for a bar framework in d-space. Show that infinitesimal rigidity and static rigidity are equivalent for bar frameworks in d-space.

(c) Give an example, for each $d > 2$, of a framework that is not generically rigid, but satisfies $|E| = d|V| - d(d+1)/2$ and $|E'| \leq d|V'| - d(d+1)/2$ for all subsets $|V'| \geq d$.

1.31. (Bolker & Roth, 1980) Show that generic realizations of the graph $K_{7,7}$ in 4-space are not infinitesimially rigid. (Hint: 14 points lie on a quadric in 4-space.) What is the dimension of its space of self-stresses?

*1.32. (a) Show that the cone framework $G*v_0(\mathbf{p})$ is infinitesimally rigid in d-space if and only if the central projection $G(\Pi(\mathbf{p}))$ from apex $\mathbf{p}_0$ is infinitesimally rigid in $(d-1)$-space.

1.33. A *simplicial 4-manifold* is a collection of tetrahedra such that each triangular face is shared by at most two tetrahedra, and the *link* of each vertex v_0 (the edges that form a triangular face of a tetrahedron with v_0) forms a triangulated 2-sphere. The manifold is *strongly connected* if each pair of tetrahedra is connected by a path of tetrahedra and shared triangles. Show that a strongly connected 2-manifold satisfies the lower bound (Barnette, 1973)

$$|E| \geq 4|V| - 10$$

by the following sequence of steps (Kalai, 1987):

(a) The star of each vertex (the edges of triangles sharing this vertex) is generically rigid in 3-space. (Hint: use Exercise 1.32.)

(b) If two infinitesimally rigid frameworks in 4-space share four vertices, their union is also infinitesimally rigid.

(c) The graph of a strongly connected simplicial 4-polytope is generically rigid in 4-space.

1.34. (a) (Gromov, 1986; Kalai, 1987) A *simplicial 4-pseudomanifold* allows the vertex link to be any triangulated 2-surface. Use Exercise 1.28 to extend the results of Exercise 1.33 to strongly connected pseudo-manifolds.

(b) (Barnette, 1973; Kalai, 1987) Generalize Exercise 1.33 to give the lower bound

$$|E| \geqslant d|V| - \frac{d(d+1)}{2}$$

for strongly connected simplicial d-manifolds.

1.35. (a) Show that if a conic-independent geometric graph $G(\mathbf{p})$ in the plane has three joints $\mathbf{p}_1$, $\mathbf{p}_2$, and $\mathbf{p}_3$, then inserting a 3-valent joint $\mathbf{p}_0$ with no two of the lines $(\mathbf{p}_0, \mathbf{p}_1)$, $(\mathbf{p}_0, \mathbf{p}_2)$, or $(\mathbf{p}_0, \mathbf{p}_3)$ parallel creates a conic-independent geometric graph.

(b) Conversely, given a conic-independent geometric graph with a 3-valent vertex, show that deleting this joint with its bars leaves a conic-independent geometric graph.

1.36. (a) Show that if a conic-independent geometric graph has four joints $\mathbf{p}_1$, $\mathbf{p}_2$, $\mathbf{p}_3$, and $\mathbf{p}_4$, of which no three are collinear, and an edge $(\mathbf{p}_1, \mathbf{p}_4)$, then removing this edge and inserting a vertex $\mathbf{p}_0$ with edges $(\mathbf{p}_0, \mathbf{p}_1)$, $(\mathbf{p}_0, \mathbf{p}_2)$, $(\mathbf{p}_0, \mathbf{p}_3)$, and $(\mathbf{p}_0, \mathbf{p}_4)$ creates a conic-independent geometric graph for almost all positions of $\mathbf{p}_0$.

(b) Conversely, given a conic-independent geometric graph with a 4-valent vertex $\mathbf{p}_0$, show that there is an edge $(\mathbf{p}_i, \mathbf{p}_j)$ among the vertices adjacent to $\mathbf{p}_0$ such that deleting $\mathbf{p}_0$ and inserting $(\mathbf{p}_i, \mathbf{p}_j)$ leaves a conic-independent geometric graph.

1.37. Prove that if a conic-rigid (independent) geometric graph $G(\mathbf{p})$ contains two disjoint edges $(1, 2)$ and $(3, 4)$ and an additional vertex v_5, then the geometric graph with edges $(1, 2)$, $(3, 4)$ removed and an added point $\mathbf{p}_0$ joined by edges $(0, 1)$, $(0, 2)$, $(0, 3)$, $(0, 4)$, and $(0, 5)$ is conic-rigid (independent) for generic positions $\mathbf{p}_0$.

1.38. (a) (Chui & Wang, 1983) Prove that two quadratic surfaces

$$z = (A^h x^2 + B^h xy + C^h y^2 + D^h x + E^h y + F^h)$$

and

$$z = (A^i x^2 + B^i xy + C^i y^2 + D^i x + E^i y + F^i)$$

meet over a line $px + qy + r = 0$, with common tangent planes, if and only if

$$(A^h x^2 + B^h xy + D^h x + E^h y + F^h) - (A^i x^2 + B^i xy + C^i y^2 + D^i x + E^i y + F^i)$$
$$= \omega^{hi}(p^2 x^2 + 2pqxy + q^2 y^2 + 2prx + 2qry + r^2).$$

(b) Given any set of scalars ω_{ij} assigned to the interior edges of a triangulated plane disc, such that at each interior vertex

$$\sum_j \omega_{ij}(A_{ij}, B_{ij}, C_{ij}) = \mathbf{0},$$

show that there are unique scalars on the three edges of the exterior triangle such that $\sum_j \omega_{ij}(A_{ij}, B_{ij}, C_{ij}) = \mathbf{0}$ at the exterior vertices as well.

1.39. (Projective invariance of conic-rigidity; Whiteley, 1986b) Show that if a non-singular projective transformation in the plane, T, takes the points $\mathbf{p} = (\mathbf{p}_1, \ldots, \mathbf{p}_{|V|})$ to the points $\mathbf{q} = (T(\mathbf{p}_1), \ldots, T(\mathbf{p}_{|V|}))$ then the geometric graphs $G(\mathbf{p})$ and $G(\mathbf{q})$ define the same conic-rigidity matroid for any graph G.

1.40. For a plane geometric graph $G(\mathbf{p})$ define the *triple-line vector* for a directed edge (i, j) as

$$\mathbf{D}_{ij}^3 = (p_{ij}^3, 3p_{ij}^2 q_{ij}, 3p_{ij}q_{ij}^2, q_{ij}^3)$$
$$= ((y_j - y_i)^3, 3(y_j - y_i)^2(x_i - x_j), 3(y_j - y_i)(x_i - x_j)^2, (x_i - x_j)^3) \quad \text{for } i < j$$

and

$$\mathbf{D}_{ji}^3 = -\mathbf{D}_{ij}^3 \quad \text{for } i < j.$$

Define $RCUB(G, \mathbf{p})$ as the $-|E| \times 4|V|$ *cubic-rigidity matrix* for a geometric graph with the following row for an edge (i, j):

$$[0\,0\,0 \quad \dots \quad \mathbf{D}_{ij}^3 \quad \dots \quad -\mathbf{D}_{ij}^3 \quad \dots \quad 0\,0\,0]$$

(a) Prove that the solution space to $RCUB(G, \mathbf{p}) \times \mathbf{u}^t = \mathbf{0}$ has dimension at least 10.

(b) Show that the graph of a tetrahedron is cubic-rigid for almost all plane geometric graphs $G(\mathbf{p})$.

(c) Show that adding a general position 4-valent vertex to a cubic-rigid geometric graph produces a new cubic-rigid geometric graph.

(d) Conclude that for the generic geometric graph $K_m(\mathbf{p})$ in the plane, the cubic-rigidity matrix defines a matroid of rank $4|V| - 10$.

(e) Show that dependences in the cubic-rigidity matroid correspond to C_3^2-splines over a planar drawing of a 2-connected graph with triangular boundary.

1.41. (a) Prove that if a cubic-rigid (independent) geometric graph $G(\mathbf{p})$ contains two disjoint edges $(1, 2)$ and $(3, 4)$ and two additional vertice v_5, v_6, then the geometric graph with edges $(1, 2)$, $(3, 4)$ removed and with an added point $\mathbf{p}_0$ joined by edges $(0, 1)$, $(0, 2)$, $(0, 3)$, $(0, 4)$, $(0, 5)$, and $(0, 6)$ is cubic-rigid (independent) for generic positions $\mathbf{p}_0$.

(b) (Maehara, 1988) Show that the graph $K_{6,7}$ comes from the graph K_6 minus one edge, by a sequence of double-edges replacements of type (a).

(c) Show that $K_{6,7}$ is generically cubic-rigid.

(d) Show, by using a quadric surface in 4-space, that $K_{6,7}$ is not generically rigid in 4-space.

(e) Prove that the generic cubic-rigidity matroid and the generic 4-space rigidity are different.

** (f) Prove that each set of edges that is independent in the generic 4-space rigidity is independent in the generic cubic-rigidity matroid.

1.42. (a) Define the *d-fold line vector* for a directed edge (i, j) of a plane geometric graph $G(\mathbf{p})$, and the corresponding $\|E\| \times (d + 1)|V|$ *d-fold-rigidity matrix* for a geometric graph. Show that for the generic geometric graph $K_{|V|}(\mathbf{p})$ in the plane, with $|V| \geq d$, the d-fold-rigidity matrix defines a matroid of rank $(d + 1)|V| - (d + 2)(d + 1)/2$.

(b) Prove that the generic d-fold-rigidity matroid and the generic $(d + 1)$-space rigidity are different, for $d \geq 3$.

(c) Show that dependences in the d-fold-rigidity matroid correspond to C_d^{d+1}-splines over a planar drawing of a 2-connected graph with a triangular boundary.

1.43. (White & Whiteley, 1987) A *k-frame* $G(\mathbf{d})$ is a multi-graph $G = (V, E)$ (without loops) and an assignment $\mathbf{d}$ of vectors $\mathbf{d}_e \in \mathbb{R}^k$ to the edges. An *infinitesimal motion* of a k-frame is an assignment $\mathbf{u}$ of centers $\mathbf{u}_i \in \mathbb{R}^k$ to the vertices such that for each edge e joining v_i and v_j, $d_e \cdot (\mathbf{u}_i - \mathbf{u}_j) = 0$. A k-frame is *infinitesimally rigid* if every infinitesimal motion is *trivial*, with all $\mathbf{u}_i = \mathbf{c}$ for a fixed vector $\mathbf{c}$. A k-frame $G(\mathbf{d})$ is *isostatic* if $G(\mathbf{d})$ is a minimal infinitesimally rigid k-frame on its vertices (i.e. $|E| = k|V| - k$ and $G(\mathbf{d})$ is infinitesimally rigid).

Prove that for a multi-graph G, the following are equivalent:

(1) $G(\mathbf{d})$ is an isostatic k-frame for some $\mathbf{d} \in \mathbb{R}^{k|E|}$;

(2) G is the union of k edge-disjoint spanning trees;

(3) G satisfies $|E| = k|V| - k$ and $|E'| \leqslant k|V'| - k$ for all non-empty subsets E'.

1.44. (Whitely, 1988a) Given a graph and a partition of the vertices $(..., V_j, ...)$ of a multi-graph, the *contracted graph* $G^*(V^*, E^*)$ identifies all vertices of each partition class V_j and removes all loops.

    (a) Prove that a multi-graph has realizations as an infinitesimally rigid k-frame if and only if $|E^*| \geqslant k|V^*| - k$ for all contracted graphs G^* of partitions of the vertices.

    (b) Show that a multi-graph G that is $2k$-connected in an edge sense is rigid as a generic k-frame.

1.45. A non-zero 3-vector $\mathbf{L}^i = (L_1^i, L_2^i, L_3^i)$ represents a *line* in the plane: $L_1^i x + L_2^i y + L_3^i = 0$. A *bar and body framework* in the plane is a multi-graph G, and an assignment $\mathbf{L}$ of lines $\mathbf{L}_{ij}$ to the edges of the graph. The *rigidity matrix* for the framework is the rigidity matric for the corresponding 3-frame. (Visualize the lines as bars joining large rigid bodies for the vertices joined by bars along the lines of the edges.)

    (a) Prove that for a multi-graph G, the following conditions are equivalent:

        (1) $G(\mathbf{L})$ is an isostatic bar and body framework in the plane for some $\mathbf{L} \in \mathbb{R}^{3|E|}$;

        (2) G is the union of three edge-disjoint spanning trees;

        (3) G satisfies $|E| = 3|V| - 3$ and $|E'| \leqslant 3|V'| - 3$ for all non-empty subsets E'.

    (b) Show that a multi-graph G that is 6-connected in an edge sense has realizations as an infinitesimally rigid bar and body framework in the plane.

    (c) (Whiteley, 1984c) Give an interpretation of the reduced simple picture matrix as a bar and body framework in the plane.

1.46. A 6-vector $\mathbf{L}^i = (L_1^i, L_2^i, L_3^i, L_4^i, L_5^i, L_6^i)$ represents a *line* in 3-space if and only if $L_1^i L_4^i + L_2^i L_5^i + L_3^i L_6^i = 0$. (These are the Plücker coordinates of the line.) A *bar and body framework* in 3-space is a multi-graph G, and an assignment $\mathbf{L}$ of lines $\mathbf{L}_{ij}$ to the edges of the graph. The *rigidity matrix* for the framework is the rigidity matrix for the corresponding 6-frame.

    (a) (Tay, 1984) Prove that for a multi-graph G, the following conditions are equivalent:

        (1) $G(\mathbf{L})$ is an isostatic bar and body framework in 3-space for some $\mathbf{L} \in \mathbb{R}^{6|E|}$;

        (2) G is the union of 6 edge-disjoint spanning trees;

        (3) (Tay, 1984) G satisfies $|E| = 6|V| - 6$ and $|E'| \leqslant 6|V'| - 6$ for all non-empty subsets E'.

    (b) Give an example of a multi-graph G that is 11-connected in an edge sense, but has no realizations as an infinitesimally rigid bar and body framework in 3-space.

1.47. Define a *body and hinge framework* in 3-space as a graph G and an assignment of hinge lines $\mathbf{H}^{ij}$ to the edges of G. An *infinitesimal motion* of a body and hinge framework is an assignment of *screw centers* $\mathbf{S}^i = (S_1^i, S_2^i, S_3^i, S_4^i, S_5^i, S_6^i)$ to the vertices such that, for each hinge, $\mathbf{S}^i - \mathbf{S}^j = \omega^{ij}\mathbf{H}^{ij}$ for a scalar ω^{ij}. A body and hinge framework is infinitesmially rigid if the only infinitesimal motions are trivial, with all $\mathbf{S}^i$ the same.

  (a) Show that a hinge between two rigid bodies in 3-space is equivalent to five bars passing through the line of the hinge.

  (b) (Tay & Whiteley, 1983; Whiteley, 1988a) Show that a body and hinge framework in 3-space is independent if and only if $5|E'| \leq 6|V'| - 6$ for all non-empty subsets of edges.

  (c) (Tay & Whiteley, 1983; Whiteley, 1988a) Show that a body and hinge framework in 3-space is generically infinitesimally rigid if and only if $5|E^*| \geq 6|V^*| - 6$ for all contracted graphs G^* from partitions of the vertices.

1.48. (a) Show that the faces and edges of a spherical polyhedron define a graph that is infinitesimally rigid as a generic body and hinge framework in 3-space.

  (b) (Crapo & Whiteley, 1982) Show that the faces and edges of a spherical polyhedron form an infinitesimally flexible body and hinge framework in 3-space if and only if the edges and vertices form a bar framework with a non-trivial self-stress.

1.49. (a) Show that a multi-graph $G = (F, E)$ has realizations as an independent picture of a simple picture if and only if the graph, with any added edge, is the union of three edge-disjoint spanning trees.

  (b) Show that a multi-graph $G = (F, E)$ has realizations as an independent picture of a sharp simple picture if and only if the graph is the union of four edge-disjoint trees, and no three subtrees span the same vertices.

1.50. (a) Extend the definitions of section 1.6 to elementary pictures, and general pictures in $(d-1)$-space of scenes in d-space.

  (b) Show that a hypergraph H of $(d+1)$-tuples has realizations as an independent $(d-1)$-picture if and only if $|F'| \leq |V'| - d$ for every non-empty subset of faces F'.

  (c) Show that a generic picture in $(d-1)$-space of an incidence structure $S = (V, F; I)$ has sharp scenes in d-space if and only if $|I'| \leq |V'| + d|F'| - (d+1)$ for all subsets I' with at least two faces.

1.51. (Projective invariance of scene analysis) Show that the rank of the picture matrix is unchanged by a projective transformation of the picture.

1.52. (a) Show that for simple scenes in the plane, the independent sets are characterized by $|E'| \leq 2|V'| - 3$ for all non-empty subsets.

  (b) Show that a plane geometric graph $G(\mathbf{p})$ at a generic point $\mathbf{p}$ is independent for parallel redrawings if and only if

$$|E| = 2|V| - 3 \quad \text{and} \quad |E'| \leq 2|V'| - 3 \quad \text{for all proper subsets.}$$

  (c) Give an isomorphism between parallel redrawings of a plane geometric graph and infinitesimal motions of the corresponding bar framework, which takes trivial parallel redrawings to trivial infinitesimal motions.

1.53. (a) For a configuration of points and planes in space, satisfying a fixed incidence structure $S = (V, F; I)$, define the *parallel redrawing matrix*.

(b) (Projective invariance of parallel redrawing; Kallay, 1982) Show that the rank of the parallel redrawing matrix in 3-space is unchanged by a projective transformation of the points and planes of a configuration.

1.54. (a) (Whiteley, 1986a) Describe the parallel redrawing polymatroid for geometric graphs in 3-space, and characterize the independent sets for generic geometric graphs.

(b) (Whiteley, 1986a) Show that a non-trivial parallel redrawing of a geometric graph in 3-space induces a set of non-trivial infinitesimal motions of the corresponding bar framework.

(c) Give an example of a geometric graph in 3-space with only trivial parallel drawings, but with non-trivial infinitesimal motions as a bar framework.

(d) (Shephard, 1963) Show that the graph of a triangulated spherical polyhedron has only trivial parallel redrawings in 3-space.

(e) (Shephard, 1963) Generalize this to graphs of spherical polyhedra where any two vertices are joined by a path of triangular faces and edges.

1.55. (a) (Whiteley, 1990b) Choose an incidence structure $I \subset V \times F$ of vertices and faces that must be satisfied by spatial points and planes:

$$A^h x_j + B^h y_j + z_j + C^h = 0.$$

For each spatial realization, this creates a linear constraint on local changes to the face planes and the vertex points:

$$A^h (x_j)' + B^h (y_j)' + (z_j)' + (A^h)' x_j + (B^h)' y_j + (C^h)' = 0.$$

Over the set of incidences, this creates an $|I| \times (3|V| + 3|F|)$ matrix $RI(\mathbf{Q}, \mathbf{q})$. Show that for the incidences of a spherical polyhedron, and any realization with all faces and vertices distinct, the rows of $RI(\mathbf{Q}, \mathbf{q})$ are independent.

1.56. (a) Show that a collinear triangle is a geometric circuit for infinitesimal plane rigidity.

(b) A *simple plane framework* is a plane framework which can be created from a single bar by adding a sequence of 2-valent joints, with the two added bars not collinear. Show that a simple plane framework creates an independent rigidity matrix.

(c) Show that the plane framework on $K_{3,3}$ is a geometric circuit if the six points lie on a plane conic section, with no three collinear.

(d) Show that the projection of a triangular prism in 3-space, with all faces distinct non-vertical planes, is a geometric circuit as a plane framework.

1.57. Give the geometric conditions for frameworks on each of the graphs in Figure 1.21 to be geometric circuits.

1.58. (a) Show that for the graph of an octahedron with aa', bb', cc' not bars, the spatial framework $G(\mathbf{p})$ is dependent if and only if the six lines ab', $b'c$, ca', $a'b$, bc', and $c'a$ lie on a projective line complex in 3-space.

(b) Show that for the graph of an octahedron, the spatial framework $G(\mathbf{p})$ is dependent if and only if the four planes abc, $a'b'c$, $a'bc'$, and $ab'c'$ are concurrent in a point d.

Figure 1.21.

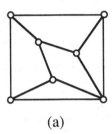 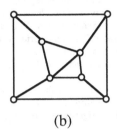 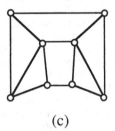

(a)   (b)   (c)

(c) (Whiteley, 1986b) Show that for the graph of an octahedron, the plane
geometric graph $G(\mathbf{p})$ is conic-dependent if and only if the six lines ab',
$b'c$, ca', $a'b$, bc', and $c'a$ are tangent to a conic section.

References

Alfeld, P. (1987). A case study of multivariate piecewise polynomials, in G. Farin (ed.),
Geometric Modeling: Algorithms and New Trends, pp. 149–60. SIAM Publications.

Asimow, L. & Roth, B. (1978). Rigidity of graphs, *Trans. Amer. Math. Soc.* **245**, 279–89.

Barnette, D. (1973). A proof of the lower bound conjecture for convex polytopes, *Pacific J. Math.* **46**, 349–54.

Barnette, D. (1982). Generating the triangulations of the projective plane, *J. Comb. Theory Ser. B* **33**, 22–230.

Billera, L. (1986). Homology theory of smooth splines: generic triangulations and a conjecture of Strang, *Trans. Amer. Math. Soc.* **310**, 325–40.

Bolker, E. & Crapo, H. (1977). Bracing rectangular frameworks I, *SIAM J. Appl. Math.* **36**, 473–90.

Bolker, E. & Roth, B. (1980). When is a bipartite graph a rigid framework?, *Pacific J. Math.* **90**, 27–44.

Brylawski, T. (1985a). Coordinatizing the Dilworth truncation, in L. Lovász & A. Recski (eds), *Matroid Theory*, (Szeged, 1982), pp. 61–95. North-Holland, New York.

Brylawski, T. (1985b). Matroid constructions, in N. White (ed.), *Matroid Theory*, Cambridge University Press.

Cauchy, A. (1831). Deuxième memoire sur les polygons et les polyédres, *J. École Polytechnique* **XVI**, 87–98.

Chui, C. K. & Wang, R. H. (1983). On smooth multivariate spline functions. *Math. of Comp.* **41**, 131–42.

Connelly, R. (1987). Basic concepts of rigidity, preprint chapter, Cornell University, Ithaca, New York.

Crapo, H. (1985). The combinatorial theory of structures, in L. Lovász & A. Recski (eds), *Matroid Theory* (Szeged, 1982), pp. 107–213, North-Holland, New York.

Crapo, H. (1988). On the generic rigidity of plane frameworks, preprint, INRIA B.P. 105, 78153 Le Chesnay Cedex, France.

Crapo, H. & Whiteley, W. (1982). Stresses on frameworks and motions of panel structures: a projective geometric introduction, *Structural Topology* **6**, 42–82.

Crapo, H. & Whiteley, W. (1988). Plane stresses and projected polyhedra I: The basic pattern, preprint, Champlain Regional College, 900 Riverside Drive, St Lambert, Québec, J4P-3P2.

Cremona, L. (1890). *Graphical Statics* (English translation), Oxford University Press.

Dehn, M. (1916). Über die Starrheit konvexer Poleder, *Math. Ann.* **77**, 466–73.

Edmonds, J. (1970). Submodular functions, matroids and certain polyhedra, in R. Guy *et al.* (eds), *Combinatorial Structures and their Applications*, pp. 69–87, Gordon & Breach, New York.

Fogelsanger, A. (1988). Generic rigidity of triangulated homology 2-cycles, Ph.D. Thesis, Cornell University, Ithaca, New York.

Gluck, H. (1975). Almost all simply connected surfaces are rigid, in *Geometric Topology*, Lecture Notes in Mathematics 438, pp. 225–39. Springer-Verlag, Berlin and New York.

Graver, J. (1984). A combinatorial approach to infinitesimal rigidity, preprint, Department of Mathematics, Syracuse University, Syracuse, New York 13210.

Gromov, M. (1986). *Partial Differential Relations*, Springer, Berlin, Heidelberg, and New York.

Henneberg, L. (1911). *Die Graphische Statik*, Leipzig.

Imai, H. (1985). On the combinatorial structure of line drawings of polyhedra, *Discrete Appl. Math.* **10**, 79–82.

Kalai, G. (1987). Rigidity and the lower bound theorem I. *Invent. Math.* **88**, 125–51.

Kallay, M. (1982). Indecomposible polytopes, *Israel J. Math.* **41**, 235–43.

Laman, G. (1970). On graphs and the rigidity of plane skeletal structures, *J. Engineering Math.* **4**, 331–40.

Lovasz, L. & Yemini, Y. (1982). On generic rigidity in the plane, *SIAM J. Alg. Discrete Meth.* **3**, 91–8.

Maehara, H. (1988). On Graver's Conjecture, Ryuku University, Nishihara, Okinawa, Japan.

Maxwell, J. C. (1864). On reciprocal diagrams and diagrams of forces. *Phil. Mag.* (4) **27**, 250–61.

Pym, J. S. & Perfect, H. (1970). Submodular functions and independence structures, *J. Math. Analysis Appl.* **30**, 1–30.

Rankine, W. J. M. (1863). On the application of barycentric perspective to the transformation of structures. *Phil. Mag.* (4) **26**, 387–8.

Recski, A. (1984). A network approach to rigidity of skeletal structures II, *Discrete Appl. Math.*, **8**, 63–8.

Recski, A. (1989). *Matroid Theory and its Applications*, Springer-Verlag, Berlin, and Akademiai Kiadó, Budapest.

Roth, B. & Whiteley, W. (1981). Tensegrity frameworks, *Trans. Amer. Math. Soc.* **265**, 419–45.

Schumaker, L. L. (1979). On the dimension of spaces of piecewise polynomials in two variables, in W. Schemp & K. Zeller (eds), *Multivariate Approximation Theory*, pp. 396–412. Birkauser, Basel.

Shephard, G. C. (1963). Decomposible convex polyhedra, *Mathematika* **10**, 89–95.

Smilansky, Z. (1987). Decomposibility of polytopes and polyhedra, *Geom. Dedicata* **24**, 29–49.

Strang, G. (1973). The dimension of piecewise polynomial spaces and one-sided approximations, in *Proceedings of the Conference on Numerical Solution of Differential Equations* (Dundee, 1973), Lecture Notes in Mathematics 365, pp. 144–52. Springer-Verlag, New York.

Sugihara, K. (1984). An algebraic and combinatorial approach to the analysis of line drawings of polyhedra, *Discrete Appl. Math.* **9**, 77–104.

Sugihara, K. (1985). Detection of structural inconsistency in systems of equations with degrees of freedom, *Discrete Appl. Math.* **10**, 297–312.

Sugihara, K. (1986). *Machine Interpretation of Line Drawings*, M.I.T. Press, Cambridge, Mass.

Tay, T.-S. (1984). Rigidity of multi-graphs I: Linking rigid bodies in *n*-space, *J. Comb. Theory Ser. B* **36**, 95–112.

Tay, T.-S. (1991). On generically dependent bar and joint frameworks in space, in *Structural Topology* (to appear).

Tay, T.-S. & Whiteley, W. (1983). Recent progress in the generic rigidity of frameworks, *Structural Topology* **9**, 31–8.

Tay, T.-S. & Whiteley, W. (1985). Generating isostatic frameworks, *Structural Topology* **11**, 21–69.

White, N. & Whiteley, W. (1983). The algebraic geometry of stresses in frameworks, *SIAM J. Alg. Discrete Meth.* **4**, 481–511.

White, N. & Whiteley, W. (1987). The algebraic geometry of motions in bar and body frameworks, *SIAM J. Alg. Discrete Meth.*, **8**, 1–32.

Whiteley, W. (1982). Motions, stresses and projected polyhedra, *Structural Topology* **7**, 13–38.

Whiteley, W. (1983). Cones, infinity and one-story buildings, *Structural Topology* **8**, 53–70.

Whiteley, W. (1984a). Infinitesimal motions of a bipartite framework, *Pacific J. Math.* **110**, 233–55.

Whiteley, W. (1984b). Infinitesimally rigid polyhedra I: Statics of frameworks, *Trans. Amer. Math. Soc.* **285**, 431–65.

Whiteley, W. (1984c). A correspondence between scene analysis and motions of frameworks,

Discrete Appl. Math. **9**, 269–295.

Whiteley, W. (1986a). Parallel redrawings of configurations in 3-space, preprint, Champlain Regional College, 900 Riverside Drive, St Lambert, Québec J4P-3P2.

Whiteley, W. (1986b). The analogy between algebraic splines and hinged panel structures, preprint, Champlain Regional College, 900 Riverside Drive, St Lambert, Québec J4P-3P2.

Whiteley, W. (1987). Infinitesimally rigid polyhedra II: Modified spherical frameworks, *Trans. Amer. Math. Soc.* **306**, 115–39.

Whiteley, W. (1988a). Unions of matroids and rigidity of frameworks. *SIAM J. Discrete Math.* **1**, 237–55.

Whiteley, W. (1988b). A matroid on hypergraphs, with applications to scene analysis and geometry, *Discrete Comp. Geometry* **4**, 75–95.

Whiteley, W. (1990a). The geometry of C_2^1 splines, preprint, Champlain Regional College, 900 Riverside Drive, St Lambert, Québec J4P-3P2.

Whiteley, W. (1990b). Determination of spherical polyhedra, preprint, Champlain Regional College, 900 Riverside Drive, St Lambert, Québec J4P-3P2.

Whiteley, W. (1991a). Vertex splitting in isostatic frameworks, *Structural Topology* **16**, 23–30.

Whiteley, W. (1991b). The combinations of bivariate splines, in P. Gritzmann & B. Sturmfels (eds), *Applied Geometry and Discrete Mathematics, The Victor Klee Feitschrift*, AMS DIMACS Series, in press.

Whiteley, W. (1991c). A matrix for splines, in *Progress in Approximation Theory*, Niven & Pinkus (eds), Academic Press, in press.

Whiteley, W. (1991d). Weaving, sections and projections of polyhedra, *Discrete Appl. Math.*, in press.

2

Perfect Matroid Designs

M. DEZA

2.1. Introduction

A *perfect matroid design* (*PMD*) is a matroid such that all flats of rank i have the same cardinality α_i. Some other names used for PMDs are: uniformly designed lattices, geometric lattices of triangular type, combinatorially homogeneous matroids, geometric designs, matroid designs, and Murty–Young–Edmonds designs (MYEDs). PMDs were introduced in Murty (1970) and the first study of them was given in Murty, Young & Edmonds (1970). The sequence $\alpha = (\alpha_0, \alpha_1, ..., \alpha_{r-1}, \alpha_r)$ of flat sizes for a PMD of rank r is called its α-*sequence*. By the same process of geometrization as in Brylawski (1986) p. 171, it is easy to obtain from any PMD with α-sequence $(\alpha_0, \alpha_1, ..., \alpha_{r-1}, \alpha_r)$ a PMD with α-sequence $(\alpha_0 - \alpha_0 = 0, \alpha_1 - \alpha_0, ..., \alpha_{r-1} - \alpha_0, \alpha_r - \alpha_0)$ and further a PMD with α-sequence

$$\left(\frac{0}{\alpha_1 - \alpha_0} = 0, \frac{\alpha_1 - \alpha_0}{\alpha_1 - \alpha_0} = 1, ..., \frac{\alpha_{r-1} - \alpha_0}{\alpha_1 - \alpha_0}, \frac{\alpha_r - \alpha_0}{\alpha_1 - \alpha_0} \right)$$

i.e. a combinatorial geometry. Without loss of generality (as far as the problems considered in this chapter are concerned) we shall suppose throughout that $\alpha_0 = 0$, $\alpha_1 = 1$.

Though PMDs are a very special case of matroid a PMD is nevertheless a common generalization of three highly symmetric finite objects – finite geometries, designs, and commutative Moufang loops. Also, two more reasons for being interested in PMDs come from combinatorics. One reason is that the existence of each PMD implies the existence of many designs, and it will be seen that Proposition 2.2.1 below can be a way to look for new ones. The other reason is that almost all PMDs correspond to the case of maximality for some inequality generalizing many results from coding theory and the extremal theory of finite sets (see Proposition 2.3.1 below).

Given a PMD $G(S)$ of rank r, let $G(S):1$ (which is called 1-*truncation of*

$G(S)$) denote the PMD of rank $r - 1$ having the set of all flats of $G(S)$ except hyperplanes as the lattice of flats. By induction for any $1 < i \leqslant r - 1$ define the i-truncation $G(S){:}i$ of $G(S)$ as the 1-truncation of $G(S){:}(i - 1)$; define the 0-truncation $G(S){:}0$ as $G(S)$ itself. We will call the *contraction* of the PMD $G(S)$ *on a flat* F' the PMD with the lattice $\{F \mid F' \subseteq F \text{ and } F \text{ is a flat of } G(S)\}$ of flats; the contraction can be seen as 'truncation from the bottom'. We will call the *restriction* (Brylawski, 1986, p. 131) of PMD $G(S)$ *on a flat* F' the PMD with lattice $\{F' \cap F \mid F \text{ is a flat of } G(S)\}$ of flats. Any i-truncation, $0 \leqslant i \leqslant r - 1$, of a PMD, any contraction, and any restriction of a PMD on one of its flats are PMDs.

The known examples of PMDs are:

(1) i-truncations $AG(d, q){:}i$, $1 \leqslant i \leqslant r - 1$, of affine geometries of dimension d ($AG(d, q){:}0$, i.e. $AG(d, q)$ itself included) having rank $d - i + 1$ and $\alpha = (0, 1 = q^0, q = q^1, ..., q^{d-i-1}, q^d)$;

(2) i-truncations $PG(d, q){:}i$, $1 \leqslant i \leqslant r - 1$, of projective geometries of dimension d ($PG(d, q){:}0$, i.e. $PG(d, q)$ itself included) having rank $d - i + 1$ and

$$\alpha = \left(0, 1 = \frac{q-1}{q-1}, q+1 = \frac{q^2-1}{q-1}, ..., \frac{q^{d-i}-1}{q-1}, \frac{q^{d+1}-1}{q-1}\right);$$

(3) Steiner systems $S(t, k, v)$ having

$$\alpha = (0, 1, 2, ..., t-1, k, v);$$

(4) PMDs of rank 4 with $\alpha = (0, 1, 3, 9, 3^n)$ arising from finite commutative Moufang (but not associative) loops; we will call them *triffids* (they are surveyed in section 2.5).

The special case $S(k, k, v)$ of a Steiner system is a truncation of a Boolean algebra. Any PMD whose α-sequence is the same as that of $AG(d, q){:}i$ ($d \geqslant 3$, $q > 3$) of $PG(d, q){:}i$ ($d \geqslant 3$) or $S(t, k, v)$ is isomorphic to $AG(d, q){:}i$ or $PG(d, q){:}i$ or $S(t, k, v)$, respectively.

For $d \geqslant 3$ $AG(d, g)$ exists if and only if q is a power of a prime; the same is true for $PG(d, q)$, $d \geqslant 3$. The classification of all $S(t, k, v)$ is a big open problem. The study of triffids started only recently. Obviously, it will be difficult to achieve the classification of all PMDs. The natural approach to this central problem will be, for example, the study of PMDs of rank 4 (because the rest is too hard) and, in particular, PMDs with $\alpha = (0, 1, 3, \alpha_3, \alpha_4)$ (those being analogs of $S(3, k, v)$) or PMDs with $\alpha = (0, 1, 3, 9, 3^n)$ (triffids and $AG(d, 3){:}(d - 3)$).

This chapter contains five more sections. Section 2.2 gives some properties of the lattice of flats of a PMD, some parameters and links among them, some inequalities for parameters, and classification of extreme cases. Section 2.3 gives information about the possible properties of a hyperplane family of being a 'large' intersection family and a subscheme of a Johnson association scheme. PMDs of rank 4 are considered in section 2.4. Section 2.5 gives a short survey on triffids. Section 2.6 mentions some generalizations and analogs of PMDs.

2.2. Some Lattice Properties and Parameters of PMDs

In this section we consider the lattice L of all flats of a PMD. In particular, we give related designs, the Sperner property of L, some rank invariants (especially Whitney numbers), and all known necessary conditions for the existence of a PMD in terms of the α-sequence.

A PMD of rank 3 is precisely a BIBD (balanced incomplete block design) with $\lambda = 1$; in other terms, it is an $S(2, k = \alpha_2, v = \alpha_3)$ or $(b, v, r, k, 1)$-configuration. PMDs of higher rank correspond to very special designs. Propositions 2.2.1–2.2.3 below are from Murty, Young & Edmonds (1970).

2.2.1. Proposition. *Given a PMD; then*

(i) *the set of all j-flats forms the set of blocks of a BIBD on the set of all 1-flats, for $1 < j < r$ (i.e. a PMD is an MD; see section 2.6 below);*

(ii) *the set of independent subsets of cardinality j is the set of blocks of a BIBD, for $1 < j \leqslant r$;*

(iii) *the set of circuits of cardinality j is the (possibly vacuous) set of blocks of a BIBD, for $2 < j \leqslant r + 1$.*

Also, each of the three sets considered above is the set of blocks of some $S(t, k, v)$, $2 \leqslant t \leqslant \beta$, where β is a maximal integer m such that every m-subset of S is independent.

The next property of the lattice of flats of a PMD considered here is its *Sperner property*, i.e. the property that the largest Sperner family (i.e. antichain) of this lattice is the set of all i-flats for some i, $1 \leqslant i \leqslant r - 1$. This follows from a theorem of Baker given in Welsh (1976), Chapter 16.5. Moreover, using Deza, Erdös & Frankl (1978) one can see that for α_r sufficiently large (with respect to α_{r-1}) this largest antichain is exactly the set of all hyperplanes.

Now we will consider some lower bounds for $|S| = \alpha_2$. Given a PMD $G(S)$ of rank r with $\alpha = (0, 1, \ldots, \alpha_{r-1}, \alpha_r)$ that is not $S(k, k, v)$, denote α_r by v and α_{r-1} by k; let c be the minimum circuit cardinality of $G(S)$.

2.2.2. Proposition. $v \geqslant k + \min(k, c, r)$.

Any PMD, other than $S(k, k, v)$, satisfies $k \geqslant r \geqslant c$ and so has $v \geqslant k + c$. We have $k = c = r = v/2$ for $AG(3, 2)$ (it is $S(3, 4, 8)$) and for Witt design $S(5, 6, 12)$.

The next property of a PMD is that for any i-flat F^i and l-flat F^l with $F^i \subset F^l$ the set $F^l - F^i$ is partitioned by sets of the form $F^{i+1} - F^i$ where F^{i+1} is an $(i+1)$-flat such that $F^i \subset F^{i+1} \subset F^l$. This is a general property of matroids (see Brylawski, 1986). Denote by $t(i, j, l)$ the number of j-flats F^j such that $F^i \subset F^j \subset F^l$; the following result comes directly from this property.

2.2.3. Proposition. *The number* $t(i, j, l)$ *is independent of the choice of* F^i, F^l; *moreover*

$$t(0, 1, i) = \alpha_i, \quad t(i, i+1, l) = \frac{\alpha_l - \alpha_i}{\alpha_{i+1} - \alpha_i},$$

$$t(0, i, r) = \sum_{j=1}^{i-1} \frac{\alpha_r - \alpha_j}{\alpha_i - \alpha_j}$$

and

$$t(i, k, l) = t(i, f, l)t(f, j, l)/t(i, f, j)$$

for
$$0 \leqslant i \leqslant f \leqslant j \leqslant l \leqslant r.$$

The number $t(0, i, r)$, i.e. the number of all i-flats, is the Whitney number $W_{r-i}^{(2)}$ of the second kind of our PMD. Denote by L the lattice of all flats of $G(S)$, by $\mu(x, y)$ the Möbius function of the lattice, and by $r(x)$ the rank of $x \in L$. The Whitney numbers $W_j^{(1)}$, $W_j^{(2)}$ of the first and second kind are defined by

$$W_j^{(1)} = \sum_{x \in L} \mu(\varnothing, x)\delta(r - r(x), r - j),$$

$$W_j^{(2)} = \sum_{x \in L} \delta(r - r(x), r - j).$$

Example 8.5.9 of White (1987) discusses *log concavity* of the sequence $W_i^{(2)}$, i.e. the inequality $W_i^{(2)} > + (W_{i-1}^{(2)} W_{i+1}^{(2)})^{\frac{1}{2}}$.

2.2.4. Proposition. (*Brini, 1980*) *Let* $x, y \in L$ *and* $x \subseteq y$; *denote* $r(x) = i$, $r(y) = l$. *Then*

$$\mu(x, y) = (-1)^{l-i} \sum_{j=i+1}^{l} [t(i, j-1, j) - t(i+1, j-1, j)].$$

So, the functions α_i, $t(i, k, l)$, $\mu(x, y)$ and both Whitney numbers are rank invariants; another rank invariant considered in Brini (1980) will be mentioned at the end of the next section.

This section gives necessary conditions for the existence of a PMD in many different terms. In terms of the α-sequence, which is our basic rank invariant, independent necessary conditions for the sequence $(0, 1, \alpha_2, ..., \alpha_r)$ to be the α-sequence of a PMD are (Murty, Young & Edmonds, 1970):

(1) $\displaystyle\prod_{f=i+1}^{j} \frac{\alpha_l - \alpha_{f-1}}{\alpha_j - \alpha_{f-1}}$ is a non-negative integer for $0 \leqslant i \leqslant j \leqslant l \leqslant r$;

(2) $(\alpha_i - \alpha_{i-1})$ divides $\alpha_{i+1} - \alpha_i$ for $2 \leqslant i \leqslant r - 1$;

(3) $(\alpha_i - \alpha_{i-1})^2 \leqslant (\alpha_{i+1} - \alpha_i)(\alpha_i - \alpha_{i-2})$ for $1 \leqslant i \leqslant r - 1$.

These conditions are not sufficient.

An example is provided (R. Wilson, unpublished, indicated in Murty, Young & Edmonds, 1970) by the sequence (0, 1, 3, 7, 43); moreover, Wilson showed that any PMD $G(S)$ with $\alpha_2 \geqslant 3$, $(\alpha_3 - \alpha_2) = (\alpha_2 - 1)^2$ has the property $\alpha_r(\alpha_2 - 2) = (\alpha_2 - 1)^f - 1$ for some f. In Ray-Chaudhuri & Singhi (1977) the inequality $\alpha_j - \alpha_i \geqslant (\alpha_2 - 1)^{j-i}(\alpha_{j-1} - \alpha_{i-1})$ was proved for $2 \leqslant i < j \leqslant r$, and the cases of equality were classified.

2.3. Intersection Properties of the Hyperplane Family

Consider a PMD $G(S)$ with hyperplane family M^{r-1} and parameters $\alpha = (\alpha_0 = 0, \alpha_1 = 1, \alpha_2, ..., \alpha_r)$; denote $\alpha_r = v$, $\alpha_{r-1} = k$, $\{\alpha_0, \alpha_1, ..., \alpha_{r-2}\} = T$. Denote by $A(T, k, v)$ any set of k-subsets of a given v-set such that the size of any pairwise intersection is a member of the given set T of integers. Of course, M^{r-1} is an $A(T, k, v)$. We will give below information known about M^{r-1} with respect to the following possible special properties of $A(T, k, v)$s:

(1) a *maximal* $A(T, k, v)$ is an $A(T, k, v)$ that is not a proper subset of an $A(T, k, v)$;

(2) more specially, a *largest* $A(T, k, v)$ is an $A(T, k, v)$ with the largest number of k-sets;

(3) a *complete* $A(T, k, v)$ is an $A(T, k, v)$ such that any number from T occurs as the size of a pairwise intersection of k-sets;

(4) an $A(T, k, v)$-*subscheme* is an $A(T, k, v)$ that is a subscheme of the Johnson association scheme of all k-subsets of a v-set.

Here, an association scheme on a finite set X *with n classes* (see, for example, Bose, 1963; Hamada & Tamari, 1975) is a partition $X \times X = N_0 \cup N_1 ... \cup N_n$ such that

(1) $(x, y) \in N_0$ iff $x = y$;

(2) $(x, y) \in N_i$ iff $(x, y) \in N_i$;

(3) for any $x \in X$ the number $m_i = |\{y \in X | (x, y) \in N_i\}|$ depends only on i;

(4) for any $x, y \in X$ with $(x, y) \in N_i$ the number $P^i_{jk} = |\{z \in X | (x, y) \in N_j$, $(y, z) \in N_k\}|$ depends only on i, j, k.

We say that the pair (x, y) is *i-associated* if $(x, y) \in N_i$. For $n = 2$ the pairs in each class can be seen as the edges of a strongly regular graph (introduced by Bose, 1963) and any such graph gives rise to an association scheme with two classes.

The *Johnson scheme* $J(k, v)$ is defined by $X = S(k, k, v)$ and $(x, y) \in N_i$ if $|x \cap y| = k - i$.

A PMD $G(S)$ can be characterized (if $v > v_0(k)$) in the class of all $A(T, k, v)$s by the following extremal property.

2.3.1. Proposition. (*Deza, Erdős & Frankl, 1978; Deza, 1978*) *Let* $A = A(T, k, v)$; *then there exists a polynomial* $v_0(T, k)$ *of argument* k *such that* $v > v_0(T, k)$ *implies*:

(i) $|A| \leqslant \prod\limits_{i \in T} \dfrac{v - i}{k - i}$;

(ii) $|A| = \prod\limits_{i \in T} \dfrac{v - i}{k - i}$

if and only if A *is the set of all hyperplanes of some PMD with* $\alpha = (\alpha_0, \alpha_1, ..., \alpha_r)$.

For the special case $T = \{0, 1, 2, ..., r - 2\}$ this proposition is just the Schönheim–Johnson bound for packings and codes (see, for example, Deza, 1978); for this T we have $v_0(T, k) = k$ and the corresponding PMD is a Steiner system $S(t = r - 1, k, v)$. From a remark of P. Frankl (1983) it follows that $v_0(T, k) \leqslant (k - \alpha_{r-2})^2 (k - \alpha_{r-2} + 1) + k$ for the case $T = \{0, 1, 2, ..., r - 3, \alpha_{r-2}\}$, i.e. if $G(S)$ is an 'extension' of an $S(r - 2, k, v)$. For the special case $T = \{0, 1, 2, ..., r - 3, r - 1\}$ of the above case Frankl gave a better bound $v_0(T, k) \leqslant (k - r + 1)^2 + k$; therefore $v_0(T, k) = k$ for all PMDs with $\alpha = \{0, 1, 3, 9, 3^n\}$, $n > 4$. The smallest $v_0(T, k)$ is unknown for truncations of $PG(d, q)$ and $AG(d, q)$ and for the triffid with $\alpha = \{0, 1, 3, 9, 81\}$.

Frankl (1983) has also shown that $|A(T, k, v)| = O(v^{|T|-1})$ unless there is a PMD with $\alpha = (\alpha_0, \alpha_1, ..., \alpha_{r-1})$. Let us also note that using the family of elliptic quadrics over $PG(d, q)$, $q > 2$, Deza, Frankl & Hirschfeld (1985) showed that for $v > v_0(q)$,

$$A(\{0, 1, 2, q + 1\}, q^2 + 1, v) > c_q \left(\frac{v}{q^2 + 1}\right)\left(\frac{v - 1}{q^2}\right)\left(\frac{v - 2}{q^2 - 1}\right)\left(\frac{v - q - 1}{q^2 - q}\right),$$

where c_q is a positive constant with $c_q \to 1$ as $q \to \infty$, i.e. the corresponding families come fairly close to PMDs.

So M^{r-1} is a largest (and, in particular, maximal) $A(T, k, v)$ for *any* v if $G(S)$ is an $S(t, k, v)$. M^{r-1} is a maximal $A(T, k, v)$ for any v if $G(S)$ is some i-truncation of either affine or projective space; it follows from the more general Theorem 3a of Deza & Singhi (1980) that M^{r-1} is a maximal $A(T, k, v)$ if $v \leqslant k^2/\alpha_{r-2}$. Hence M^{r-1} is maximal whenever v is sufficiently 'large' or 'small' with respect to k. The hyperplane family M^{r-1} of a triffid (the only remaining example of a known PMD) is also maximal; the maximality was proved in Theorem 3b of Deza & Singhi (1980) for a class of PMDs of rank 4 including any PMD with $\alpha = (0, 1, 3, k, v)$. So, we conjecture that the hyperplane family of any PMD with $\alpha = (0, 1, \alpha_2, ..., \alpha_{r-2}, k, v)$ is a maximal $A(T, k, v)$.

M^{r-1} can be a non-complete $A(T, k, v)$ but all such cases in the class of all known examples of PMDs are classified as follows.

2.3.2. Proposition. *Let $r > 3$. Then*

(i) *the i-truncation of a projective geometry $PG(d, q)$ has a complete M^{r-1} if and only if $i \geqslant (d + 1)/2$;*

(ii) *th i-truncation of an affine geometry $AG(d, q)$ has a complete M^{r-1} if and only if $i \geqslant d/2$;*

(iii) *each triffid has a complete M^{r-1};*

(iv) *the Steiner system $S(t, k, v)$ has an incomplete M^{r-1} only in the following cases (if an $S(t, k, v)$ exists):*

    (a) *$(t, k, v) = (2, n + 1, n^2 + n + 1)$ and is a projective plane of order n;*

    (b) *$t = 3$ and $(k, v) = (6, 22), (4, 8), (12, 112)$;*

    (c) *$(t, k, v) = (4, 7, 23), (5, 8, 24)$;*

    (d) *$(t, k, v) = (t, t + 1, 2t + 3)$ and $t + 3$ is a prime number;*

    (e) *$(t, k, v) = (t, t + 1, 2t + 2)$ and $t + 2$ is a prime number.*

The most difficult part, (iv), of the above proposition is given in Gross (1974); parts (i) and (ii) are given in Deza & Singhi (1980). We note that a triffid with $\alpha = (0, 1, 3, 9, 3^4)$ is complete, but $AG(4, 3)$ is not.

Any PMD with v sufficiently large (with respect to k) has a complete M^{r-1}; more exactly, Proposition 2.3.1 implies the following.

2.3.3. Proposition. *Let the PMD $G(S)$ have an incomplete M^{r-1} and, more exactly, $|H_1 \cap H_2| \neq \alpha_j$ for some hyperplanes H_1, H_2 and some $j, 0 \leqslant j \leqslant r - 2$. Then $v < v_0(T - \{\alpha_j\}, k)$.*

Given an i-flat E and a u-flat F such that $E \subset F, 0 \subset F, 0 \leqslant i < u < r$; denote by $c(i, u)$ the number of all u-flats F_1 such that $F_1 \cap F = E$.

2.3.4. Proposition. *(Deza & Singhi, 1980) The number $c(i, u)$ is independent of the choice of E and F; moreover, we have a recurrence*

$$c(i, u) = t(i, u, r) - \sum_{i < w \leqslant u} t(i, w, u)c(w, u).$$

As an immediate corollary we have the following test of completeness for M^{r-1}: it is complete if and only if all numbers $c(i, r - 1), 0 \leqslant i \leqslant r - 1$ are positive. From the above proposition one can see that

$$c(i, i+1) = (v - \alpha_i)/(\alpha_{i+1} - \alpha_i) - 1 = (v - \alpha_{i+1})/(\alpha_{i+1} - \alpha_i) > 0$$

for any $0 \leqslant i \leqslant r - 2$ and, in particular, $c(r - 2, r - 1) > 0$ always. Also, it is easy to check that $c(r - 3, r - 1) = 0$ if and only if the contraction of our PMD on a $(r - 3)$-flat is a projective plane; in particular, Gross (1974) gives that $S(t, k, v)$ has $c(r - 3, r - 1) = 0$ if and only if $(t, k, v) = (3, 6, 22), (4, 7, 23),$ (5, 8, 24), (3, 4, 8) or (3, 12, 112), if it exists.

PMDs of rank 4 having $c(1, 3) = 0$ are classified (under the name of finite regular locally projective spaces of dimension 3) by Doyen & Hubaut (1971). Each of them is either a projective space, or an affine space, or a Lobachevski space (in terms of Doyen & Hubaut, 1971) of type $\alpha_2^2 - \alpha_2 + 1$ or $\alpha_2^3 + 1$, i.e. each top rank 3 interval of the lattice of flats is a projective plane of order $q = \alpha_2^2$ or $q = \alpha_2^3 + \alpha_2$. Concerning PMDs of rank 4 having $c(0, 3) = 0$: there is a conjecture in Cameron (1980) that each of them is a truncation of a Boolean algebra or projective space of rank 4 or 5.

The number $c(i, u)$ was generalized by Brini (1980) in the following direction – given an i-flat E and a u-flat F, such that $E \subset F$, $0 \leqslant i < u < r$, denote by $c(i, j, u)$ the number of all j-flats F_1 such that $F_1 \cap F = E$. Of course, $c(i, u)$ is just $c(i, u, u)$. Then $c(i, j, u)$ is $\gamma(r - 1, r - j, r - u)$ in the notation of Brini (1980); in that paper it was proved that $c(i, j, v)$ is independent of the choice of E and F and, moreover, that the following holds.

2.3.5. Proposition. $c(i, j, u) = \displaystyle\sum_{p = r - u}^{r - i} (-1)^{p - i} t(i, p, u) t(p, j, r) \prod_{q = i + 1}^{p} [t(i, q - 1, q)$
$- t(i + 1, q - 1, q)].$

The remainder of this section will give information known about PMDs in which M^{r-1} is a subscheme of a Johnson association scheme. Any PMD of rank 3 is exactly a BIBD with $\lambda = 1$; its line graph is strongly regular and hence M^{r-1} will be a subscheme.

2.3.6. Proposition. *Let $r > 3$. Then*

(i) (*Hamada & Tamari, 1975*) *any i-truncation of $PG(d, q)$ is a subscheme;*

(ii) (*Hamada & Tamari, 1975*) *the i-truncation of $AG(d, q)$ is a subscheme if and only if $i = 0$, i.e. $G(S)$ is $AG(d, q)$ itself;*

(iii) (a) (*Cameron, 1975*) *$S(t, k, v)$ is a subscheme if it has at most $\lfloor t/2 \rfloor + 1$ different sizes of pairwise intersections of blocks (hyperplanes);*

(b) (*anonymous referee*) *$S(t, k, v)$ is a subscheme if it has at most $\lfloor (t + 1)/4 \rfloor + 1$ such sizes and if the complement of any block is a block;*

(iv) (a) (*Cameron, 1975*) *if $S(3, k, v)$ is a subscheme then $v \leqslant 2 + k(k - 1)(k - 2)/2$;*

(b) (*Atsumi, 1979*) *if $S(t, k, v)$ is a subscheme then $v \leqslant k^4 \dbinom{k}{\lfloor k/2 \rfloor}$.*

All known $S(t, k, v)$-subschemes are inversive (Möbius) planes

$I(q) = S(3, q + 1 = k, q^2 + 1 = v)$ with three classes; $S(5, 6, 12)$ is so because of case (iii)(b) above, and $S(3, 4, 8)$, $S(3, 6, 22)$, $S(4, 5, 11)$ (the unique $S(4, 5, v)$-subscheme), $S(4, 7, 23)$, $S(5, 8, 24)$, and $S(3, 12, 122)$ (if it exists) are so because of case (iii)(a) of the above theorem.

The triffid with $\alpha = (0, 1, 3, 9, 81)$ (i.e. the smallest one) is not a subscheme but its M^{r-1} has a partition into two association schemes (Deza & Hamada, 1980).

2.4. PMDs of Rank 4

The central problem in PMD studies is to find all possible α-sequences, $\alpha = (0, 1, \alpha_2, ..., \alpha_r)$. For $r = 3$ the concepts of PMD (with $\alpha = (0, 1, \alpha_2, \alpha_3)$) and BIBD $S(2, \alpha_2, \alpha_3)$ coincide: so one can use the deep but still incomplete information obtained on BIBDs. Below we survey the known information on the next, much more difficult and messy, case of rank 4. Throughout section 2.4 we shall consider, a PMD $G(S)$ of rank 4 with $\alpha = (0, 1, \alpha_2 = l, \alpha_3 = k, \alpha_4 = v)$. In sections 2.4 and 2.5 we shall denote 1-, 2- and 3-flats by the terms points, lines, and planes. Denote $\lambda = (v - l)/(k - l)$.

The existence of our PMD $G(S)$ with $\alpha = (0, 1, l, k, v)$ implies the existence of the four following related BIBDs:

(1) $D_1 = S(2, l, k)$, in which points and blocks are points and lines of $G(S)$ contained in a given plane, i.e. D_1 is the restriction of $G(S)$ on this plane;

(2) $D_2 = S\left(2, \dfrac{k-1}{l-1}, \dfrac{v-1}{l-1}\right)$, which is the geometrization of the contraction of $G(S)$ on some point $x \in S$;

(3) $D_3 = S(2, l, v)$, in which points and blocks are points and lines of $G(S)$, i.e. D_3 is the 1-truncation of $G(S)$;

(4) $D_4 = S_\lambda(2, k, v)$, in which points and blocks are points and planes of $G(S)$.

(A) The intersection of any two blocks of D_4 has size 0, 1, or l.

The existence of D_1, D_2, D_3, D_4 and condition (A) are necessary for the existence of $G(S)$.

Proposition 2.3.1(ii) gives (for the case of rank 4) that the existence of D_4 satisfying (A) is sufficient for the existence of $G(S)$, if $v > v_0$ for some polynomial $v_0(T, k)$ of k.

The necessary conditions of Murty, Young & Edmonds (1970) (of existence of a PMD in terms of divisibility of some numbers related to α) take, for our PMD, the form:

(B) All the numbers $\dfrac{k-1}{l-1}, \dfrac{k(k-1)}{l(l-1)}, \dfrac{v-l}{k-l} = \lambda, \dfrac{(v-1)(v-l)}{(k-1)(k-l)}, \dfrac{v(v-1)(v-l)}{k(k-1)(k-l)}$
are integers.

The existence of some BIBDs with the parameters of D_1, D_2, D_3, and D_4, and the validity of (B), do not together imply the condition (A) on a BIBD with the parameters of D_4; an example (given by R. M. Wilson, 1982) is provided by $(l, k, v) = (3, 19, 307)$.

Denote by $(b_i, v_i, r_i, k_i, \lambda_i)$ the parameters of the BIBDs D_i, $1 \leqslant i \leqslant 4$. Fischer's (1966) properties of a BIBD give the following necessary condition for the existence of $G(S)$:

(C) $b_i k_i = v_i r_i$, $r_i(k_i - 1) = \lambda_i(v_i - 1)$, and $r_i \geqslant k_i$ for $i = 1, 2, 3, 4$.

The case $r_i = k_i$ corresponds to a *symmetric* BIBD D_i, $\lambda_1 = \lambda_2 = \lambda_3 = 1$; so if D_1, D_2, or D_3 is symmetric then it is a projective plane.

2.4.1. Proposition.

(i) D_1 *is symmetric if and only if* $G(S)$ *is* $PG(d, q)$: $(d - 3)$, $d \geqslant 3$;
(ii) D_2 *is symmetric if* $G(S)$ *is* $PG(3, q)$, $AG(3, q)$, *or* $S(3, 6, 22)$;
(iii) D_3 *is not symmetric;*
(iv) D_4 *is symmetric if and only if* $G(S)$ *is* $PG(3, q)$.

(i) is given in Ray-Chaudhuri & Singhi (1977), (iv) follows from Dembowski & Wagner (1960). It can be shown that if $G(S)$ has rank at least 5 then D_2 is symmetric if and only if $G(S)$ is $PG(d, q)$ or $AG(d, q)$.

If the property (A) holds (that is, blocks of D_4 have only 0, 1, or l as the size of pairwise intersections), Deza & Singhi (1980) gives that: (1) there exists a point $x \in S$ such that $a \cap b \neq \{x\}$ for any blocks a, b of D_4 (i.e. planes of $G(S)$) if and only if D_2 is symmetric; (2) any $a \cap b$ is non-empty if and only if

$$(k - 1)(k\lambda - k + l) + l\left(\lambda^2 - \frac{\lambda(\lambda - 1)(l - 1)}{k - 1}\right)\left(\frac{v}{k} - l\right) = 0.$$

Using (C) one can easily calculate the parameters $(b_i, v_i, r_i, k_i, \lambda_i)$ of D_i, $1 \leqslant i \leqslant 4$.

Now, denote $t = (k - 1)/(l - 1)$, $u = kt/l$ and recall that $\lambda = (v - l)/(k - l)$. We shall call the triplet (l, k, v) of integers, $0 < l < k < v$, *admissible* if it satisfies conditions (B) and (C). It is easy to check that (l, k, v) is admissible if and only if $2 \leqslant l \leqslant t \leqslant \lambda$ and all the following numbers are integers: t, λ, $t(t - 1)/l$, $\lambda(\lambda - 1)/t$, $\lambda(\lambda - 1)(\lambda t - \lambda + 1)/u$ (it will be equivalent to require that t, λ, u, b_2, b_4 are integers). The extreme cases of the inequalities $2 \leqslant l \leqslant t \leqslant \lambda$ are given below:

2.4.2. Proposition.

(i) $l = 2$ *if and only if* $G(S)$ *is* $S(3, k, v)$;
(ii) $l = t > 2$ *if and only if* $G(S)$ *is* $PG(d, q = t - 1)$: $(d - 3)$ (recall that $G(S)$: i denotes truncation);

(iii) $\lambda = t$ *if and only if D_2 is a projective plane of order $t - 1$* (i.e. we are in case (ii) of 2.4.1);

(iv) *(Buekenhout, 1969) $l = t - 1 > 3$ if and only if $G(S)$ is $AG(d = \log_{t-1} v, q = t - 1)$: $(d - 3)$;*

(v) *(Hall, 1960) $l = t - 1 = 3$ if and only if $G(S)$ is a triffid* (see section 2.5) *or $AG(\log_3 v, 3)$.*

For case (i) of Proposition 2.4.2 we recall that the existence of $S(3, k, v)$ will imply the existence of $S(2, k - 1, v - 1)$. Hanani (1963) gives that $S(3, 4, v)$ exists if and only if $v \equiv 2, 4 \pmod 6$, $v \geq 8$. For any k such that $k - 1$ is a power of a prime there exists an inversive plane $I(k - 1)$ of order $k - 1$ that is an $S(3, k, (k - 1)^2 + 1)$. In case (ii) of Proposition 2.4.2 the dimension d of $PG(d, q)$ is defined by $v = (q^{d+1} - 1)/(q - 1)$, and $PG(d, q)$ (for our case $d > 2$) exists if and only if $q = t - 1$ is a power of a prime. In case (iv) of Proposition 2.4.2 we have $v = l^d$ and our $G(S)$ exists if and only if $q = t - 1$ is a power of a prime. Case (v) is considered in section 2.5. One can see, for example, that 'smallest' parameters (l, k, v), such that BIBDs with the parameters of D_1, D_2, D_3 exist but a PMD is unknown, are (3, 13, 183), (3, 13, 313), (3, 15, 183), (3, 15, 675), (3, 19, 307), (3, 19, 1027), (4, 25, 1201), and (4, 25, 8404). In all these cases the existence of a $G(S)$ requires the existence of a $D_4 = S_\lambda(2, l, v)$ with only (0, 1, 3) (for $l = 3$) and (0, 1, 4) (for $l = 4$) as sizes of pairwise intersections of blocks.

2.5. Triffids

The triffid is a special PMD of rank 4 with α-sequence $(0, 1, 3, 9, 3^n)$. The motivation to study it comes from combinatorics (automorphisms and extensions of an $S(2, 3, v)$), algebra (groups generated by 3-transpositions, weak associativity, generalization of mean by quasigroups), and even from algebraic geometry (cubic hypersurfaces). In this section we shall first present our objects in their different equivalent forms: special loops, matroids, Steiner triple systems, and also quasigroups and groups. Afterwards, we shall give (in appropriate terms of loops) some of their parameters, bounds on parameters, and the cases of equality and small parameters.

A *commutative Moufang loop* (CML) is a generalization of an Abelian group in the following direction – the law of associativity is replaced by a weaker property (the commutative version of the Moufang identity):

$$(x \cdot x) \cdot (y \cdot z) = (x \cdot y) \cdot (x \cdot z).$$

(A *quasigroup* is a set with a binary operation $\circ$ such that both left and right translations are bijections; a *loop* is a quasigroup with an element 1, such that $1 \circ x = x \circ 1 = x$ for all elements x.)

Below we consider only the finite *exponent 3 CMLs* (exp. 3 CMLs), i.e. CMLs in which each element, except 1, has order 3. They are important for us because they provide a class of PMDs. Also, Bruck (1946) gives that for any CML E the central quotient $E/Z(E)$ (see the definition of center, $Z(E)$, below) is an exp. 3 CML and that any finite CML is a direct product of some Abelian group and some CML in which each element has order a power of 3. Any of our CMLs, i.e. a finite exp. 3 CML, has order 3^n for some n.

The simplest example of a CML is $Z_3 \times Z_3 \times \ldots \times Z_3 = Z_3^n$, an Abelian elementary 3-group where Z_3 is the group on $\{0, 1, 2\}$ with addition modulo 3. Zassenhaus (p. 423 of Bol, 1937) constructed a non-associative CML on the set of all 81 sequences $x = (x_1, x_2, x_3, x_4)$ with $x_i = 0, 1, 2$. He defined a product by

$$x \cdot y = (x_1 + y_1, x_2 + y_2, x_3 + y_3, x_4 + y_4 + (x_3 - y_3)(x_1 y_2 - x_2 y_1))$$

(all sums above are modulo 3). Denote this CML by E_4 and E_n the CML $Z_3^{n-4} \times E_4$ of order 3^n. We will see below that E_n is 'close' to Z_3^n.

A *triffid* on a given CML E is a PMD of rank 4 such that its *points* (1-flats) are just elements of E, its *lines* (2-flats) are all 3-subsets $L(x, y) = \{x, y, (x \cdot y)^2\}$ for any $x, y \in E$ with $y \neq 1, x, x^2$, and its planes (or 3-flats or hyperplanes since our PMD has rank 4) are 9-subsets

$$H(x, y, z) = \begin{pmatrix} x, & y, & (x \cdot y)^2 \\ z, & t, & (z \cdot t)^2 \\ (x \cdot z)^2, & (y \cdot t)^2, & (x \cdot t)^2 \end{pmatrix}$$

for any $x, y, z \in E$ that are not in a line (here $t = [(x \cdot y)^2 \cdot (x \cdot z)^2]^2$). So, a triffid is a PMD with $\alpha = (0, 1, 3, 9, 3^n)$. Any PMD with such an α comes from some CML E and in the special case $E = Z_3^n$ this PMD is just an $(n - 3)$-truncation $AG(n, 3)$: $(n - 3)$ of an affine geometry over $GF(3)$. Any hyperplane (i.e. plane) of a triffid is an affine plane $AG(2, 3)$.

The set of all lines $L(x, y)$ of a triffid is an $S(2, 3, 3^n)$ such that any triangle (i.e. 3 points not in a line) generates an affine plane. We call such an $S(2, 3, v)$ a *Hall triple system* (*HTS*). Any HTS is the set of lines of some triffid. The result in Buekenhout (1969) gives that any PMD with $a = (0, 1, \lambda, \lambda^2, v)$ such that $\lambda \geqslant 4$ and any triangle generates an affine plane is $AG(n, \lambda)$: $(n - 3)$, so λ is a prime power. The triffids, i.e. the case $\lambda = 3$ (and also $S(3, 4, 10)$ for $\lambda = 2$, where all planes are $AG(2, 2)$s) are exceptional with respect to this result.

So, CML and HTS are equivalent concepts. They are also equivalent to each of the following 3 structures:

(1) $S(2, 3, 3^n)$ such that any symmetry •—•—• is an automorphism, or, in other words, for any point a from our set E of 3^n points there exists an involution moving exactly $E - \{a\}$ (this was proven in Hall, 1960);

(2) a *symmetric distributive quasigroup* $(E, \circ)$, i.e. a quasigroup such that all translations are automorphisms and $x \circ y = z$ is independent of permutations of x, y, z; we say that it has *identity* e if $e \circ e = e$,

$e \circ x = x \circ e = x$ (for any $x \in E$) for some $e \in E$; one can get a CML $(E, \cdot)$ from such a quasigroup $(E, e, \circ)$ by defining $x \cdot y = (e \circ x) \circ (e \circ y)$;

(3) a *group* (G, A), i.e. a group G generated by a subset A such that $x^2 = 1 = (xy)^3$ and $xyx \in A$ for any $x, y \in A$ and such that the commutative center of G is just $\{1\}$ (these groups introduced by Fischer, 1966, are considered in Beneteau, 1978); one can get such a group (G, A) from a symmetric distributive quasigroup $(E, \circ)$ by choosing as generating set A the set of all automorphisms $\tilde{x} = \tilde{x}(y) = x \circ y$ of this quasigroup.

We will now consider the following parameters of the CML E:

(1) the *order* $|E| = 3^n$;
(2) the *dimension* $d(E)$, i.e. the size of any minimal subset generating E;
(3) the *order* $|Z(E)|$ of the *center* (the *associative center* $Z(E)$ of the CML E is the Abelian 3-group $\{z \in E \mid (x \cdot y) \cdot z = x \cdot (y \cdot z), \forall x, y \in E\}$);
(4) the *nilpotency class* $k(E)$ (the central quotient $E/Z(E)$ is a CML $E^{(1)}$, $E^{(1)}/Z(E^{(1)})$ is a CML $E^{(2)}$, etc.; $k(E)$ is the smallest number k such that $E^{(k-1)}$ is an Abelian 3-group, with $E^{(0)} = E$);
(5) the $\mathrm{rank}_3(E)$ is the *3-rank*, i.e. the rank over $GF(3)$, of the incidence matrix (lines–points) of the Steiner triple system $S(2, 3, 3^n)$ of all lines of E.

2.5.1. Proposition.
(i) (*Beneteau, 1980a; Sandu, 1979*) $d(E) \leqslant n$ with equality if and only if $E = Z_3^n$;
(ii) (*trivial*) $|Z(E)| \leqslant |E|$ with equality if and only if $E = Z_3^n$, $|k(E)| \geqslant 1$ with equality if and only if $E = Z_3^n$;
(iii) (*Doyen, Hubaut & Vandensavel, 1978*) $\mathrm{rank}_3(E) \leqslant |E| - n + 1$ with equality if and only if $E = Z_3^n$.

It is easy to see that $|E_4| = 3^4$, $Z(E_4) = \{(0, 0, 0, 0, 0), (0, 0, 0, 1), (0, 0, 0, 2)\}$, E_4 is generated by $\{(1, 0, 0, 0), (0, 1, 0, 0), (0, 0, 1, 0)\}$, $k(E_4) = 2$, and so $|E_n| = 3^n$, $|Z(E_n)| = |Z(Z_3^{n-4}) \times Z(E_4)| = 3^{n-3}$, $d(E_n) = n - 1$, $k(E_n) = 2$. Hamada (1981) gives $\mathrm{rank}_3(E_n) = 3^n - n$. So E_n is very close to an Abelian 3-group; moreover, it is closest to Z_3^n in the following sense – for any non-associative CML E with $|E| = 3^n$, Beneteau (1980a) gives $|Z(E)| \leqslant 3^{n-3}$ with equality if and only if $E = E_n$.

The order $|E|$ of a non-associative CML E is 3^n for some $n \geqslant 4$; one exists (for example, E_n) for any such n.

2.5.2. Proposition.
(i) (*Hall, 1960*) E_4 is the unique non-associative CML of order 3^4;
(ii) (*Beneteau, 1980a; Smith, 1982*) E_5 is the unique non-associative CML of order 3^5;
(iii) (*Beneteau, 1980b*) there exist exactly three non-associative CMLs of order

3^6: $E_6, I_5,$ and $I_4,$ of dimensions 5, 5, and 4 (see also Kepka & Nemec,1981; Roth & Ray-Chaudhuri, 1984).

For the dimension $d(E)$ of any non-associative CML E we have $3 \leqslant d(E) \leqslant \log_3 |E| - 1$, $d(E_n) = n - 1$, and $d(E) \geqslant 4$ for $E \neq E_4$.

2.5.3. Proposition. (*Beneteau, 1980b*) *Let* $d(E) = 4$; *then* $|E| \leqslant 3^{12}$ *and for* $|E| \leqslant 3^8$ *there exist exactly eight CMLs with* $d(E) = 4$; *for each of the orders* 3^4, 3^5, 3^6, 3^7, 3^{12} *there exists exactly one such CML* (Z_3^4, E_5, I_4, J_4, L_4).

There exists for any m the *free* CML of dimension m, denoted L_m, such that any CML E with $d(E) = m$ is a representation (homeomorphic image) of L_m. So $|E| \leqslant |L_{d(E)}|$ and $k(E) \leqslant k(L_{d(E)})$; $L_3 = E_4$. In Manin (1974) the problem arises of determining $|L_m|$; it is known only that $|L_m| = 3^4$, 3^{12}, 3^{49}, 3^{220}, 3^{1028}, 3^{4592} for $m = 3, 4, 5, 6, 7, 8$ (Beneteau, 1980b; Smith, 1982). So, for example, the bound $|E| \leqslant 3^{12}$ for CMLs with $d(E) = 4$, given in Proposition 2.5.3, follows from $|E| \leqslant |L_{d(E)}|$ and $|L_4| = 3^{12}$.

The dimension, $d(E)$, is also linked with the order of the *Frattini subloop* $\Phi(E)$ of CML E by the equality $|\Phi(E)| = |E|/3^{d(E)}$. This subloop $\Phi(E)$ is defined as the intersection of all maximal subloops of E; also, $\Phi(E)$ is the set of all *non-generators*, i.e. elements that can be taken away from any subset generating E. Further, we remark that any CML is *diassociative*, i.e. any two elements generate an Abelian subgroup.

$|Z(E)|$ is a power of 3, because $Z(E)$ is an Abelian 3-group. For any non-associative CML E with $|E| = 3^n$ we have

$$1 \leqslant \log_3 |Z(E)| \leqslant n - 3.$$

The left bound is given in Bruck (1958), the right bound in Beneteau (1980a). Roth & Ray-Chandhuri (1984) gives that each of numbers 1, 2, 3, ..., $n - 3$ is $\log_3 |Z(E)|$ for some CML E with $|E| = 3^n$ (of course, we have to take $n \neq 5$). $|Z(E_n)| = 3^{n-3}$ and $|Z(L_m)| = 3^{(m/3)}$.

We now consider the central nilpotency class $k(E)$ of the CML E. $k(Z_3^n) = 1$ and $k(E_n) = 2$. Beneteau (1978) gives $3^4 \leqslant |E| \leqslant 3^{d+(d/3)}$ for E with $k(E) = d$ and $3^8 \leqslant |E| \leqslant 3^{d+(d/3)+4d+1/5)}$ for E with $k(E) = 3$, $d(E) = d$ (all four bounds are exact.) All CMLs E with $k(E) = 2$ are classified in Bruck (1946) and also in Kepka & Nemec (1981) and in Roth & Ray-Chaudhuri (1984).

2.5.4. Proposition.

(i) (*Bruck, 1958*) $k(E) \leqslant d(E) - 1$ (*Bruck–Slaby*);
(ii) (*Beneteau, 1978; Malbos, 1978; Smith, 1978*): $k(L_m) = m - 1$ if $m \geqslant 2$ (*Beneteau–Malbos–Smith*).

Here we describe the construction given in Malbos (1978) for $k(L_m) = m - 1$.

Let V be a vector space over $GF(3)$ and $\Lambda * V = \bigoplus_{i=0}^{\infty} \Lambda^{2i+1} V$ the submodule of odd sums $\Lambda^{2i+1} V$ of an exterior algebra ΛV of V. Denote $E_M = (\Lambda * V) \times (\Lambda * V)$ and define a product $\circ$ on E_M by

$$(x_1, x_2) \circ (y_1, y_2) = (x_1 + y_1 - x_1 y_1 (x_2 - y_2), x_2 + y_2 - (x_1 y_2 + x_2 y_1)(x_2 - y_2)).$$

$(E_M, \circ)$ is a CML and for any linearly independent $2m$-subset $\{x_1, ..., x_m, y_1, .., y_m\}$ of $V (m > 1)$ the m-subset $\{(x_i, y_i) | i = 1, 2, ..., m\}$ of E_M generates a subloop $\tilde{E}$ of $(E_M, \circ)$ of class $m - 1$. So $m - 1 = k(\tilde{E}) = d(\tilde{E}) - 1$ and $k(\tilde{E}) \leqslant k(L_m) \leqslant m - 1$ implies $k(L_m) = m - 1$.

There exists for any m and for any $k, k \leqslant m - 1$, the *free* CML of dimension m and nilpotency class k; denote it $L_{m,k}$. Then $|E| \leqslant |L_{d(E),k(E)}|$ and also $L_{m,m-1} = L_m$. Now Beneteau & Lacaze (1980) gives

$$\log_3 |L_{m,k}| \leqslant m + \binom{m}{3} + 4\binom{m+1}{5} \frac{\binom{n}{2}^{k-2} - 1}{\binom{n}{2} - 1};$$

moreover $|L_{m,2}| = 3^{m + (m/3)}$. Also in Beneteau & Lacaze (1980) it is proved that the order of the automorphism group $|\mathrm{Aut}\, E|$ of any 3-CML E of dimension m divides the number $(|E|^m / 3^{(m + 1/2)}) \prod_{i=1}^{m} (3^i - 1)$ and that in the special case $E = L_{m,k}$ the order is equal to that number.

There is a developed theory of infinite CMLs. An example of an infinite CML is the loop of octonions. Infinite CMLs have connections with differential geometry (3-webs), topological algebra (CMLs on topological spaces, Lie loops), and algebraic geometry (the cubic hypersurfaces of Manin).

Basic work on CMLs was carried out by Bruck (1946, 1958). HTSs and links with groups (G, A) were given by Hall (1960). Young (1973) observed the link between CMLs and HTSs, and PMDs. Symmetric distributive quasigroups (equivalent to CMLs) were studied by Beloussov and his students (see Sandu, 1979, and some other papers in the same book) and (in terms of symmetric medial spaces) by Soublin (1971) and, recently, Malbos (1978). A promising (but not yet completely developed) link of CMLs with cubic hypersurfaces, and many other developments, were given by Manin (1974). In recent years CMLs have been studied intensively by Beneteau (for example, Beneteau, 1978, 1980a, b; Beneteau & Lacaze, 1980), Smith (for example, 1978, 1982), Kepka and Nemec (for example, Kepka, 1978; Kepka & Nemec, 1981), and Roth & Ray-Chaudhuri (1984); some new observations on CMLs are given in Deza & Hamada (1980), Deza & Singhi (1980), and Hamada (1981).

2.6. Some Close Relatives of PMDs and Further Directions

An *equicardinal matroid* $M(S)$ is a matroid such that all its hyperplanes have the same cardinality k. It was introduced in Murty (1970) and, in fact, the PMD was introduced later just as a special case of equicardinal matroid. Kestenband & Young (1978) gives information on such matroids and, especially, on the case when $|S| - k$ is a prime power p^n. These equicardinal matroids (which are a direct generalization of $PG(d, q)$ having $|S| - k = q^d$) are shown in Kestenband & Young (1978) to be 'almost' PMDs; they are classified for the cases $n = 1, 2$ and for the case when $M(S)$ has rank 3.

A *matroid design* (*MD*, for short) is a matroid whose hyperplanes form a BIBD, i.e. a (b, v, r, k, λ) configuration on 1-flats. Any *PMD* is an *MD*. Any BIBD with $\lambda = 1$ is an MD (moreover, a PMD) of rank 3; it is easy to see that any non-trivial BIBD with $\lambda = 2$ is never an MD. Theorem 3 in section 12.3 of Welsh (1976) gives the necessary conditions for a BIBD with $\lambda > 2$ to be an MD. Kantor (1969) showed that a symmetric BIBD, i.e. a (v, k, λ)-configuration, is an MD if and only if either $\lambda = 1$, or $(q - 1)$ $(v, k, \lambda) = (q^{n+1} - 1, q^n - 1, q^{n-1} - 1)$ for some prime power q and the BIBD is the point-hyperplane design of $PG(n, q)$. Section 12.3 of Welsh (1976) also contains information on *base designs*, i.e. matroids whose circuits are the blocks of a BIBD. Also, finding circuit designs, i.e. matroids whose bases are the blocks of a BIBD, is, of course the dual problem to finding MDs.

Another close relative of PMDs of rank 4 is a *planar M-space* (Buekenhout & Deherder, 1971). A *planar space* (or respectively a *linear space*) is exactly a combinatorial geometry of rank 4 (or respectively 3). A *planar M-space* is a planar space such that any of its planes is isomorphic to the given linear space M. Any $PG(d, q)$: $(d - 3)$, $AG(d, q)$: $(d - 3)$, or triffid is a planar M-space. An example of a non-trivial (i.e. with at least two planes) planar M-space with two different sizes of lines is known, but only one; Buekenhout & Deherder (1971) show that the number of such planar M-spaces is finite for any given M.

Brylawski (1979) considers uniform matroids, i.e. matroids such that all upper intervals (in the lattice L of all flats) of the same rank have the same Whitney numbers of the second kind: $W^{(2)}_{[x,S]}(u, v) = W^{(2)}_{[y,S]}(u, v)$ for any flats x, y of the same rank and flats u, v with $v \subseteq u$. Any PMD is a uniform matroid. Theorems 4.5 and 4.7 of Brylawski (1979) give a criterion for a matroid to be a PMD in terms of rank invariants and calculate the characteristic polynomial of a PMD as a function of its α-sequence. Among other types of uniformity, considered in Brylawski (1979), we mention *latticially homogeneous* matroids, i.e. $[\varnothing, x] = [\varnothing, y]$ for all bottom intervals (of the lattice L of all flats) with x, y of the same rank (these are a special case of PMD), and *latticially uniform* matroids, i.e. $[x, S] = [y, S]$ for all upper intervals of L with x, y of the same rank (these are a special case of uniform matroids and not necessarily PMDs). Another generalization of the lattice of a PMD is considered in Neumaier (1981).

PMDs also appear in the following lattice-theoretic context (Theorem 9.1 of Doubilet, Rota & Stanley, 1972). For any upper semimodular lattice L of *triangular type* (i.e. the number of all maximal chains in any segment $[x, y]$ of L depends only on the ranks of x and y) there are PMDs $L_1, L_2, ..., L_r$ and an upper semimodular atomic lattice L_{r+1} of triangular type, such that L is isomorphic to the lattice obtained by identifying the top of L_i with the bottom of L_{i+1} for $1 \leqslant i \leqslant r$.

The papers by Deza (1978) and by Deza, Erdös & Frankl (1978), in which was given Proposition 2.3.1 (a characterization of all PMDs with large α_r as the case of equality in an upper bound for $|A(T, k, v = \alpha_r)|$; here T is a set of integers) contain also several generalizations of the inequality such that the equality will correspond to a generalization of a PMD. Denote by $A_\lambda(T, k, v)$ any family of k-subsets of a given v-set S such that any $(\lambda + 1)$-wise intersection of them is a number from the set $T = \{l_0, l_1, ..., l_{r-2}\}$ of integers, $0 \leqslant l_0 < l_1 < ... < l_{r-2}$. Denote by $M_\lambda(T, k, v)$ a family of k-subsets of S, called 'hyperplanes', such that for any $(r-2)$-flat (of given PMD of rank $r-1$ with $\alpha = (l_0, l_1, ..., l_{r-2}, v))$ and point outside it there are exactly λ 'hyperplanes' containing this $(r-2)$-flat and point. Any $S_\lambda(t, k, v)$ is an $M_\lambda(\{0, 1, ..., t-1\}, k, v)$; also we obtain $M_\lambda(T, k, v)$, with $\lambda = \prod_{i=1}^{j} (v - l^{(i)})/(k - l^{(i)})$, from a PMD with $\alpha = (l_0, l_1, ..., l_{r-2}, l^{(1)}, l^{(2)}, ..., l^{(j)}, k, v)$, $l_{r-2} < l^{(1)} < ... l^{(j)} < k$, by deleting all flats of sizes $l^{(1)}, ..., l^{(j)}$. Deza (1978) gives, for $v > v_0(\lambda, T, k)$, that $|A_\lambda(T, k, v)| \leqslant \lambda \prod_{l \in T} (v - l)/(k - l)$ with equality if and only if $A_\lambda(T, k, v)$ is an $M_\lambda(T, k, v)$; also that any $M_\lambda(T, k, v)$ with $v > v_0$ is *complete*, i.e. any number from T is the size of some $(\lambda + 1)$-wise intersection of sets from it.

Let A be an $A(T, k, v)$, $T = \{l_0, l_1, ..., l_{r-2}\}$, $0 \leqslant l_0 < ... < l_{r-2}$, such that A is a subset of the hyperplane family of some PMD $G'(S)$ with $\alpha = (l_0, l_1, ..., l_{r-2}, l^{(1)}, ..., l^{(j)}, k, v)$. Then $|A| \leqslant \prod_{l \in T} (v - l)/(k - l))$ (in contrast with Proposition 2.3.1 we have no condition $v > v_0$); the case of equality is a special PMD $M(S)$ which can be seen as an $S(t = r - 1, k, v)$ *in the PMD $G'(S)$*. In the special case when the hyperplane family of $G'(S)$ is partitioned by such $A(T, k, v)$s we will call $G'(S)$ *i-resolvable* (with $i = r - 1$). A 1-resolvable PMD is just an *affine resolvable* PMD ($M\{l_0\}, k, v)$ if $l_0 = 0$ forms a partition of S). In design theory some cases of i-resolvability of $S(k, k, v)$ are also considered. Another useful concept in relation to the resolvability of a PMD is affine resolvability of the configuration dual to the BIBD of hyperplanes and points (Deza, 1978).

Finally, in Deza & Frankl (1984) a new notion, *squashed design*, is introduced. Squashed designs are more general than PMDs; they also include some geometric structures on the set of permutations. Deza & Frankl (1984), extending Proposition 2.3.1, characterized squashed designs as families $A(T, k, v)$ of maximum size and belonging to some meet semi-lattice ($v > v_0(T, k)$) F of subsets of S. If F is the lattice of flats of some PMD, then it is a design within this PMD as above.

References

Atsumi, T. (1979). An extension of Cameron's result on block schematic Steiner systems, *J. Comb. Theory Ser. A* **27**, 388–91.

Beneteau, L. (1978). Problèmes de majorations dans les quasigroupes distributifs et les groupes de Fischer, in *Actes Colloque 'Algebra Appliquée et Combinatoire'*, pp. 22–34. Univ. Sci. et Med. de Grenoble.

Benteau, L. (1980a). Topics about Moufang loops and Hall triple systems, *Simon Stevin* **54**, 2, 107–24.

Beneteau, L. (1980b). Une classe particulière de matroides parfaits, in *Actes Colloque France-Canada de Combinatoire* (Montreal, 1979), North-Holland, Amsterdam. Discrete Math. 8, pp. 229–32.

Beneteau, L. & Lacaze, J. (1980). Groups d'automorphismes des boucles de Moufang commutatives, *Europ. J. Comb.*, 1–4, 299–309.

Bol, G. (1937). Gewebe und Gruppen, *Math. Ann.* **114**, 414–31 (especially p. 423).

Bose, R. C. (1963). Strongly regular graphs, partial geometries and partially balanced designs, *Pacific J. Math.* **13**, 389–419.

Brini, A. (1980). A class of rank-invariants for perfect matroid designs, *Europ. J. Comb.* **1-1**, 33–8.

Bruck, R. H. (1946). Contributions to the theory of loops, *Trans. Amer. Math. Soc.* **60**, 245–354.

Bruck, R. H. (1958). *A Survey of Binary Systems*. Springer, Berlin, Göttingen, and Heidelberg.

Brylawski, T. H. (1979). Intersection theory for embedding of matroids into uniform geometries, *Stud. Appl. Math.* **61**, 211–44.

Brylawski, T. H. (1986). Constructions, in N. White (ed.), *Matroid Theory*, Cambridge University Press.

Buekenhout, F. (1960). Characterization des espaces affine basée sur la notion de droit, *Math. Z.* **111**, 367–71.

Buekenhout, F. & Deherder, R. (1971). Espaces linéaires finis à plans isomorphes, *Bull. Soc. Math. Belg.* **23**, 348–59.

Cameron, P. J. (1975). Two remarks on Steiner systems, *Geom. Dedicata* **4**, 403–18.

Cameron, P. J. (1980). Extremal results and configuration theorems for Steiner systems, in C. Lindner & A. Rosa (eds), *Topics on Steiner systems*, Ann. Discrete Math. 7, pp. 43–63. North-Holland, Amsterdam.

Dembowski, P. & Wagner, A. (1960). Some characterizations of finite projective spaces, *Arch. Math.* **11**, 465–69.

Deza, M. (1978). Pavage généralisé parfait comme généralisation de matroide-configurations et de simples t-configurations, in *Problèmes Combinatoires*, Coll. Int. CNRS, No. 260, pp. 97–100. Paris-Orsay.

Deza, M. & Frankl, P. (1984). Injection Geometries, *J. Comb. Theory Ser. B* **36**, 31–70.

Deza, M. & Hamada, N. (1980). The geometric structure of a matroid design derived from some commutative Moufang loop and a new MDPB association scheme, Technical Report No. 18, Statistical Research Group, Hiroshima University.

Deza, M. & Singhi, N. M. (1980). Some properties of perfect matroid designs, in *Proceedings of the Symposium on Combinatorial Mathematics and Optimal Design*, Ann. Discrete Math. 6, pp. 57–76. North-Holland, Amsterdam.

Deza, M., Erdös, P. & Frankl, P. (1978). Intersection properties of systems of finite sets, *Proc.*

London Math. Soc. (3) **36**, 369–84.

Deza, M., Frankl, P. & Hirschfeld, J. W. P. (1985). Sections of varieties over finite fields as large intersection families, *Proc. London Math. Soc.* (3) **50**, 405–25.

Doubilet, P., Rota, G. C. & Stanley, R. (1972). On the foundation of combinatorial theory (VI): The idea of generating functions, in *Proceedings of thje Sixth Berkeley Symposium of Mathematical Statistics and Probability* Vol. 2, pp. 267–318. University of California Press, Berkeley.

Doyen, J., Hubaut, X. & Vandensavel, M. (1978). Ranks of incidence matrices of a Steiner triple system, *Math. Z.* **163**, 251–59.

Doyen, J. & Hubaut, X. (1971). Finite regular locally projective spaces, *Math. Z.* **119**, 83–8.

Fischer, B. (1966). Finite groups generated by 3-transformations, *Invent. Math.* **13**, 232–76.

Frankl, P. (1983). Personal communication.

Gross, B. H. (1974). Intersection triangles and block intersection numbers of Steiner systems, *Math. Z.* **139**, 87–104.

Hall, M. (1960). Automorphisms of Steiner triple systems, *IBM J. Res. Dev.* **4**, 460–72.

Hamada, N. (1981). The geometric structure and the p-rank of an affine triple system, derived from a nonassociative Moufang loop with the maximum associative center, *J. Comb. Theory Ser A* **30**, 285–97.

Hamada, N. & Tamari, F. (1975). Duals of balanced incomplete block designs derived from an affine geometry, *Ann. Stat.* **3**, 926–38.

Hanani, H. (1963). On some tactical configurations, *Can. J. Math.* **15**, 702–22.

Kantor, W. M. (1969). Characterizations of finite projective and affine spaces, *Can. J. Math.* **21**, 64–75.

Kepka, T. (1978). Distributive Steiner quasigroups of order 3^5, *Comment. Math. Carolin.* **19-2**, 389–401.

Kepka, T. & Nemec, P. (1981). Commutative Moufang loops and distributive groupoids of small order, *Czechoslovak Math. J.* **31** (106), 633–69.

Kestenband, B. C. & Young, H. P. (1978). *J. Comb. Theory Ser. A* **24**, 211–34.

Malbos, J.-P. (1978). Sur la classe de nilpotence des boucles commutatives de Moufang et des espaces médiaux. *C.R. Acad. Sc. Paris Sér. A* **287**, 691–93.

Manin, J. I. (trans. M. Hazewinkel) (1974). *Cubic Forms*, North-Holland, Amsterdam.

Murty, U. S. R. (1970). Equicardinal matroids and finite geometries, in *Proceedings of the International Conference on Combinatorial Structures and Applications*, pp. 289–91. Gordon & Breach, New York.

Murty, U. S. R., Young, H. P. & Edmonds, J. (1970). Equicardinal matroids and matroid-designs, in *Proceedings of the Second Chapel Hill Conference on Combinatorial Mathematics and Applications*, pp. 498–547. Chapel Hill, North Carolina.

Neumaier, A. (1981). Distance matrices and n-dimensional designs, *Europ. J. Comb.* **2-2**, 165–72.

Ray-Chaudhuri, D. K. & Singhi, N. M. (1977). A characterization of line hyperplane design of projective space and some extremal theorems for matroid-designs, in H. J. Zassenhaus (ed.), *Number Theory and Algebra*, pp. 289–301. Academic Press, New York.

Roth, R. & Ray-Chaudhuri, D. K. (1984). Hall triple systems and commutative Moufang exponent 3 loops: The case of nilpotence class 2, *J. Comb. Theory Ser. A* **36**, 129–62.

Sandu, N. I. (1979). On centrally nilpotent commutative Moufang loops, in *Quasigroups and Loops (in Russian)*, pp. 145–55. Stiintza, Kishiniev.

Smith, J. D. H. (1978). On the nilpotence class of commutative Moufang loops, *Math. Proc. Cam. Phil. Soc.* **84**, 387–404.

Smith, J. D. H. (1982). Commutative Moufang loops and Bessel functions, *Invent. Math.* **67**, 173–86.

Soublin, J.-P. (1971). Etude algébrique de la notion de moyenne, *J. Math. Pures et Appl. Sér.* 9 **50**, 53–264.

Young, H. P. (1973). Affine triple systems and matroid designs, *Math. Z.* **132**, 343–59; *Math. Rev.* **50**, 142.

Welsh, D. J. A. (1976). *Matroid Theory*, Academic Press, London and New York.

White, N. (ed.) (1987). *Combinatorial Geometries*, Cambridge University Press.

Wilson, R. M. (1982). Personal communication.

3

Infinite Matroids

JAMES OXLEY

The many different axiom systems for finite matroids given in Chapter 2 of White (1986) offer numerous possibilities when one is attempting to generalize the theory to structures over infinite sets. Some axiom systems that are equivalent when one has a finite ground set are no longer so when an infinite ground set is allowed. For this reason, there is no single class of structures that one calls infinite matroids. Rather, various authors with differing motivations have studied a variety of classes of matroid-like structures on infinite sets. Several of these classes differ quite markedly in the properties possessed by their members and, in some cases, the precise relationship between particular classes is still not known.

The purpose of this chapter is to discuss the main lines taken by research into infinite matroids and to indicate the links between several of the more frequently studied classes of infinite matroids.

There have been three main approaches to the study of infinite matroids, each of these being closely related to a particular definition of finite matroids. This chapter will discuss primarily the independent-set approach. Some details of the closure-operator approach will also be needed, but a far more complete treatment of this has been given by Klee (1971) and by Higgs (1969a, b, c). The third approach, via lattices, will not be considered here. This approach is taken by Maeda & Maeda (1970) and they develop it in considerable detail.

Throughout this chapter, the Axiom of Choice will be assumed.

3.1. Pre-independence Spaces and Independence Spaces

The first class of infinite matroids that we consider is obtained essentially by deleting the references to finite sets in the independence axioms (i1)–(i3) of Chapter 2 of White (1986). We note, however, that the finiteness of I_1 and I_2 in (i3), which is implicit when S is finite, is made explicit when $|S|$ is unrestricted.

A *pre-independence space* $M_P(S)$ is a set S together with a collection $\mathscr{I}$ of subsets of S (called *independent sets*) such that

(i1) $\mathscr{I} \neq \varnothing$.
(i2) A subset of an independent set is independent.
(i3′) (Finite augmentation) If I_1 and I_2 are finite members of $\mathscr{I}$ with $|I_2| > |I_1|$, then there exists x in $I_2 - I_1$ such that $I_1 \cup x \in \mathscr{I}$.

Generalizing the terminology of finite matroid theory, we call a subset X of S *dependent* if $X \notin \mathscr{I}$. A *circuit* of $M_P(S)$ is a minimal dependent set, and a *basis* of $M_P(S)$ is a maximal independent set. The notation $Y \subset\subset X$ indicates that Y is a finite subset of X.

Although pre-independence spaces seem to be natural objects for study, they have received little attention in their own right primarily because they fail to possess many of the fundamental properties of finite matroids.

3.1.1. Example. Let $S = \mathbb{R}$, the set of real numbers, and $\mathscr{I}$ be the set of all countable subsets of S. Then $\mathscr{I}$ is the collection of independent sets of a pre-independence space $M_P(S)$. However, $M_P(S)$ has no circuits and no bases.

In the face of such examples, it is natural to strengthen one's axiom system. As with finite matroids, a principal example of a pre-independence space is obtained from a vector space V. In this case, we let S be an arbitrary subset of V, and $\mathscr{I}$ be the collection of subsets of S that are linearly independent in V. Such a pre-independence space satisfies the following additional condition.

(I4) (Finite character) If $X \subseteq S$ and every finite subset of X is in $\mathscr{I}$, then X is in $\mathscr{I}$.

Much of the work done on infinite matroids has been algebraically motivated and for this reason (i1), (i2), and (i3′) have frequently been augmented by (I4). We shall call a pre-independence space satisfying (I4) an *independence space*. Such structures are also commonly referred to as *finitary matroids* (Bean, 1976; Higgs, 1969a; Klee, 1971). In this section we examine the properties of independence spaces. In the next section we shall add different conditions to (i1), (i2), and (i3′) with different consequences.

3.1.2. Example. It is easy to extend the relevant arguments from Chapter 4 of White (1987) to show that the set of partial transversals of an arbitrary family of subsets of a set S is the set of independent sets of a pre-independence space on S. However, this pre-independence space need not be an independence space. For instance, if $\mathscr{X} = (X_1, X_2, X_3, \ldots)$, and $X_i = \{1, i + 1\}$ for all i, then every finite subset of $\mathbb{Z}^+$ is a partial transversal of $\mathscr{X}$, yet $\mathbb{Z}^+$ itself is not.

Two immediate consequences of (I4) are that every dependent set in an independence space contains a circuit and that this circuit is finite. Thus an independence space is uniquely determined by its collection of circuits. We leave it to the reader to show that the independence axioms, (i1), (i2), (i3'), and (I4), are cryptomorphic to each of the circuit axioms, the strong circuit axioms, and the basis axioms stated below.

3.1.3. Circuit axioms for independence spaces. *An independence space $M(S)$ is a set S together with a collection $\mathscr{C}$ of subsets (called circuits) such that $\mathscr{C}$ satisfies (c1)–(c3) (Chapter 2 of White, 1986) together with*
(C4) *Every circuit is finite.*

3.1.4. Strong circuit axioms for independence spaces. *These are the same as 3.1.3, except that (c3) is replaced by (c3.1) of Chapter 2 of White (1986).*

The bases of an independence space also behave similarly to their counterparts in a finite matroid. Indeed, an easy consequence of Zorn's Lemma is that every independent subset of an independence space is contained in a basis. Hence every independence space is determined by its collection of bases. To obtain a definition of an independence space in terms of bases, one augments the basis axioms (b1)–(b3) of Chapter 2 of White (1986) by the following form of the finite character condition.

(B4) If X is not contained in a basis, then some finite subset of X is not contained in a basis.

It is not difficult to extend the axioms (cl1)–(cl4) of White (1986) to give a closure-operator definition for independence spaces. We leave the reader to check the details of this (see Exercise 3.1).

3.1.5. Example. Let S be the set of edges of a graph Γ and let $\mathscr{C}(\Gamma)$ be the collection of edge-sets of cycles of Γ. It was noted in Chapters 1 and 6 of White (1986) that, when Γ is finite, $\mathscr{C}(\Gamma)$ is the set of circuits of a (finite) matroid on S. Since every cycle in an infinite graph is finite, we can extend this immediately to get that $\mathscr{C}(\Gamma)$ is the set of circuits of an independence space $M_\Gamma(S)$ regardless of whether Γ is finite or infinite.

If $M(S)$ is an independence space having $\mathscr{I}$ as its collection of independent sets, then for $X \subseteq S$, let $\mathscr{I}|X$ be defined, as for finite matroids, by

$$\mathscr{I}|X = \{Y \subseteq X : Y \in \mathscr{I}\}. \tag{3.1}$$

Clearly $\mathscr{I}|X$ is the collection of independent sets of an independence space $M(X)$ on X. We call $M(X)$ the *restriction* of $M(S)$ to X. One can also define the operation of contraction for independence spaces in the same way as for finite matroids. However, in order to establish that this operation is well defined and that it gives an independence space, we shall require three lemmas.

3.1.6. Lemma. *Let B_1 and B_2 be bases of an independence space $M(S)$ and suppose that $x \in B_1 - B_2$. Then there is an element y of $B_2 - B_1$ such that $(B_1 - x) \cup y$ is a basis of $M(S)$.*

Proof. This result is an easy consequence of the proof of Proposition 2.1.1 of White (1986). $\square$

3.1.7. Lemma. *Suppose that $M(S)$ is an independence space and $Y \subseteq X \subseteq S$. If B_1 and B_2 are bases of $M(S - X)$, then $B_1 \cup Y \in \mathscr{I}$ if and only if $B_2 \cup Y \in \mathscr{I}$.*

Proof. Suppose $B_1 \cup Y \in \mathscr{I}$, but $B_2 \cup Y \notin \mathscr{I}$. Consider $M((S - X) \cup Y)$. This is an independence space having $B_1 \cup Y$ as a basis. Moreover, $M((S - X) \cup Y)$ has a basis B such that $B_2 \subseteq B \subseteq B_2 \cup Y$. Now $B = B_2 \cup Y'$ where $Y' \subsetneqq Y$. Choose x in $Y - Y'$. Then $x \in (B_1 \cup Y) - B$ and so, by Lemma 3.1.6, there is an element y of $B - (B_1 \cup Y)$ such that $((B_1 \cup Y) - X) \cup y$ is a basis of $M((S - X) \cup Y)$. As $Y' \subseteq Y$, we have that $y \in B_2 - B_1$ and so $B_1 \subsetneqq B_1 \cup y \subseteq S - X$, and $B_1 \cup y$ is independent in $M(S - X)$. This contradicts the fact B_1 is a basis of $M(S - X)$ and thereby completes the proof of the lemma. $\square$

We require one further lemma before defining contraction for independence spaces. If S and I are sets and $\mathscr{X} = (X_i : i \in I)$ is a family of subsets of S, then a *choice function* for $\mathscr{X}$ is a mapping $\phi : I \to S$ such that $\phi(i) \in X_i$ for all i in I. If $J \subseteq I$, then $\phi|_J$ denotes the mapping from J into S defined by $\phi|_J(j) = \phi(j)$ for all j in J.

3.1.8. Lemma. *(Rado's selection principle) Let $(X_i : i \in I)$ be a family of finite subsets of a set S. For each finite subset J of I, let ϕ_J be a choice function for $(X_i : i \in J)$. Then there is a choice function ϕ for $(X_i : i \in I)$ such that if $J \subset\subset I$, then there is a set K for which $J \subseteq K \subset\subset I$ and $\phi|_J = \phi_K|_J$.*

Mirsky's book (1971) contains several applications of this result together with a short proof of it using Tychonoff's theorem. We shall not reproduce these here.

3.1.9. Proposition. *Suppose that $M(S)$ is an independence space, $X \subseteq S$, and B is a basis of $M(S - X)$. Let*

$$\mathscr{I}.X = \{Y \subseteq X : Y \cup B \in \mathscr{I}\}. \tag{3.2}$$

Then $\mathscr{I}.X$ is the set of independent sets of an independence space $M.X$, the contraction of $M(S)$ to X.

Proof. By Lemma 3.1.7, $\mathscr{I}.X$ does not depend on the basis B chosen for $M(S - X)$. It is clear that $\mathscr{I}.X$ satisfies (i1) and (i2). Moreover, if $Y \subseteq X$ and every finite subset of Y is in $\mathscr{I}.X$, then every finite subset of $Y \cup B$ is in $\mathscr{I}$. Hence $Y \cup B \in \mathscr{I}$, and so $Y \in \mathscr{I}.X$. Thus, $\mathscr{I}.X$ satisfies (I4).

To show that $\mathscr{I}.X$ satisfies (i3'), we shall use Rado's selection principle. Suppose that U and T are finite members of $\mathscr{I}.X$ and $|T| > |U|$. Then $B \cup T$ and $B \cup U$ are in $\mathscr{I}$. Moreover, for every finite subset B_0 of B, we have that $M(B_0 \cup T \cup U)$ is a finite matroid, and so, applying (i3) to $B_0 \cup T$ and $B_0 \cup U$, there is an element t of $T - U$ such that $B_0 \cup U \cup t \in \mathscr{I}$. Thus the set $S_{B_0} = \{t \in T - U : B_0 \cup U \cup t \in \mathscr{I}\}$ is finite and non-empty. Let I be the set of finite subsets of B and suppose that $J = \{B_1, B_2, \ldots, B_n\} \subset\subset I$. Let $B' = \bigcup_{i=1}^{n} B_i$ and choose an element t' from $S_{B'}$. We define a choice function ϕ_J for $(S_{B_1}, S_{B_2}, \ldots, S_{B_n})$ by $\phi_J(B_i) = t'$ for all i in $\{1, 2, \ldots, n\}$. Let ϕ be a choice function for $(S_{B_0} : B_0 \in I)$ satisfying the conclusion of Lemma 3.1.8. We show next that ϕ maps every element of I to the same element t_0 of $T - U$. It will follow from this and (I4) that $B \cup U \cup t_0 \in \mathscr{I}$, and hence that $U \cup t_0 \in \mathscr{I}.X$, as required. To show that the image of ϕ is a single element of $T - U$, suppose that $B_0, B_0' \in I$, and let $J = \{B_0, B_0'\}$. Then there is a set K such that $J \subseteq K \subset\subset I$ and $\phi_K|_J = \phi|_J$. But $\phi_K(B_0) = \phi_K(B_0')$ and so $\phi(B_0) = \phi(B_0')$. $\square$

It is well known that all bases of a vector space are equicardinal. This result was extended to independence spaces by Rado (1949).

3.1.10. Proposition. *If B_1 and B_2 are bases of an independence space $M(S)$, then $|B_1| = |B_2|$.*

The proof of this result uses the following infinite extension of Rado's theorem on independent transversals (see Chapter 4 of White, 1987). Note that, since every restriction of an independence space $M(S)$ to a *finite* set X is a finite matroid, one can define the *rank* $r(X)$ of X to be the rank of the matroid $M(X)$.

3.1.11. Proposition. *Let $M(S)$ be an independence space and $\mathscr{X} = (X_i : i \in I)$ be a family of finite subsets of S. The following statements are equivalent.*

(i) $r\left(\bigcup_{j \in J} X_j\right) \geqslant |J|$ *for every $J \subset\subset I$.*

(ii) *Every finite subfamily of $\mathscr{X}$ has an independent transversal.*

(iii) *$\mathscr{X}$ has an independent transversal.*

Proof. The fact that (i) implies (ii) follows from the finite case of Rado's theorem. Moreover, it is clear that (iii) implies (i). We shall complete the proof of Proposition 3.1.11 by using Rado's selection principle to show that (ii) implies (iii).

Suppose $L \subset\subset I$. Then, by (ii), $(X_i : i \in L)$ has an independent transversal. Thus there is an injective choice function ϕ_L for $(X_i : i \in L)$ such that $\phi_L(L) \in \mathscr{I}$.

Let ϕ be a choice function for $(X_i: i \in I)$ satisfying the conclusion of Lemma 3.1.8. Then if $i_1, i_2 \in I$ and $J' = \{i_1, i_2\}$, there is a finite subset K' of I such that $J' \subseteq K'$ and $\phi_{K'}|_{J'} = \phi|_{J'}$. As $\phi_{K'}(i_1) \neq \phi_{K'}(i_2)$, it follows that $\phi(i_1) \neq \phi(i_2)$, so ϕ is an injection and hence $\phi(I)$ is a transversal of $\mathscr{X}$. To see that $\phi(I) \in \mathscr{I}$, suppose that $Y \subset\subset \phi(I)$. Then $Y = \phi(J)$ for some $J \subset\subset I$. Now I has a finite subset K such that $K \supseteq J$ and $\phi_K|_J = \phi|_J$. Since $\phi_K(K) \in \mathscr{I}$, we have that $Y = \phi(J) = \phi_K(J) \in \mathscr{I}$. Thus every finite subset of $\phi(I)$ is in $\mathscr{I}$, and so $\phi(I) \in \mathscr{I}$, as required. $\square$

Proof of Proposition 3.1.10. The required result will follow if we can show that for $U, T \in \mathscr{I}$ and $|U| < |T|$, there is an element t of $T - U$ such that $U \cup t \in \mathscr{I}$; that is, if we can show that the finiteness restriction on I_1 and I_2 in (i3') can be dropped.

Suppose that $U, T \in \mathscr{I}$ and $|U| < |T|$, but $U \cup t \notin \mathscr{I}$ for all t in $T - U$. Then it is easy to show that for each element t of $T - U$ there is a unique circuit C_t such that $t \in C_t \subseteq U \cup t$. Now, if $t \in T - U$, let $X_t = C_t - t$, and if $t \in T \cap U$, then let $X_t = \{t\}$. We shall show next that the family $\mathscr{X} = (X_t: t \in T)$ satisfies Proposition 3.1.11(i). Suppose $T' \subset\subset T$. Then $T' \in \mathscr{I}$. Now let g be the closure operator of the finite matroid $N = M\left(T' \cup \left(\bigcup_{t \in T'} X_t\right)\right)$. Then, for all t in $T' - U$, the set C_t is a circuit of N and so $T' \subseteq g\left(\bigcup_{t \in T'} X_t\right)$. Hence $r\left(\bigcup_{t \in T'} X_t\right) \geqslant r(T') = |T'|$. Thus, by Proposition 3.1.11, $\mathscr{X}$ has a transversal; that is, there is an injection from T into a subset of U. Hence $|T| \leqslant |U|$; a contradiction. This completes the proof of Proposition 3.1.10. $\square$

It follows from Proposition 3.1.10 that in an independence space $M(S)$ one can define the *rank* $r(X)$ of an *arbitrary* subset X of S to be the common cardinality of all bases of $M(X)$.

The preceding discussion has shown that a large number of fundamental properties of finite matroids are shared by independence spaces. In particular, the basic operations of restriction and contraction can be defined for independence spaces. Another important and powerful tool for finite matroids that one would naturally wish to extend to independence spaces is the operation of orthogonality. We shall show, however, that this cannot be done. Indeed, the lack of a satisfactory theory of orthogonality for independence spaces has been an important motivating factor in the study of other classes of infinite matroids.

Suppose that $M(S)$ is an independence space and let $\mathscr{I}^*$ be defined as follows.

$$\mathscr{I}^* = \{X: S - X \text{ contains a basis of } M(S)\}. \tag{3.3}$$

We leave the reader to check that $\mathscr{I}^*$ is the set of independent sets of a pre-independence space $M^*(S)$. Evidently, every independent set of $M^*(S)$ is contained in a basis of $M^*(S)$. The term 'cofinitary matroid' has often been used to refer to such pre-independence spaces $M^*(S)$ (Bean, 1976; Klee, 1971). When S is finite $\mathscr{I}^*$, of course, is the collection of independent sets of the matroid orthogonal to $M(S)$. However, as the next example shows, when S is infinite, $M^*(S)$ may fail to satisfy the finite character condition.

3.1.12. Example. Let S be an infinite set and k be a positive integer. If $\mathscr{I}_k = \{X \subseteq S: |X| \leqslant k\}$, then $\mathscr{I}_k$ is the set of independent sets of an independence space $M^k(S)$. However, although every finite subset of S is in $\mathscr{I}_k^*$, the set S itself is not.

This example plays a central role in the proof of the next theorem. Let S be an arbitrary infinite set and $\mathscr{S}$ be the set of independence spaces on S. An *orthogonality function* Δ on $\mathscr{S}$ is a mapping from $\mathscr{S}$ into $\mathscr{S}$ such that for all $\mathscr{I}$ in $\mathscr{S}$ we have that Δ is an involution, that is,

$$\Delta(\Delta \mathscr{I}) = \mathscr{I} \tag{3.4}$$

and

$$(\Delta \mathscr{I})|X = (\mathscr{I} \cdot X)^* \quad \text{for all } X \subset\subset S. \tag{3.5}$$

The second of these conditions expresses agreement between Δ and the usual orthogonality for finite methods.

3.1.13. Theorem. *There is no orthogonality function on the collection $\mathscr{S}$ of independence spaces on an infinite set S.*

Proof. Assume that there is an orthogonality function Δ on $\mathscr{S}$. Now if k is a positive integer and $X \subset\subset S$, then $S - X$ is infinite, so $\mathscr{I}_k|(S - X)$ contains a basis of $M^k(S)$. Thus $\mathscr{I}_k.X = \{\varnothing\}$, hence $X \in (\mathscr{I}_k.X)^*$ and so, by (3.5), $X \in \Delta \mathscr{I}_k$. It follows that $\Delta \mathscr{I}_k$ contains all finite subsets of S and hence $\Delta \mathscr{I}_k = 2^S$. Therefore, if j and k are distinct positive integers, then, by (3.4), $\mathscr{I}_j = \Delta(\Delta \mathscr{I}_j) = \Delta(\Delta \mathscr{I}_k) = \mathscr{I}_k$; a contradiction. This completes the proof of Theorem 3.1.13. $\square$

Rado (1966) raised the problem of developing a non-trivial theory of infinite matroids in which the finite character condition does not feature. The motivation for discarding this condition is increased by Example 3.1.12, which suggests that the problem with attempting to define orthogonality for independence spaces may arise because the class of independence spaces is too restricted. The rest of this chapter will be concerned with solving Rado's problem.

3.2. B-matroids

In this section we discuss the properties of basis-matroids (B-matroids), a class of pre-independence spaces introduced by Higgs (1969a) whose members have many of the properties of finite matroids. It is shown that this class contains the class of independence spaces and is closed under restriction and contraction. Moreover, unlike the class of independence spaces, the class of B-matroids is closed under the natural orthogonality function.

If $\mathscr{L}$ is a collection of subsets of a set S, then an $\mathscr{L}$-*subset* of S is a subset of S that is a member of $\mathscr{L}$.

A *B-matroid* $M_B(S)$ is a set S together with a collection $\mathscr{I}$ of subsets of S such that $\mathscr{I}$ satisfies (i1) and (i2) together with

(I_B1) If $T \subseteq X \subseteq S$ and $T \in \mathscr{I}$, then there is a maximal $\mathscr{I}$-subset of X containing T.

(I_B2) For all $X \subseteq S$, if B_1 and B_2 are maximal $\mathscr{I}$-subsets of X and $x \in B_1 - B_2$, then there is an element y of $B_2 - B_1$ such that $(B_1 - x) \cup y$ is a maximal $\mathscr{I}$-subset of X.

It is easy to check that (I_B2) implies (i3) and hence that every B-matroid is a pre-independence space. Thus, if $M_B(S)$ is a B-matroid, the members of $\mathscr{I}$ are called independent sets, and a maximal member of $\mathscr{I}$ is called a basis. An immediate consequence of Lemma 3.1.6 is that every independence space is a B-matroid. But, as the following example shows, not every B-matroid is an independence space.

3.2.1. Example. Let S be the edge set of the infinite graph Γ shown in Figure 3.1 and let $\mathscr{I}$ consist of those subsets of S that do not contain the edge set of any cycle or two-way infinite path in Γ. It is straightforward to check that $\mathscr{I}$ satisfies (i1), (i2), (I_B1), and (I_B2). Hence $\mathscr{I}$ is the collection of independent sets of a B-matroid $M_B(S)$ on S. However, $\mathscr{I}$ does not satisfy (I4) because every finite subset of a two-way infinite path is in $\mathscr{I}$, yet the path itself is not. Therefore, $M_B(S)$ is not an independence space.

Figure 3.1.

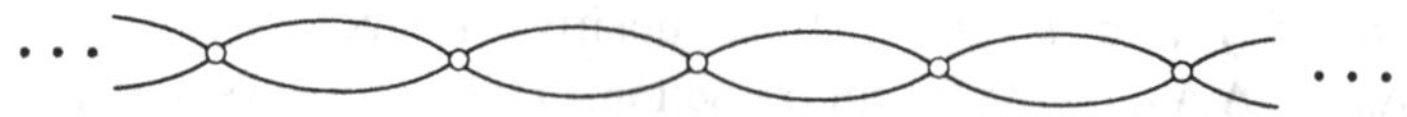

This example is one member of a class of B-matroids introduced by Higgs (1969c) (see Exercise 3.19).

If $M_B(S)$ is a B-matroid having $\mathscr{I}$ as its collection of independent sets, then for $X \subseteq S$, let $\mathscr{I}|X$ be defined as in (3.1). It is easy to check that $\mathscr{I}|X$ is the set of independent sets of a B-matroid, $M_B(X)$, the restriction of $M_B(S)$ to X.

We show next that the operation of contraction can be defined for

B-matroids as in (3.2). First note that the argument that was used to prove Lemma 3.1.7 can be applied in this case to give that $\mathscr{I}.X$ does not depend on the basis chosen for $\mathscr{I}|(S-X)$.

3.2.2. Proposition. *If $M_B(S)$ is a B-matroid and $X \subseteq S$, then $\mathscr{I}.X$ is the set of independent sets of a B-matroid $M_B.X$ on X.*

Proof. Let B be a basis of $M_B(S-X)$. Clearly $\mathscr{I}.X$ satisfies (i1) and (i2). Now suppose $Z \subseteq Y \subseteq X$ and Z is an $\mathscr{I}.X$-subset. Then $Z \cup B \in \mathscr{I}$ and so $Z \cup B$ is contained in a maximal $\mathscr{I}$-subset B' of $S-(X-Y)$. Now $B' = Z' \cup B$ where $Z' \supseteq Z$ and $Z' \cap (S-Y) = \varnothing$. Evidently Z' is a maximal $\mathscr{I}.X$-subset of Y containing Z. Thus $\mathscr{I}.X$ satisfies (I_B1).

We now check that $\mathscr{I}.X$ satisfies (I_B2). Let B_1 and B_2 be maximal $\mathscr{I}.X$ subsets of a subset Y of X and suppose $x \in B_1 - B_2$. Then $B_1 \cup B$ and $B_2 \cup B$ are maximal $\mathscr{I}$-subsets of $S-(X-Y)$. Since $x \in (B_1 \cup B) - (B_2 \cup B)$, there is an element y of $(B_2 \cup B) - (B_1 \cup B)$ such that $((B_1 \cup B) - x) \cup y$ is a maximal $\mathscr{I}$-subset of $S-(X-Y)$. Clearly $y \in B_2 - B_1$. Moreover, $((B_1 \cup B) - x) \cup y = ((B_1 - x) \cup y) \cup B$, and so $(B_1 - x) \cup y$ is a maximal $\mathscr{I}.X$-subset of Y. Hence $\mathscr{I}.X$ satisfies (I_B2) and the proposition is proved. $\qquad\square$

In the preceding section it was suggested that one of the problems with attempting to define orthogonality for independence spaces may be that the class of independence spaces is too restricted. In this section we shall confirm this by showing that (3.3) is an involution on the class of B-matroids. In order to achieve this objective we shall consider certain aspects of the closure-operator approach to infinite matroids as developed by Klee (1971). We defer to the exercises consideration of an equivalent approach that was adopted by Higgs (1969a, b, c).

An *operator* f on a set S is a function from 2^S to 2^S satisfying

(cl1) $X \subseteq f(X)$ for all $X \subseteq S$.
(cl2) $X \subseteq Y \subseteq S$ implies $f(X) \subseteq f(Y)$.

We shall concentrate primarily on operators satisfying the following two additional conditions.

(cl3) $f(f(X)) = f(X)$ for all $X \subseteq S$.
(cl4') If X, $Y \subseteq S$ and $p \in f(Y) - f(Y-X)$, then $x \in f((Y-x) \cup p)$ for some x in X.

Such operators will be called *idempotent-exchange-* or *IE-operators*.

It is straightforward to check that (cl1)–(cl3) and (cl4') provide another cryptomorphic axiom system for finite matroids to add to the many presented in White (1986).

For an operator f on a set S, let f^* be defined, for all $X \subseteq S$, by

$$f^*(X) = X \cup \{x: x \notin f(S - (X \cup x))\}.$$

3.2.3. Proposition. *If f is an operator on a set S, then f^* is an operator on S and $(f^*)^* = f$. Moreover, if f is an IE-operator, so is f^*.*

The proof of this is left as an exercise.

Again, for f an operator on S and $X \subseteq S$, the *restriction* f_X and *contraction* f^X of f to X are defined, for all $Y \subseteq X$, by

$$f_X(Y) = f(Y) \cap X, \text{ and } f^X(Y) = f(Y \cup (S - X)) \cap X \text{ respectively.}$$

3.2.4. Proposition. *If f is an operator on a set S and $X \subseteq S$, then both f_X and f^X are operators on X, and $(f^*)_X = (f^X)^*$ and $(f_X)^* = (f^*)^X$. Furthermore, if f is an IE-operator, so are both f_X and f^X.*

Proof. It is not difficult to check that both f_X and f^X are operators on X and that if f is an IE-operator on S, then both f_X and f^X are IE-operators on X. Now, if $Y \subseteq X$, then

$$(f^*)_X(Y) = f^*(Y) \cap X = (Y \cup \{y \in S: y \notin f(S - (Y \cup y))\}) \cap X$$

$$= Y \cup \{y \in X: y \notin f((S - X) \cup (X - (Y \cup y)))\}$$

$$= Y \cup \{y \in X: y \notin f^X(X - (Y \cup y))\} = (f^X)^*(Y).$$

Thus $(f^*)_X = (f^X)^*$. If we replace f by f^* in this equation, then, by Proposition 3.2.3 we get $f_X = ((f^*)^X)^*$. Hence, by Proposition 3.2.3 again, $(f_X)^* = (f^*)^X$, and the proof of the proposition is complete. $\qquad\square$

Let f be an operator on a set S and suppose $X \subseteq S$. Then X is *independent* if $x \notin f(X - x)$ for all x in X; otherwise X is *dependent*. We call X *spanning* or *non-spanning* according as $f(X) = S$ or $f(X) \subsetneqq S$; and X is a *basis* if X is both independent and spanning. A minimal dependent set is a *circuit* and a maximal non-spanning set is a *hyperplane*. Clearly when S is finite and f is an IE-operator, f is the closure operator of a (finite) matroid on S and then the definitions above are consistent with the usage of these terms in White (1986). We leave the routine proofs of the next two results as exercises.

3.2.5. Proposition. *If f is an IE-operator on S, then the sets of bases, maximal independent sets, and minimal spanning sets are identical.*

3.2.6. Proposition. *If f is an operator, then*

(i) *the f-independent sets are precisely the complements of the f^*-spanning sets;*
(ii) *the f-circuits are precisely the complements of the f^*-hyperplanes; and*
(iii) *the f-bases are precisely the complements of the f^*-bases.*

The next proposition shows that B-matroids can be axiomatized in this operator framework. This important fact will be used in the discussion of orthogonality for B-matroids. We require the following lemma.

3.2.7. Lemma. *Let f be an IE-operator on a set S and $\mathscr{I}$ be the collection of f-independent sets. If $X \subseteq S$ and B is a maximal $\mathscr{I}$-subset of X, then $f(B) = f(X)$. Moreover, if $I \subseteq S$ and $I \in \mathscr{I}$, then $f(I) = I \cup \{x: I \cup x \notin \mathscr{I}\}$.*

Proof. As B is a maximal $\mathscr{I}$-subset of X, by Propositions 3.2.4 and 3.2.5, $f_X(B) = X$. Thus $X \subseteq f(B)$ and so, by (cl2) and (cl3), $f(X) \subseteq f(f(B)) = f(B)$. But, by (cl2), $f(B) \subseteq f(X)$, hence $f(B) = f(X)$ as required. If $I \in \mathscr{I}$ and $I \cup x \notin \mathscr{I}$, then $y \in f((I \cup x) - y)$ for some element y of $I \cup x$. If $y = x$, then $x \in f(I)$. But if $y \neq x$, we still get $x \in f(I)$ by (cl4') because $y \notin f(I - y)$. Therefore, $f(I) \supseteq I \cup \{x: I \cup x \notin \mathscr{I}\}$. Now, if $x \in f(I) - I$, then $I \cup x$ is not f-independent; that is, $I \cup x \notin \mathscr{I}$. Hence $f(I) \subseteq I \cup \{x: I \cup x \notin \mathscr{I}\}$, and the required result follows. $\square$

3.2.8. Proposition. *Suppose that $M_B(S)$ is a B-matroid having $\mathscr{I}$ as its collection of independent sets and let f be defined, for all subsets X of S, by*

$$f(X) = X \cup \{x: I \cup x \notin \mathscr{I} \text{ for some } I \subseteq X \text{ such that } I \in \mathscr{I}\}. \qquad (3.6)$$

Then f is the unique IE-operator on S having $\mathscr{I}$ as its collection of independent sets, and, since $M_B(S)$ is a B-matroid, $\mathscr{I}$ satisfies $(I_B 1)$. Conversely, if f is an IE-operator on S such that the set $\mathscr{I}$ of f-independent sets satisfies $(I_B 1)$, then $\mathscr{I}$ is the set of independent sets of a B-matroid on S.

Proof. Suppose that $M_B(S)$ is a B-matroid and that f is defined as in (3.6). Then clearly f satisfies (cl1) and (cl2). We show next that for all $X \subseteq S$,

$$f(X) = \begin{cases} X \cup \{x: X \cup x \notin \mathscr{I}\}, & \text{if } X \in \mathscr{I}, \\ f(I_X), & \text{if } I_X \text{ is a maximal } \mathscr{I}\text{-subset of } X. \end{cases} \qquad (3.7)$$

If $X \in \mathscr{I}$ and $I \cup x \notin \mathscr{I}$ for some $I \subseteq X$, then $X \cup x \notin \mathscr{I}$. Thus (3.6) and (3.7) agree when $X \in \mathscr{I}$. If $X \notin \mathscr{I}$ and I_X is a maximal $\mathscr{I}$-subset of X, then, by (cl2), $f(X) \supseteq f(I_X)$. We now show that the reverse inclusion also holds. If $x \in X - I_X$, then $I_X \cup x \notin \mathscr{I}$, so $x \in f(I_X)$ and hence, $f(I_X) \supseteq X$. Next assume that $x \in f(X) - X$. Then $I \cup x \notin \mathscr{I}$ for some $\mathscr{I}$-subset I of X. If $I_X \cup x \notin \mathscr{I}$, then $x \in f(I_X)$, so suppose that $I_X \cup x \in \mathscr{I}$. Then $I_X \cup x$ is a maximal $\mathscr{I}$-subset of $X \cup x$. Moreover, I is contained in a maximal $\mathscr{I}$-subset B of $X \cup x$, and $x \notin B$. Thus $x \in (I_X \cup x) - B$, and so, by $(I_B 2)$, there is an element y of $B - (I_X \cup x)$ such that $((I_X \cup x) - x) \cup y = I_X \cup y$ is a maximal $\mathscr{I}$-subset of $X \cup x$. But $I_X \cup y \subseteq X$ and so the choice of I_X is contradicted. Thus (3.7) is established.

From (3.7), it is easy to see that f satisfies (cl3). To show that f satisfies (cl4'), suppose $X \subseteq Y$ and $p \in f(Y) - X$ and $p \notin f(Y - X)$. Let I_{Y-x} be a maximal $\mathscr{I}$-subset of $Y - X$. Then I_{Y-X} is contained in a maximal $\mathscr{I}$-subset I_Y of Y, and, as $p \in f(Y) = f(I_Y)$, we have $I_Y \cup p \notin \mathscr{I}$ and so I_Y is a maximal $\mathscr{I}$-subset of $Y \cup p$. Now $p \notin f(Y - X) = f(I_{Y-X})$, hence $I_{Y-X} \cup p \in \mathscr{I}$, and so $I_{Y-X} \cup p \subseteq I'_Y$, a maximal $\mathscr{I}$-subset of $Y \cup p$. If there is an element x of X such that $x \notin I'_Y$, then $x \in f(Y) \subseteq f(Y \cup p) = f(I'_Y) = f((Y - x) \cup p)$; that is, (cl4′) holds. Thus assume that $X \subseteq I'_Y$. Then $I'_Y = I_{Y-X} \cup X \cup p \supsetneqq I_{Y-X} \cup X \supseteq I_Y$. This is a contradiction since both I_Y and I'_Y are maximal $\mathscr{I}$-subsets of $Y \cup p$. We conclude that f satisfies (cl4′).

The fact that $\mathscr{I}$ is precisely the collection of f-independent sets follows without difficulty from (3.7). Moreover, by Lemma 3.2.7, f is the unique IE-operator on S having $\mathscr{I}$ as its collection of independent sets.

To prove the converse, let f be an IE-operator on S having $\mathscr{I}$ as its collection of independent sets and suppose that $\mathscr{I}$ satisfies (I_B1). Evidently $\mathscr{I}$ satisfies (i1) and (i2). Therefore to show $\mathscr{I}$ is the collection of independent sets of a B-matroid, it remains only to check that (I_B2) holds. Note that, by Lemma 3.2.7 and (I_B1), f is defined as in (3.7) for all subsets of S. Suppose $X \subseteq S$ and let B_1 and B_2 be maximal $\mathscr{I}$-subsets of X. Assume that $x \in B_1 - B_2$ and, for all y in $B_2 - B_1$, the set $(B_1 - x) \cup y \notin \mathscr{I}$. Then, by (3.7), $B_2 - B_1 \subseteq f(B_1 - x)$. Hence $f(B_1 - x) = f((B_1 - x) \cup B_2) \supseteq f(B_2) \supseteq X$. Hence $x \in f(B_1 - x)$; a contradiction of the fact that B_1 is f-independent. This completes the proof of Proposition 3.2.8. $\qquad\square$

If $M_B(S)$ is a B-matroid and f is defined as in (3.6), then we call f the *closure operator* of $M_B(S)$.

3.2.9. Proposition. *Let f be the closure operator of a B-matroid $M_B(S)$ having $\mathscr{I}$ as its collection of independent sets. Then f_X and f^X are the closure operators of $M_B(X)$ and $M_B.X$ respectively.*

Proof. It is clear that $\mathscr{I}|X$ is the collection of f_X-independent sets, hence by Proposition 3.2.8, f_X is the closure operator of $M_B(X)$. Now suppose that $Y \in \mathscr{I}.X$, but Y is not f^X-independent. Then $Y \cup B \in \mathscr{I}$ for some basis B of $M_B(S - X)$, and $y \in f^X(Y - y)$ for some y in Y. Thus $y \in f((Y - y) \cup (S - X))$ and, as $S - X \subseteq f(B) \subseteq f((Y - y) \cup B)$, it follows by (cl2) and (cl3) that $f((Y - y) \cup (S - X)) \subseteq f((Y - y) \cup B)$. Therefore $y \in f((Y - y) \cup B)$. This contradicts the fact that $Y \cup B \in \mathscr{I}$. Thus, if $Y \in \mathscr{I}.X$, then Y is f^X-independent. On the other hand, if Y is f^X-independent and $Y \notin \mathscr{I}.X$, then $Y \cup B \notin \mathscr{I}$ where B is a basis of $M_B(S - X)$. As $B \in \mathscr{I}$, by (I_B1), $Y \cup B$ contains a maximal $\mathscr{I}$-subset $Y' \cup B$ where $Y' \subsetneqq Y$. By Lemma 3.2.7, $f(Y' \cup B) = f(Y \cup B)$. Thus, if $y \in Y - Y'$, then $y \in f(Y' \cup B) \subseteq f((Y - y) \cup B) \subseteq f((Y - y) \cup (S - X))$; that is, Y is not f^X-independent. This contradiction implies that if Y is f^X-independent, then $Y \in \mathscr{I}.X$. We conclude that the set of f^X-independent sets equals $\mathscr{I}.X$ and hence that f^X is the closure operator of $M_B.X$. This completes the proof of Proposition 3.2.9. $\qquad\qquad\qquad\qquad\qquad\qquad\qquad\qquad\qquad\qquad\qquad\qquad\qquad\square$

We are now in a position to prove that the class of B-matroids is closed under the natural orthogonality function (3.3).

3.2.10. Theorem. *Suppose that $M_B(S)$ is a B-matroid having f as its closure operator. Then f^* is the closure operator of a B-matroid on S and its collection $\mathscr{I}^*$ of independent sets is given by $\mathscr{I}^* = \{X : S - X$ contains a basis of $M_B(S)\}$.*

Proof. By Proposition 3.2.3, f^* is an IE-operator on S. Now, by Propositions 3.2.3 and 3.2.6(i), the f^*-independent sets are precisely the complements of the f-spanning sets. In addition, by (3.7), a set is f-spanning if and only if it contains an f-basis. Therefore $\mathscr{I}^*$ is as given. It follows from Proposition 3.2.8 that to complete the proof we need to show that $\mathscr{I}^*$ satisfies (I_B1). Suppose that $Y \subseteq X \subseteq S$ and $Y \in \mathscr{I}^*$. Then $S - Y$ is f-spanning and so $X - Y$ is f^X-spanning. But, by Proposition 3.2.9, f^X is the closure operator of the B-matroid $M_B.X$ and so, by (3.7), $X - Y$ contains an f^X-basis B. Thus $X - B \supseteq Y$ and, by Proposition 3.2.6(iii), $X - B$ is an $(f^X)^*$-basis. Therefore, since $(f^X)^* = (f^*)_X$ by Proposition 3.2.4, it follows that $X - B$ is a maximal f^*-independent subset of X containing Y. Thus $\mathscr{I}^*$ satisfies (I_B1), and so, by Proposition 3.2.8, Theorem 3.2.10 is proved. $\qquad\qquad\qquad\square$

The attention in this chapter has been concentrated on two classes of infinite matroids, the classes of independence spaces and of B-matroids. The motivation for the study of independence spaces was algebraic. The last result of this chapter provides additional motivation for the study of B-matroids, showing that this is a particularly natural class of infinite matroids to examine.

Let S be an arbitrary infinite set and suppose that on every subset W of S we have a distinguished class $\mathscr{D}_W$ of pre-independence spaces so that

(3.1) and (3.2) give well-defined operations of restriction and contraction on $\mathscr{D}_W$ such that if $Z \subseteq W$, the restriction or contraction of a member of $\mathscr{D}_W$ is in $\mathscr{D}_Z$; $\qquad\qquad\qquad\qquad\qquad\qquad\qquad\qquad$ (3.8)

and

the function defined by (3.3) is an involution on $\mathscr{D}_W$. $\qquad\qquad\qquad$ (3.9)

Evidently, if for all $W \subseteq S$, we let $\mathscr{D}_W$ be the set of B-matroids on W, then (3.8) and (3.9) hold. In fact, for every choice of $\mathscr{D}_W$ subject to these conditions, the members of $\mathscr{D}_W$ are B-matroids.

3.2.11. Theorem. *If S is an infinite set, then the largest class of pre-independence spaces defined on S and all its subsets such that (3.8) and (3.9) are satisfied is the class of B-matroids.*

For the proof of this result the reader is referred to Oxley (1978a). Note, however, that there is an error in Oxley (1978a) in that only Theorem 3.2.11 above is proved, although a stronger result is stated there. Whether this stronger result is true is an unsolved problem (see Exercise 3.22).

Exercises

The more difficult exercises are marked with an asterisk.

Section 3.1

3.1. (a) Extend the closure axioms (cl1)–(cl4) of Chapter 2 of White (1986) to give a closure-operator definition of an independence space.

   (b) Show that, by adding (B4) to (b1)–(b3) of White (1986), one gets an axiom system that is cryptomorphic to the system (i1), (i2), (i3′), and (I4).

   (c) Show that (b) holds with (b3) replaced by (b3.1) of White (1986).

3.2. (a) Let I be a set and $\{S_i : i \in I\}$ be a collection of pairwise disjoint sets. For all i in I, let $M_i(S_i)$ be an independence space having $\mathscr{I}_i$ as its collection of independent sets. Show that $\left\{ \bigcup_{i \in I} U_i : U_i \in \mathscr{I}_i \right\}$ is the collection of independent sets of an independence space on $\bigcup_{i \in I} S_i$. This independence space is called the *direct sum* of the independence spaces $M_i(S_i)$ $(i \in I)$.

   *(b) (Las Vergnas, 1971; Mason, 1970; Bean, 1976) Let $M(S)$ be an independence space for which $M^*(S)$ is also an independence space. Show that $M(S)$ is the direct sum of a collection of finite matroids.

3.3. Let $M(S)$ be an independence space and suppose that $X \subset\subset S$. If the restriction of a pre-independence space is defined as in (3.1), show that $\mathscr{I}.X = (\mathscr{I}^*|X)^*$.

3.4. (Las Vergnas, 1971) Let $M(S)$ be an independence space and define $\mathscr{I}^+$ to be the collection of subsets X of S such that for all $Y \subset\subset X$ there is a basis of $M(S)$ disjoint from Y.

   (a) Show that $\mathscr{I}^+$ is the collection of independent sets of an independence space on S.

   (b) Prove that $((\mathscr{I}^+)^+)^+ = \mathscr{I}^+$, and $\mathscr{I}^+|X = (\mathscr{I}.X)^*$ for all $X \subset\subset S$.

   (c) Give an example to show that it is not necessarily the case that $(\mathscr{I}^+)^+ = \mathscr{I}$.

   *(d) Let f be the closure operator of $M(S)$; that is, for all $X \subseteq S$,

$$f(X) = X \cup \{x : I \cup x \notin \mathscr{I} \text{ for some } I \subseteq X \text{ such that } I \in \mathscr{I}\}.$$

   Show that $(\mathscr{I}^+)^+ = \mathscr{I}$ if and only if for all finite subsets X of S and all elements x of $S - f(X)$, there is a set Y containing X such that $S - Y$ is finite and $x \notin f(Y)$.

*3.5. (Piff, 1971) Let $\mathscr{X}$ be a family of subsets of a set S and suppose that every element of S is in finitely many members of $\mathscr{X}$. Use Rado's selection principle to show that the set of partial transversals of $\mathscr{X}$ is the collection of independent sets of an independence space on S.

*3.6. (Piff, 1971) Prove that if $M(S)$ is an independence space such that for all $X \subset\subset S$ the matroid $M(X)$ is graphic, then there is a graph Γ such that $M(S)$ is isomorphic to M_Γ.

*3.7. Show that the following are equivalent.

(a) The Axiom of Choice.

(b) (Zorn's lemma) If every chain in a non-empty partially ordered set possesses an upper bound, then the set has at least one maximal element.

(c) (Tukey's lemma) If $\mathscr{X}$ is a non-empty collection of sets satisfying (i2) and (I4), then $\mathscr{X}$ has a maximal member.

Section 3.2

3.8. Higgs's approach to infinite matroids (Higgs, 1969a) uses the idea of a derived set operator. The series of problems below links this approach to that taken above. For an operator f on a set S, we define $\partial_f: 2^S \rightarrow 2^S$ by

$$\partial_f(X) = \{x: x \in f(X - x)\}.$$

(a) Show that ∂_f satisfies the following conditions.

(D1) If $X \subseteq Y \subseteq S$, then $\partial_f(X) \subseteq \partial_f(Y)$.
(D2) If $X \subseteq S$ and $x \in \partial_f(X)$, then $x \in \partial_f(X - x)$.

(b) Show that for all $X \subseteq S$, $f(X) = X \cup \partial_f(X)$.

(c) Let ∂ be a function on 2^S satisfying (D1) and (D2). Define $f: 2^S \rightarrow 2^S$ by $f(X) = X \cup \partial(X)$. Show that f is an operator on S and that $\partial_f = \partial$.

(d) Characterize IE-operators f in terms of properties of the corresponding derived set operators ∂_f.

(e) If $X \subseteq S$ and f is an operator on S, find ∂_{f^*}, ∂_{f_x}, and ∂_{f^x} in terms of ∂_f.

(f) For an operator f, characterize f-independent sets, f-spanning sets and f-bases in terms of ∂_f.

3.9. Axiomatize B-matroids in terms of their collections of bases.

3.10. Show that Proposition 3.2.5 does not hold for arbitrary operators by giving an example of an operator whose sets of bases, maximal independent sets, and minimal spanning sets are all distinct.

3.11. Show that an IE-operator is not uniquely determined by the pair consisting of its collection of independent sets and its collection of spanning sets.

3.12. Consider the following conditions on an operator f on a set S.

(cl3′) If $X, Y \subseteq S$ and $X \subseteq f(Y)$, then $f(X \cup Y) \subseteq f(Y)$.

(C) If $Y \subseteq S$ and $p \in f(Y)$, then there is a minimal subset U of Y such that $p \in f(U)$ and U is independent.

(H) If $Y \subseteq S$ and $p \in S - f(Y)$, then there is a maximal superset V of Y such that $p \notin f(V)$ and $V \cup p$ is spanning.

Show that

(a) f satisfies (cl3′) if and only if f satisfies (cl3);

(b) f satisfies (C) if and only if f^* satisfies (H);

(c) if f satisfies (cl4′) and p, Y, and U are as in (C), then $U \cup p$ is a circuit;

(d) if f satisfies (cl3) and p, Y, and V are as in (H), then V is a hyperplane.

3.13. Find an example of a B-matroid $M_B(S)$ and an operator g on S so that g is not the closure operator of $M_B(S)$, yet $M_B(S)$ and g have the same collections of independent sets.

3.14. If f is the closure operator of a B-matroid, show that f satisfies (C) and (H).

3.15. Let f be an IE-operator on a set S and suppose that f satisfies (C). Show that the collection $\mathscr{C}$ of f-circuits satisfies (c1), (c2), and (c3.1) and that, for all $X \subseteq S$, $f(X) = X \cup \{x : x \in C \subseteq X \cup x \text{ for some } C \text{ in } \mathscr{C}\}$.

3.16. (Higgs, 1969a) Let f be an IE-operator on a set S and suppose that f satisfies both (C) and (H).
   (a) (Unsolved) Is f the closure operator of a B-matroid on S?
   (b) (Unsolved) If $X \subseteq S$, does f_X satisfy (H)?
   (c) If $X \subseteq S$, does f_X satisfy (C)?
   (d) Show that if (a) holds, then so does (b).

3.17. Let f be an IE-operator on a set S. Suppose f satisfies (C) and B is an f-basis. If $x \in S - B$, use the result of Exercise 3.15 to prove that there is a unique circuit C such that $x \in C \subseteq B \cup x$, and that, if $y \in B$, then $(B \cup x) - y$ is a basis if and only if $y \in C$.

*3.18. (Oxley, 1978a) Use the results of Exercises 3.14, 3.15, and 3.17 to show that if B_1 and B_2 are bases of a B-matroid and $b_1 \in B_1 - B_2$, then there is an element b_2 of $B_2 - B_1$ such that both $(B_1 - b_1) \cup b_2$ and $(B_2 - b_2) \cup b_1$ are bases.

*3.19. (Higgs, 1969c) Let Γ be a graph that may be finite or infinite, and let $\mathscr{K}(\Gamma)$ consist of all finite cycles of Γ together with all two-way infinite paths in Γ. Use the results of Exercises 3.14 and 3.15 to show that $\mathscr{K}(\Gamma)$ is the set of circuits of a B-matroid on the edge set of Γ if and only if Γ has no subgraph homeomorphic to the graph in Figure 3.2.

Figure 3.2.

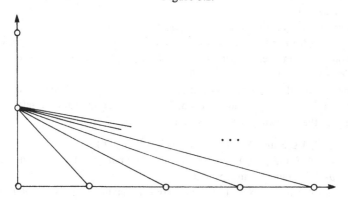

*3.20. (Matthews and Oxley, 1977) Let Γ be a graph that may be finite or infinite, and let $\mathscr{N}(\Gamma)$ be the set of subgraphs of Γ that are homeomorphic to one of the five graphs shown in Figure 3.3 (where an arrow denotes a one-way infinite path). Use the results of Exercises 3.14 and 3.15 to show that $\mathscr{N}(\Gamma)$ is the set of circuits of a B-matroid on the set of edges of Γ. A detailed discussion of the properties of the finite matroids that arise in this way may be found in Chapter 4 of this volume.

*3.21. (Higgs, 1969b) Show that if the Generalized Continuum Hypothesis holds, then all bases of a B-matroid are equicardinal.

Figure 3.3.

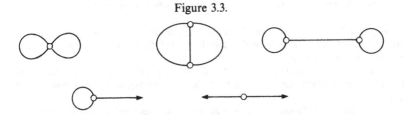

3.22. Let S be an infinite set and $\mathscr{D}$ be a class of pre-independence spaces defined on S and all its subsets such that (3.8) and (3.9) are satisfied.

(a) Show that $\mathscr{D}$ need not equal the class of B-matroids defined on all subsets of S.

(b) (Unsolved) Does (a) remain true if we insist that $\mathscr{D}$ contains the class of all those independence spaces that are defined on some subset of S?

References

Asche, D. S. (1966). Minimal dependent sets, *J. Austral. Math. Soc.* **6**, 259–62.

Bean, D. W. T. (1976). A connected finitary co-finitary matroid is finite, in *Proceedings of the Seventh Southeastern Conference on Combinatorics, Graph Theory and Computing,* Congressus Numerantium 17, pp. 115–19. Utilitas Mathematica, Winnipeg.

Bleicher, M. N. & Marczewski, E. (1962). Remarks on dependence relations and closure operators, *Colloq. Math.* **9**, 209–11.

Bleicher, M. N. & Preston, G. B. (1961). Abstract linear dependence relations, *Publ. Math. Debrecen* **8**, 55–63.

Brualdi, R. A. (1969). Comments on bases in dependence structures, *Bull. Austral. Math. Soc.* **1**, 161–7.

Brualdi, R. A. (1970). Admissible mappings between dependence spaces, *Proc. London Math. Soc.* (3) **21**, 296–312.

Brualdi, R. A. (1971). On families of finite independence structures, *Proc. London Math. Soc.* (3) **22**, 265–93.

Brualdi, R. A. & Scrimger, E. B. (1968). Exchange systems, matchings and transversals, *J. Comb. Theory* **5**, 244–57.

d'Ambly, C. G. (1970). Whitneysche Abhängigkeitsstrukturen, in *Combinatorial Mathematics and its Applications I,* Colloq. Math. Soc. János Bolyai 4, pp. 25–37. North-Holland, Amsterdam.

Dlab, V. (1965a). Axiomatic treatment of bases in arbitrary sets, *Czechoslovak Math. J.* **15** (90), 554–64.

Dlab, V. (1965b). The rôle of the 'finite character property' in the theory of dependence, *Comment. Math. Carolin.* **6**, 97–104.

Glazek, K. (1979). Some old and new problems in independence theory, *Colloq. Math.* **42**, 127–89.

Györi, E. (1975). Végtelen matroidok összehasonlítása (Comparison of infinite matroids), *Mat. Lapok* **26** (34), 311–17.

Györi, E. (1978). On the structures induced by bipartite graphs and infinite matroids, *Discrete Math.* **22**, 257–61.

Higgs, D. A. (1969a). Matroids and duality, *Colloq. Math.* **20**, 215–20.

Higgs, D. A. (1969b). Equicardinality of bases in B-matroids, *Can. Math. Bull.* **12**, 861–62.

Higgs, D. A. (1969c). Infinite graphs and matroids, in *Recent Progress in Combinatorics, Proceedings of the Third Waterloo Conference on Combinatorics,* pp. 245–53. Academic Press, New York.

Hughes, N. J. S. (1963). Steinitz' exchange theorem for infinite bases, *Compositio Math.* **15**, 113–18.

Hughes, N. J. S. (1965). Steinitz' exchange theorem for infinite bases II, *Compositio Math.* **17**, 152–5.

Klee, V. (1971). The greedy algorithm for finitary and confinitary matroids, in *Combinatorics, Proceedings of Symposia in Pure Mathematics* 19, pp 137–52. American Mathematical Society, Providence, Rhode Island.

Las Vergnas, M. (1971). Sur la dualité en théorie des matroïdes, in *Théorie des Matroïdes*, Lecture Notes in Mathematics 211, pp. 67–85. Springer-Verlag, Berlin, Heidelberg, and New York.

Las Vergnas, M. (1972). Problèmes de couplages et problèmes hamiltoniens en théorie des graphes, Thesis L'Université Paris VI.

Maeda, F. & Maeda, S. (1970). *Theory of Symmetric Lattices*, Springer-Verlag, Berlin, Heidelberg, and New York.

Mason, J. H. (1970). A characterization of transversal independence spaces, in *Combinatorial Structures and Their Applications, Proceedings of the Calgary International Conference*, pp. 257–9. Gordon & Breach, New York.

Matthews, L. R. & Oxley, J. G. (1977). Infinite graphs and bicircular matroids. *Discrete Math.* **19**, 61–5.

McDiarmid, C. J. H. (1975). An exchange theorem for independence spaces, *Proc. Amer. Math. Soc.* **47**, 513–14.

Mirsky, L. (1971). *Transversal Theory. An Account of Some Aspects of Combinatorial Mathematics*, Academic Press, London and New York.

Oxley, J. G. (1978a). Infinite matroids, *Proc. London Math. Soc.* (3) **37**, 259–72.

Oxley, J. G. (1978b). Infinite matroids and duality, in *Problèmes Combinatoires et Théorie des Graphes*, Colloques Internationaux du C.N.R.S 260, pp. 325–6. Centre National de la Recherche Scientifique, Paris.

Perfect, H. (1969). Independence spaces and combinatorial problems, *Proc. London Math. Soc.* **19**, 17–30.

Piff, M. J. (1971). Properties of finite character of independence spaces, *Mathematika* **18**, 201–8.

Rado, R. (1949). Axiomatic treatment of rank in infinite sets, *Can. J. Math.* **1**, 337–43.

Rado, R. (1966). Abstract linear dependence, *Colloq. Math.* **14**, 257–64.

Robert, P. (1967). Sur l'axiomatique des systèmes générateurs, des rangs etc., *C.R. Acad. Sci. Paris Sér. A* **265**, 649–51.

Robertson, A. P. & Weston, J. D. (1959). A general basis theorem, *Proc. Edinburgh Math. Soc.* (2) **11**, 139–41.

Simões-Pereira, J. M. S. (1974). Matroids, graphs and topology, in *Proceedings of the Fifth Southeastern Conference on Combinatorics, Graph Theory and Computing*, Congressus Numerantium 10, pp. 145–55. Utilitas Mathematica, Winnipeg.

Wagstaff, S. S. Jr. (1973). Infinite matroids, *Trans. Amer. Math. Soc.* **175**, 141–53.

Welsh, D. J. A. (1976). *Matroid Theory*, London Mathematical Society Monographs 8, Academic Press, London, New York, and San Francisco.

White, N. (ed.) (1986). *Theory of Matroids*, Cambridge University Press.

White, N. (ed.) (1987). *Combinatorial Geometries*, Cambridge University Press.

4

Matroidal Families of Graphs

J. M. S. SIMÕES-PEREIRA

4.1. Introduction

The connections between graph theory and matroid theory can be traced back to the study of graphic matroids, which were introduced by Whitney (1935) and have been extensively investigated (see Chapters 1, 2, and 6 of White, 1986). We recall that a matroid is graphic if it is isomorphic to the polygon matroid of some graph.

In this chapter we present some recent results that give a new setting to the relations between graphs and matroids. In the light of this setting, the polygon matroid appears as the simplest and best known object among an uncountably infinite collection of similar objects.

The fundamental concept that we want to introduce is the concept of a 'matroidal family of graphs'. The precise definition is given below. According to this definition, the collection of all polygons is a matroidal family of graphs, the simplest among the non-trivial ones. Another example of a matroidal family of graphs is the collection of all bicycles, where a bicycle is a connected graph with two independent cycles and no vertex of degree less than two, that is to say, a bicycle is a graph homeomorphic to one of the graphs J_{00}, $J_{0.0}$, or $J_{0\text{-}0}$ pictured in Figure 4.1.

As we shall see, there are uncountably many matroidal families of graphs; the subject is virtually unexplored and this chapter is just a brief introduction to this fascinating new field.

Figure 4.1.

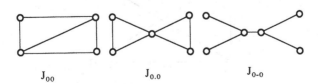

J_{00} $J_{0.0}$ $J_{0\text{-}0}$

4.2. Matroidal Families of Graphs: a Definition

Throughout this chapter, Γ denotes a graph, $V(\Gamma)$ its vertex set, and $E(\Gamma)$ its edge set. We shall consider simple graphs, unless otherwise stated. If Γ' and Γ are graphs, then $\Gamma' \cong \Gamma$ means that they are isomorphic, $\Gamma' = \Gamma$ means that they are coincident, $\Gamma' \lesssim \Gamma$ means that Γ' is isomorphic to a subgraph of Γ, and $\Gamma' \subseteq \Gamma$ means that Γ' is a subgraph of Γ. Sometimes the distinction between $\lesssim$ and $\subseteq$ is convenient. Behzad & Chartrand (1971) is our graph theory reference.

A matroidal family of graphs is a non-empty collection P of finite, connected graphs with the following property: given an arbitrary graph Γ, the edge sets of the subgraphs of Γ that are isomorphic to some member of P are the circuits of a matroid on $E(\Gamma)$. The matroid defined in this way by the matroidal family P on the edge set of the graph Γ will be denoted by $P(\Gamma)$.

The assumption of connectedness in the definition is important. If we drop the requirement of connectedness, then, for each positive integer q, the collection P of all graphs (connected or not) with q edges is a matroidal family, $P(\Gamma)$ being a uniform matroid.

As examples, we present now the four matroidal families of graphs that were discovered first; they will be denoted throughout by P_0, P_1, P_2, and P_3. The family P_0 consists of one graph only, namely the complete graph on two points, K_2. The family P_1 consists of all polygons or cycles C_n, for $n \geq 3$. (Polygon and cycle are here essentially synonyms but 'cycle' conveys the idea of a polygon that is a proper subgraph of some graph.) The family P_2 consists of all bicycles. The family P_3 consists of the even polygons of length at least four and the bicycles with no even polygon.

In Figure 4.2, $\Gamma = K_4$ and $E(\Gamma) = \{e_1, e_2, ..., e_6\}$. The circuits of $P_0(\Gamma)$ are the sets $\{e_i\}$ for $i = 1, 2, ..., 6$; the circuits of $P_1(\Gamma)$ are $\{e_1, e_2, e_5\}$, $\{e_3, e_4, e_5\}$, $\{e_2, e_3, e_6\}$, $\{e_1, e_4, e_6\}$, $\{e_1, e_2, e_3, e_4\}$, $\{e_1, e_3, e_5, e_6\}$, and $\{e_2, e_4, e_5, e_6\}$; the circuits of $P_2(\Gamma)$ are the sets $E(\Gamma) - \{e_i\}$ for $i = 1, 2, ..., 6$; the circuits of $P_3(\Gamma)$ are $\{e_1, e_2, e_3, e_4\}$, $\{e_1, e_3, e_5, e_6\}$, and $\{e_2, e_4, e_5, e_6\}$.

Among matroidal families of simple graphs, P_0 is the only one that is finite, as proved in Simões-Pereira (1973). If Γ is simple, $P_1(\Gamma)$ is the polygon

Figure 4.2.

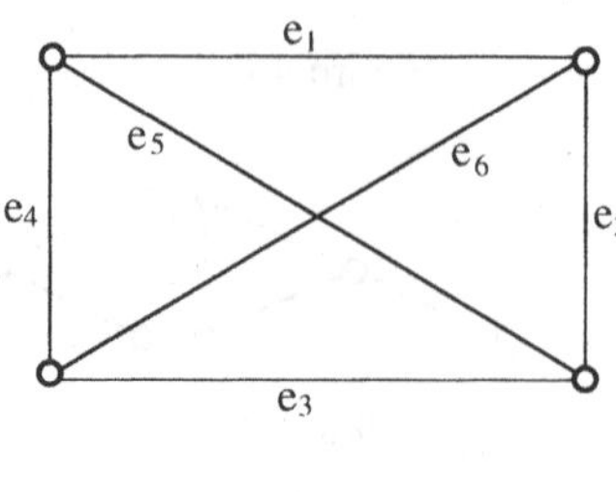

matroid of Γ and $P_2(\Gamma)$ is called the bicircular matroid of Γ. If we allow multiple edges in Γ, then, in the definition of P_1 we should write $n \geqslant 2$ instead of $n \geqslant 3$; if we also allow loops, then we should write $n \geqslant 1$. Similar considerations apply to P_2 and P_3. Note that generalizing definitions or results in this chapter to non-simple graphs may not be straightforward.

The matroidal families P_1 and P_2 were first singled out by a theorem proved in Simões-Pereira (1972). To state it, recall that two graphs are homeomorphic if both can be obtained from a third graph by a (possibly empty) sequence of edge subdivisions. Define a family P of graphs to be closed under homeomorphism if it has the following property: if a graph Γ'' is homeomorphic to a graph Γ' and $\Gamma' \in P$, then $\Gamma'' \in P$.

4.2.1. Theorem. *The only matroidal families of graphs closed under homeomorphism are P_1 and P_2.*

The proof is based on the next three lemmas. For many results on matroidal families, proofs are graph-theoretic. Some lemmas forbid certain subgraphs or configurations in members of P; other lemmas forbid or require certain graphs as members of P provided that other graphs belong to P or that P satisfies some prescribed conditions. To prove most of the lemmas we take two graphs Γ' and Γ'' (possibly, $\Gamma' \cong \Gamma''$) of a matroidal family P and, by partly superposing them, we form a graph Γ; we may write $\Gamma = \Gamma' \cup \Gamma''$. The graphs Γ' and Γ'' are now subgraphs of Γ and their edge sets are circuits of $P(\Gamma)$. For any edge $e \in E(\Gamma') \cap E(\Gamma'')$, $E(\Gamma') \cup E(\Gamma'') - \{e\}$ contains a circuit of $P(\Gamma)$. By conveniently choosing Γ' and Γ'', the edges of Γ' and Γ'' that coincide in Γ, and the edge e, we determine new members of P or are led to a contradiction. In the latter case we deduce that Γ' and Γ'' cannot be in P.

4.2.2. Lemma. *If P is a matroidal family distinct from P_0, then none of its members has a pendant edge.*

Proof. Suppose the contrary. Let $J \in P$ have a pendant edge e. Take $\Gamma' \cong \Gamma'' \cong J$ and form Γ from Γ' and Γ'' so that Γ' and Γ'', as subgraphs of Γ, have e as their only common edge and e is a bridge of Γ (Figure 4.3). Since $E(\Gamma')$ and $E(\Gamma'')$ are circuits of $P(\Gamma)$, the (circuit) definition of a matroid requires the existence of another circuit contained in $E(\Gamma') \cup E(\Gamma'') - \{e\} = E(\Gamma) - \{e\}$. This circuit is either properly contained in one of the circuits $E(\Gamma')$ or $E(\Gamma'')$, contrary to the definition, or it is, as a graph, disconnected, with edges in $E(\Gamma') - \{e\}$ and edges in $E(\Gamma'') - \{e\}$. Since all graphs in P are connected, this is also a contradiction and thus the lemma is proved. $\square$

Figure 4.3. —— , Γ'; ---- , Γ''.

4.2.3. Lemma. *In any matroidal family P distinct from P_1 and closed under homeomorphism, there is a member that contains, as a proper induced subgraph, a pendant triangle with an edge incident to one of its vertices.*

Proof. Take any $J \in P$ and let $(u, v) \in E(J)$. Let Γ' and Γ'' be obtained from J by substituting for (u, v) paths (u, w, x, v) and (u, w, y, x, v), respectively. Let Γ have Γ' and Γ'' as subgraphs with all edges coincident except for the triangle (w, y), (y, x), (x, w) as pictured in Figure 4.4. Let $e = (u, w)$. Since $E(\Gamma')$ and $E(\Gamma'')$ are circuits of $P(\Gamma)$, the (circuit) definition of a matroid requires the existence of another circuit contained in $E(\Gamma') \cup E(\Gamma'') - \{e\} = E(\Gamma) - \{(u, w)\}$. This circuit contains the edge (w, x) and at least one of the edges (w, y) or (y, x) otherwise it is properly contained in another circuit. By Lemma 4.2.2, it contains in fact (w, y) and (y, x). Since, as a graph, it must be connected, either it is the triangle, and $P = P_1$, or it also contains the edge (x, v). This completes the proof.   □

Figure 4.4.

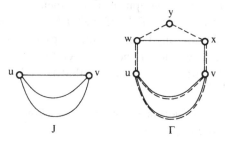

J Γ

4.2.4. Lemma. *If P is a matroidal family distinct from P_1 and closed under homeomorphism, then the square with a diagonal, the graph formed by two triangles sharing one vertex, and the graph formed by two triangles linked by an edge belong to P.*

Proof. We denote these graphs by J_{00}, $J_{0.0}$, and $J_{0\text{-}0}$, respectively (Figure 4.1). Let $J \in P$ contain the induced subgraph referred to in Lemma 4.2.3. Let (v, x) be the edge in J incident with the pendant triangle (w, y), (y, x), (x, w). Let $\Gamma' \cong \Gamma'' \cong J$ and form Γ so that Γ' and Γ'', as subgraphs of Γ, have all edges coincident except those in the pendant triangles. Let $e = (v, x)$ (Figure 4.5a).

Figure 4.5.

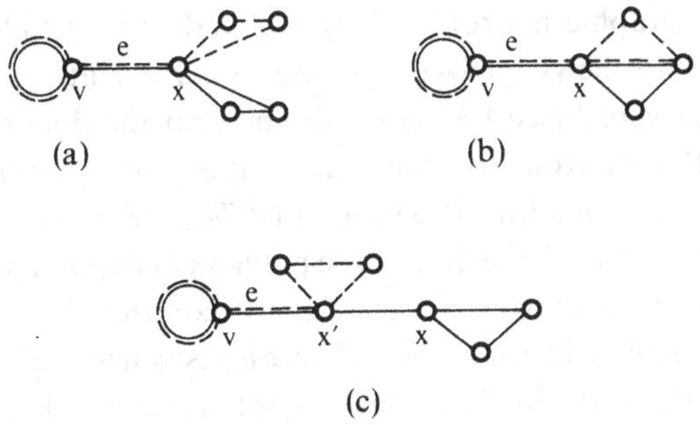

(a)   (b)

(c)

Reasoning similar to that used in the preceding lemmas shows that $J_{0.0} \in P$; a slightly different construction of Γ (Figure 4.5b) shows that $J_{00} \in P$; finally, substituting a path (v, x'), (x', x) for the edge (v, x) in Γ' (which is possible because P is closed under homeomorphism), forming Γ from Γ' and Γ'' as indicated in Figure 4.5c, and setting $e = (v, x')$ shows that $J_{0\text{-}0} \in P$. This completes the proof of the lemma.

We can now prove Theorem 4.2.1: As a consequence of Lemma 4.2.4, the bicycles belong to any matroidal family P closed under homeomorphism and distinct from P_1. By Lemma 4.2.2, any other graph of P contains at least one cycle. By the definition of a matroidal family and the (circuit) definition of a matroid, it cannot contain only one cycle; if it contains two or more, then it contains a bicycle. Hence $P = P_2$ and thus the theorem is proved.

$\square$

Each of the matroidal families P_0, P_1, P_2, and P_3 can be partitioned into one or three classes of homeomorphic graphs. This is one of several properties shared only by these four families.

4.2.5. Property. *No member of a matroidal family P distinct from P_0, P_1, P_2, and P_3 has fewer than three independent cycles.*
In P_0, P_1, P_2, and P_3 no member has more than two independent cycles.

4.2.6. Property. *No member of a matroidal family P distinct from P_0, P_1, P_2, and P_3 has a vertex of degree less than three.*
In P_1, P_2, and P_3 each member has vertices of degree two.

4.2.7. Property. *If P is a matroidal family distinct from P_0, P_1, P_2, and P_3, then no two members of P are homeomorphic.*

Property 4.2.7 is an immediate consequence of Property 4.2.6. For the proofs of Properties 4.2.5 and 4.2.6, we refer the reader to Simões-Pereira (1975a).

A matroid is called bicircular if it is isomorphic to the bicircular matroid of some graph. In view of Theorem 4.2.1, bicircular matroids are, in a sense, an alternative to graphic matroids. They are studied in detail in Matthews (1977a; 1978a). Matthews proved, among other results, that bicircular matroids are transversal (see Exercise 4.4) and that the dual of a bicircular matroid is a pathic matroid together with a (possibly empty) set of loops. Pathic matroids, as defined in McDiarmid (1974), are in fact an alternative to the cocycle matroid, itself the dual of the polygon matroid. The connectivity of $P_2(\Gamma)$ and its relation to a certain type of connectivity of Γ has been studied in Wagner (1985). The bases of $P_2(\Gamma)$ are investigated in Shull *et al.* (1989).

Unlike $P_2(\Gamma)$, the matroid $P_3(\Gamma)$ has not yet, as far as I know, received a comprehensive analysis. It emerged on an equal footing with $P_2(\Gamma)$ in Simões-Pereira (1975a) and was probably perceived in Duchamp (1972). It appears in the work of Doob (1973) where $P_2(\Gamma)$ and $P_3(\Gamma)$ are related with the matroid represented by the eigenspace that corresponds to the eigenvalue -2 of the adjacency matrix of the line graph of Γ. $P_3(\Gamma)$ appears also in Tutte (1976) within the context of chain groups and in Zaslavsky (1982a) within the context of signed graphs. By allowing multiple edges, loops, and even free loops (which are edges with no endpoints), Zaslavsky (1982a) obtains $P_1(\Gamma), P_2(\Gamma)$, and $P_3(\Gamma)$ as particular instances of his voltage-graphic matroids.

$P_2(\Gamma)$ appears also in Zaslavsky (1987; 1989) and $P_3(\Gamma)$ in Wagner (1988) where it is called the factor matroid.

4.3. Countably Many Matroidal Families

Andreae (1978) gave the first description of matroidal families distinct from P_0, P_1, P_2, and P_3. Theorem 4.3.1 below is a generalization of his result. Although this theorem may be proved by graph-theoretic methods, we prefer to give a proof due to Lorea (1979), which is not essentially graph-theoretic.

4.3.1. Theorem. *Let n and r be integers, $n \geq 0$ and $-2n + 1 \leq r \leq 1$. Let $P_{n,r}$ be the set of all graphs Γ such that:*

(i) $n|V(\Gamma)| + r = |E(\Gamma)|$ *and* $|V(\Gamma)| \geq 2$; *and*
(ii) Γ *is minimal with respect to* $\lesssim$ *among graphs with property* (i).
Then $P_{n,r}$ is a matroidal family.

Proof. Let Γ be an arbitrary graph. Define an integer-valued function f on the power set of $E(\Gamma)$, by setting

$$f(X) = 0 \qquad \text{for } X = \varnothing$$
$$f(X) = n|V'(X)| + r - 1 \quad \text{for } X \neq \varnothing, \ X \subseteq E(\Gamma),$$

where $V'(X)$ is the set of vertices of Γ incident to edges in X. Such a function is non-decreasing and submodular. It therefore defines a matroid, written $P_{n,r}(\Gamma)$, on $E(\Gamma)$, whose independent sets are the sets $X \subseteq E(\Gamma)$ such that, $\forall\, Y \subseteq X$, $f(Y) \geqslant |Y|$ (see Welsh, 1976, Corollary 8.1.1).

Given Γ we now claim that the circuits of $P_{n,r}(\Gamma)$ are the minimal, nonempty subsets Z of $E(\Gamma)$ for which

$$n|V'(Z)| + r = |Z|. \tag{4.1}$$

To prove this claim it is enough to show that
(a) if Z satisfies (4.1), then it is dependent
(b) if Z is a circuit, then it satisfies (4.1).

To show (a): from (4.1) we get $f(Z) < |Z|$ and therefore Z is dependent in $P_{n,r}(\Gamma)$.

To show (b): suppose Z is a circuit. We must have $f(Z) < |Z|$, that is,

$$n|V'(Z)| + r \leqslant |Z|. \tag{4.2}$$

By hypothesis, Z is a circuit; this therefore excludes the possibility that $f(Z) \geqslant |Z|$. Now, if $|Z| = 1$, then the inequality (4.2) and $f(Z) = n|V'(Z)| + r - 1 \geqslant 0$ imply $n|V'(Z)| + r - 1 = 0 = |Z| - 1$, which proves (4.1). If $|Z| > 1$, then let $Z' \subseteq Z$, $|Z'| = |Z| - 1$. The set Z' is independent, hence $f(Z') \geqslant |Z'|$ and, therefore,

$$n|V'(Z')| + r \geqslant |Z|. \tag{4.3}$$

Inequalities (4.2) and (4.3) yield

$$|Z| \geqslant n|V'(Z)| + r \geqslant n|V'(Z')| + r \geqslant |Z|$$

which proves (4.1) and, therefore, our claim.

To achieve the proof of the theorem it is enough to show that the subgraphs of Γ induced by the edge subsets Z are connected. Suppose the contrary. Let $Z_1, Z_2, \ldots, Z_k$, with $k \geqslant 2$, be the connected components of the subgraph of Γ induced by Z. Since Z is a circuit, $Z_1, Z_2, \ldots, Z_k$ are independent, therefore we have $n|V(Z_i)| + r - 1 \geqslant |Z_i|$ for $i = 1, 2, \ldots, k$. It follows that $n|V(Z)| + r - 1 \geqslant |Z| - (k-1)(r-1)$, which contradicts $n|V(Z)| + r = |Z|$, because $r - 1 \leqslant 0$. This completes the proof. $\square$

Note that, in proving this theorem, we became aware of the following fact stated here as a proposition for later reference.

4.3.2. Proposition. *If Γ is such that, with n and r as in Theorem 4.3.1, $|E(\Gamma)| \geqslant n|V(\Gamma)| + r$, $|V(\Gamma)| \geqslant 2$, then there is a connected graph $\Gamma' \subseteq \Gamma$ such that $|E(\Gamma')| = n|V(\Gamma')| + r$, $|V(\Gamma')| \geqslant 2$.*

The matroidal families P_0, P_1, and P_2 are of the form $P_{n,r}$. It is an easy exercise to verify that $P_{1,1} = P_2$, $P_{1,0} = P_1$ and, by Lemma 4.2.2, $P_{n,1-2n} = P_0$ for all $n \geqslant 0$.

Figure 4.6.   Figure 4.7.

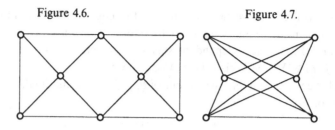

Given $P_{n,r}$, it does not seem feasible to draw a picture that visualizes all the graphs in $P_{n,r}$. The reader should, however, pause to check that some well known graphs belong to some of these families. For instance, $P_{2,-2}$ contains all the wheels; it contains also, for example, $K_{3,4}$ and the graph pictured in Figure 4.6. For $n \geqslant 1$, $P_{n,-n}$ contains K_{2n} and $P_{n,0}$ contains K_{2n+1}. The graph $K_{4,4}$ and the graph in Figure 4.7 belong to $P_{2,0}$, which contains all connected, 4-regular graphs (Lemma 4.4.3). Finally, no $P_{n,r}$ contains $K_{3,3}$ (Exercise 4.9).

4.4. Uncountably Many Matroidal Families

Schmidt (1979) was able to use certain subsets of the families $P_{n,r}$ to construct uncountably many matroidal families. He called these subsets partly closed sets.

A collection of graphs $Q \subseteq P_{n,r}$ is called partly closed if the following holds: if X, $Y \in Q$, $W = X \cup Y$ is different from X and Y, and $|E(W)| = n|V(W)| + r + 1$, then, for every edge $e \in E(X \cap Y)$, there exists $Z \in Q$ such that $Z \lesssim W - \{e\}$. As an example, the set of all even polygons is partly closed. We denote it by T. We have $T \subseteq P_{1,0}$. For more examples, see Lemma 4.4.5.

In what follows, the notations are those of the preceding section and all proofs are essentially due to Schmidt (1979).

4.4.1. Theorem. Let $Q \subseteq P_{n,r}$, $r \leqslant 0$, be a partly closed set and define $P_{n,r+1,Q} = Q \cup P'$, where P' is the set of all graphs of $P_{n,r+1}$ that do not contain a subgraph isomorphic to a graph in Q. Then $P_{n,r+1,Q}$ is a matroidal family.

Proof. It is enough to see that:

(1) By the definitions, $P_{n,r+1,Q}$ contains only minimal graphs with respect to $\lesssim$; and
(2) If X, $Y \in P_{n,r+1,Q}$, $W = X \cup Y$, and W is different from X and Y, then, for every edge $e \in E(X \cap Y)$ there exists $Z \in P_{n,r+1,Q}$ such that $Z \lesssim W - \{e\}$.

To prove (2) note first that Proposition 4.3.2 and the minimality of X and Y allow us to write

$$
\begin{aligned}
|E(W)| &= |E(X \cup Y)| \\
&= |E(X)| + |E(Y)| - |E(X \cap Y)| \geqslant n|V(X)| + r + n|V(Y)| \\
&\qquad\qquad\qquad\qquad + r - n|V(X \cap Y)| - r + 1 \\
&= n(|V(X)| + |V(Y)| - |V(X \cap Y)|) + r + 1 \\
&= n|V(W)| + r + 1. \qquad\qquad\qquad\qquad\qquad (4.4)
\end{aligned}
$$

We distinguish now:

CASE 1. $X, Y \in Q$. If, in $|E(W)| \geqslant n|V(W)| + r + 1$, the equality holds, then (2) follows from the definition of a partly closed collection. Otherwise, take $e \in E(X \cap Y)$. From Proposition 4.3.2, it follows that there exists $Z' \in P_{n,r+1}$ such that $Z' \lesssim W - \{e\}$. Now either $Z' \in P_{n,r+1,Q}$ or there exists $Z \in Q, Z \lesssim Z'$ hence $Z \lesssim W - \{e\}$.

CASE 2. $X, Y \in P_{n,r+1}$. By definition of a matroidal family, we find $Z' \in P_{n,r+1}$, $Z' \lesssim W - \{e\}$. Then we use the argument of Case 1.

CASE 3. $X \in Q, Y \notin Q$. We have $|E(W)| \geqslant n(|V(X)| + |V(Y)| - |V(X \cap Y)|) + r + 2$, which is the situation found in Case 1.

This completes the proof of the theorem. □

Recalling that the set T of all even polygons is partly closed and contained in $P_{1,0}$, it is easy to see that $P_3 = P_{1,1,T}$. As a further exercise, the reader is urged to verify that $P_{n,r+1} = P_{n,r+1,\varnothing}$ and that $P_{n,-2n+1} = P_{n,-2n+2,U}$ where $U = P_{n,-2n+1}$.

To show that there are uncountably many matroidal families of the form $P_{n,r+1,Q}$ we have only to describe a sufficiently large collection of partly closed sets and show that the families they yield are distinct. This can be achieved by the following lemmas.

4.4.2. Lemma. *Let the pairs of integers (n, r) and (n', r') be as in Theorem 4.3.1. Then $P_{n,r} \neq P_{n',r'}$ if and only if $(n, r) \neq (n', r')$ and at least one of the inequalities $2n + r > 1$ or $2n' + r' > 1$ holds.*

Proof. For the 'only if' part we have trivially $(n, r) \neq (n', r')$. Moreover $2n + r \leqslant 1$ implies $2n + r = 1$, hence $P_{n,r} = P_{n,1-2n} = P_0$. Since the same is true for (n', r'), if neither of the inequalities in the lemma holds, then we would have $P_{n,r} = P_{n',r'}$.

For the 'if' part, suppose, without loss of generality, that $2n + r > 1$, that is, $P_{n,r} \neq P_0$. It follows that $P_{n,r}$ is infinite. Since $(n, r) \neq (n', r')$, we get $P_{n,r} \neq P_{n',r'}$, which proves the lemma. □

4.4.3. Lemma. *For every $n \geqslant 1$, $P_{n,0}$ contains all connected, $2n$-regular graphs.*

Proof. For any $2n$-regular graph Γ, $|E(\Gamma)| = n|V(\Gamma)|$. Moreover, if Γ is connected and $2n$-regular and Γ' is a proper subgraph of Γ, then there exists $e \in E(\Gamma) - E(\Gamma')$ incident with $v \in V(\Gamma')$. Hence $|E(\Gamma')| < n|V(\Gamma')|$, which implies the minimality of Γ, that is to say, no connected, $2n$-regular graph contains properly another such graph. This completes the proof of the lemma. $\square$

4.4.4. Lemma. *For every $n \geqslant 2$, there are infinitely many non-isomorphic, 3-connected, $2n$-regular graphs.*

Proof. Consider a path of length $q \geqslant 2$, say $(x_0, x_1), (x_1, x_2), ..., (x_{q-1}, x_q)$ and add to it $2n - 2$ pendant edges at each vertex $x_1, x_2, ..., x_{q-1}$. Let J denote the tree so obtained. The tree J has all vertices of degree $2n$ except for the set V' of its endvertices. Form a graph J' by adding new edges to J linking endvertices in such a way that the subgraph C induced by V' is a cycle. J' is 3-connected. Let $J_1, J_2, ..., J_{2n-2}$ be copies of J' and form with them a graph Γ in such a way that, for $i \neq k$, $J_i \cap J_k = C_i = C_k$ where $C_i \subseteq J_i$ is the replica of C. This graph Γ, obtained as a join of all J_i for $i = 1, 2, ..., 2n - 2$ is $2n$-regular and 3-connected. Since distinct values of q obviously yield non-isomorphic graphs Γ, the lemma is proved. $\square$

4.4.5. Lemma. *Let $Q \subseteq P_{n,0}$, $n \geqslant 2$, be a set of 3-connected, $2n$-regular graphs. Then Q is partly closed.*

Proof. With X, Y, and W as in Theorem 4.4.1, we show that the equation $|E(W)| = n|V(W)| + 1$ cannot be satisfied, which implies the lemma. For this purpose, suppose that this equation holds. We have then, by (4.4), $|E(X \cap Y)| = |E(X)| + |E(Y)| - |E(W)| = n|V(X \cap Y)| - 1$. With no loss of generality, we now distinguish two cases, both leading to contradictions:

CASE 1. $|V(X \cap Y)| = |V(X)| = |V(Y)|$. Then we get $X \cap Y$ by deleting just one edge from X, incident with, say, vertices u and v in $V(X)$. As u and v are the only vertices in $X \cap Y$ of degree less than $2n$, they are connected by an edge in Y too. All graphs here are presumed to be simple. It follows that $X = Y$, that is, X and Y are coincident, a contradiction.

CASE 2. $|V(X \cap Y)| < |V(X)|$. From the $2n$-regularity of X, it follows that $X \cap Y$ has exactly two vertices, say u and v, of degree less than $2n$. It follows also that $\{u, v\}$ separates X because $V(X \cap Y) - \{u, v\} \neq \varnothing$ and $V(X) - V(Y) \neq \varnothing$, and no edge of X may join a vertex of $X - Y$ with a vertex of $X \cap Y$ other than u or v. This contradicts the 3-connectivity of X. $\square$

4.4.6. Lemma. *If* $Q, Q' \subseteq P_{n,r}$ *are partly closed and* $Q \neq Q'$, *then* $P_{n,r+1,Q} \neq P_{n,r+1,Q'}$.

Proof. The proof of this lemma is trivial.   □

From the preceding lemmas, we immediately obtain the following.

4.4.7. Theorem. *There are uncountably many matroidal families of graphs. More precisely, for every* $n \geq 2$, *there are uncountably many matroidal families of the form* $P_{n,1,Q}$ *where* Q *is an infinite set of 3-connected, 2n-regular graphs.*

Partly closed collections of graphs do not look like a natural concept. But they are. To see this, let us give the following generalization which is due to Walter (1982):

For $n \geq 0$, $-2n + 1 \leq r \leq 1$, a collection $Q \subseteq P_{n,r}$ is *s-partly closed* if the following holds: if $X, Y, \in Q$, $W = X \cup Y$ is different from X and Y, and $|E(W)| \leq n|V(W)| + s$, then, for every edge $e \in E(X \cap Y)$, there exists $Z \in Q$ such that $Z \leq W - \{e\}$.

The proof of the following result can be found in Walter (1982).

4.4.8. Theorem. *Let* $n \geq 0$, $-2n + 1 \leq r \leq s$, *and* $Q \subseteq P_{n,r}$. *Let* $N_{n,s}$ *be the set of those graphs in* $P_{n,s}$ *that do not contain a subgraph isomorphic to a graph in* Q. *Then* $Q \cup N_{n,s}$ *is a matroidal family if and only if* Q *is s-partly closed and* $s \leq 1$.

Another result that is worth mentioning is that the existence of uncountably many matroidal families of graphs can also be proved without requiring the graphs to be simple. This generalization of Theorem 4.4.7 is due to Prüss (1984) whose main result is a characterization of matroidal families in terms of what he calls matroidal functions: finitely additive, submodular, non-decreasing functions that map graphs onto non-negative integers and are zero for edgeless graphs.

4.5. Digraphs and Infinite Graphs

Some of the results obtained for matroidal families of graphs have analogues for digraphs. Research in this area was started by Duchamp (1972), and Matthews (1978b) has extensively developed it. Matthews' proofs are analogous to those used for graphs as exemplified in section 4.2. Loops and multiple edges are however allowed.

A matroidal family of digraphs is defined as is a matroidal family of graphs but with 'digraph' instead of 'graph' and 'weakly connected' instead of 'connected'. From a given matroidal family P of graphs, we can obtain a matroidal family of digraphs by taking, for each Γ in P, all possible orientations of the edges of Γ. Obviously, only those families not obtained in this way deserve a study in terms of digraphs. In this context, a matroidal family P of digraphs is called proper if there is some digraph $\Gamma \in P$ and some edge e of Γ such that the digraph obtained from Γ by reversing the orientation of e is not in P.

We refer the reader to Matthews (1977c; 1978b) for a proof that there exist no proper matroidal families of digraphs closed under homeomorphism and for a description of all proper matroidal families of digraphs closed under antihomeomorphism. Note that two digraphs Γ' and Γ'' are homeomorphic if both can be obtained from a third one by a (possibly empty) sequence of operations consisting of the substitution of a directed path $(u, x), (x, v)$ for a directed edge (u, v), x being a new vertex; they are antihomeomorphic if both can be obtained from a third one by a (possibly empty) sequence of operations consisting of the addition of two new vertices u' and v' and the substitution of an antidirected path $(u, u'), (v', u'), (v', v)$ for the directed edge (u, v).

Matthews (1979) has also investigated a generalization of the definition of a matroidal family of graphs where connectedness is not required and infinite, non-simple graphs are allowed as members of the family. He concentrated his efforts on families closed under homeomorphism and used a definition of infinite matroids in terms of circuits that satisfy the (finite) matroid axioms, namely:

(C1) No circuit contains properly another circuit; and
(C2) If C' and C'' are distinct circuits and $e \in C' \cap C''$, then there is a circuit C such that $C \subseteq C' \cup C'' - \{e\}$.

His main result is a description of all those generalized matroidal families that can be partitioned into a finite number of classes of homeomorphic graphs. He proves that a generalized matroidal family of graphs has only a finite number of homeomorphic classes when all graphs in the family are finite (but not necessarily connected) or are connected (but not necessarily finite).

It is an open question whether there are generalized matroidal families closed under homeomorphism with infinitely many classes of homeomorphic graphs. Matthews (1977c) conjectures that there are none. Theorem 4.4.7 leads me to conjecture that there are infinitely many. Let our disagreement stimulate the reader to try to settle the question.

Exercises

4.1. Show that P_0, P_2 and P_3 are in fact matroidal families of graphs.
4.2. (Matthews, 1978b)
      (a) Show that, allowing non-simple graphs as members, the family consisting of a single graph with j loops at a single vertex is matroidal.

(b) Find other finite matroidal families of non-simple graphs.

4.3. (Matthews, 1977a) Consider Γ as a road network with a vehicle traveling along the edges of Γ with U-turns and reversals at vertices forbidden. Let a *position* of the vehicle be an edge of Γ together with the orientation of the edge that specifies in which direction the vehicle is moving. Give an intuitive description of the connection of $P_2(\Gamma)$ in terms of positions.

4.4. (Matthews, 1977a) For each $v \in V(\Gamma)$, let S_v be the set of edges incident with v. Prove that $P_2(\Gamma)$ is isomorphic to the transversal matroid $M(\mathscr{I})$ of the family $\mathscr{I} = (S_v: v \in V(\Gamma))$.

4.5. (Bryant & Perfect, 1977) Define a circuit space as a matroid in which each base is contained in a circuit. Prove that, for Γ simple or non-simple, if $P_2(\Gamma)$ is a circuit space, then no base has more than two pendant edges, no vertex of Γ has degree one, and Γ is connected.

4.6. (Simões-Pereira, 1978) Let P be a matroidal family distinct from P_0, P_1, P_2, and P_3 and let $m \geqslant 3$ be the minimum degree of the vertices in any graph of P. Prove that, for some $n \geqslant m$, the complete bipartite graph $K_{m,n}$ belongs to P.

4.7. (Matthews, 1977c; Andreae, 1978) Prove that, if a wheel belongs to a matroidal family P, then all wheels belong to P.

4.8. Prove that any matroidal family containing the wheels contains also the graph pictured in Figure 4.7.

4.9. Prove that there is no matroidal family of the form $P_{n,r}$ that contains the complete bipartite graph $K_{3,3}$.

4.10. (Matthews, 1978b) Show that the digraphs pictured in Figure 4.8 form a proper

Figure 4.8.

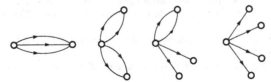

matroidal family of digraphs.

4.11. Let Π_3 be the collection of graphs whose members are the polygons and the two-way infinite path. Let Π_4 be the collection of graphs whose members are the bicycles, the two-way infinite path, and the graphs formed by a cycle and a one-way infinite path starting at a vertex of the cycle.
(Bean, 1967; Higgs, 1969b; Klee, 1971) Show that Π_3 and Π_4 are matroidal families as defined in section 4.5.

4.12. (Schmidt, 1979) Show that there are infinitely many matroids that cannot have the form $P(\Gamma)$ for any graph Γ and matroidal family P.

Research Exercises

4.13. Characterize the matroids that are isomorphic to some $P(\Gamma)$ where Γ is a graph and P a matroidal family (see Prüss, 1984).

4.14. Prove or disprove the following conjecture:
For each pair of integers n, m with $n \geqslant m \geqslant 3$, there is at least one matroidal family that contains the complete bipartite graph $K_{m,n}$.

References

Andreae, T. (1978). Matroidal families of finite connected nonhomeomorphic graphs exist, *J. Graph Theory* **2**, 149–53.

Bean, D. W. T. (1967). Infinite exchange systems, Ph.D. Thesis, McMaster University, Hamilton, Ontario.

Behzad, M. & Chartrand, G. (1971). *Introduction to the Theory of Graphs*, Allyn & Bacon, Boston.

Bryant, V. & Perfect, H. (1977). Some characterization theorems for circuit spaces associated with graphs, *Discrete Math.* **18**, 109–24.

Doob, M. (1973). An interrelation between line graphs, eigenvalues and matroids, *J. Comb. Theory Ser. B* **15**, 40–50.

Duchamp, A. (1972). Sur certains matroides définis à partir des graphes, *C. R. Acad. Sci. Sér. A, Paris* **274**, 9-A, 11.

Higgs, D. A. (1969a). Infinite graphs and matroids, in W. T. Tutte (ed.) *Recent Progress in Combinatorics*, pp. 245–53. Academic Press, New York.

Higgs, D. A. (1969b). Matroids and duality, *Colloq. Math.* **20**, 215–20.

Klee, V. (1971). The greedy algorithm for finitary and cofinitary matroids, in *Combinatorics*, Amer. Math. Soc. Publication XIX, pp. 137–52. American Mathematical Society, Providence, Rhode Island.

Lorea, M. (1979). On matroidal families, *Discrete Math.* **28**, 103–6.

Matthews, L. R. (1977a). Bicircular matroids, *Quart. J. Math. Oxford* (2) **28**, 213–28.

Matthews, L. R. (1977b). Matroids on edge sets of directed graphs, in *Optimization and Operations Research* (Bonn, 1977), Lecture Notes in Economics and Mathematical Systems 157, pp. 193–9. Springer-Verlag, Berlin.

Matthews, L. R. (1977c). Matroids from graphs and directed graphs, Ph.D. Thesis, Oxford.

Matthews, L. R. (1978a). Properties of bicircular matroids, in *Problèmes Combinatoires et Théorie des Graphes* (Orsay 1976), Editions du C.N.R.S. 260, pp. 289–90. Centre National de la Recherche Scientifique, Paris.

Matthews, L. R. (1978b). Matroids from directed graphs, *Discrete Math.* **24**, 47–61.

Matthews, L. R. (1979). Infinite subgraphs as matroid circuits, *J. Comb. Theory Ser. B* **27**, 260–73.

McDiarmid, C. J. H. (1974). Path-partition structures of graphs and digraphs, *Proc. London Math. Soc.* (3) **29**, 750–68.

Prüss, V. (1984). A characterization of matroidal families of multigraphs, *Discrete Math.* **52**, 101–5.

Schmidt, R. (1979). On the existence of uncountably many matroidal families, *Discrete Math.* **27**, 93–7.

Shull, R., Orlin, J. B., Shuchat, A. & Gardner, M. L. (1989). The structure of bases in bicircular matroids, *Discrete Appl. Math.* **23**, 267–83.

Simões-Pereira, J. M. S. (1972). On subgraphs as matroid cells, *Math. Z.* **127**, 315–22.

Simões-Pereira, J. M. S. (1973). On matroids on edge sets of graphs with connected subgraphs as circuits, *Proc. Amer. Math. Soc.* **38**, 503–6.

Simões-Pereira, J. M. S. (1974). Matroids, graphs and topology, in *Proceedings of the Fifth Southeastern Conference on Combinatorics, Graph Theory and Computing*, Congressus Numerantium 10, pp. 145–55. Utilitas Mathematica, Winnipeg.

Simões-Pereira, J. M. S. (1975a). On matroids on edge sets of graphs with connected subgraphs as circuits – II, *Discrete Math.* **12**, 55–78.

Simões-Pereira, J. M. S. (1975b). Subgraphs as circuits and bases of matroids, *Discrete Math.* **12**, 79–88.

Simões-Pereira, J. M. S. (1978). A comment on matroidal families, in *Problèmes Combinatoires et Théorie des Graphs* (Orsay 1976), Editions du C.R.N.S. 260, pp. 385–7. Centre Nationale de la Recherche Scientifique, Paris.

Tutte, W. T. (1976). On chain-groups and the factors of graphs, Research Report CORR 76/45,

University of Waterloo.

Wagner, D. K. (1985). Connectivity in bicircular matroids, *J. Comb. Theory Ser. B* **39**, 308–24.

Wagner, D. K. (1988). Equivalent factor matroids of graphs, *Combinatorica* **8**, 373–7.

Walter, M. (1982). Construction of matroidal families by partly closed sets, *Discrete Math.* **41**, 309–15.

Welsh, D. J. A. (1976). *Matroid Theory*, Academic Press, London.

White, N. (ed.) (1986). *Theory of Matroids*, Cambridge University Press.

Whitney, H. (1935). On the abstract properties of linear dependence, *Amer. J. Math.* **57**, 509–33.

Zaslavsky, T. (1982a). Signed graphs, *Discrete Appl. Math.* **4**, 47–74.

Zaslavsky, T. (1982b). Bicircular geometry and the lattice of forests of a graph, *Quart. J. Math. Oxford* (2) **33**, 493–511.

Zaslavsky, T. (1987). The biased graphs whose matroids are binary, *J. Comb. Theory Ser. B* **42**, 337–47.

Zaslavsky, T. (1989). Biased graphs. *I*. Bias, balance and gains, *J. Comb. Theory Ser. B* **47**, 32–52.

5

Algebraic Aspects of Partition Lattices

IVAN RIVAL and MIRIAM STANFORD

5.1. Introduction

Two results of the last decade highlight the algebraic advances in the theory of partition lattices. The first answers affirmatively a celebrated conjecture about the place of (finite) partition lattices in lattice theory itself. The second is a compactness result according to which every partition on a finite set can be expressed algebraically in terms of a surprisingly small subset of all of the partitions on that set.

Briefly the results are these.

In a now famous footnote to his paper 'Lattices, equivalence relations, and subgroups', Whitman (1946) conjectured that

every finite lattice is lattice embeddable in a finite partition lattice.

The conjecture resisted proof for many years. In the late 1970s a positive solution was announced by Pudlak and Tůma (1977) and in 1980, they published their proof of it. The proof is long and complicated.

It follows from a result of Wille (1976) on simple lattices that only a few simple lattices have at most three generators: the 1-element chain **1**, the 2-element chain **2**, and the 5-element modular, non-distributive lattice M_3.

Figure 5.1. **1**, the 1-element chain; **2**, the 2-element chain; M_3, the 5-element modular non-distributive lattice; Π_4, the lattice of partitions on a 4-element set; **PG**(2), the lattice of subspaces of a field of characteristic 2.

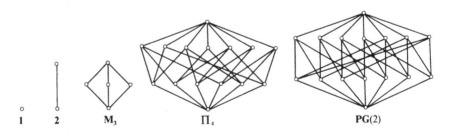

1   **2** M_3 Π_4 **PG**(2)

It is well known too that the lattice of subspaces of a finite projective geometry is 4-generated. In the early 1970s Wille put forth a bold conjecture that

every finite partition lattice is four-generated.

This was proved by Strietz (1975). The proof is not particularly difficult once a 4-generating set is discovered.

Our aim in this chapter is to survey the background to these two results, as well as related results on the algebraic side of the theory of partitions.

5.2. The Lattice of All Partitions

Let S be a non-empty set. A *partition* on S is a family of non-empty, disjoint subsets of S whose union is all of S. These subsets are called *blocks*. Of course, partitions correspond to 'equivalence relations', a term preferred by some. (For a variety of simple examples, see Chapter 1, Examples and basic concepts, in Volume I of this series (Crapo, 1986).)

Let $\Pi(S)$ stand for the set of all partitions on S. Let θ, $\psi \in \Pi(S)$. For a, $b \in S$, we write $a\,\theta\,b$ if a and b are in the same block of θ. Write $\theta \leqslant \psi$ if, for each a, $b \in S$, $a\,\theta\,b$ implies $a\,\psi\,b$. Then $(\Pi(S), \leqslant)$ (or simply $\Pi(S)$) becomes an ordered set; in fact, $\Pi(S)$ is a lattice with operations *meet (infimum)* $\wedge$ and *join (supremum)* $\vee$ defined by

$$a(\theta \wedge \psi)b \text{ if and only if } a\,\theta\,b \text{ and } a\,\psi\,b$$

and

$a(\theta \vee \psi)b$ if and only if there is a positive integer n, a sequence $a = a_0, a_1, a_2, ..., a_n = b$ of elements of S, and a sequence $\theta_1, \theta_2, ..., \theta_n$ of partitions, each equal to either θ or ψ, such that $a_0\theta_1 a_1, a_1\theta_2 a_2, ..., a_{n-1}\theta_n a_n$.

These operations can be extended to any subset of $\Pi(S)$. Thus, for a family $(\theta_i | i \in I)$ of partitions,

$$a\left(\bigwedge_{i \in I} \theta_i\right)b \text{ if and only if } a\theta_i b \text{ for each } i \in I$$

and,

$a\left(\bigvee_{i \in I}\right)b$ if and only if there is a positive integer n, a sequence $a = a_0, a_1, ..., a_n = b$ of elements of S, and a sequence $\theta_{i_1}, \theta_{i_2}, ..., \theta_{i_n}$, each a member of the family $(\theta_i | i \in I)$, such that $a_0\theta_{i_1} a_1, a_1\theta_{i_2} a_2, ..., a_{n-1}\theta_{i_n} a_n$.

In other words, $\Pi(S)$ is a *complete lattice.* (For an introductory account of lattices, see Chapter 3, Lattices, in Volume I of this series (Faigle, 1986), and, for more elaborate and extensive accounts refer to Birkhoff (1967), Crawley & Dilworth (1973), or Grätzer (1978).) $\Pi(S)$ has a *bottom,* denoted 0, and a *top,* denoted 1, prescribed by

$$a\,0\,b \text{ if and only if } a = b \text{ and, } a\,1\,b \text{ for all } a, b \in S.$$

Besides the self-evident symmetry and duality built into the very definition of lattices, they have a further, rather far-reaching quality: a split personality. The salient points are these. First, there is a natural order-theoretic aspect to a lattice. Indeed, the relation $\leqslant$ defined by

$$\theta \leqslant \psi \text{ if and only if } \theta \vee \psi = \psi$$

is an *order* on $\Pi(S)$, that is, a reflexive, antisymmetric, and transitive relation. Thus, a lattice may be considered as an ordered set in which every pair of elements has a supremum and an infimum. This is the order-theoretic aspect and it has natural consequences for the choice of the mappings and subobjects that are appropriate for its theoretical development. Second, there is the algebraic aspect which emphasizes the definition of a lattice as an algebra, that is, as a set $\Pi(S)$ with two binary operations $\vee$ and $\wedge$ defined by these postulates:

$$\theta \vee \theta = \theta = \theta \wedge \theta;$$
$$\theta \vee \psi = \psi \vee \theta, \theta \wedge \psi = \psi \wedge \theta;$$
$$\theta \vee (\psi \vee \tau) = (\theta \vee \psi) \vee \tau, \theta \wedge (\psi \wedge \tau) = (\theta \wedge \psi) \wedge \tau;$$
$$\theta \vee (\psi \wedge \theta) = \theta = \theta \wedge (\psi \vee \theta).$$

This definition, based entirely on the formal properties of the two binary operations $\vee$ and $\wedge$, is the algebraic aspect. The two aspects are formally equivalent yet each leads, according to its respective tradition, to different questions, techniques, and theories (cf. Rival, 1978). Our needs in this chapter are mostly served by following the algebraic viewpoint. Thus, the appropriate morphisms here are mappings preserving both operations $\vee$ and $\wedge$. (Such mappings are, *a fortiori*, order-preserving.) Thus, a *lattice embedding* of a lattice K into a lattice L is a one-to-one map f of K to L such that, for each $a, b \in K$, $f(a \wedge b) = f(a) \wedge f(b)$ and $f(a \vee b) = f(a) \vee f(b)$. An *order embedding* of K into L is a one-to-one, order-preserving map of K to L whose inverse, too, is order-preserving (cf. Figure 5.4). Every finite lattice K can be order embedded into the lattice 2^n of all subsets of an n-element, where $n = |K|$. To find familiar classes of finite lattices into which every finite lattice can be lattice embedded is a much, much harder problem!

Our primary interest here will be in the algebraic aspect, yet we should certainly exploit the convenience of the order-theoretic aspect for its facility in combinatorial themes.

For ease in working we denote the partition $(\{a, b\}, \{c, d\})$, say, of $\{a, b, c, d\}$ by $(a, b)(c, d)$, or simply $(ab)(cd)$, and the partition $(\{a, b\}, \{c\}, \{d\})$ simply by its non-trivial block (ab). Thus, for instance (cf. Figure 5.2),

Figure 5.2. $\Pi(\{a, b, c, d\})$, the lattice of partitions on the 4-element set $\{a, b, c, d\}$.

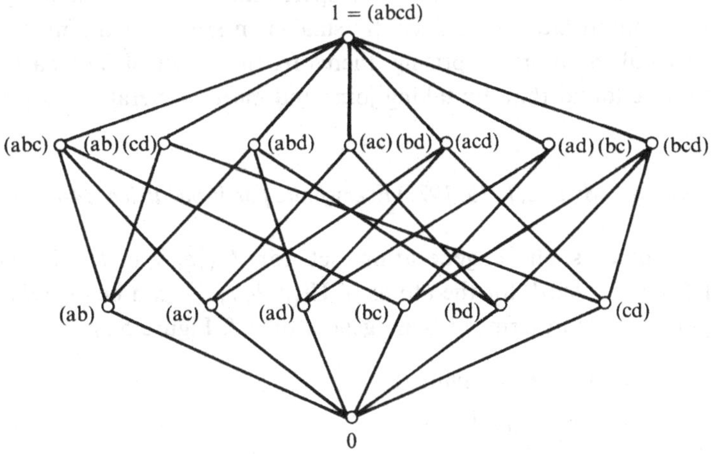

$$0 = (ab) \wedge (ac) = (ab) \wedge (cd) = (ab)(cd) \wedge (ad)(bc) = \text{etc.},$$
$$(abc) = (ab) \vee (bc) = (ab) \vee (ac) = \text{etc.},$$
$$(ac)(bd) = (ac) \vee (bd),$$
$$1 = (abc) \vee (cd) = (abd) \vee (bcd) = \text{etc.}$$

Say that θ *covers* ψ in $\Pi(S)$, and write $\theta \succ \psi$, if $\theta > \tau \geqslant \psi$ in $\Pi(S)$ implies $\tau = \psi$. Each partition $\theta \succ 0$ is called an *atom* of $\Pi(S)$. Every partition is a join of atoms (see Exercise 5.2). A partition θ is *compact* if $\theta \leqslant \bigvee\limits_{i \in I} \theta_i$ implies that $\theta \leqslant \bigvee\limits_{i \in F} \theta_i$ for a finite subset F of I. In fact, every atom is *compact* (see Exercise 5.3). Thus $\Pi(S)$ is *compactly generated*, that is, every element of $\Pi(S)$ is the join of compact elements.

$\Pi(S)$ is *semimodular*, that is, for every $\theta, \psi \in \Pi(S)$, θ non-comparable to ψ, if $\theta \succ \theta \wedge \psi$ then $\theta \vee \psi \succ \theta$ (see Exercise 5.5).

In summary, the partition lattice $\Pi(S)$ is a *geometric lattice*, that is, a complete, semimodular lattice in which every element is the join of atoms. Moreover, $\Pi(S)$ is *complemented*, that is, for each $\theta \in \Pi(S)$ there is $\bar{\theta} \in \Pi(S)$ satisfying $\theta \wedge \bar{\theta} = 0$ and $\theta \vee \bar{\theta} = 1$. $\bar{\theta}$ is a *complement* of θ (see Exercise 5.6).

This is an important result in the algebraic theory of partition lattices (Ore, 1942, cf. Grätzer, 1978).

5.2.1. Proposition. $\Pi(S)$ *is simple*.

A partition lattice $\Pi(S)$ is *simple* if it has no non-trivial congruence relations besides the *identity* relation ω, defined by $\theta \equiv \psi(\omega)$ if and only if $\theta = \psi$, and the *all* relation ι defined by $\theta \equiv \psi(\iota)$ for all $\theta, \psi \in \Pi(S)$.

Every element in $\Pi(S)$ is the join of atoms; thus the atoms form a generating set. There would seem little reason to suspect that there are generating sets for finite partition lattices significantly smaller in size than the number of atoms. It is all the more surprising, therefore, that a set of four partitions can always be found that, by taking joins and meets, generates every other partition.

5.2.2. Theorem. *(Strietz, 1975; 1977) Every finite partition lattice is 4-generated.*

For instance, such a generating set for $\mathbf{\Pi}(\{a, b, c, d)\}$ is $\{(abc), (ab)(cd), (ad)(bc), (acd)\}$, for the atoms of $\Pi(\{a, b, c, d\})$ can be expressed as lattice polynomials in terms of these generators (cf. Figure 5.2):

$$(ab) = (abc) \wedge (ab)(cd)$$
$$(ac) = (abc) \wedge (acd)$$
$$(ad) = (acd) \wedge (ad)(bc)$$
$$(bc) = (abc) \wedge (ad)(bc)$$
$$(cd) = (acd) \wedge (ab)(cd)$$
$$(bd) = (abd) \wedge (bcd) = ((ab) \vee (ad)) \wedge ((bc) \vee (cd))$$
$$= (((abc) \wedge (ab)(cd)) \vee ((acd) \wedge (ad)(bc)))$$
$$\wedge (((abc) \wedge (ad)(bc)) \vee ((acd) \wedge (ab)(cd))).$$

Here are the general ideas. For convenience we write Π_n for $\Pi(S)$, where $S = \{1, 2, ..., n\}$. As every element of Π_n is the join of atoms a subset of Π_n is a generating set if it generates the atoms of Π_n. As above, let (i, j) stand for the atom of Π_n whose non-trivial block consists of i and j. In fact, a subset of Π_n is a generating set if it generates all atoms $(1, i)$ and $(2, j)$, where $2 \leqslant i \leqslant n$ and $3 \leqslant j \leqslant n$, for these in turn, generate all atoms, since

$$(i, j) = ((1, i) \vee (1, j)) \wedge ((2, i) \vee (2, j))$$

as long as i and j are both distinct from 1 and 2. For $n \geqslant 16$ satisfying $n \equiv 1 \pmod 3$ this is a generating system for Π_n.

$\theta_1 = (1, 2),$
$\theta_2 = (1, 2, 11), (3, 4, 12), (5, 6, 8), (7, 9, 10), ..., (k + 3, k + 4, k + 5),$
$\theta_3 = (1, 4, 6), (3, 5, 9), (2, 7, 12), ..., (k - 2, k, k + 5), ..., (n - 5, n - 3), (n - 2, n),$
$\theta_4 = (2, 3), (4, 8), (1, 5, 10), (6, 7, 13), ..., (k - 1, k + 1, k + 6), ..., (n - 4, n - 2), (n - 1),$

where $k \equiv 1 \pmod 3$ and $10 \leqslant k \leqslant n - 6$. Let $T = \{1, 2, ..., m\} \subseteq S$ satisfy $10 \leqslant m \leqslant n$. Then the generators $\theta_1, \theta_2, \theta_3, \theta_4$ restricted to T determine generators $\theta'_1, \theta'_2, \theta'_3, \theta'_4$ of Π_m. Strietz (1977) exhibits 4-element generating sets for $\Pi_4, \Pi_5, \Pi_6, \Pi_7, \Pi_8$, and Π_9 separately, each isomorphic to $1 + 1 + 1 + 1$ (see Figure 5.3).

Figure 5.3.

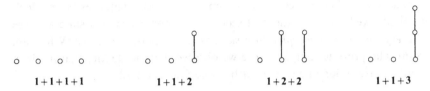

$1+1+1+1$ $1+1+2$ $1+2+2$ $1+1+3$

Strietz (1975) has also considered the structure of the generating sets as ordered sets. Every partition lattice Π_n, $n \geqslant 4$, contains a generating set isomorphic to $1+1+1+1$ (see Figure 5.3). Π_4 contains no generating set isomorphic to $1+1+2$ although Π_n does for each $n \geqslant 10$. Moreover, every Π_n, $n \geqslant 4$, has a generating set isomorphic to $1+2+2$, and every Π_n, $n \geqslant 5$, has a generating set isomorphic to $1+1+3$. In fact, Strietz (1977) shows the following.

5.2.3. Theorem. *If the ordered set P is a generating set of Π_n, $n \geqslant 10$, then P contains an isomorphic copy of $1+1+1+1$ or $1+1+2$.*

The generating set $\{\theta_1, \theta_2, \theta_3, \theta_4\}$ described above is isomorphic to $1+1+2$, where $\theta_1 < \theta_2$. For $n \geqslant 10$, on replacing θ_1 with

$$\theta'_1 = (1, 2, 3)$$

and observing that $\theta'_1 \vee \theta_2 = \theta_1$ it follows that $\{\theta'_1, \theta_2, \theta_3, \theta_4\}$ is also a generating set of Π_n isomorphic to $1+1+1+1$.

Here is another basic result. It concerns lattice identities, that is, identities of the form

$$x \wedge (y \vee z) = (x \wedge y) \vee (x \wedge z) \qquad \textit{(distributivity)}$$
$$x \wedge (y \vee (x \wedge z)) = (x \wedge y) \vee (x \wedge z) \qquad \textit{(modularity)}$$
$$x = y \qquad \textit{(triviality)}.$$

5.2.4. Proposition. *(Sachs, 1961) Any lattice identity that holds in every finite partition lattice must hold in every lattice.*

In other words, there are no non-trivial lattice identities that are satisfied by all finite partition lattices. A proof of this result can be pieced together from these steps.

Let S be a set (not necessarily finite) and let $\Pi_f(S)$ stand for the sublattice of $\Pi(S)$ consisting of all finite joins of atoms of $\Pi(S)$. A *directed down* set (alias *ideal*) in $\Pi(S)$ is a subset D such that $x \leqslant y$ and $y \in D$ implies $x \in D$ and $x, y \in D$ implies $x \vee y \in D$. The set of all directed down sets ordered by set inclusion is a lattice too. If there is a lattice identity that holds in every finite partition lattice then it must hold in every partition lattice as well (see Exercises 5.9–5.11). The concluding step uses a fundamental result on which we shall dwell longer in the next section. For now we simply state and use it: *every lattice is isomorphic to a sublattice of a partition lattice* (Whitman, 1946). The proof is complete once we observe that an identity that holds in a lattice also holds in every sublattice (see Exercise 5.12).

5.3. Lattice Embeddings

We say that K is *lattice embeddable* in L if there is a lattice embedding of the lattice K into the lattice L. For convenience, we also identify a lattice embedding f of K into L with the *sublattice* $f(K)$ of L. But notice that, for a one-to-one order-preserving map f of K to L, $f(K)$ may be a lattice (with respect to the induced ordering of L), in which case we call it an *order sublattice*. f need not, of course, be an embedding (see Figure 5.4). (This is the order-theoretic versus the algebraic aspect.)

The first important result on lattice embeddings of partition lattices is as follows (Birkhoff, 1935).

Figure 5.4. (a) An order embedding that is not a lattice embedding; (b) a lattice embedding.

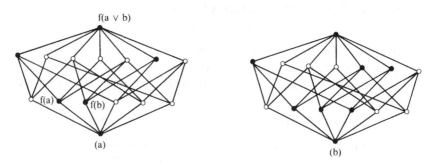

(a)   (b)

5.3.1. Theorem. *Every partition lattice is lattice embeddable in the lattice of subgroups of some group.*

The most fundamental and far-reaching result on lattice embeddings of partition lattices is due to Whitman (1946).

5.3.2. Theorem. *Every lattice is lattice embeddable in a partition lattice.*

It was actually conjectured by Birkhoff much earlier (Birkhoff, 1935). The attractive consequence is this.

5.3.3. Corollary. *Every lattice is lattice embeddable in the lattice of subgroups of a group.*

This corollary was, at one time, advertized (not entirely unconvincingly), in order to market lattices, along side groups, rings, and other mathematical produce. Up to the mid 1970s the preeminent unsolved problem in lattice theory itself was an apparently technical point that seemed to arise from Whitman's efforts to polish and improve his lattice embedding theorem. Indeed, it is in a footnote to his paper (Whitman, 1946) that this conjecture seems to be first recorded:

every finite lattice is lattice embeddable in a finite partition lattice.

Over the next three decades efforts to settle the conjecture were focussed on the two following approaches.

(1) *Construct a class of finite lattices (as much like finite partition lattices as possible) into which every finite lattice can be lattice embedded.*
(2) *Construct classes of finite lattices (as many as possible) that can be lattice embedded into finite partition lattices.*

In spirit, at least, the first approach aimed to settle the conjecture positively by converging on finite partition lattices and the second aimed to settle it by the inevitable march to a counterexample. The first was destined to play the major role. Nevertheless, the second has shed light on the 'effectiveness' of lattice embeddings into finite partition lattices.

Every element of a (finite) partition lattice is the join of atoms. Every (finite) lattice is order embeddable in a (finite) lattice, say 2^n, in which every element is the join of atoms (cf. Exercise 5.1). The first major result is recorded in Finkbeiner (1960) who credits it as an unpublished result of Dilworth.

5.3.4. Theorem. *Every finite lattice is lattice embeddable in a finite, semimodular lattice.*

Actually Finkbeiner showed considerably more. An integer-valued function r defined on a finite lattice L is a *(semimodular) rank* function if

$$r(0_L) = 0,$$

where 0_L is the bottom of L, and, for each $a, b \in L$,

$$r(a) < r(b) \text{ whenever } a < b \text{ in } L,$$

and

$$r(a \vee b) + r(a \wedge b) \leqslant r(a) + r(b).$$

Actually, every lattice has a rank function. The principal result in Finkbeiner (1960) is this:

5.3.5. Theorem. *Every finite lattice L with rank function r can be lattice embedded in a finite, semimodulator lattice K such that $r = h|L$, where $h|L$ is the restriction of the height function h for K to L.*

These semimodular lattices need not be geometric. Although it was not published until much later (Crawley & Dilworth, 1973), Dilworth proved (apparently around 1950) the following:

5.3.6. Theorem. *Every finite lattice can be lattice embedded in a finite geometric lattice.*

The proof entails and presumes several basic facts about closure operators on a set. A *closure operator* on a set S is a map of the set of all subsets of S to itself, which assigns to each $A \subseteq S$ a subset $A^c \subseteq S$ satisfying

$$A \subseteq A^c, \quad A^{cc} = A, \text{ and } A \subseteq B \text{ implies } A^c \subseteq B^c.$$

Call a subset A of S *closed* if $A^c = A$. A closure operator on S satisfies the *exchange property* if, for each $A \subseteq S$, and for each $a, b \in S$,

$$a \in (A \cup \{b\})^c \text{ and } a \notin A^c \text{ imply } b \in (A \cup \{a\})^c.$$

Recently Grätzer & Kiss (1986) have succeeded in blending the results of Dilworth and of Finkbeiner:

5.3.7. Theorem. *Every finite lattice L with rank function r can be lattice embedded in a finite, geometric lattice K such that $r = h|L$, where h is the height function for K.*

In a series of papers Hartmanis (1956, 1959, 1961) sought to exploit a geometric analogy. A partition on a set S is, after all, a family of subsets such that every element of S belongs to exactly one of these subsets and every subset contains at least one element of S. A *geometry* on S is, by analogy, a family of subsets ('lines') of S such that every pair of distinct elements ('points') of S is contained in a unique subset and each subset contains at least two elements of S. The analogy may be pursued. A *partition of type n* ($n \geqslant 1$) on a set S is a family of subsets such that any n distinct elements of S are contained in a unique set and each subset contains at least n elements of S. Let $LP_n(S)$ stand for the set of all partitions of type n on S. $LP_n(S)$ is a subset of $\Pi(S)$ from which it inherits an order. Note that $LP_1(S) = \Pi(S)$ and $LP_2(S)$ is a geometry on S (cf. Tůma, 1980). The principal result is this (Hartmanis, 1961):

5.3.8. Theorem. *For each $n \geqslant 2$, and for every finite lattice L, there is a finite set S such that L can be lattice embedded in $LP_n(S)$.*

In a now famous paper, 'Algebras whose congruence lattices are distributive', seminal to the modern theory of universal algebra, Jónsson (1967) proved, *inter alia*, the following.

5.3.9. Theorem. *Every (finite) lattice is lattice embeddable in a (finite) subdirectly irreducible lattice.*

A lattice is *subdirectly irreducible* if it has a unique smallest congruence relation larger than the identity relation ω. Every simple lattice is, *a fortiori*, subdirectly irreducible.

This, and another basic property of finite partition lattices, is the motivation for yet another result of this type. Poguntke & Rival (1976) have proved the following.

5.3.10. Theorem. *Every finite lattice can be embedded in a finite, 4-generated, simple lattice.*

Proof. Let L be a finite lattice and let $T(L)$ stand for the set of all triples (x_i, y_i, z_i) of elements of L such that $z_i \succ y_i \succ x_i$. Let

$$L' = L \cup \bigcup_i \{a_i, b_i\},$$

where $\{a_1, b_1\}. \{a_2, b_2\}, \ldots$ is a sequence of disjoint pairs of elements (disjoint also from L), and with order induced by L and the covering relations $z_i \succ a_i \succ x_i$ and $z_i \succ b_i \succ x_i$. Then L' is a finite, simple lattice containing L as a sublattice.

The next step in the proof (see Exercises 5.18, 5.19) relies on the following classical result.

5.3.11. Proposition. *(Sorkin, 1954; Dean, 1956) Every finite lattice is lattice embeddable in a finite, 3-generated lattice.*

The last step in the proof is to construct a lattice

$$L''' = L'' \cup \{c, d\},$$

where c, d are distinct elements, disjoint from L'', with order induced by L'' and $1 \succ c \succ b$ and $c \succ d \succ 0$ (see Exercises 5.20, 5.21).   □

Of course, the culmination of this approach is the theorem of Pudlák & Tůma (1977, 1980).

5.3.12. *Every finite lattice can be lattice embedded in a finite partition lattice.*

Proof. The complicated proof relies basically on graph-theoretical and combinatorial techniques.

Here is the opening gambit which, interestingly, turns on an order-theoretical aspect of lattices. For a lattice L and elements $a \leqslant b$ in L, let

$$L_{a,b} = \{x \in L \mid a \nleqslant x \text{ or } b \leqslant x\}.$$

$L_{a,b}$ is an order sublattice of L (see Exercise 5.22). Define a map $f_{a,b}$ of L to $L_{a,b}$ by

$$f_{a,b}(x) = \begin{cases} x & \text{if } a \nleqslant x \\ x \vee b & \text{if } a \leqslant x. \end{cases}$$

$f_{a,b}$ is a join homomorphism of L onto $L_{a,b}$ (see Exercises 5.23–5.25). The rest of the proof relies on techniques developed by Pudlák and Tůma to establish that, whenever L is a finite lattice, lattice embeddable in a finite partition lattice, and $a < b$ in L, then $L_{a,b}$ is also lattice embeddable in a finite partition lattice. The starting point, at least, is that every lattice 2^n is lattice embeddable in a finite partition lattice (see Exercise 5.26).

Their result can also be cast in the language of groups thus.

5.3.13. Theorem. *Every finite lattice is isomorphic to a sublattice of the subgroup lattice of a finite group.*

The theorem of Pudlák and Tůma settles the long standing conjecture of Whitman which was, after all, the best known and the most sought after in lattice theory. Nevertheless, it is likely that the earlier results, while apparently concluding less, will survive too for they often invoke important and interesting techniques and provide 'effective' lattice embeddings in lattices that are structurally like partition lattices.

The Pudlák–Tůma theorem guarantees that for every finite lattice L there is a finite set S such that L can be lattice embedded in $\Pi(S)$. How big is S? In their proof this set S is large. There is, however, a body of results, some of early vintage, others fairly recent, that address this natural question.

Hales (1970) concludes that *every finite sublattice F of a free lattice can be lattice embedded in a finite partition lattice* by extending meet-homomorphisms of F, element by element, into successively larger partition lattices. The argument proceeds by induction on the length of elements of F and applies an 'extension lemma' of Jónsson (1953).

This early paper of Jónsson, in turn, establishes several striking properties about the lattice embeddings of lattices into partition lattices (albeit infinite ones). Notice that, for $\theta, \psi \in \Pi(S)$, since

$$\theta \wedge \psi = \theta \cap \psi,$$

the meet $\theta \wedge \psi$ is easy to compute. In contrast, the join $\theta \vee \psi$ in $\Pi(S)$ is not so easy to express, for $\theta \vee \psi$ is the union of the sequence

$$\theta, \; \theta \circ \psi, \; \theta \circ \psi \circ \theta, \; \theta \circ \psi \circ \theta \circ \psi, \; \ldots$$

where the 'composition' $\theta \circ \psi$ is the partition (or equivalence relation) on S defined by $a \, \theta \circ \psi \, b$ if there is c satisfying $a \, \theta \, c$ and $c \, \psi \, b$. Say a lattice embedding f of a lattice L into $\Pi(S)$ has

$\quad\quad$ *type* 1 if $f(x) \vee f(y) = f(x) \circ f(y)$,

$\quad\quad$ *type* 2 if $f(x) \vee f(y) = f(x) \circ f(y) \circ f(x)$,

$\quad\quad$ *type* 3 if $f(x) \vee f(y) = f(x) \circ f(y) \circ f(x) \circ f(y)$,

$\quad\quad$ for each $x, \, y \in L$.

In fact (cf. Exercise 5.27):

5.3.14. Theorem. *A lattice has a type 2 lattice embedding in a partition lattice if and only if it is modular. The lattice of subspaces of a projective plane has a lattice embedding of type 1 if and only if the plane is Desarguesian.*

Finally, we have

5.3.15. Theorem. *Every lattice has a lattice embedding of type* 3.

These results of 1953 did not, of course, necessarily bear on lattice embeddings to *finite* partition lattices.

For a finite lattice L, let $n(L)$ denote the least positive integer such that there is a lattice embedding of L to a partition lattice $\Pi(S)$, where $|S| = n(L)$. Call a sublattice L of $\Pi(S)$ a $\{0, 1\}$-*sublattice* if it contains the bottom 0 and the top 1 of L. It follows from Exercise 5.28 that, if L is a sublattice of $L_1 \times L_2 \times \ldots \times L_m$, then

$$n(L) \leqslant n(L_1) + n(L_2) + \ldots + n(L_m) - m + 1.$$

Let $\mathbf{L}(i, j)$ be the lattice constructed from chains of size $i + 2$ and $j + 2$ by identifying only their bottom and top elements (see Figure 5.5). Ehrenfeucht, Faber, Fajtlowicz & Mycielski (1973) prove

$$n(\mathbf{L}(i, j)) = i + j - 1 + \lceil 2\sqrt{(i + j - 2)} \, \rceil.$$

For the modular, non-distributive lattice $\mathbf{M}_n$ with n atoms, they show that

$$n(\mathbf{M}_n) \leqslant p,$$

where p is the smallest prime satisfying $p \geqslant n$.

Building on Dilworth's embedding theorem for geometric lattices, Peele (1982) has shown that certain geometric lattices are lattice embeddable in finite partition lattices. Peele defines an index called the *nullity* of a geometric lattice L as the number of atoms of L minus the rank of the top of L. (The

Figure 5.5.

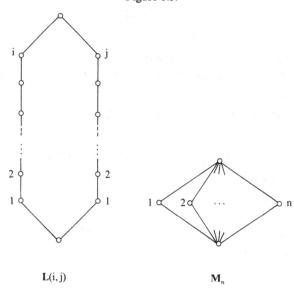

L(i, j) M_n

rank is the least number of atoms whose join is the top. The nullity of 2^n is zero and the nullity of M_3 is one.) Let L be any lattice obtained from 2^n by succesively removing elements covered by the top element while retaining the original order. Every such lattice L with m atoms and of nullity one is lattice embeddable in a partition lattice $\Pi(S)$ satisfying $|S| = 2m^3 - 5m^2 + 4m$. (This estimate is, apparently, not best-possible, e.g. $m = 4$.)

5.4. Postscript

The longstanding lattice embedding problem for partition lattices has been solved. Attention has since focussed on lattices of congruence relations of universal algebras. Grätzer & Schmidt (1963) characterized these lattices (cf. Pudlák, 1976). The result was in some sense analogous to Whitman's original lattice embedding theorem, for it used (possibly infinite) algebras to characterize even finite (congruence) lattices. The main open problem now is this.

Is every finite lattice isomorphic to the congruence lattice of a finite algebra?

Among lattices M_n, for example, it is known that for n equal to a prime power plus one, $n = 7$ (Feit, 1983) or $n = 11$ (Pálfy), M_n is isomorphic to the congruence lattice of a finite algebra. M_{13} is the smallest undecided case at this time. There is an attractive equivalent formulation due to Pálfy & Pudlák (1980): *every finite lattice is isomorphic to the congruence lattice of a finite algebra if and only if every finite lattice is isomorphic to an interval in the subgroup lattice of a finite group.*

Exercises

Show that the following hold.

5.1. For every finite lattice K there is an order embedding of K into 2^n, where $n = |K|$.

5.2. Every partition in $\Pi(S)$ is the join of a family of partitions each of which contains precisely one non-trivial block (that is, a block with more than one element of S), and this non-trivial block itself consists of precisely two elements of S.

5.3. Let $\theta, \theta_1, \theta_2, \ldots$ be elements of $\Pi(S)$. If θ is an atom and $\theta \leqslant \bigvee_{i \in I} \theta_i$ then there

   is a finite subset F of I such that $\theta \leqslant \bigvee_{i \in F} \theta_i$.

5.4. Let $U(\psi) = \{\theta \in \Pi(S) | \psi \leqslant \theta \leqslant 1\}$ and let T be the set whose elements are the blocks of ψ. Then, as lattices, $U(\psi) \cong \Pi(T)$.

5.5. Let $\theta, \psi \in \Pi(S)$ satisfy $\theta \succ 0$ and $\theta \nleqslant \psi$ (that is, $\theta \wedge \psi = 0$). Then $\theta \vee \psi \succ \psi$.

5.6. For each $\theta \in \Pi(S)$ there is $\bar{\theta} \in \Pi(S)$ satisfying $\theta \wedge \bar{\theta} = 0$ and $\theta \vee \bar{\theta} = 1$.

5.7. $\Pi(S)$ is *relatively complemented*, that is, for each $\theta \leqslant \tau \leqslant \psi$ in $\Pi(S)$ there is $\tau' \in \Pi(S)$ such that $\tau \wedge \tau' = \theta$ and $\tau \vee \tau' = \psi$.

5.8. Every 4-element generating set of $\Pi(\{a, b, c, d\})$ is an antichain.

5.9. $\Pi(S)$ is isomorphic to the lattice of all directed down sets of $\Pi_f(S)$.

5.10. Every lattice identity that holds in a lattice also holds in the lattice of all of its directed down sets.

5.11. If there is a lattice identity that holds in every finite partition lattice then this lattice identity must hold in every $\Pi_f(S)$ too.

5.12. An identity that holds in a lattice also holds in every sublattice.

5.13. The height function

$$h(a) = \max\{|C| \, | \, C \text{ is a chain in } L \text{ and } a \text{ is the top of } C\},$$

   for a semimodular lattice L, is a rank function.

5.14. The set of all closed subsets of a set S, with respect to a closure operator on S, and ordered by set inclusion, is a complete lattice.

5.15. The lattice of all closed subsets of a set S with respect to a closure operator on S satisfying the exchange axiom is geometric.

5.16. The set $LP_n(S)$ of all partitions of type n on S is a complete lattice.

Exercises 5.17–5.21 refer to Theorem 5.3.10 *et seq.*

5.17. L' is a finite, simple lattice containing L as a sublattice.

5.18. Every finite lattice is lattice embeddable in a finite, 3-generated, subdirectly irreducible lattice L''.

5.19. In a finite, subdirectly irreducible lattice L'' with top 1 and bottom 0, there are elements $1 > a \succ b > 0$ such that the smallest congruence relation on L'' that *identifies* a and b is the smallest non-trivial congruence relation on L''.

5.20. L''' is 4-generated and contains L'' as a sublattice.

5.21. The smallest non-trivial congruence relation on L''' that identifies a and b also identifies 0 and 1 in L'''. Therefore, L''' is a simple lattice.

Exercises 5.22–5.26 refer to Theorem 5.3.12 *et seq.*

5.22. $L_{a,b}$ is an order sublattice of L.

5.23. $f_{a,b}$ is an onto map and a *join homomorphism*, that is, for each $x, y \in L$, $f_{a,b}(x \vee y) = f_{a,b}(x) \vee f_{a,b}(y)$.

5.24. Every join homomorphism f of a lattice K to a lattice L such that $f(a) = f(b)$, for some $a < b$ in K, is the composition of the join homomorphism $f_{a,b}$ of K to $K_{a,b}$ and a join homomorphism g of $K_{a,b}$ to L.

5.25. Let $\mathscr{L}$ be a class of lattices containing the lattices 2^n of all subsets of an n-element set ordered by inclusion ($n \geqslant 1$) and containing, for every $L \in \mathscr{L}$ and every $a < b$ in L, the lattices $L_{a,b}$. Then $\mathscr{L}$ is the class of all finite lattices.

5.26. Every lattice 2^n is lattice embeddable in a finite partition lattice.

5.27. If f is a lattice embedding of type 2 then L is modular. (Show that, for a, b, $c \in L$ with $a \leqslant c$, $f((a \vee b) \wedge c) \subseteq f(a \vee (b \wedge c))$.)

5.28. Let K and L be $\{0, 1\}$-sublattices of $\Pi(S)$ and $\Pi(T)$ respectively. Then the set of all partitions of $\Pi(S \cup T)$ of the form $\theta \in K$, $\psi \in L$, or $\theta \circ \psi$ is a $\{0, 1\}$-sublattice isomorphic to the direct product $K \times L$. (Each partition $\theta \in \Pi(S)$ can also be considered as a partition of $\Pi(S \cup T)$ by adjoining to its blocks a singleton block for each element of $T - S$.)

References

Aigner, M. (1974). Uniformität des Verbandes der Partitionen, *Math. Ann.* **207**, 1–22.

Birkhoff, G. (1935). On the structure of abstract algebras, *Proc. Camb. Phil. Soc.* **31**, 433–54.

Birkhoff, G. (1940, 1948, 1967). *Lattice Theory*, Amer. Math. Soc. Colloq. Publ. XXV.

Birkhoff, G. (1970). What can lattices do for you?, in Abbott (ed.) *Trends in Lattice Theory*, pp. 1–40. Van Nostrand.

Crapo, H. (1986). Examples and basic concepts, in N. White (ed.), *Theory of Matroids*, pp. 1–28. Cambridge University Press.

Crapo, H. H. & Rota, G. C. (1969). *On the Foundations of Combinatorial Theory: Combinatorial Geometries*, MIT Press, Cambridge, Mass.

Crawley, P. & Dilworth, R. P. (1973). *Algebraic Theory of Lattices*, Prentice-Hall, Englewood Cliffs, New Jersey.

Dean, R. A. (1956). Component subsets of the free lattice on n generators, *Proc. Amer. Math. Soc.* **7**, 220–6.

Dilworth, R. P. (1961). Structure and decomposition theory of lattices, in *Proceedings of a Symposium on Pure Mathematics*, pp. 3–16. American Mathematical Society, Providence, Rhode Island.

Ehrenfeucht, A., Faber, V., Fajtlowicz, S. & Mycielski, J. (1973). Representations of finite lattices as partition lattices on finite sets, in *Proceedings of the University of Houston Lattice Theory Conference*, pp. 17–35. University of Houston.

Faigle, U. (1986). Lattices, in N. White (ed.), *Theory of Matroids*, pp. 45–61. Cambridge University Press.

Feit, W. (1983). An interval in the subgroup lattice of a finite group which is isomorphic to M_7, *Algebra Universalis* **17**, 220–1.

Finkbeiner, D. T. (1951). A general dependence relation for lattices, *Proc. Amer. Math. Soc* **2**, 756–9.

Finkbeiner, D. T. (1960). A semimodular imbedding of lattices, *Can. J. Math.* **12**, 582–91.

Goralčik, P. (1975). Problem 1.3., in *Universal Algebra*, Colioq. Math. Soc. János Bolyai **17**, p. 604. North-Holland, Amsterdam.

Goralčik, P. (1980). On an elegant algebraic result (in Czechoslovakian), *Pokroky Mat. Fyz. Astronom.* **25**, 139–44.

Grätzer, G. (1978). *General Lattice Theory*, Academic Press, New York.

Grätzer, G. & Kiss, E. W. (1986). A construction of semimodular lattices, *Order* **2**, 351–65.

Grätzer, G. & Schmidt, E. T. (1963). Characterization of congruence lattices of abstract algebras, *Acta. Sci. Math. Szeged* **24**, 34–59.

Hales, A. W. (1970). Partition representation of free lattices, *Proc. Amer. Math. Soc.* **24**, 517–20.

Hall, M. (1943). Projective planes, *Trans. Amer. Math. Soc.* **54**, 239.

Hartmanis, J. (1956). Two embedding theorems for finite lattices, *Proc. Amer. Math. Soc.* **7**, 571–7.

Hartmanis, J. (1959). Lattice theory of generalized partitions, *Can. J. Math.* **11**, 97–106.

Hartmanis, J. (1961). Generalized partitions and lattice embedding theorems, in *Proceedings of a Symposium on Pure Mathematics*, pp. 22–30. American Mathematical Society, Providence, Rhode Island.

Hsü, C. J. (1981). On the characterization of the partition lattice $LP_n(S)$ as a geometric lattice, *Chinese J. Math.* **9**, 37–46.

Jónsson, B. (1953). On the representation of lattices, *Math. Scand.* **1**, 193–206.

Jónsson, B. (1956). Universal relational systems, *Math. Scand.* **4**, 193–208.

Jónsson, B. (1962). Algebraic extensions of relational systems, *Math. Scand.* **11**, 179–205.

Jónsson, B. (1967). Algebras whose congruence lattices are distributive, *Math. Scand.* **21**, 110–21.

Lin, C. (1973). The lattice of all partitions on a set, *Tamkang J. Math.* **4**, 117–22.

Ore, O. (1942). Theory of equivalence relations, *Duke Math. J.* **9**, 573–627.

Pálfy, P. P. & Pudlák, P. (1980). Congruence lattices of finite algebras and intervals in subgroup lattices of finite groups, *Algebra Universalis* **11**, 22–7.

Peele, R. (1982). On finite partition representations of lattices, *Discrete Math.* **42**, 267–80.

Poguntke, W. & Rival, I. (1976). Finite four-generated simple lattices contain all finite lattices, *Proc. Amer. Math. Soc.* **55**, 22–4.

Pudlák, P. (1976). A new proof of the congruence lattice representation theorem. *Algebra Universalis* **6**, 269–75.

Pudlák, P. (1977). Distributivity of strongly representable lattices, **7**, 85–92.

Pudlák, P. (1979). Symmetric embedding of finite lattices into finite partition lattices. *Comment. Math. Carolin.* **20**, 183–7.

Pudlák, P. & Tůma, J. (1976). Yeast graphs and fermentation of algebraic lattices, in *Lattice Theory* (Szeged, 1974), Colloq. Math. Soc. János Bolyai **14**, pp. 301–41. North-Holland, Amsterdam.

Pudlák, P. & Tůma, J. (1977). Every finite lattice can be embedded in the lattice of all equivalences over a finite set, *Comment. Math. Carolin.* **18**, 409–14.

Pudlák, P. and Tůma, J. (1980). Every lattice can be embedded in a finite partition lattice, *Algebra Universalis* **10**, 74–95.

Pudlák, P. & Tůma, J. (1983). Regraphs and congruence lattices, *Algebra Universalis* **17**, 339–43.

Rival, I. (1978). Lattices and partially ordered sets, in *Problèmes Combinatoires et Théorie des Graphs* (Orsay, 1976), Editions du C.N.R.S. **260**, pp. 349–51. Centre National de la Recherche Scientifique, Paris.

Rousseau, R. (1983). Representations of a sublattice of the partition lattice on a lattice, *Discrete Math.* **47**, 307–14.

Sachs, D. (1961). Identities in finite partition lattices, *Proc. Amer. Math. Soc.* **12**, 944–5.

Sachs, D. (1961). Partition and modulated lattices, *Pacific J. Math.* **11**, 325–45.

Sasaki, U. & Fujiwara, S. (1952). The characterization of partition lattices, *J. Sci. Hiroshima Univ.* **15**, 189–201.

Sorkin, J. I. (1954). On the embedding of latticoids in lattices (in Russian), *Dokl. Akad. Nauk. S.S.S.R.* **95**, 931–4.

Strietz, H. (1975) Finite partition lattices are four-generated, in *Proceedings of the Lattice Theory*

Conference (Universität Ulm), pp. 257–59.

Strietz, H. (1977). Über Erzeugendenmengen endlicher Partitionenverbände, *Studia Scient. Math. Hung.* **12**, 1–17.

Tůma, J. (1980). A structure theorem for lattices of generalized partitions, in *Lattice Theory* (Szeged), Colloq. Math. Soc. János Bolyai **33**, pp. 737–57. North-Holland, Amsterdam.

Tůma, J. (1985). A simple geometric proof of a theorem on M_n, *Comment. Math. Carolin.* **26**, 233–9.

Werner, H. (1976). Which partition lattices are congruence lattices?, in *Lattice Theory* (Szeged, 1974), Colloq. Math. Soc. János Bolyai 14, pp. 433–53. North-Holland, Amsterdam.

White, N. (ed.) (1986). *Theory of Matroids*, Cambridge University Press.

Whitman, P. M. (1946). Lattices, equivalence relations, and subgroups, *Bull. Amer. Math. Soc.* **52**, 507–22.

Wille, R. (1967). Verbandstheoretische Kennzeichnung n-stufiger Geometrien, *Arch. Math.* **18**, 465–8.

Wille, R. (1976). A note on simple lattices, in *Lattice Theory* (Szeged, 1974), Colloq. Math. Soc. János Bolyai 14, pp. 455–62. North-Holland, Amsterdam.

6

The Tutte Polynomial and Its Applications

THOMAS BRYLAWSKI[†] and JAMES OXLEY[‡]

6.1. Introduction

The theory of numerical invariants for matroids is one of the many aspects of matroid theory having its origins within graph theory. Indeed, most of the fundamental ideas in matroid invariant theory were developed for graphs by Veblen (1912), Birkhoff (1912–13), Whitney (1933c), and Tutte (1947; 1954) when considering colorings and flows in graphs. The applications of matroid invariant theory now extend well beyond graphs, reaching into such fields as coding theory, percolation theory, electrical network theory, and statistical mechanics. In addition, many new graph-theoretic applications of the theory have been found. The purpose of this chapter is to review the many diverse applications of matroid invariant theory. In White (1987), Chapters 7 and 8 deal with several fundamental examples of matroid invariants, and we shall make frequent reference to these chapters here, particularly the former (Zaslavsky, 1987).

A matroid *isomorphism invariant* is a function f on the class of all matroids such that

$$f(M) = f(N) \quad \text{whenever } M \cong N. \tag{6.1}$$

The starting point for our chapter is the observation that several numbers associated with a matroid $M(E)$, such as its number of bases, its number of independent sets, and its number of spanning sets, are isomorphism invariants satisfying the following two basic recursions. Recall that if $T \subseteq E$, then $M(T)$ denotes the submatroid of M on T.

† Supported in part by ONR grant N00014-86-K-0449.

‡ Supported in part by NSF grant No. DMS-8500494 and by a grant from the Louisiana Education Quality Support Fund through the Board of Regents.

For every element e of M,

$$f(M) = f(M - e) + f(M/e) \text{ if } e \text{ is neither a loop nor an isthmus,} \quad (6.2)$$
$$f(M) = f(M(e))f(M - e) \text{ otherwise.} \quad (6.3)$$

If $\mathscr{K}$ is a class of matroids closed under isomorphism and the taking of minors, and f is a function on $\mathscr{K}$ satisfying (6.1), (6.2), and (6.3), then f is called a *Tutte–Grothendieck* or *T–G invariant*.

Theorem 7.2.4 of White (1987) notes that the characteristic polynomial $p(M; \lambda)$ of a matroid M satisfies the deletion–contraction formula

$$p(M; \lambda) = p(M - e; \lambda) - p(M/e; \lambda)$$

for all elements e that are neither loops nor isthmuses of M. This prompts consideration of matroid isomorphism invariants satisfying (6.3) and the following generalization of (6.2).

For some fixed non-zero numbers σ and τ,

$$f(M) = \sigma f(M - e) + \tau f(M/e) \quad (6.4)$$

provided e is neither a loop nor an isthmus.

Such invariants will be called *generalized T–G invariants*.

The fundamental result of this theory is that every T–G invariant f is an evaluation of a certain two-variable polynomial $t(x, y)$ where, for an isthmus I and a loop L,

$$f(I) = x \quad \text{and} \quad f(L) = y. \quad (6.5)$$

From this result, it is straightforward to deduce a characterization of all generalized T–G invariants. The precise statements of these results will be given in the next section, which contains a review of the basic results in matroid invariant theory. A more detailed development of this theory appears in Brylawski (1982). The primary focus of this chapter is on the applications of the theory. In particular, we concentrate mainly on those applications that are related to graphs and coding theory, preferring to treat a few applications in detail rather than to give a superficial treatment of all the applications. We hope that the extensive list of references will compensate in part for our failure to provide encyclopedic treatment of every application.

6.2. The Tutte Polynomial

In this section we state the fundamental results characterizing all T–G and generalized T–G invariants. We also investigate several closely related but coarser matroid invariants.

For an arbitrary matroid $M(E)$ having rank and nullity functions r and n respectively, the *rank generating polynomial* $S(M; x, y)$ of M is defined by

$$S(M; x, y) = \sum_{X \subseteq E} x^{r(E) - r(X)} y^{n(X)} = \sum_{X \subseteq E} x^{r(E) - r(X)} y^{|X| - r(X)}. \quad (6.6)$$

Thus

$$S(M; x, y) = \sum_i \sum_j a_{ij} x^i y^j \tag{6.7}$$

where a_{ij} is the number of submatroids of M of rank $r(M) - i$ and nullity j. Brylawski (1982) calls $S(M; x, y)$ the *corank-nullity polynomial* of M, the *corank* of a set X in M being $r(E) - r(X)$. Note that this usage differs from that in Welsh (1976) where 'corank' means rank in the dual matroid. Clearly, $S(M; x, y)$ is an isomorphism invariant for the class of all matroids. Moreover, one can easily check that if I is an isthmus and L is a loop, then

$$S(I; x, y) = x + 1 \quad \text{and} \quad S(L; x, y) = y + 1. \tag{6.8}$$

6.2.1. Lemma. *$S(M; x, y)$ is a T–G invariant for the class of all matroids.*

Proof. Let e be an element of the matroid $M(E)$. Clearly

$$S(M; x, y) = \sum_{\substack{X \subseteq E \\ e \notin X}} x^{r(E)-r(X)} y^{n(X)} + \sum_{\substack{X \subseteq E \\ e \in X}} x^{r(E)-r(X)} y^{n(X)}. \tag{6.9}$$

Consider the first term on the right-hand side. Clearly this equals $\sum_{X \subseteq E-e} x^{r(E)-r(X)} y^{n(X)}$. Moreover,

$$r(E) = \begin{cases} r(E - e) + 1 & \text{if } e \text{ is an isthmus,} \\ r(E - e) & \text{otherwise.} \end{cases}$$

Thus

$$\sum_{\substack{X \subseteq E \\ e \notin X}} x^{r(E)-r(X)} y^{n(X)} = \begin{cases} x \sum_{X \subseteq E-e} x^{r(E-e)-r(X)} y^{n(X)} & \text{if } e \text{ is an isthmus,} \\ \sum_{X \subseteq E-e} x^{r(E-e)-r(X)} y^{n(X)} & \text{otherwise.} \end{cases}$$

Hence

$$\sum_{\substack{X \subseteq E \\ e \notin X}} x^{r(E)-r(X)} y^{n(X)} = \begin{cases} x S(M - e; x, y) & \text{if } e \text{ is an isthmus,} \\ S(M - e; x, y) & \text{otherwise.} \end{cases} \tag{6.10}$$

Now consider the second term on the right-hand side of (6.9). This equals $\sum_{Y \subseteq E-e} x^{r((E-e)\cup e)-r(Y \cup e)} y^{n(Y \cup e)}$. Let r' and n' denote the rank and nullity functions of M/e. Then, for all $Y \subseteq E - e$,

$$r'(Y) = \begin{cases} r(Y \cup e) & \text{if } e \text{ is a loop,} \\ r(Y \cup e) - 1 & \text{otherwise;} \end{cases}$$

and

$$n'(Y) = \begin{cases} n(Y \cup e) - 1 & \text{if } e \text{ is a loop,} \\ n(Y \cup e) & \text{otherwise.} \end{cases}$$

Thus

$$\sum_{\substack{X \subseteq E \\ e \in X}} x^{r(E)-r(X)} y^{n(X)} = \begin{cases} y \displaystyle\sum_{Y \subseteq E-e} x^{r'(E-e)-r'(Y)} y^{n'(Y)} & \text{if } e \text{ is a loop,} \\ \displaystyle\sum_{Y \subseteq E-e} x^{r'(E-e)-r'(Y)} y^{n'(Y)} & \text{otherwise.} \end{cases}$$

Hence

$$\sum_{\substack{X \subseteq E \\ e \in X}} x^{r(E)-r(X)} y^{n(X)} = \begin{cases} yS(M/e; x, y) & \text{if } e \text{ is a loop,} \\ S(M/e; x, y) & \text{otherwise.} \end{cases} \tag{6.11}$$

On substituting from (6.10) and (6.11) into (6.9), we obtain

$$S(M; x, y) = \begin{cases} S(M - e; x, y) + S(M/e; x, y) & \text{if } e \text{ is neither an isthmus} \\ & \text{nor a loop,} \\ (x + 1)S(M - e; x, y) & \text{if } e \text{ is an isthmus,} \\ (y + 1)S(M/e; x, y) & \text{if } e \text{ is a loop.} \end{cases}$$

But $S(I; x, y) = x + 1$ and $S(L; x, y) = y + 1$. Moreover, if e is a loop, then $M/e = M - e$. The lemma now follows easily. $\qquad\qquad\square$

The next theorem, the main result of this section, extends the preceding lemma by showing that not only is $S(M; x, y)$ a T–G invariant but, more importantly, it is essentially the universal T–G invariant. The sets of isomorphism classes of matroids and non-empty matroids will be denoted by $\mathcal{M}$ and $\mathcal{M}'$, respectively. Note that (ii) and (iii) in this theorem are no more than restatements of the fundamental recursions (6.2) and (6.3).

6.2.2. Theorem. (*Brylawski, 1972b*) *There is a unique function t from $\mathcal{M}$ into the polynomial ring $\mathbb{Z}[x, y]$ having the following properties*:

(i) $t(I; x, y) = x$ *and* $t(L; x, y) = y$.

(ii) *(Deletion–contraction) If e is an element of the matroid M and e is neither a loop nor an isthmus, then*

$$t(M; x, y) = t(M - e; x, y) + t(M/e; x, y).$$

(iii) *If e is a loop or an isthmus of the matroid $M(E)$, then*

$$t(M; x, y) = t(M(e); x, y)t(M - e; x, y).$$

Furthermore, let R be a commutative ring and suppose that f is any function from $\mathcal{M}'$ into R. If f satisfies (6.2) and (6.3) whenever $|E| \geq 2$, then, for all matroids M,

$$f(M) = t(M; f(I), f(L)).$$

Proof. By Lemma 6.2.1, if $t(M; x, y) = S(M; x - 1, y - 1)$, then (i)–(iii) hold. Now the only non-empty matroids that cannot be decomposed using (ii) or

(iii) are I and L, and, for these matroids, $t(M; x, y)$ is fixed by (i). Hence an easy induction argument establishes that t is unique. Finally, the last part of the theorem can be proved using another straightforward induction argument, the details of which are left to the reader. ☐

We shall call the function $t(M; x, y)$ the *Tutte polynomial* of M. Evidently $t(M; x, y)$ can be written as $\sum_i \sum_j b_{ij} x^i y^j$ where $b_{ij} \geq 0$ for all i and j. We shall usually abbreviate this double summation to $\sum b_{ij} x^i y^j$. It follows immediately from the last proof that

$$t(M; x, y) = S(M; x - 1, y - 1). \tag{6.12}$$

Hence

$$t(M; x, y) = \sum_{X \subseteq E} (x - 1)^{r(E) - r(X)} (y - 1)^{n(X)}. \tag{6.13}$$

The Tutte polynomial can be calculated directly from this summation, or alternatively, it can be determined by using the recursions 6.2.2(ii) and (iii). The second of these techniques is illustrated as follows.

6.2.3. Example. Let M be $U_{2,4}$, the rank 2 uniform matroid on a set of four elements, that is, the 4-point line. In the calculation below, we shall abbreviate $t(N; x, y)$ throughout as (N), and represent each matroid affinely, where a loop e is written as **e**. By repeated application of 6.2.2(i), (ii), and (iii) we have:

$$
\begin{aligned}
&= \ (\quad b \ ) \ + \ (\bullet ab) \ + \ (\bullet ab) \ + \ (\{\mathbf{a}, \mathbf{b}\}) \\
&= \ x(\bullet a) \ + \ 2(\bullet ab) \ + \ y(\mathbf{a}) \\
&= \ x^2 \ + \ 2(\bullet a) \ + \ 2(\mathbf{a}) \ + \ y^2 \\
&= \ x^2 \ + \ 2x \ + \ 2y \ + \ y^2.
\end{aligned}
$$

Thus

$$t(U_{2,4}; x, y) = x^2 + 2x + 2y + y^2.$$

Evidently $t(U_{2,4}; x, y)$ is symmetric with respect to x and y. Since $U_{2,4}$ is a self-dual matroid, this observation is a special case of the next result, the proof of which follows easily from (6.13) by using the fact that if $X \subseteq E$, then its rank in $M^*(E)$ is $|X| - r(E) + r(E - X)$, where r is the rank function of $M(E)$.

6.2.4. Proposition. *For all matroids M,*

$$t(M^*; x, y) = t(M; y, x).$$

An easy induction argument beginning with 6.2.2(iii) establishes the next result.

6.2.5. Proposition. *For matroids* $M_1(E_1)$ *and* $M_2(E_2)$ *where* E_1 *and* E_2 *are disjoint,*

$$t(M_1 \oplus M_2; x, y) = t(M_1; x, y)t(M_2; x, y).$$

Evidently 6.2.2(iii) is a special case of 6.2.5. A consequence of this observation and the preceding result is that one can use the direct sum formula 6.2.5 in place of 6.2.2(iii) in defining the Tutte polynomial.

The following characterization of generalized T–G invariants is a straightforward extension of Theorem 6.2.2. Its proof is left to the reader.

6.2.6. Corollary. (*Oxley & Welsh, 1979b*) *Let* σ *and* τ *be non-zero elements of a field F. Then there is a unique function* t' *from* $\mathcal{M}$ *into the polynomial ring* $F[x, y]$ *having the following properties:*

(i) $t'(I; x, y) = x$ *and* $t'(L; x, y) = y$.

(ii) *If e is an element of the matroid M and e is neither a loop nor an isthmus, then*

$$t'(M; x, y) = \sigma t'(M - e; x, y) + \tau t'(M/e; x, y).$$

(iii) *If e is a loop or an isthmus of the matroid M, then*

$$t'(M; x, y) = t'(M(e); x, y)t'(M - e; x, y).$$

Furthermore, this function t' *is given by*

$$t'(M; x, y) = \sigma^{|E| - r(E)} \tau^{r(E)} t(M; x/\tau, y/\sigma).$$

We defer to the exercises consideration of a still more general invariant which admits a multiplicative constant on the right-hand side of 6.2.6(iii).

6.2.7. Example. Suppose that every element of a matroid $M(E)$ has, independently of all other elements, a probability $1 - p$ of being deleted from M and assume that $0 < p < 1$. We call the resulting restriction minor $\omega(M)$ of M a *random submatroid* of M, corresponding in the obvious way to a random graph on m vertices when M is the cycle matroid of the complete graph K_m. If we let $\Pr(M)$ denote the probability that $\omega(M)$ has the same rank as M, then, evidently, $\Pr(I) = p$ and $\Pr(L) = 1$. Moreover,

$$\Pr(M) = \begin{cases} (1 - p)\Pr(M - e) + p\Pr(M/e) & \text{if } e \text{ is neither a loop nor an isthmus,} \\ \Pr(M(e))\Pr(M - e) & \text{otherwise.} \end{cases}$$

It follows by Corollary 6.2.6 that

$$\Pr(M) = (1 - p)^{|E| - r(E)} p^{r(E)} t(M; 1, 1/(1 - p)).$$

Section 7.3 of White (1987) is devoted to the beta invariant for matroids. This invariant is a member of the class of matroid isomorphism invariants that satisfy the additive recursion (6.2) but not necessarily the multiplicative recursion (6.3). Such matroid isomorphism invariants will be called $(T\text{–}G)$ *group invariants*. The next result shows that the theory already developed can be used to characterize invariants of this type.

6.2.8. Proposition. *If A is an Abelian group, then there is a unique function g from $\mathcal{M}'$ into A such that*

(i) $g(M) = g(M - e) + g(M/e)$ *provided e is neither a loop nor an isthmus of the matroid M; and*

(ii) $g(U_{i,i} \oplus U_{0,j}) = \alpha_{ij}$ *for all i and j such that $i + j > 0$.*

Moreover, if $t(M; x, y) = \sum_i \sum_j b_{ij} x^i y^j$, then $g(M) = \sum_i \sum_j b_{ij} \alpha_{ij}$.

Proof. Evidently, if we define $g(M) = \sum_i \sum_j b_{ij} \alpha_{ij}$ for all matroids M, then g satisfies (i). Moreover, as $t(U_{i,i} \oplus U_{0,j}; x, y) = x^i y^j$, g satisfies (ii). Thus there is at least one function satisfying the required conditions. To obtain that such a function is unique, we argue by induction noting that the only non-empty matroids that cannot be decomposed using (i) are those consisting entirely of loops and isthmuses. As the value of the function is fixed on such matroids by (ii), the required result follows. $\square$

We shall now illustrate the use of the last result. If M is a matroid, then $i_{r-j}(M)$ will denote the number of independent sets of M having $r(M) - j$ elements.

6.2.9. Proposition. *For a non-negative integer k, the isomorphism invariant $i_{r-k}(M)$ is a group invariant and, if $t(M; x, y) = \sum_i \sum_j b_{ij} x^i y^i$, then*

$$i_{r-k}(M) = \sum_i \sum_j b_{ij} \binom{i}{k}.$$

Proof. Let e be an element of M and suppose that e is neither a loop nor an isthmus. Partition the set $\mathcal{I}_{r(M)-k}$ of independent sets of M having $r(M) - k$ elements into subsets $\mathcal{I}'_{r(M)-k}$ and $\mathcal{I}''_{r(M)-k}$ consisting, respectively, of those members of $\mathcal{I}_{r(M)-k}$ that contain e and those that do not. Evidently $\mathcal{I}''_{r(M)-k}$ is in one-to-one correspondence with the set of independent sets of $M - e$ having $r(M - e) - k$ elements. Moreover, $\mathcal{I}'_{r(M)-k}$ is in one-to-one correspondence with the set of independent sets of M/e having $r(M/e) - k$ elements. It follows that $i_{r-k}(M) = i_{r-k}(M - e) + i_{r-k}(M/e)$. Thus $i_{r-k}(M)$ is a group invariant. Now clearly $i_{r-k}(U_{i,i} \oplus U_{0,j}) = \binom{i}{i-k} = \binom{i}{k}$ and so, by Proposition 6.2.8, we conclude that for all matroids M,

$$i_{r-k}(M) = \sum_i \sum_j b_{ij} \binom{i}{k}. \qquad \square$$

By (6.6), $i_{r-k}(M)$ is the coefficient of x^k in $S(M; x, y)$. Indeed, the preceding proposition is just a special case of the result (see Exercise 6.4) that every coefficient of $S(M; x, y)$ is a group invariant.

Theorem 6.2.2 and Corollary 6.2.6 give characterizations of T–G invariants and group invariants that are neatly expressible in terms of the Tutte polynomial. In order to address the problem of precisely which matroid isomorphism invariants can be determined from the Tutte polynomial, a function f from $\mathcal{M}$ into a set Ω will be called a *Tutte invariant* if it has the property that $f(M) = f(N)$ whenever M and N have the same Tutte polynomial. Thus all generalized T–G and group invariants are examples of Tutte invariants. There are many other examples that are not of one of these types. For instance, since

$$t(M; x, y) = S(M; x-1, y-1) = \sum_{X \subseteq E} (x-1)^{r(E)-r(X)}(y-1)^{n(X)},$$

$$r(M) \text{ is the highest power of } x \text{ in } t(M; x, y), \qquad (6.14)$$

while

$$n(M) \text{ is the highest power of } y \text{ in } t(M; x, y). \qquad (6.15)$$

Thus rank and nullity are Tutte invariants. Clearly, since $|E| = r(M) + n(M)$, cardinality of the ground set is also a Tutte invariant. The next result characterizes all Tutte invariants, although it is essentially just a restatement of the definition. We leave the straightforward proof as an exercise.

6.2.10. Proposition. *Let Ω be a set and f be a function from $\mathcal{M}$ into Ω such that $f(M) = f(N)$ whenever $t(M; x, y) = t(N; x, y)$. Then $f(M)$ is a function of the coefficients b_{ij} of $t(M; x, y)$.*

Notice that this proposition does not assert that we can find an explicit formula for a particular Tutte invariant in terms of the Tutte polynomial coefficients. Nevertheless, most of the examples of Tutte invariants that we shall consider will have such an explicit formula. For example, by (6.14) and (6.15), we have

$$r(M) = \max\{i: b_{ij} > 0 \text{ for some } j\}, \text{ and} \qquad (6.16)$$

$$n(M) = \max\{j: b_{ij} > 0 \text{ for some } i\}. \qquad (6.17)$$

Having stated the characterizations of T–G, generalized T–G, group, and Tutte invariants, we now give a number of the more basic applications of these results. For a matroid M, we denote by $b(M)$, $i(M)$, and $s(M)$, the numbers of bases, independent sets, and spanning sets, respectively, of M. It was asserted in section 6.1 that these three numbers are T–G invariants. We now prove this.

6.2.11. Proposition.

(i) $b(M) = t(M; 1, 1) = S(M; 0, 0)$;
(ii) $i(M) = t(M; 2, 1) = S(M; 1, 0)$;
(iii) $s(M) = t(M; 1, 2) = S(M; 0, 1)$; and
(iv) $2^{|E|} = t(M; 2, 2) = S(M; 1, 1)$.

Proof. We begin by proving (i). Let e be an element of the matroid M and suppose that e is neither a loop nor an isthmus. Partition the set of bases of M into subsets $\mathscr{B}'$ and $\mathscr{B}''$ consisting, respectively, of those bases containing e and those bases not containing e. Now $\mathscr{B}''$ is equal to the set of bases of $M - e$, while the set of bases of M/e is $\{B - e: B \in \mathscr{B}'\}$. Thus $|\mathscr{B}''| = b(M - e)$ and $|\mathscr{B}'| = b(M/e)$. Therefore if e is neither a loop nor an isthmus of M, then $b(M) = b(M - e) + b(M/e)$. On the other hand, if e is a loop or an isthmus, then it is clear that $b(M) = b(M(e))b(M - e)$. Thus $b(M)$ satisfies (6.2) and (6.3). But clearly $b(I) = b(L) = 1$. Therefore, by Theorem 6.2.2, $b(M) = t(M; 1, 1)$. Moreover, by (6.12), $t(M; 1, 1) = S(M; 0, 0)$. Thus (i) holds.

The proof technique used above can also be used to prove (ii) and we leave this to the reader. To prove (iii), note that $s(M) = i(M^*)$. Thus, by (ii), $s(M) = t(M^*; 2, 1)$. But by Proposition 6.2.4, $t(M^*; 2, 1) = t(M; 1, 2)$, and (iii) follows.

Finally, we have by (6.12) and (6.8) that $t(M; 2, 2) = S(M; 1, 1) = \sum_{X \subseteq E} 1^{r(E)-r(X)} 1^{n(X)} = 2^{|E|}$. $\square$

Since $b(M) = t(M; 1, 1)$ and $t(M; x, y) = \sum_i \sum_j b_{ij} x^i y^j$, it follows that $\sum_i \sum_j b_{ij} = b(M)$. In section 6.6A, we shall describe how, by ordering the ground set of M, we can interpret each of the coefficients b_{ij} as counting a particular set of bases of M.

The characteristic polynomial, the Möbius function, and the beta invariant were all considered in detail in Chapter 7 of White (1987). We now relate each of these functions to the Tutte polynomial. Recall that for a matroid M having a lattice of flats L, the characteristic polynomial $p(M; \lambda)$ of M is defined by the equation

$$p(M; \lambda) = \sum_{F \in L} \mu_M(\varnothing, F) \lambda^{r(M)-r(F)} \qquad (6.18)$$

and satisfies the identity

$$p(M; \lambda) = \sum_{X \subseteq S} (-1)^{|X|} \lambda^{r(M)-r(X)}. \qquad (6.19)$$

We noted in section 6.1 that $p(M; \lambda)$ is a generalized T–G invariant. Using Corollary 6.2.6, the characterization of such invariants, we obtain

$$p(M; \lambda) = (-1)^{r(M)} t(M; 1 - \lambda, 0) = (-1)^{r(M)} S(M; -\lambda, -1). \qquad (6.20)$$

By (6.18), $\mu(M) = \mu_M(\varnothing, E) = p(M; 0)$. Hence the Möbius function $\mu(M)$ is a generalized T–G invariant and

$$\mu(M) = (-1)^{r(M)} t(M; 1, 0) = (-1)^{r(M)} S(M; 0, -1). \tag{6.21}$$

6.2.12. Proposition. $\beta(M)$ *is a group invariant whose value is* b_{10}. *Moreover, if* $|E| \geqslant 2$, *then* $b_{10} = b_{01}$.

Proof. As noted earlier, $\beta(M)$ satisfies the additive recursion 6.2.5 and hence is a group invariant. Now, by Theorems 7.3.2(b) and 7.3.4 of White (1987),

$$\beta(U_{i,i} \oplus U_{0,j}) = \begin{cases} 1 & \text{if } i = 1 \text{ and } j = 0, \\ 0 & \text{otherwise.} \end{cases}$$

Hence by Proposition 6.2.8, $\beta(M) = b_{10}$. The fact that $b_{10} = b_{01}$ for $|E| \geqslant 2$ follows by a straightforward induction argument using the additive recursion 6.2.8(i). □

The identity $b_{10} = b_{01}$ noted in the preceding proposition is one of a number of identities that hold for the coefficients b_{ij} of the Tutte polynomial. The next result (Brylawski, 1972b; 1982) completely characterizes all identities of the form $\sum_i \sum_j \alpha_{ij} b_{ij} = \gamma$ where γ and all the α_{ij} are constants.

6.2.13. Theorem. *The following identities form a basis for the affine linear relations that hold among the coefficients* b_{ij} *in the Tutte polynomial*

$$t(M; x, y) = \sum_{i \geqslant 0} \sum_{j \geqslant 0} b_{ij} x^i y^j$$

where M *is a rank* r *geometry having* m *elements none of which is an isthmus.*

  (i) $b_{ij} = 0$ *for all* $i > r$ *and all* $j \geqslant 0$;
 (ii) $b_{r0} = 1$; $b_{rj} = 0$ *for all* $j > 0$;
(iii) $b_{r-1,0} = m - r$; $b_{r-1,j} = 0$ *for all* $j > 0$;
 (iv) $b_{ij} = 0$ *for all* i *and* j *such that* $1 \leqslant i \leqslant r - 2$ *and* $j \geqslant m - r$;
  (v) $b_{0,m-r} = 1$; $b_{0j} = 0$ *for all* $j > m - r$;
 (vi) $\displaystyle\sum_{s=0}^{k} \sum_{t=0}^{k-s} (-1)^t \binom{k-s}{t} b_{st} = 0$ *for all* k *such that* $0 \leqslant k \leqslant m - 3$.

Moreover, (vi) *holds for all matroids* $M(E)$ *such that* $|E| > k$. *Hence,*

 (vii) $b_{00} = 0$ *if* $|E| \geqslant 1$;
(viii) $b_{10} = b_{01}$ *if* $|E| \geqslant 2$;
 (ix) $b_{20} - b_{11} + b_{02} = b_{10}$ *if* $|E| \geqslant 3$; *and*
  (x) $b_{30} - b_{21} + b_{12} - b_{03} = b_{11} - 2b_{02} + b_{10}$ *if* $|E| \geqslant 4$.

Next we consider what sorts of matroid properties are Tutte invariants. First we note four elementary examples of such invariants. These assertions are easily proved by induction.

6.2.14. Example. If $b_{r(M),t} > 0$, then t is the number of loops of M and $b_{r(M),t} = 1$.

6.2.15. Example. If $b_{s,n(M)} > 0$, then s is the number of isthmuses of M and $b_{s,n(M)} = 1$.

6.2.16. Example. The number of rank 1 flats of M is $r(M) + b_{r(M)-1,t}$, where t is the number of loops of M.

6.2.17. Example. Provided $r(M)$ and $n(M)$ are both positive, the number of connected components of M is $\min\{j: b_{0j} > 0\} = \min\{i: b_{i0} > 0\}$. If exactly one of $r(M)$ and $n(M)$ is zero, then M has $|E|$ components.

To see the sorts of properties that are not Tutte invariants, we look at two non-isomorphic matroids having the same Tutte polynomial.

6.2.18. Example. Let M_1 and M_2 be the matroids for which affine representations are shown in Figure 6.1. Their Tutte polynomials are equal because $M_1 - e \cong M_2 - e$ and $M_1/e \cong M_2/e$. We now list various properties that M_1 and M_2 do not share. Each such property is an example of a matroid isomorphism invariant that is not a Tutte invariant, and hence cannot be determined from the Tutte polynomial.

(i) $M_1 \cong M_\Gamma$ for a planar graph Γ, so M_1 is both graphic and cographic and hence is unimodular and binary. On the other hand, M_2 has a 4-point line as a restriction so it is not even binary.

(ii) M_2 is transversal, but M_1 is not even a gammoid.

(iii) Although M_1 and M_2 have the same number of flats of rank 1 and the same number of 4- and 3-element lines, M_1 has two 2-element lines

Figure 6.1.

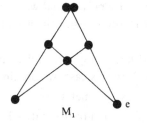
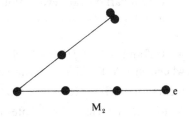

whereas M_2 has three. Hence M_1 and M_2 have different numbers of flats and different numbers of hyperplanes.

(iv) M_1 and M_2 each have one 2-element circuit and six 3-element circuits, but M_1 has five 4-element circuits and M_2 has six such circuits. Thus M_1 and M_2 have different numbers of circuits.

By Proposition 6.2.4, M_1^* and M_2^* have the same Tutte polynomial. Moreover, unlike M_1 and M_2, each of M_1^* and M_2^* is a geometry. We defer to the exercises (Exercise 6.9) consideration of an elementary counting argument which has as a consequence that for any number N there are at least N geometries all of which have the same Tutte polynomial.

Returning to the basic Tutte recursion (6.2), namely $f(M) = f(M - e) + f(M/e)$, we note that there are several slightly different techniques which are used to show that, when $f(M)$ enumerates the family $\mathscr{F}(M)$,

$$|\mathscr{F}(M)| = |\mathscr{F}(M - e)| + |\mathscr{F}(M/e)|.$$

(1) There is a bijection $b: \mathscr{F}(M) \to \mathscr{F}(M - e) \cup \mathscr{F}(M/e)$. Thus the members of $\mathscr{F}(M)$ are partitioned into two classes corresponding to the analogous families in the deletion and the contraction. This technique was used in the proof of 6.2.11(i).

(2) There is an injection $i: \mathscr{F}(M/e) \to \mathscr{F}(M - e)$ such that $|\mathscr{F}(M)| = \sum_{x \in \mathscr{F}(M - e)} m(x)$, where $m(x)$ is 2 when x is in the image of i and is 1 otherwise. This idea is used, for example, to prove that the number of acyclic orientations of a graph is a T–G invariant (6.3.17), where, in this case, $m(x)$ counts the number of ways to orient the edge e while maintaining the property of being acyclic.

(3) There are two surjections $\pi_1: \mathscr{F}(M) \to \mathscr{F}(M/e)$ and $\pi_2: \mathscr{F}(M - e) \to \mathscr{F}(M/e)$, such that, for all x in $\mathscr{F}(M/e)$, $|\pi_1^{-1}(x)| = |\pi_2^{-1}(x)| + 1$. Here we think of partitioning both $\mathscr{F}(M)$ and $\mathscr{F}(M - e)$ into $f(M/e)$ blocks such that corresponding blocks have one more member in M than in $M - e$. An example occurs in the calculation of the number of different score vectors that arise from the orientations of a graph (see Proposition 6.3.19).

Many T–G invariants are evaluations of the Tutte polynomial when y, or dually x, is 0, 1, or 2. We now list some salient features of such invariants f.

(1) $y = 0$ if and only if $f(M) = 0$ whenever M has a loop, or, equivalently, whenever parallel elements can be ignored. In this case, we obtain the recursion $f(G) = f(G - e) + f(\overline{G/e})$ for any geometry G and any non-isthmus e. Here $\overline{G/e}$ denotes the simplification of the matroid G/e.

(2) $y = 1$ if and only if loops can be ignored so that $f(M) = f(\tilde{M})$, where $\tilde{M}$ is obtained from M by deleting its loops. Here, if A is a set of parallel elements of M that is not a component, then

$$f(M) = f(M - A) + |A| f(M/A).$$

(3) $y = 2$ if and only if the recursion $f(M) = f(M - e) + f(M/e)$ holds not only when e is a non-loop or a non-isthmus, but also if e is a loop.

In some of the applications of T-G techniques, it is convenient to work with a four-variable version of the Tutte polynomial. This is defined on the class $\mathscr{M}_P$ of *pointed matroids*, that is, matroids M_d having a distinguished point d. It is not difficult to modify the proof of Theorem 6.2.2 to establish the next result and we leave the details to the reader. Evidently, if e is an element of M_d other than d, then $M_d - e$ and M_d/e are members of $\mathscr{M}_P$, the distinguished point of each being d.

6.2.19. Proposition. *There is a unique function t_P from $\mathscr{M}_P$ into the polynomial ring $\mathbb{Z}[x', x, y', y]$ having the following properties:*

(i) $t_P(M_d(d)) = x'$ *if $M_d(d)$ is an isthmus, and*

$t_P(M_d(d)) = y'$ *if $M_d(d)$ is a loop.*

(ii) *If e is an element of a member M_d of $\mathscr{M}_P$ and $e \neq d$, then*

$$t_P(M_d) = t_P(M_d - e) + t_P(M_d/e).$$

(iii) *If e is a loop or an isthmus of a member M_d of $\mathscr{M}_P$ and $e \neq d$, then*

$$t_P(M_d) = t_P(M_d - e)t(M_d(e)). \text{ In particular, } t_P(M_d(e)) = t(M_d(e)).$$

The polynomial $t_P(M_d; x', x, y', y)$ is called the *pointed Tutte polynomial*. The following proposition, which summarizes some of the basic properties of this polynomial, is not difficult to prove and is left to the reader as an exercise.

6.2.20. Proposition. *Suppose that $M_d \in \mathscr{M}_P$. Then*

(i) *for some f and g in $\mathbb{Z}[x, y]$,*

$$t_P(M_d; x', x, y', y) = x'f(x, y) + y'g(x, y).$$

Moreover, for this f and g,

(ii) $t(M_d; x, y) = xf(x, y) + yg(x, y)$;

(iii) $t_P(M_d^*; x', x, y', y) = t_P(M_d; y', y, x', x) = x'g(y, x) + y'f(y, x)$;

(iv) *if d is neither a loop nor an isthmus of M_d, then*

$$t(M_d - d; x, y) = (x - 1)f(x, y) + g(x, y) \text{ and}$$

$$t(M_d/d; x, y) = f(x, y) + (y - 1)g(x, y); \text{ and}$$

(v) *if d is a loop or an isthmus of M_d, then*

$$t(M_d - d; x, y) = t(M_d/d; x, y) = \begin{cases} f(x, y) & \text{if } d \text{ is an isthmus,} \\ g(x, y) & \text{if } d \text{ is a loop.} \end{cases}$$

We have seen that the rank generating polynomial is fundamental in the class of T–G invariants. Another related polynomial that arises in applications of T–G techniques is the *cardinality-corank polynomial* $S_{KC}(M; x, y)$. This is defined by

$$S_{KC}(M; x, y) = \sum_{X \subseteq E} x^{|X|} y^{r(E) - r(X)}. \tag{6.22}$$

We leave the reader to prove that this polynomial is a generalized T–G invariant, as stated in the following.

6.2.21. Proposition. $S_{KC}(M; x, y) = x^r t\left(M; \dfrac{x+y}{x}, x+1\right).$

6.3. T–G Invariants in Graphs

In this section we shall review the occurrence of T–G invariants in graphs. Although the most important such invariants occur in the context of colorings and flows, a number of others arise, for example, in connection with acyclic and totally cyclic orientations, score vectors, and network reliability. Here, most of our attention will be devoted to colorings and flows. We begin with the former.

6.3.A. Colorings

Iet Γ be a graph and λ be a positive integer. A *proper vertex coloring of Γ with λ colors* or a *proper λ-coloring of Γ* is a function f from $V(\Gamma)$ into $\{1, 2, ..., \lambda\}$ such that if $uv \in E(\Gamma)$, then $f(u) \neq f(v)$. The number of such colorings will be denoted $\chi_\Gamma(\lambda)$. It was noted in Chapter 7 of White (1987) that $\chi_\Gamma(\lambda)$ is a polynomial in λ. This polynomial is called the *chromatic polynomial* of Γ. The next result relates $\chi_\Gamma(\lambda)$ to the Tutte polynomial of M_Γ.

6.3.1. Proposition. *For a graph Γ having $k(\Gamma)$ connected components,*

$$\chi_\Gamma(\lambda) = \lambda^{k(\Gamma)} p(M_\Gamma; \lambda) = \lambda^{k(\Gamma)} (-1)^{|V(\Gamma)| - k(\Gamma)} t(M_\Gamma; 1 - \lambda, 0).$$

Proof. Let $f(\Gamma; \lambda) = \lambda^{-k(\Gamma)} \chi_\Gamma(\lambda)$. Ostensibly f depends on the particular graph Γ. However, we shall show that in fact f depends only on M_Γ and that if we let $f(M_\Gamma; \lambda) = f(\Gamma; \lambda)$, then this matroid function is a well defined generalized T–G invariant for which $\sigma = 1$ and $\tau = -1$. If M_Γ is I or L, then

Γ is formed from a single isthmus or a single loop by adjoining isolated vertices. It follows easily that $f(I; \lambda) = \lambda - 1$ and $f(L; \lambda) = 0$ so that f is well defined if $|E(\Gamma)| = 1$. Assume that f is a well defined matroid function when $|E(\Gamma)| < n$ and let $|E(\Gamma)| = n$.

Now suppose that e is an edge of Γ having endpoints u and v. Assume that e is neither a loop nor an isthmus of Γ. Then we can partition the set of proper λ-colorings of $\Gamma - e$ into those in which u and v are colored alike and those in which they are colored differently. But the first subset is one-to-one correspondence with the set of proper λ-colorings of Γ/e, and the second subset is in one-to-one correspondence with the set of proper λ-colorings of Γ. Evidently Γ and Γ/e have the same number of components. Moreover, as e is not an isthmus, Γ and $\Gamma - e$ have the same number of components. Thus $f(\Gamma - e; \lambda) = f(\Gamma; \lambda) + f(\Gamma/e; \lambda)$, and so, by the induction assumption,

$$f(\Gamma; \lambda) = f(M_{\Gamma-e}; \lambda) - f(M_{\Gamma/e}; \lambda).$$

If e is a loop of Γ, then $f(\Gamma; \lambda) = f(L; \lambda) = 0$, and so

$$f(\Gamma; \lambda) = f(L; \lambda) f(M_{\Gamma-e}; \lambda).$$

Finally, if e is an isthmus of Γ, then the number of ways to properly λ-color Γ equals the number of ways to properly λ-color $\Gamma - e$ so that u and v are colored differently. But in a proper λ-coloring of $\Gamma - e$, once a color is assigned to u, there are λ possible colors that can be assigned to v. Of these, $\lambda - 1$ are different from the color assigned to u. Thus the number of ways to properly λ-color $\Gamma - e$ so that u and v are colored differently is $\dfrac{\lambda - 1}{\lambda} \chi_{\Gamma-e}(\lambda)$. Hence $\chi_\Gamma(\lambda) = \dfrac{\lambda - 1}{\lambda} \chi_{\Gamma-e}(\lambda)$. Since $f(I; \lambda) = \lambda - 1$ and $k(\Gamma - e) = k(\Gamma) + 1$, we conclude that $f(\Gamma; \lambda) = f(I; \lambda) f(\Gamma - e; \lambda)$, and so, by the induction assumption,

$$f(\Gamma; \lambda) = f(I; \lambda) f(M_{\Gamma-e}; \lambda).$$

On comparing the equations for $f(\Gamma; \lambda)$ when e is a loop, an isthmus, and neither a loop nor an isthmus, we conclude by induction that f is well defined as a matroid function. Moreover, the same equations imply that f is a generalized T–G invariant for which $\sigma = 1$ and $\tau = -1$. Since $f(I; \lambda) = \lambda - 1$ and $f(L; \lambda) = 0$, it follows by Corollary 6.2.6 that

$$f(M_\Gamma; \lambda) = (-1)^{|V(\Gamma)| - k(\Gamma)} t(M_\Gamma; 1 - \lambda, 0).$$

Thus

$$\chi_\Gamma(\lambda) = \lambda^{k(\Gamma)} (-1)^{|V(\Gamma)| - k(\Gamma)} t(M_\Gamma; 1 - \lambda, 0).$$

Since $p(M; \lambda) = (-1)^{r(M)} t(M; 1 - \lambda, 0)$ for all matroids M, the rest of the proposition follows easily. $\qquad\square$

The chromatic polynomial for graphs was introduced by Birkhoff (1912–13) as a tool for attacking the persistent Four Color Map problem. This problem remained unsolved for another sixty years after Birkhoff's paper, and the eventual resolution of the problem did not use chromatic polynomials. Appel & Haken (1976) were able to reduce the Four Color problem to one that involved checking a very large, but finite, number of cases. Using a computer to do the case checking, they were then able to prove the following result, which is known as the Four Color theorem.

6.3.2. Theorem. *Let Γ be a loopless planar graph. Then Γ has a proper 4-coloring.*

Accounts of the history of the Four Color problem and of the methods used to prove the last theorem can be found in Biggs, Lloyd & Wilson (1976) and in Woodall & Wilson (1978). In terms of chromatic polynomials, the assertion of the theorem is that, for a loopless planar graph Γ, $\chi_\Gamma(4) > 0$.

In general, if Γ is a planar graph, we can construct a geometric dual Γ^* of it (see Chapter 6 of White, 1986). Consider now the number of proper λ-colorings of Γ^*. In section 6.3.B, we shall show how to interpret this number in the graph Γ.

6.3.B. Flows

Let θ be some fixed orientation of an arbitrary graph Γ and let Γ_θ denote the associated directed graph. Let H be an additively written Abelian group with identity element 0 and order $|H|$. An *H-flow* on Γ_θ is an assignment of weights from H to the directed edges of Γ_θ so that, at each vertex v of Γ_θ, the sum in H of the weights of the edges directed into v equals the sum of the weights of the edges directed out from v. If none of the edges receives zero weight, the H-flow is called *nowhere zero*. We denote by $\chi^*_{\Gamma_\theta}(H)$ the number of nowhere-zero H-flows on Γ_θ.

6.3.3. Lemma. $\chi^*_{\Gamma_\theta}(H)$ *does not depend on the orientation θ of Γ.*

Proof. Suppose that the orientation θ' is obtained from θ by reversing the direction of a single edge e. Then, by replacing the weight of e by its additive inverse in H, we determine a bijection between the sets of nowhere-zero H-flows on Γ_θ and $\Gamma_{\theta'}$. Thus $\chi^*_{\Gamma_\theta}(H) = \chi^*_{\Gamma_{\theta'}}(H)$ and an obvious extension of this yields the required result.   □

In view of the preceding lemma, we shall abbreviate $\chi^*_{\Gamma_\theta}(H)$ as simply $\chi^*_\Gamma(H)$. The next result should be compared with Proposition 6.3.1.

6.3.4. Proposition. *For a graph Γ having $k(\Gamma)$ connected components,*

$$\chi_\Gamma^*(H) = p(M_\Gamma^*; |H|) = (-1)^{|E(\Gamma)| - |V(\Gamma)| + k(\Gamma)} t(M_\Gamma; 0, 1 - |H|).$$

Proof. Let $g(M_\Gamma; H) = \chi_\Gamma^*(H)$. We shall prove that g is well defined and that it is a generalized T–G invariant for which $\sigma = -1$ and $\tau = 1$. Evidently $g(I; H) = 0$ and $g(L; H) = |H| - 1$. So g is well defined if $|E(\Gamma)| = 1$. Assume it is well defined for $|E(\Gamma)| < n$ and let $|E(\Gamma)| = n$.

Now suppose that e is an edge of Γ having endpoints u and v. Assume that e is not a loop or an isthmus of Γ. Partition the set W of nowhere-zero H-flows on Γ/e into subsets W' and W'' where W' consists of those members of W that are also nowhere-zero H-flows on $\Gamma - e$. Thus $|W'| = \chi_{\Gamma - e}^*(H)$. Moreover, W'' is in one-to-one correspondence with the set of nowhere-zero H-flows on Γ. To see this, we note that a member of W'' fails as a nowhere-zero H-flow on $\Gamma - e$ precisely because, at each of the vertices u and v in $\Gamma - e$, the sum of the weights of the edges directed into the vertex does not equal the sum of the weights of the edges directed out. We may assume, without loss of generality, that the resultant flow into u is n. Then the resultant flow out of v is also n and, by directing e from u to v and assigning it the weight n, we obtain a nowhere-zero H-flow on Γ. Since every nowhere-zero H-flow on Γ is uniquely obtainable in this way, it follows that $|W''| = \chi_\Gamma^*(H)$ and so $\chi_\Gamma^*(H) = \chi_{\Gamma/e}^*(H) - \chi_{\Gamma - e}^*(H)$. Thus, by the induction assumption,

$$\chi_\Gamma^*(H) = g(M_{\Gamma/e}; H) - g(M_{\Gamma - e}; H).$$

If e is a loop in Γ, then, corresponding to every nowhere-zero H-flow on $\Gamma - e$, we may take the weight of e to be any one of the non-zero elements of H. This gives a nowhere-zero H-flow on Γ, and every such flow on Γ arises in this way. Thus if e is a loop in M_Γ, then $\chi_\Gamma^*(H) = g(L; H)\chi_{\Gamma - e}^*(H)$ and so, by the induction assumption,

$$\chi_\Gamma^*(H) = g(L; H)g(M_{\Gamma - e}; H).$$

If e is an isthmus of Γ, then $\chi_\Gamma^*(H) = 0 = g(I; H)$ and so

$$\chi_\Gamma^*(H) = g(I; H)\chi_{\Gamma - e}^*(H).$$

On comparing the equations for $\chi_\Gamma^*(H)$ when e is a loop, an isthmus, and neither a loop nor an isthmus, we conclude that g is well defined. The same equations imply that g is a generalized T–G invariant for which $\sigma = -1$ and $\tau = 1$. The proposition now follows immediately from Corollary 6.2.6 and 6.20. $\qquad\square$

A consequence of the last result is that $\chi_\Gamma^*(H)$ does not depend on the particular Abelian group H but only on its order. Thus, if $|H| = n$, we shall denote $\chi_\Gamma^*(H)$ by $\chi_\Gamma^*(n)$. In particular, $\chi_\Gamma^*(\mathbb{Z}_n) = \chi_\Gamma^*(n)$. The last result implies that $\chi_\Gamma^*(\lambda)$ is a polynomial in λ. We call this the *flow polynomial* of Γ.

Let Γ^* be a geometric dual of the planar graph Γ. On combining Propositions 6.3.1 and 6.3.4, we deduce the following result.

6.3.5. Corollary. *The number of proper λ-colorings of Γ equals the product of $\lambda^{k(\Gamma)}$ and the number of nowhere-zero $\mathbb{Z}_\lambda$-flows on Γ^*.*

The next result comes from combining this corollary with the Four Color theorem (6.3.2).

6.3.6. Corollary. *Let Γ be a planar graph having no isthmuses. Then Γ has a nowhere-zero $\mathbb{Z}_4$-flow.*

An immediate consequence of Proposition 6.3.4 is that if H is an Abelian group of order k, then Γ has a nowhere-zero H-flow if and only if Γ has a nowhere-zero $\mathbb{Z}_k$-flow. The next result shows that the existence of the latter corresponds to the existence of a nowhere-zero $\mathbb{Z}$-flow for which the weights lie in $[-(k-1), k-1]$. This result holds in the more general context of flows in unimodular matroids (Tutte, 1965), but we prove it here only for graphs.

6.3.7. Proposition. *Let k be an integer exceeding one, Γ be a graph, and θ be a fixed orientation of Γ. Then the following statements are equivalent.*
   (i) *Γ_θ has a nowhere-zero $\mathbb{Z}_k$-flow.*
   (ii) *Γ_θ has a nowhere-zero $\mathbb{Z}$-flow with weights in $[-(k-1), k-1]$.*

Proof. In this proof, the symbols $0, 1, 2, ..., k-1$ will be used to denote integers *and* to denote the members of $\mathbb{Z}_k$, and it will be convenient to switch between these. Suppose Γ_θ has a nowhere-zero $\mathbb{Z}$-flow with weights in $[-(k-1), k-1]$. Then, by regarding these weights as elements of $\mathbb{Z}_k$, we obtain a nowhere-zero $\mathbb{Z}_k$-flow in Γ_θ. Thus (ii) implies (i).

Now suppose that ψ is a nowhere-zero $\mathbb{Z}_k$-flow on Γ_θ. Define a function ϕ on $E(\Gamma_\theta)$ by, for each e in $E(\Gamma_\theta)$, taking $\phi(e)$ to be an integer in $[-(k-1), k-1]$ such that, regarded as a member of $\mathbb{Z}_k$, $\phi(e)$ equals $\psi(e)$. This does not uniquely determine ϕ since there are two choices for $\phi(e)$ for each edge e. We call e positive if $\phi(e)$ is positive, and negative otherwise. Evidently $\phi(e)$ is non-zero.

Now, for each vertex v of Γ, define the weight $w(v)$ of v by

$$w(v) = \sum_{e \in N^+(v)} \phi(e) - \sum_{e \in N^-(v)} \phi(e)$$

where $N^+(v)$ is the set of edges directed into v, and $N^-(v)$ is the set of edges directed out from v. We call v positive, negative, or zero according to whether its weight is positive, negative, or zero. Clearly $w(v) \equiv 0 \pmod{k}$ for all vertices

v. We choose the function ϕ so that $\sum\limits_{v \in V(\Gamma)} |w(v)|$ is a minimum.

A path P in Γ from a vertex u to a vertex v will be called positive provided that the positive edges of P are precisely those whose orientation in Γ_θ agrees with the direction of traversal of P. We show next that there is no positive path in Γ_θ with a negative initial vertex and a positive final vertex. If there is such a path, P', then let ϕ' be defined as follows for all edges e of Γ:

$$\phi'(e) = \begin{cases} \phi(e) - k & \text{if } e \text{ is a positive edge of } P', \\ \phi(e) + k & \text{if } e \text{ is a negative edge of } P', \\ \phi(e) & \text{otherwise.} \end{cases}$$

Evidently, regarded as members of $\mathbb{Z}_k$, $\phi'(e)$ and $\phi(e)$ are equal for all edges e. Moreover, using the fact that $w(v) \equiv 0 \pmod{k}$ for all vertices v, it is easy to show that if w' is the weight function associated with ϕ', then

$$\sum\limits_{v \in V} |w'(v)| < \sum\limits_{v \in V} |w(v)|.$$

This contradiction to the choice of ϕ establishes that the path P' does not exist.

Let V_P and V_N be the sets of positive and negative vertices, respectively, of $V(\Gamma)$. Let V_1 be V_N together with all the vertices u for which there is a positive path from a member of V_N to u. Then $V - V_1 \supseteq V_P$ and

$$\sum\limits_{v \in V - V_1} w(v) = \sum\limits_{v \in V_P} w(v). \tag{6.23}$$

But

$$\sum\limits_{v \in V - V_1} w(v) = \sum\limits_{v \in V - V_1} \left(\sum\limits_{e \in N^+(v)} \phi(e) - \sum\limits_{e \in N^-(v)} \phi(e) \right). \tag{6.24}$$

If both endpoints of e are in $V - V_1$, the net contribution of e to the right-hand side of (6.24) is zero. Moreover, by the definition of V_1, no positive edge has its tail in V_1 and its head in $V - V_1$, and no negative edge has its tail in $V - V_1$ and its head in V_1. Thus the right-hand side of (6.24) is non-positive. Hence, by (6.23), $\sum\limits_{v \in V_P} w(v) \leqslant 0$. But, as every vertex in V_P is positive, $\sum\limits_{v \in V_P} w(v) \geqslant 0$ with equality only if $V_P = \varnothing$. We conclude that Γ has no positive vertices. A similar argument shows that Γ has no negative vertices. Thus ϕ is a $\mathbb{Z}$-flow on Γ having all its weights in $[-(k-1), k-1]$ and Proposition 6.3.7 is proved. $\square$

If Γ is a graph, a $\mathbb{Z}$-flow with weights in $[-(k-1), k-1]$ is called a k-flow.

6.3.C. Tutte's 5-flow Conjecture

In sections 6.3.A and 6.3.B, we noted that certain fundamental T–G invariants occur in the study of both colorings and flows in graphs. In each of these areas, the focal point of much of the research has been one very difficult problem. For colorings this problem was, for many years, the Four Color problem. Now that this has been solved, attention has turned to the more general conjecture of Hadwiger, which we shall discuss later in the chapter (section 6.4). For flows, the outstanding problem has been to prove or disprove the following conjecture of Tutte (1954).

6.3.8. Conjecture. *Every graph without isthmuses has a nowhere-zero 5-flow.*

Appearing with this conjecture was the following weakening of it.

6.3.9. Conjecture. *There is some integer k such that every graph without isthmuses has a nowhere-zero k-flow.*

Both conjectures remained unresolved for over twenty years until Jaeger (1976b) proved that every graph without isthmuses has a nowhere-zero 8-flow. This result was sharpened by Seymour (1981) and, in this section, we shall prove his result, which is still the best partial result toward Conjecture 6.3.8.

6.3.10. Theorem. *Every graph without isthmuses has a nowhere-zero 6-flow.*

As an example of a graph having no nowhere-zero 4-flow, Tutte (1954) cited the Petersen graph P_{10} (see Figure 6.2). The reader can check this by showing that the flow polynomial of P_{10} is

$$\chi^*_{P_{10}}(\lambda) = (\lambda - 1)(\lambda - 2)(\lambda - 3)(\lambda - 4)(\lambda^2 - 5\lambda + 10).$$

In 1966, Tutte (1966a) advanced a variant of 6.3.8, namely that P_{10} is the

Figure 6.2.

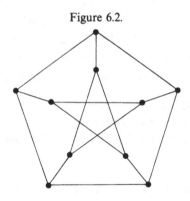

unique minimal obstruction to the existence of a nowhere-zero 4-flow. More precisely, he proposed the following.

6.3.11. Conjecture. *If a graph without isthmuses has no nowhere-zero 4-flow, then it has a subgraph contractible to P_{10}.*

Let Γ be a graph without isthmuses. To prove Theorem 6.3.10, we shall show that Γ has a nowhere-zero $(\mathbb{Z}_3 \times \mathbb{Z}_2)$-flow. We begin with the following simple observation, the proof of which is left to the reader.

6.3.12. Lemma. *The following statements are equivalent for a graph Γ.*

(i) Γ *has a nowhere-zero 2-flow.*
(ii) *Every vertex of Γ has even degree.*
(iii) $E(\Gamma)$ *is a disjoint union of circuits.*

We show next that if a minimal counterexample to Theorem 6.3.10 exists, then it is simple and 3-connected.

6.3.13. Lemma. *Suppose that k is an integer exceeding 2. Let Γ be a graph that, among all graphs Δ with no isthmuses and no nowhere-zero k-flows, has $|V(\Delta)| + |E(\Delta)|$ minimum. Then Γ is simple and 3-connected.*

Proof. Evidently Γ is loopless and 2-connected. Moreover, if $\{e_1, e_2\}$ is a circuit of Γ, then $E(\Gamma) \neq \{e_1, e_2\}$, so $\Gamma - e_1$ has no isthmuses. The choice of Γ now implies that $\Gamma - e_1$ has a nowhere-zero k-flow, and, since $k \geqslant 3$, this k-flow can easily be modified to give a nowhere-zero k-flow on Γ. We conclude that Γ is simple.

Now suppose that Γ is not 3-connected. Then, as Γ has at least four vertices, it follows by Lemma 6.3.3 of White (1986) that Γ has a representation as a generalized circuit, each part of which is a block. This means that, for some $m \geqslant 2$, Γ has subgraphs $\Gamma_1, \Gamma_2, .., \Gamma_m$ so that the following conditions hold:

(1) Each Γ_i is connected, loopless, has no cut vertices, and has a non-empty edge set; and, if $m = 2$, both Γ_1 and Γ_2 have at least three vertices.
(2) The edge sets of $\Gamma_1, \Gamma_2, ..., \Gamma_m$ partition the edge set of Γ, and each Γ_i shares exactly two vertices, its *contact vertices*, with $\bigcup_{j \neq i} \Gamma_j$.
(3) If each Γ_i is replaced by an edge joining its contact vertices, the resulting graph is a circuit.

If none of $\Gamma_1, \Gamma_2, ..., \Gamma_m$ consists of a single edge, then each has no isthmuses and therefore, by the choice of Γ, each has a nowhere-zero k-flow. It follows that Γ has such a flow; a contradiction. Thus, we may suppose that Γ_1 consists of a single edge. Hence $m \geq 3$. Therefore Γ is the series connection of two graphs Δ_1 and Δ_2, neither of which has an isthmus. Thus $M(\Gamma) = S(M(\Delta_1), M(\Delta_2)) = [P(M^*(\Delta_1), M^*(\Delta_2))]^*$ (White, 1986, p. 180), so $M^*(\Gamma) = P(M^*(\Delta_1), M^*(\Delta_2))$. Now, by Theorem 7.2.9 of White (1987),

$$p(M^*(\Gamma); \lambda) = p(M^*(\Delta_1); \lambda)p(M^*(\Delta_2); \lambda),$$

that is,

$$\chi_\Gamma^*(\lambda) = \chi_{\Delta_1}^*(\lambda)\chi_{\Delta_2}^*(\lambda).$$

Since each of Δ_1 and Δ_2 has a nowhere-zero k-flow, so does Γ; a contradiction.
$\square$

The proof of Theorem 6.3.10 will use the following function defined on the set of subsets of $E(\Gamma)$. For $X \subseteq E(\Gamma)$, we take the *S-closure* $\langle X \rangle$ of X to be the smallest subset Y of $E(\Gamma)$ with the following properties:

(1) $X \subseteq Y$; and
(2) there is no circuit C of Γ such that $0 < |C - Y| \leq 2$.

Evidently if Y_1 and Y_2 both satisfy (1) and (2), then so does $Y_1 \cap Y_2$. Hence $\langle X \rangle$ is well defined. It is not difficult to check that $\langle X \rangle$ can be obtained constructively as follows. If C is a circuit of Γ with $0 < |C - X| \leq 2$, then let $X' = X \cup C$. Repeat this procedure with X' replacing X and continue in this manner until no further elements can be added. The resulting set is $\langle X \rangle$.

We leave it to the reader (Exercise 6.32) to check that S-closure is a closure operator, that is, $X \subseteq \langle X \rangle$; $\langle \langle X \rangle \rangle = \langle X \rangle$; and if $X_1 \subseteq X_2$, then $\langle X_1 \rangle \subseteq \langle X_2 \rangle$.

If f is a $\mathbb{Z}_n$-flow on Γ, then the *support* $S(f)$ of f is the set $\{e \in E(\Gamma): f(e) \neq 0\}$.

6.3.14. Lemma. *Let Γ be a graph and X be a subset of $E(\Gamma)$ such that $\langle X \rangle = E(\Gamma)$. Then there is a $\mathbb{Z}_3$-flow f on Γ with $E(\Gamma) - X \subseteq S(f)$.*

Proof. We argue by induction on $|E(\Gamma) - X|$. If this is zero, the result is immediate so assume that $|E(\Gamma) - X| > 0$. Then $X \neq \langle X \rangle$, so there is a circuit C with $0 < |C - X| \leq 2$. Certainly $\langle X \cup C \rangle = E(\Gamma)$ and so, by the induction assumption, there is a $\mathbb{Z}_3$-flow g on Γ such that $E(\Gamma) - (C \cup X) \subseteq S(g)$. Evidently there is a $\mathbb{Z}_3$-flow h on Γ so that $S(h) = C$. As $|C - X| \leq 2$, we can choose n from $\mathbb{Z}_3$ so that, for all e in $C - X$, $n \neq -g(e)/h(e)$. Let $f = g + nh$. Then, for e in $E(\Gamma) - (X \cup C)$, $f(e) = g(e) \neq 0$. Moreover, for e in $C - X$, $f(e) = g(e) + nh(e)$. This sum is non-zero by the choice of n, and we conclude that, for all e in $E(\Gamma) - X$, $f(e)$ is non-zero.
$\square$

By Lemma 6.3.13, we know that we may assume that Γ is simple and 3-connected. The next lemma focusses on such graphs. The proof of Theorem 6.3.10 will be obtained by combining this lemma with Lemma 6.3.14.

6.3.15. Lemma. *A simple 3-connected graph Γ has a collection $C_1, C_2, ..., C_m$ of disjoint circuits such that $\langle C_1 \cup C_2 \cup ... \cup C_m \rangle = E(\Gamma)$.*

To prove this lemma, we shall use the following technical but elementary result.

6.3.16. Lemma. *Let Δ be a non-null simple graph in which each vertex has degree at least two. Then Δ has a block Δ' with at least three vertices so that at most one vertex of Δ' is adjacent in Δ to some vertex of Δ not in Δ'.*

Proof. Let $bc(\Delta)$ be the graph having as its vertices the blocks and cut vertices of Δ; the edges of $bc(\Delta)$ join a cut vertex to a block if the block contains the cut vertex. Evidently $bc(\Delta)$ is a forest. Let v be a pendant vertex of $bc(\Delta)$. Then v corresponds to a block of Δ. Since Δ is simple and has no vertices of degree less than 2, this block has at least three vertices. We take this block to be Δ'.   □

Proof of Lemma 6.3.15. A subset X of $E(\Gamma)$ will be called *connected* if the subgraph of Γ consisting of X and all incident vertices is connected. Now certainly Γ has a circuit C. Moreover, as Γ is simple, $\langle C \rangle$ is connected. Thus we can choose a maximum positive integer m so that there are disjoint circuits $C_1, C_2, ..., C_m$ with $\langle C_1 \cup C_2 \cup ... C_m \rangle$ connected.

Let U be the set of vertices of Γ incident with $\langle C_1 \cup C_2 \cup ... \cup C_m \rangle$ and let Δ be the subgraph of Γ obtained by deleting U. If Δ is the null graph, then $U = V(\Gamma)$ and so, as $\langle C_1 \cup C_2 \cup ... \cup C_m \rangle$ is connected, $\langle\langle C_1 \cup C_2 \cup ... \cup C_m \rangle\rangle = E(\Gamma)$. Thus $\langle C_1 \cup C_2 \cup ... \cup C_m \rangle = E(\Gamma)$ and the lemma holds.

Now suppose that Δ is non-null. No vertex v of Δ is adjacent in Γ to two distinct vertices, say u_1 and u_2, of U; otherwise, since $\langle C_1 \cup C_2 \cup ... \cup C_m \rangle$ is connected, there would be a path joining u_1 and u_2 using only edges of $\langle C_1 \cup C_2 \cup ... \cup C_m \rangle$. This path together with the edges vu_1 and vu_2 forms a circuit contradicting the definition of $\langle C_1 \cup C_2 \cup ... \cup C_m \rangle$. Thus, as every vertex of the simple 3-connected graph Γ has degree at least 3, every vertex of Δ has degree at least 2. Therefore, by Lemma 6.3.16, Δ has a block Δ' with at least three vertices and with at most one vertex adjacent in Δ to some vertex of Δ not in Δ'. Since Γ is 3-connected and $|V(\Delta')| \geq 3$, there are at least 3 vertices of Δ' that are adjacent in Γ to vertices not in Δ'. Hence there are distinct vertices b_1 and b_2 of Δ' both of which are adjacent in Γ to vertices in U. As Δ' is a block with at least three vertices, it has a circuit, say C_{m+1}, using both b_1 and b_2. Let e_1 and e_2 be edges of Γ joining b_1 and b_2 respectively to U. Then $\{e_1, e_2\} \subseteq \langle C_1 \cup C_2 \cup ... \cup C_{m+1} \rangle$ and so $\langle C_1 \cup C_2 \cup ... C_{m+1} \rangle$ is connected. This contradicts the maximality of m and completes the proof of the lemma.   □

We are now ready to prove Theorem 6.3.10.

Proof of Theorem 6.3.10. Let Γ be a graph that, among all graphs Δ with no isthmuses and no nowhere-zero 6-flows, has $|V(\Delta)| + |E(\Delta)|$ minimum. By Lemma 6.3.13, we may assume that Γ is simple and 3-connected. Then, by Lemma 6.3.15, Γ has a set $\{C_1, C_2, ..., C_m\}$ of disjoint circuits so that $\langle C_1 \cup C_2 \cup ... \cup C_m \rangle = E(\Gamma)$. By Lemma 6.3.14, there is a $\mathbb{Z}_3$-flow f_1 on Γ with $E(\Gamma) - (C_1 \cup C_2 \cup ... \cup C_m) \subseteq S(f_1)$. By Lemma 6.3.12, there is a $\mathbb{Z}_2$-flow f_2 on Γ with $S(f_2) = C_1 \cup C_2 \cup ... \cup C_m$. Then the $(\mathbb{Z}_3 \times \mathbb{Z}_2)$-flow f defined, for all e in $E(\Gamma)$, by $f(e) = (f_1(e), f_2(e))$ is nowhere zero. We conclude, by Propositions 6.3.4 and 6.3.7, that Γ has a nowhere-zero 6-flow. $\qquad\square$

6.3.D. Orientations of Graphs

As we have seen, T–G invariants are important in the study of colorings and flows in graphs. Another area of graph theory in which there have been numerous applications of T–G techniques is in the consideration of certain special types of orientations of graphs. Several of these are considered below and some further examples are considered in section 6.3.G and the exercises. An *acyclic orientation* of a graph Γ is an orientation of Γ in which there are no directed cycles. Let $a(\Gamma)$ denote the number of such orientations of Γ.

6.3.17. Proposition. *(Stanley, 1973b)* $a(\Gamma) = (-1)^{|V(\Gamma)|} \chi_\Gamma(-1) = t(M_\Gamma; 2, 0)$.

Proof. Suppose that the edge e of Γ is neither a loop nor an isthmus and let u and v be the endpoints of e. Partition the set $\mathscr{A}$ of acyclic orientations of $\Gamma - e$ into subsets $\mathscr{A}'$ and $\mathscr{A}''$ where $\mathscr{A}'$ consists of those members θ of $\mathscr{A}$ for which $(\Gamma - e)_\theta$ contains a directed path from u to v or from v to u. It is straightforward to show that if $\theta \in \mathscr{A}'$, then $(\Gamma - e)_\theta$ cannot contain both a directed path from u to v and a directed path from v to u, as otherwise $(\Gamma - e)_\theta$ certainly contains a directed cycle. Therefore, for each orientation θ in $\mathscr{A}'$, there is precisely one orientation of e that will extend θ to an acyclic orientation of Γ. On the other hand, if $\theta \in \mathscr{A}''$, then each of the two orientations of e extends θ to an acyclic orientation of Γ. Since every acyclic orientation of Γ can be uniquely obtained from a member of $\mathscr{A}$ by assigning an orientation to e, it follows that

$$a(\Gamma) = |\mathscr{A}'| + 2|\mathscr{A}''|.$$

But

$$a(\Gamma - e) = |\mathscr{A}'| + |\mathscr{A}''|.$$

Moreover, it is easy to see that

$$a(\Gamma/e) = |\mathscr{A}''|.$$

We conclude that if e is neither a loop nor an isthmus of Γ, then

$$a(\Gamma) = a(\Gamma - e) + a(\Gamma/e).$$

Now suppose that e is an isthmus of Γ. Then $a(\Gamma) = 2a(\Gamma - e)$ unless Γ consists of a single edge, in which case $a(\Gamma) = 2$. Finally, if Γ has a loop, then Γ has no acyclic orientations, that is, $a(\Gamma) = 0$.

We may now apply Theorem 6.2.2 to obtain that

$$a(\Gamma) = t(M_\Gamma; 2, 0).$$

Thus, by Proposition 6.3.1,

$$a(\Gamma) = (-1)^{|V(\Gamma)|}\chi_\Gamma(-1),$$

as required.   □

The preceding proposition showed that the number of acyclic orientations of a graph is a T–G invariant. Certain proper subsets of the set of acyclic orientations can also be associated with T–G invariants. The next result gives one such example and another example is given in Exercise 6.35. A vertex v in a directed graph is a *source* if no edge is directed toward v, and a *sink* if no edge is directed away from v. We shall denote by $N_v(\Gamma)$ the number of acyclic orientations of Γ in which v is the unique source.

6.3.18. Proposition. (*Greene & Zaslavsky, 1983*) $N_v(\Gamma)$ *is* $(-1)^{r(M_\Gamma)}\mu(M_\Gamma)$ *if* Γ *is connected, and is 0 otherwise. Thus* $N_v(\Gamma)$ *does not depend on the choice of the vertex* v.

Proof. The proof of this result differs slightly from the usual pattern in that, instead of establishing the deletion–contraction formula for an arbitrary edge e, we show it only for certain special choices of e. In particular, we assume that the edge e has v as an endpoint. Let v' be the other endpoint of e. If Γ has e as its only edge, it is clear that

$$N_v(\Gamma) = \begin{cases} 0, & \text{if } e \text{ is a loop,} \\ 1, & \text{if } e \text{ is an isthmus.} \end{cases} \tag{6.25}$$

Now suppose that Γ has at least two edges and that e is still a loop or an isthmus. Then

$$N_v(\Gamma) = \begin{cases} 0, & \text{if } e \text{ is a loop,} \\ N_v(\Gamma/e), & \text{if } e \text{ is an isthmus.} \end{cases} \tag{6.26}$$

Next assume that e is neither a loop nor an isthmus. Then we can partition the set $\mathscr{S}$ of acyclic orientations of Γ in which v is the unique source into subsets $\mathscr{S}'$ and $\mathscr{S}''$, where $\theta \in \mathscr{S}'$ provided that the only edge of Γ_θ directed into v' is e. Evidently

$$|\mathscr{S}'| = N_v(\Gamma/e) \quad \text{and} \quad |\mathscr{S}''| = N_v(\Gamma - e),$$

hence

$$N_v(\Gamma) = N_v(\Gamma - e) + N_v(\Gamma/e). \tag{6.27}$$

Now although (6.27) has not been established for an arbitrary edge of Γ, it is clear that by repeated application of (6.25), (6.26), and (6.27), one can determine the value of $N_v(\Gamma)$ for any graph Γ having a distinguished vertex v. Moreover, as (6.26) and (6.27) are Tutte–Grothendieck recursions, a straightforward induction argument establishes that $N_v(\Gamma) = t(M_\Gamma; 1, 0)$. It follows, by (6.21), that $N_v(\Gamma) = (-1)^{r(M_\Gamma)}\mu(M_\Gamma)$, as required.   □

Our last result for oriented graphs concerns score vectors. If the graph Γ has vertex set $\{v_1, v_2, ..., v_n\}$ and θ is an orientation of Γ, then the *score vector* of Γ_θ is the ordered n-tuple $(s_1, s_2, ..., s_n)$ where s_i, the *score* of v_i, is the number of edges of Γ_θ that are directed away from v_i. We shall denote the number of distinct score vectors of Γ by $s(\Gamma)$.

6.3.19. Proposition. (*Stanley, 1980*) $s(\Gamma) = t(M_\Gamma; 2, 1) = i(M_\Gamma)$.

The proof of this proposition will use the following result.

6.3.20. Lemma. *Let e be an edge of Γ joining v_1 and v_2. Suppose that $(s_2, s_2, s_3, s_4, ..., s_n)$ and $(s'_1, s'_2, s_3, s_4, ..., s_n)$ are score vectors of Γ with $s_2 < s'_2$. Then $(s_1 - 1, s_2 + 1, s_3, s_4, ..., s_n)$ is a score vector for Γ.*

Proof. Let θ and θ' be orientations of Γ having $(s_1, s_2, s_3, s_4, ..., s_n)$ and $(s'_1, s'_2, s_3, s_4, ..., s_n)$, respectively, as their score vectors. If, in Γ_θ, the edge e is directed from v_1 to v_2, then reversing the orientation of e gives an orientation of Γ having $(s_1 - 1, s_2 + 1, s_3, s_4, ..., s_n)$ as its score vector. Therefore we may assume that e is directed from v_2 to v_1 in Γ_θ. Now consider the set V_1 of vertices v such that there is a directed path in Γ_θ from v_1 to v. We distinguish two cases:

$$(1) \quad v_2 \in V_1, \quad \text{and} \quad (2) \quad v_2 \notin V_1.$$

In case (1), on taking a directed path from v_1 to v_2 and adding the edge e, we obtain a directed cycle containing the edge e. Reversing the directions of all the edges in this cycle except e gives an orientation having $(s_1 - 1, s_2 + 1, s_3, s_4, ..., s_n)$ as its score vector.

In case (2), $v_2 \in V(\Gamma) - V_1$. Now, by definition, every edge in Γ_θ joining a vertex in V_1 to a vertex in $V(\Gamma) - V_1$ must be directed from the vertex in $V(\Gamma) - V_1$ to the vertex in V_1. Therefore, for any orientation of Γ and, in particular, for $\Gamma_{\theta'}$, the sum of the scores of the vertices in $V(\Gamma) - V_1$ cannot exceed the sum of the scores of these vertices in Γ_θ. But since $s'_2 > s_2$, this is a contradiction and the proof of the lemma is complete.   □

Proof of Proposition 6.3.19. The equality of $t(M_\Gamma; 2, 1)$ and $i(M_\Gamma)$ follows immediately from Proposition 6.2.11. We now show that $t(M_\Gamma; 2, 1) = s(\Gamma)$. Let e be an edge of Γ which is neither a loop nor an isthmus and assume that e joins the vertices v_1 and v_2. We shall show that

$$s(\Gamma) = s(\Gamma - e) + s(\Gamma/e). \tag{6.28}$$

Suppose that $(s'_1, s'_2, s_3, s_4, ..., s_n)$ is a score vector for $\Gamma - e$ and that $\mathcal{S}$ is the set of score vectors of $\Gamma - e$ having $(s_3, s_4, ..., s_n)$ as the last $n - 2$ entries. Then, by Lemma 6.3.20, there are integers s_1, s_2, and k such that

$$\mathcal{S} = \{(s_1 - j, s_2 + j, s_3, s_4, ..., s_n): 0 \leqslant j \leqslant k\}.$$

Now, given an orientation of $\Gamma - e$, we can orient the edge e in two different ways to obtain orientations of Γ. Thus if $\mathcal{S}'$ is the set of score vectors for Γ having $(s_3, s_4, ..., s_n)$ as the last $n - 2$ entries, then

$$\mathcal{S}' = \{(s_1 + 1 - j, s_2 + j, s_3, s_4, ..., s_n): 0 \leqslant j \leqslant k + 1\}.$$

Hence $|\mathcal{S}'| = |\mathcal{S}| + 1$. Since the only score vector for Γ/e having $(s_3, s_4, ..., s_n)$ as the last $n - 2$ entries is $(s_1 + s_2, s_3, s_4, ..., s_n)$, we conclude that (6.28) holds.

To complete the proof, it only remains to notice that if Γ has a loop, then $s(\Gamma) = 0$, while if e is an isthmus of Γ, then $s(\Gamma) = 2s(\Gamma - e)$. Since the value of s on an isthmus is 2, the proposition follows immediately on applying Theorem 6.2.2. □

6.3.E. Reliability and Percolation

Classical percolation theory was introduced by Broadbent & Hammersley (1957) to model the flow of liquid through a random medium. As such, the classical theory is a branch of random graph theory. Another closely related branch of random graph theory is the study of the reliability of a network. Here one is interested in determining the probability that, in a random subgraph of the network, two distinguished vertices are joined by a path. In this section we show how the Tutte polynomial is useful in the study of the matroid generalizations of these graph problems.

In matroid reliability and percolation problems, every element e_i of a matroid $M(E)$ has, independently of all other elements, a probability $1 - p_i$ of being deleted from M where, except when otherwise stated, $0 < p_i < 1$. Then, writing q_i for $1 - p_i$, the probability $\Pr(A)$ that a subset A of E consists of precisely those elements that are retained is given by

$$\Pr(A) = \prod_{e_i \in A} p_i \prod_{e_i \notin A} q_i. \tag{6.29}$$

The standard problem in this area is to find ways to efficiently compute the probability $\Pr(\mathscr{F})$ that the set of retained elements is in some family $\mathscr{F}$. Evidently $\Pr(\mathscr{F}) = \sum_{A \in \mathscr{F}} \Pr(A)$. Usually the family $\mathscr{F}$ is ascending, that is, if $A \in \mathscr{F}$ and $B \supseteq A$, then $B \in \mathscr{F}$. For example, when $\mathscr{F}$ is the family $\mathscr{S}$ of spanning sets of M and all the retention probabilities p_i equal some constant p, then $\Pr(\mathscr{F})$ is the probability $\Pr(M)$ that a random submatroid of $M(E)$ has the same rank as M. We saw in Example 6.2.7 that this probability is

$$q^{|E| - r(M)} p^{r(M)} t(M; 1, 1/q). \tag{6.30}$$

Toward the end of this section, we shall present a procedure for modifying the matroid M so as to adapt the last formula to the case when the retention probabilities p_i are different.

Next we consider how to put the problem of computing network reliability into this framework. Given two distinguished vertices s and t in a graph Γ, we are interested in determining the probability that a random subgraph of Γ contains a path between s and t. To do this, we first form a new graph $\bar{\Gamma}$ from Γ by adjoining a basepoint edge d between s and t. Let $\mathscr{D}$ be the family of subsets A of $E(\Gamma)$ for which $A \cup d$ contains a cycle of $\bar{\Gamma}$ containing d. Equivalently, $A \in \mathscr{D}$ if and only if d is not an isthmus in the subgraph of $\bar{\Gamma}$ induced by $A \cup d$. We shall develop a formula for $\Pr(\mathscr{D})$ for an arbitrary pointed matroid $M_d(E \cup d)$, where, in this more general context, $A \in \mathscr{D}$ if and only if d is not an isthmus of $M_d(A \cup d)$. Our formula for $\Pr(\mathscr{D})$ will involve the pointed Tutte polynomial $t_P(M_d)$ that was introduced in Proposition 6.2.19. We shall first determine $\Pr(\mathscr{D})$ in the case when, for all elements e_i of E, the retention probability p_i equals a constant p. The number of elements in a matroid M will be denoted by $|M|$.

6.3.21. Proposition. *Let the matroid M be $M_d(E \cup d)$ and assume that every element of E has, independently of all other elements, probability $1 - p$ of being deleted from M, while the element d has probability 0 of being deleted. Then the probability that, in a random submatroid $\omega(M)$ of M, the element d is not an isthmus is given by the formulas*

(i) $\Pr(\mathscr{D}) = p^{r(M)} q^{|M| - r(M) - 1} g(1/p, 1/q)$ *and*
(ii) $\Pr(\mathscr{D}) = 1 - p^{r(M) - 1} q^{|M| - r(M)} f(1/p, 1/q)$

where $x'f(x, y) + y'g(x, y) = t_P(M_d(E \cup d); x', x, y', y)$.

Proof. We first show that $\Pr(\mathscr{D})$ obeys the weighted recursion

$$\Pr(\mathscr{D}(M)) = q \Pr(\mathscr{D}(M - e)) + p \Pr(\mathscr{D}(M/e)) \tag{6.31}$$

where e is a point of $M - d$ that is not a loop or an isthmus. By (6.29), we have

$$\Pr(\mathcal{D}(M)) = \sum_{A \in \mathcal{D}} p^{|A|} q^{|E - A|}$$

$$= \sum_{\substack{A \in \mathcal{D} \\ e \notin A}} p^{|A|} q^{|E - A|} + \sum_{\substack{A \in \mathcal{D} \\ e \in A}} p^{|A|} q^{|E - A|}.$$

Thus

$$\Pr(\mathcal{D}(M)) = q \sum_{\substack{A \in \mathcal{D} \\ e \notin A}} p^{|A|} q^{|(E - e) - A|} + p \sum_{\substack{A' \subseteq E - e \\ A' \cup e \in \mathcal{D}}} p^{|A'|} q^{|(E - e) - A'|}. \tag{6.32}$$

The first summation in (6.32) is clearly over those subsets A of $E - e$ for which d is not an isthmus of $A \cup d$ in $M - e$. Thus this summation is over those members A of $\mathcal{D}(M - e)$. On the other hand, since d is an isthmus of $M_d(A' \cup d \cup e)$ if and only if it is an isthmus of $M_d(A' \cup d \cup e)/e$, the second summation in (6.32) is over those members A' of $\mathcal{D}(M/e)$. Thus (6.31) holds. It follows that if

$$h(M) = (1/q)^{|M| - r(M)} (1/p)^{r(M)} \Pr(\mathcal{D}(M)),$$

then

$$h(M) = h(M - e) + h(M/e) \tag{6.33}$$

for all elements e of $M - d$ that are not loops or isthmuses of M. Moreover, it is routine to check that if e is an element of $M - d$, then

$$h(M) = \begin{cases} (1/q)h(M - e) & \text{if } e \text{ is a loop,} \\ (1/p)h(M - e) & \text{if } e \text{ is an isthmus.} \end{cases} \tag{6.34}$$

Finally, one easily checks that

$$h(M(d)) = \begin{cases} 1/q, & \text{if } M(d) \text{ is a loop,} \\ 0, & \text{if } M(d) \text{ is an isthmus.} \end{cases} \tag{6.35}$$

We conclude, by Proposition 6.2.19, that $h(M_d) = t_P(M_d; x', x, y', y)$ where $x' = 0$, $y' = 1/q$, $x = 1/p$, and $y = 1/q$. Therefore, as $t_P(M_d; x', x, y', y) = x'f(x, y) + y'g(x, y)$, $h(M_d) = (1/q)g(1/p, 1/q)$ and so $\Pr(\mathcal{D}(M)) = q^{|M| - r(M) - 1} p^{r(M)} g(1/p, 1/q)$. This establishes (i). A similar argument applied to $\Pr(2^E - \mathcal{D})$ gives (ii). $\qquad \square$

With M still equal to $M_d(E \cup d)$, we note that, by (6.29),

$$\Pr(\mathcal{D}(M)) = \sum_{i=0}^{|E|} a_i p^i q^{|E| - i}. \tag{6.36}$$

Thus $\Pr(\mathcal{D}(M))$ is a polynomial in p and q of constant total degree. The coefficient a_i here equals the number of i-element subsets A of E for which d is not an isthmus of $M_d(A \cup d)$. Thus, provided d is not an isthmus or a loop of M, $a_i \geqslant 0$ for all i and $a_{|E|} = 1$. Now suppose that M_d is the polygon matroid of the graph Γ and let the basepoint edge d join the distinguished vertices s and t. Let k be the least j for which a_j is non-zero. Then k equals the length of the shortest (s, t)-path in $\Gamma - d$, and the number of such shortest paths is a_k.

On substituting $1 - p$ for q in (6.36), we obtain $\Pr(\mathscr{D})$ as a polynomial in p alone. We leave as an exercise the problem of determining the coefficients of this polynomial (Exercise 6.37).

Before attacking the reliability problem in the case of unequal retention probabilities, we note a remarkable fact about the evaluation $t(M; 1/p, 1/q)$. Recall that $t_P(M_d; x', x, y', y) = x'f(x, y) + y'g(x, y)$. If M_d is viewed as simply a matroid M rather than as a pointed matroid, then d is no longer distinguished, so $x' = x, y' = y$, and $t(M; x, y) = xf(x, y) + yg(x, y)$. In Proposition 6.2.20(iv), it was noted that, if d is neither a loop nor an isthmus of M, then

$$\begin{cases} t(M - d) = (x - 1)f(x, y) + g(x, y) & \text{and} \\ t(M/d) = f(x, y) + (y - 1)g(x, y). \end{cases} \tag{6.37}$$

Clearly these formulas can be inverted in the Tutte–Grothendieck ring to give expressions for $f(x, y)$ and $g(x, y)$ in terms of $t(M - d)$ and $t(M/d)$. However, the determinant of this system equals $xy - x - y$, which is zero when $(1/x) + (1/y) = 1$. Since $p + q = 1$, $\Pr(\mathscr{D})$ is computable from the evaluations of $t(M - d)$ and $t(M/d)$ at $x = 1/p$ and $y = 1/q$ only in the most formal sense, that is, when the identity $p + q = 1$ is never invoked. On the other hand, we note that, by (6.37), $t(M/d; 1/p, 1/q) = (p/q)t(M - d; 1/p, 1/q)$. Therefore, for a given matroid M', $t(M''; 1/p, 1/q)$ is the same for any strong map image M'' of M'. To see this, we note that if M'' and M' are so related, then, for some matroid M and element d which is neither a loop nor an isthmus, $M - d = M'$ and $M/d = M''$.

The above remarks are summarized in the following proposition, the first part of which generalizes the following identity, a trivial consequence of 6.2.11(iv):

$$t(M - d; 2, 2) = t(M/d; 2, 2) = 2^{|M| - 1}.$$

6.3.22. Proposition. *Let $q(M) = t(M; 1/p, 1/q)$ where $p + q = 1$ and let d be an element of M that is neither a loop nor an isthmus. Then*

(i) $q(M/d) = (p/q)q(M - d) = f(M; 1/p, 1/q) + (p/q)g(M; 1/p, 1/q)$;
(ii) *Both $q(M - d)$ and $q(M/d)$ are independent of the modular cut of $M - d$ determined by d in M.*

These ideas are illustrated in the following.

6.3.23. Example. Let M be the polygon matroid of the graph Γ in Figure 6.3. Then, it is straightforward to check that

$$t_P(M_d; x, y) = x'(x + y + 1) + y'(y + 1). \tag{6.38}$$

Figure 6.3.

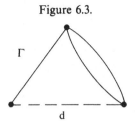

Thus, by 6.3.21(i),

$$\Pr(\mathscr{D}(M)) = p^2 q\left(\left(\frac{1}{q}\right) + 1\right)$$
$$= 2p^2 q + p^3$$
$$= 2p^2 - p^3.$$

Now, by (6.37) and (6.38),

$$t(M - d) = x^2 + xy$$

and

$$t(M/d) = y^2 + y + x.$$

Thus

$$q(M/d) = \frac{1}{q^2} + \frac{1}{q} + \frac{1}{p}$$
$$= \frac{1}{pq^2}$$
$$= \frac{p}{q}\left(\frac{1}{p^2} + \frac{1}{pq}\right)$$
$$= \frac{p}{q} q(M - d).$$

By 6.3.21(i), if M_1 and M_2 are the polygon matroids of the graphs Γ_1 and Γ_2 shown in Figure 6.4, then $q(M/d) = q(M_1/d_1) = q(M_2/d_2)$. However, $\Pr(\mathscr{D}(M_1)) = p$ while $\Pr(\mathscr{D}(M_2)) = 2p - p^2$.

Proposition 6.3.21 gives two formulas for $\Pr(\mathscr{D})$ in the case when all the retention probabilities are equal. We now turn to the general problem of determining $\Pr(\mathscr{D})$ when M_d is the polygon matroid of a graph $\bar{\Gamma}$ and the retention probabilities p_i can vary from edge to edge. In particular, we shall describe how, if p_i is equal to a k-place binary decimal, we can replace the corresponding edge e_i by an appropriate series–parallel network in which each edge has retention probability equal to $1/2$. Since $\Pr(\mathscr{D})$ will be unaffected by the presence of loops, we assume that $\bar{\Gamma}$ is loopless.

Figure 6.4.

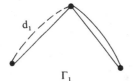
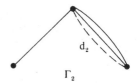

Let $0.d_{i1}d_{i2} \ldots d_{ik}$ be the binary decimal for p_i where $d_{ik} = 1$. We form the $(k + 1)$-edge series–parallel network N_i as follows. Start with the graph consisting of the edge e_i and its endpoints. Then, since $d_{ik} = 1$, add an edge e_{ik} in parallel with e_i. Assuming the edges $e_{ik}, e_{i(k-1)}, \ldots, e_{i(j+1)}$ have been added, add e_{ij} in parallel with e_i if $d_{ij} = 1$, and in series with e_i otherwise. After the edge e_{i1} has been added, form the 2-sum of this network N_i and $\bar{\Gamma}$ along the basepoint e_i. As an example, if $p_i = 0.0111001$, then N_i is as shown in Figure 6.5.

We repeat the above procedure for every edge of $\bar{\Gamma}$ other than d to obtain a new graph $\tilde{\Gamma}$ with polygon matroid $\tilde{M}_d$. Clearly $\tilde{\Gamma}$ retains the distinguished edge d. In $\tilde{\Gamma}$, we assign to each edge e_{ij} the retention probability $1/2$. Then it is straightforward to check that

$$\Pr(\mathcal{D}(M_d)) = \Pr(\mathcal{D}(\tilde{M}_d)).$$

But, since $\tilde{M}_d$ has constant retention probability, we get, by 6.3.21(i), that

$$\Pr(\mathcal{D}(M_d)) = (\tfrac{1}{2})^{|\tilde{M}_d|-1}g(\tilde{M}_d; 2, 2). \tag{6.39}$$

The techniques just described when M_d is a polygon matroid can be equally well applied to find $\Pr(\mathcal{D}(M_d))$ for an arbitrary pointed matroid. This technique can also be used in other situations where the retention probabilities can vary. For instance, by replacing each element of an arbitrary matroid M by an appropriate series–parallel network and using (6.30), we can obtain a formula for the probability that a random submatroid of M has the same rank as M.

Figure 6.5.

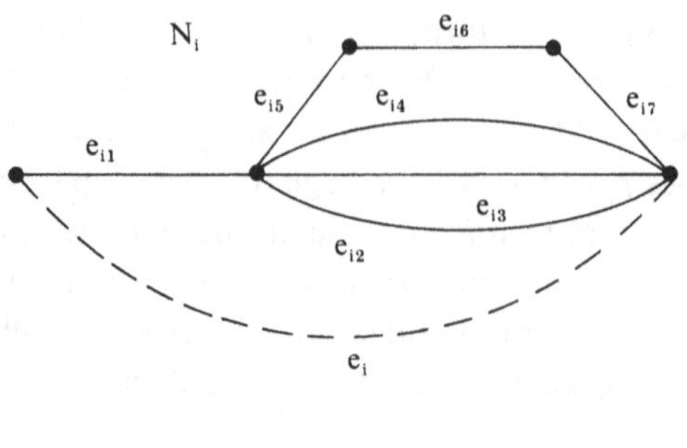

To conclude this section, we shall extend Proposition 6.3.19 on the number of score vectors that can arise when orienting a graph. The following lemma will be used in the proof of this extension.

6.3.24. Lemma. *Let k be a positive integer and M be a matroid. Let $M^{(k)}$ be the matroid that is obtained from M by replacing each non-loop element by k parallel elements and replacing each loop by k loops. Then*

$$t(M^{(k)}; x, y) =$$

$$(y^{k-1} + y^{k-2} + \dots + y + 1)^{r(M)} t\left(M; \frac{y^{k-1} + y^{k-2} + \dots + y + x}{y^{k-1} + y^{k-2} + \dots + y + 1}, y^k \right).$$

Proof. Let $f(M) = t(M^{(k)}; x, y)$. Then it is straightforward to show that f is a generalized T–G invariant. The lemma then follows easily from Corollary 6.2.6. We leave it to the reader to complete the details of this argument. $\square$

In the North American National Hockey League (NHL), teams are awarded 2 points for a victory, 1 point for a tie, and 0 points for a loss. To compute the number of possible score vectors at the end of an NHL season, it suffices to compute the number of score vectors for $\Gamma^{(2)}$ where Γ is the graph corresponding to the NHL schedule and $\Gamma^{(k)}$ is obtained from Γ by replacing each edge by k edges in parallel. A special case of the next result is that this number of score vectors can be determined directly from $t(M_\Gamma)$.

6.3.25. Proposition. *Suppose that the vertices of a loopless graph Γ correspond to teams in some league with each edge corresponding to a game that must be played between the two endpoints. Let k be a fixed positive integer so that, for each game a team plays, it may score any number of points from the set $\{0, 1, \dots, k\}$ provided that the two teams in any game score a total of k points from that game. Then, when all the games have been played, the number of possible score vectors is*

$$t(M_{\Gamma^{(k)}}; 2, 1) = k^{r(M_\Gamma)} t\left(M_\Gamma; \frac{k+1}{k}, 1 \right).$$

Proof. The fact that the number of possible score vectors is $t(M_{\Gamma^{(k)}}; 2, 1)$ follows easily from Proposition 6.3.19. The equality of this and $k^{r(M_\Gamma)} t\left(M_\Gamma; \frac{k+1}{k}, 1 \right)$ is a consequence of Lemma 6.3.24. $\square$

We note that this result can also be used to treat the case when, from each game, a team may score any number of points from the set $\{-k, -(k-1), \dots, k-1, k\}$ provided that from any one game a total of 0 points are scored. A one-to-one correspondence between the possible score vectors $(s(v_1), s(v_2), ..., s(v_m))$ of Γ in this case and the possible score vectors obtained from orienting the edges of $\Gamma^{(2k)}$ is given by

$$(s(v_1), s(v_2), ..., s(v_m)) \rightarrow (s(v_1) + k \deg(v_1), s(v_2) + k \deg(v_2), ..., s(v_m) + k \deg(v_m)).$$

Hence there are $t(M_{\Gamma^{(2k)}}; 2, 1)$ possible score vectors in this case. We know of no easy formula for determiming the number of score vectors when the sum of the points scored in each contest is allowed to vary.

6.3.F. Two-variable Coloring

In this section we generalize the relationship (6.3.1) between the chromatic polynomial of a graph Γ and the Tutte polynomial of its polygon matroid. So far we have considered only proper colorings of Γ, that is, assignments of colors to the vertices of Γ so that two adjacent vertices receive different colors. We now consider arbitrary colorings where adjacent vertices are no longer required to be colored differently. In such a coloring, an edge is called *monochromatic* if its two endpoints receive the same color. Let Γ be connected and let $c_i(\lambda, \Gamma)$, or briefly $c_i(\lambda)$, denote the number of ways to color Γ with λ colors so that exactly i edges are monochromatic. Then $c_i(\lambda)$ is a polynomial in λ. If Γ has n edges, define

$$\bar\chi(\Gamma; \lambda, v) = \frac{1}{\lambda} \sum_{i=0}^{n} c_i(\lambda)v^i. \qquad (6.40)$$

Then $\bar\chi(\Gamma; \lambda, v)$ is easily determined from the Tutte polynomial of M_Γ.

6.3.26. Proposition. *If M_Γ has rank r, then*

$$\bar\chi(\Gamma; \lambda, v) = (v - 1)^r t\left(M_\Gamma; \frac{v + \lambda - 1}{v - 1}, v\right).$$

Proof. We shall sketch three different but suggestive proofs of this identity, leaving the reader the exercise of filling in the details.

(1) We can generalize the recursion for proper colorings to obtain:

$$\bar\chi(\Gamma) = \begin{cases} (v + \lambda - 1)\bar\chi(\Gamma/e) & \text{if } e \text{ is an isthmus,} \\ v\bar\chi(\Gamma - e) & \text{if } e \text{ is a loop,} \\ \bar\chi(\Gamma - e) + (v - 1)\bar\chi(\Gamma/e) & \text{otherwise.} \end{cases} \qquad (6.41)$$

To see the third part of this, suppose e is neither a loop nor an isthmus of Γ. Then, for any fixed λ and any $i \geqslant 1$, we can partition the λ-colorings of Γ with i monochromatic edges into those in which the two endpoints v and v' of e are colored the same and those in which they are colored differently. Evidently there are $c_{i-1}(\lambda, \Gamma/e)$ members of the first class. Moreover, the number of members of the second class equals the number of λ-colorings of $\Gamma - e$ with i monochromatic edges in which v and v' are colored differently. In turn, this number is the difference between $c_i(\lambda, \Gamma - e)$ and the number of λ-colorings of $\Gamma - e$ with i monochromatic edges in which v and v' are colored the same. Since the last quantity clearly equals $c_i(\lambda, \Gamma/e)$, we have

$$c_i(\lambda, \Gamma) = c_{i-1}(\lambda, \Gamma/e) + c_i(\lambda, \Gamma - e) - c_i(\lambda, \Gamma/e).$$

Using this, the third part of (6.41) is not difficult to deduce.

(2) Following Crapo (1969), define the *coboundary polynomial* of an arbitrary matroid M having lattice of flats $L(M)$ by

$$\bar{\chi}(M; \lambda, v) = \sum_{X \in L(M)} v^{|X|} p(M/X; \lambda)$$

$$= \sum_{\substack{X, Y \in L(M) \\ X \subseteq Y}} v^{|X|} \lambda^{r(M) - r(Y)} \mu(X, Y).$$

We leave it to the reader to check that

$$\bar{\chi}(M; \lambda, v) = (v - 1)^r t\left(M; \frac{v + \lambda - 1}{v - 1}, v \right).$$

To show that $\bar{\chi}(\Gamma; \lambda, v) = \bar{\chi}(M_\Gamma; \lambda, v)$ we note that, in any coloring of Γ, the set X of monochromatic edges forms a flat in M_Γ. Hence the coloring induced on Γ/X by contracting all the edges in X is proper.

(3) This proof is based on the pervasive combinatorial idea of 'counting in two different ways'. It is quite similar, for example, to the calculations involving permutations with restricted position found in Stanley (1986, section 2.3). Let M_{KC} be the $(n + 1) \times (r + 1)$ matrix with rows indexed by $0, 1, 2, ..., n$ and columns indexed by $0, 1, 2, ..., r$ whose (i, j)-entry equals the number of subsets A of $E(\Gamma)$ of size i such that A has rank $r - j$ in M_Γ, or equivalently, the subgraph $\Gamma'[A]$ of Γ having edge set A and vertex set $V(\Gamma)$ has $j + 1$ connected components. Then

$$\sum_{j=0}^{r} M_{KC}(i, j)\lambda^{j+1} = \sum_{j=0}^{n} \binom{j}{i} c_j(\lambda), \tag{6.42}$$

since each side counts the pairs (A, c) in which A is an i-element subset of $E(\Gamma)$ and c is a λ-coloring of Γ for which each edge in A is monochromatic. Indeed, the left-hand side sums first over all such A and then, for each such subset, counts the number of λ-colorings that are monochromatic on each component of $\Gamma'[A]$. The right-hand side sums over all λ-colorings c according to their number j of monochromatic edges and then picks i of these j edges.

In matrix form, we have, from (6.42), that

$$M_{KC} \cdot \boldsymbol{\lambda} = T \cdot \mathbf{c}_\lambda \tag{6.43}$$

where

$$\lambda = (\lambda, \lambda^2, ..., \lambda^{r+1})^t,$$
$$\mathbf{c}_\lambda = (c_0(\lambda), c_1(\lambda), ..., c_n(\lambda))^t,$$

and T is the $(n+1) \times (n+1)$ matrix for which $T(i, j) = \binom{j}{i}$, the row and column indices, i and j, ranging over the set $\{0, 1, 2, ..., n\}$. It is well known that the inverse T^{-1} of T is given by $T^{-1}(i, j) = (-1)^{i+j}\binom{j}{i}$. By (6.43),

$$T^{-1} \cdot M_{KC} \cdot \lambda = \mathbf{c}_\lambda.$$

Now recall that, for a matroid $M(E)$, the cardinality-corank polynomial $S_{KC}(M; v, \lambda)$ is equal to

$$\sum_{A \subseteq E} v^{|A|}\lambda^{r(E)-r(A)}.$$

Thus

$$S_{KC}(M_\Gamma; v, \lambda) = \sum M_{KC}(i, j)v^i\lambda^j$$

and so

$$S_{KC}(M_\Gamma; v, \lambda) = \left(\frac{1}{\lambda}\right)\mathbf{v}^t \cdot M_{KC} \cdot \lambda$$

where

$$\mathbf{v}^t = (1, v, v^2, ..., v^n).$$

Thus, by (6.43),

$$S_{KC}(M_\Gamma; v, \lambda) = \left(\frac{1}{\lambda}\right)\mathbf{v}^t \cdot T \cdot \mathbf{c}_\lambda.$$

It is now not difficult to check that

$$\lambda S_{KC}(M_\Gamma; v, \lambda) = \lambda\bar{\chi}(\Gamma; v+1, \lambda). \tag{6.44}$$

But, by Proposition 6.2.21,

$$S_{KC}(M; v, \lambda) = v^r t\left(M; \frac{v+\lambda}{v}, v+1\right).$$

Substituting this into (6.44), we immediately get 6.3.26. ☐

The formula in Proposition 6.3.26 is invertible. Hence $t(M_\Gamma)$ can be computed from knowing the distribution of monochromatic edges among all λ-colorings of Γ for at least $r(M_\Gamma)$ values of λ. For the coboundary polynomial in general, it is not difficult to show that

$$t(M; x, y) = \frac{1}{(y-1)^r}\bar{\chi}(M; (x-1)(y-1), y). \tag{6.45}$$

The *dual coboundary polynomial* $\bar{\chi}^*(M; \lambda, v)$ is defined by

$$\bar{\chi}^*(M; \lambda, v) = \bar{\chi}(M^*; v, \lambda). \qquad (6.46)$$

Moreover, for a graph Γ, we let $\bar{\chi}^*(\Gamma; \lambda, v) = \bar{\chi}^*(M_\Gamma; \lambda, v)$. Then one can show that the coefficient of v^j in $\bar{\chi}^*(\Gamma; \lambda, v)$ is the number of λ-flows that are zero on precisely j edges of Γ. Using the duality of the Tutte polynomial, we then get the following link between two-variable colorings and two-variable flows in graphs.

6.3.27. Proposition. *If M_Γ has n elements and rank r, then*

$$\bar{\chi}^*(\Gamma; v, \lambda) = \frac{(v-1)^n}{\lambda^r} \bar{\chi}\left(\Gamma; \lambda, \frac{v+\lambda-1}{v-1}\right).$$

Proof.
$$\begin{aligned}
\bar{\chi}^*(\Gamma; v, \lambda) &= \bar{\chi}^*(M_\Gamma; v, \lambda) \\
&= \bar{\chi}(M_\Gamma^*; \lambda, v) \\
&= (v-1)^{n-r} t\left(M_\Gamma^*; \frac{v+\lambda-1}{v-1}, v\right) \\
&= (v-1)^{n-r} t\left(M_\Gamma; v, \frac{v+\lambda-1}{v-1}\right) \\
&= (v-1)^{n-r} \frac{(v-1)^r}{\lambda^r} \bar{\chi}\left(\Gamma; (v-1)\left(\frac{\lambda}{v-1}\right), \frac{v+\lambda-1}{v-1}\right)
\end{aligned}$$

where the last step follows by (6.45). $\qquad\qquad\square$

Clearly the last result also holds if we replace Γ by an arbitrary matroid M although, in this more general context, we no longer have the link to colorings and flows.

6.3.G. Other Graph-theoretic Tutte Invariants

In this section, we briefly list some other Tutte invariants for graphs that have appeared in the literature. Most of the results here can be proved by verifying the fundamental recursion 6.2.2(ii).

Dual to acyclic orientations we have *totally cyclic orientations*: those in which every edge of the graph Γ is contained in some directed cycle. To avoid distracting complications, the statements of these results will assume, unless otherwise stated, that Γ has no isthmuses. Then, following the results of Greene & Zaslavsky (1983) or Las Vergnas (1977) we have the following.

6.3.28. Example. The number $a^*(\Gamma)$ of totally cyclic orientations of Γ is given by

$$a^*(\Gamma) = t(M_\Gamma; 0, 2) = |\chi_\Gamma^*(-1)| = |p(M_\Gamma^*; -1)|.$$

Hence if Γ is a planar graph and Γ^* is a geometric dual of Γ, then $a^*(\Gamma) = a(\Gamma^*)$.

6.3.29. Example. Let e be a fixed edge of Γ and $a_e^*(\Gamma)$ be the number of totally cyclic orientations of Γ such that reversing the orientation of e makes the orientation acyclic. Equivalently, $a_e^*(\Gamma)$ equals the number of totally cyclic orientations of Γ such that every directed cycle uses e, that is, such that $\Gamma - e$ is acyclic. Then, provided $|E(\Gamma)| > 1$,

$$a_e^*(\Gamma) = 2\beta(M_\Gamma^*) = 2b_{10} = 2\beta(M_\Gamma).$$

Thus $a_e^*(\Gamma)$ does not depend on the edge e.

Some further results of this type can be found in the exercises for section 6.3.D. The next two examples link totally cyclic orientations and the Möbius function. They are both special cases of a more general result of Greene & Zaslavsky (1983, Theorem 8.3).

6.3.30. Example. Let Γ be a directed graph having a fixed ordering on its edges. Then the number of totally cyclic reorientations τ of Γ such that in each cycle of τ the lowest edge is not reoriented is $|\mu(M_\Gamma^*)|$.

6.3.31. Example. Let Γ be a plane graph. The number of totally cyclic orientations of Γ in which there is no clockwise cycle equals $|\mu(M_\Gamma^*)|$ if Γ has no isthmuses, and equals 0 otherwise.

When Stanley (1973b) proved his famous result (6.3.17) for acyclic orientations, he actually obtained the following stronger result in order to interpret evaluations of the chromatic polynomial at negative integers.

6.3.32. Example. Let m be a positive integer and $w(\Gamma; m)$ denote the number of pairs (θ, σ) such that θ is an acyclic orientation of Γ and σ is a function from $V(\Gamma)$ into $\{1, 2, ..., m\}$ with the property that if θ directs the edge uv of Γ from u to v then $\sigma(u) \geqslant \sigma(v)$. Suppose that Γ has k connected components and no isolated vertices. Then

$$w(\Gamma; m) = (-1)^{|V(\Gamma)|}\chi_\Gamma(-m) = m^k t(M_\Gamma; m+1, 0).$$

Let Γ be a plane graph and, for each e of Γ, let $v(e)$ be a point in the interior of e. The *medial graph* Γ_m of Γ will appear in the next result and we now describe its construction. If Γ is disconnected, its medial graph is the union of the medial graphs of its components. Now suppose that Γ is connected. The construction of Γ_m in this case is illustrated in Figure 6.6. In general, Γ_m is a plane graph with vertex set $\{v(e): e \in E(\Gamma)\}$. As one can see from Figure 6.6, two such vertices $v(e)$ and $v(f)$ are joined by one edge for every face in which e and f occur successively on the boundary. More

Figure 6.6. (a) Γ; (b) Γ_m superimposed on Γ; (c) Γ_m.

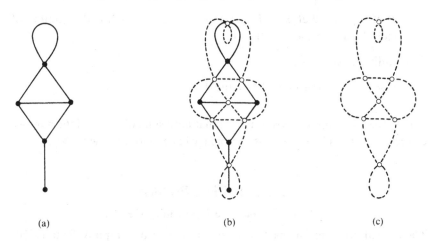

(a)   (b)   (c)

precisely, the edge set of Γ_m depends upon the set of ordered pairs (F, u) where F is a face of Γ and u is a vertex on the boundary of F. For every such pair, we add an edge $v(e)v(f)$ to Γ_m whenever e and f are edges of Γ incident with u that occur successively when one traverses the boundary of F just inside its interior.

It is not difficult to check that Γ_m is 4-regular for all plane graphs Γ. Moreover, if Γ^* is the geometric dual of Γ, then $\Gamma_m = (\Gamma^*)_m$. If Δ is an arbitrary connected, 4-regular plane graph, we can form a graph Γ for which $\Gamma_m = \Delta$ as follows. Two-color the faces of Δ so that two faces sharing an edge are colored differently. This is possible because Δ is Eulerian and hence Δ^* is bipartite; this 2-coloring of the faces of Δ is just a proper 2-coloring of the vertices of Δ^*. We construct Γ by letting its vertices consist of the faces in one of the color classes; two vertices of Γ are joined by an edge for every vertex shared by the corresponding faces of Δ. Note that if we choose the faces of the other color class to be the vertices of the graph, then this construction will produce Γ^*.

An *Eulerian partition* of a graph is a set of closed trails partitioning the edge set of the graph. Clearly every 4-regular plane graph Δ has such a partition P. Suppose that, at a vertex v of Δ, the edges, in cyclic order, are e_1, e_2, e_3 and e_4, where a loop, if it occurs, is listed once for each of its ends. We say that P has a *crossing* at v if one of the closed trails in P uses both e_1 and e_3. It is not difficult to check that Δ has a unique Eulerian partition P_0 in which there is a crossing at every vertex. Note that, since Δ is a plane graph, there are two ways of travelling along each loop.

For an arbitrary plane graph Γ, results in Jaeger (1988a), Las Vergnas (1979, 1981), Martin (1977, 1978), and Rosenstiehl & Read (1978) show that

(i) $t(M_\Gamma; x, x) = \sum_P (x-1)^{\gamma(P)-1}$, where the sum is over all Eulerian partitions

P of Γ_m such that each P has no crossings and $\gamma(P)$ is the number of closed trails in the partition P.

In addition, we have

(ii) $t(M_\Gamma; -1, -1) = (-1)^{|E(\Gamma)|}(-2)^{\gamma(P_0)-1}$.

It will follow using Proposition 6.5.4 that both sides of (ii) equal the number of vectors in $C \cap C^*$ where C is the binary code associated with M_Γ.

6.4. The Critical Problem

6.4.A. Definitions and Elementary Results

The critical problem for matroids was introduced by Crapo & Rota (1970) to provide a unified setting for a number of problems in extremal combinatorial theory including such fundamental graph-theoretic problems as Tutte's 5-flow conjecture (6.3.8) and the following celebrated conjecture of Hadwiger (1943).

6.4.1. Conjecture. *Let Γ be a loopless graph and k be a positive integer. If Γ has no proper k-coloring, then some subgraph of Γ is contractible to K_{k+1}.*

The cases $k = 1$ and $k = 2$ of this conjecture are straightforward. The conjecture was proved by Dirac (1952) when $k = 3$. For $k = 4$, Wagner (1964) proved that the conjecture is equivalent to what was then the Four Color conjecture. Now that the latter has become the Four Color theorem (6.3.2), we know that Hadwiger's conjecture is true for $k \leqslant 4$.

Several other important instances of the critical problem were noted in section 7.5 of White (1987). In this section, we shall give a more detailed discussion of the critical problem focussing attention on more recent results. Inevitably there will be some overlap between this section and Chapter 7 of White (1987). We begin here by recalling some basic definitions. A *linear functional* on $V(n, q)$, the n-dimensional vector space over $GF(q)$, is a linear map from $V(n, q)$ into $GF(q)$. If $A \subseteq V(n, q)$, then a k-tuple $(f_1, f_2, ..., f_k)$ of linear functionals on $V(n, q)$ is said to *distinguish* A if A is disjoint from $\{e: f_i(e) = 0$ for all i such that $1 \leqslant i \leqslant k\}$. Let $M(E)$ be a rank r matroid that is coordinatizable over $GF(q)$. The following fundamental result of Crapo & Rota (1970) was proved in White (1987, Theorem 7.4.1).

6.4.2. Theorem. *If $k \in \mathbb{Z}^+$ and ϕ is a coordinatization of M in $V(n, q)$, then the number of k-tuples of linear functionals on $V(n, q)$ that distinguish $\phi(E)$ equals $q^{k(n-r)}p(M; q^k)$.*

It follows from this result that, for a matroid M coordinatizable over $GF(q)$,

$$p(M; q^k) \geqslant 0 \text{ for all } k \text{ in } \mathbb{Z}^+. \tag{6.47}$$

The *critical exponent* $c(M; q)$ of M is defined by

$$c(M; q) = \begin{cases} \infty & \text{if } M \text{ has a loop,} \\ \min\{j \in \mathbb{N}: p(M; q^j) > 0\} & \text{otherwise.} \end{cases} \tag{6.48}$$

It follows from Theorem 6.4.2 that

$$c(M; q) = \min\{j \in \mathbb{N}: p(M; q^k) > 0 \text{ for all integers } k \geqslant j\}. \tag{6.49}$$

Since the kernel of a linear functional is a hyperplane, the following result follows easily from Theorem 6.4.2.

6.4.3. Corollary. *Let M be a rank r loopless matroid and ϕ be a coordinatization of M in $V(n, q)$. Then*

$$c(M; q) = \min\left\{j \in \mathbb{N}: V(n, q) \text{ has hyperplanes } H_1, H_2, \ldots, H_j \text{ such that}\right.$$

$$\left.\left(\bigcap_{i=1}^{j} H_i\right) \cap \phi(E) = \varnothing\right\}$$

$$= \min\{j \in \mathbb{N}: V(n, q) \text{ has a subspace of dimension } n - j \text{ having empty}$$
$$\text{intersection with } \phi(E)\}.$$

A noteworthy and somewhat surprising aspect of Theorem 6.4.2 and Corollary 6.4.3 is that the value of $c(M; q)$ does not depend upon the particular coordinatization ϕ. From (6.48) and Propositions 6.3.1 and 6.3.4, we deduce that when M is isomorphic to the polygon matroid of a graph Γ, $c(M; q)$ is the least integer c such that the chromatic number of Γ does not exceed q^c; when M is isomorphic to the bond matroid of Γ, $c(M; q)$ is the least integer c for which Γ has a nowhere-zero q^c-flow.

For a matroid M that is coordinatizable over $GF(q)$, the critical problem is the problem of determining the critical exponent, $c(M; q)$. This is theoretically possible for any matroid M, simply by calculating $p(M; \lambda)$. In general, however, this will require exponentially many steps. In particular then, the critical problem becomes one of efficiently determining $c(M; q)$ by, for example, recognizing M as a member of a class of matroids whose critical exponents are bounded above.

We now note some basic properties of the critical exponent. In each of these, we shall assume that M is a matroid coordinatizable over $GF(q)$.

6.4.4. Proposition. *If M is loopless and T is a subset of $E(M)$, then*

$$c(M(T); q) \leqslant c(M; q) \leqslant c(M(T); q) + c(M(E - T); q).$$

Proof. This is an immediate consequence of Corollary 6.4.3. $\qquad\square$

6.4.5. Proposition. (*Asano, Nishizeki, Saito & Oxley, 1984*) *Suppose that* $S \subseteq E(M)$ *and* $k \in \mathbb{N}$. *Then the following are equivalent.*

  (i) S *is minimal with the property that* $c(M - S; q) \leqslant k$.
  (ii) S *is minimal with the property that* $c(M/S; q) \leqslant k$.
  (iii) S *is minimal with the property that* M *has a minor with ground set* $E(M) - S$ *and critical exponent not exceeding* k.

This result is proved in Asano, Nishizeki, Saito & Oxley (1984) by using Tutte's theory of chain-groups (1965). We prove it here using a deletion–contraction argument on the characteristic polynomial. We shall require several preliminaries.

6.4.6. Lemma. *Let* S *be a subset of* $E(M)$ *for which* $c(M - S; q) \leqslant k$. *Then* S *has a subset* T *such that* $c(M/T; q) \leqslant k$.

Proof. We argue by induction on $|S|$, noting that the result is immediate if this is zero. Assume the result to be true for $|S| = n - 1$ and let $|S| = n$. Choose an element e of S. Then $p((M - e) - (S - e); q^k) = p(M - S; q^k) > 0$. Therefore, by the induction assumption, $S - e$ has a subset T such that $p((M - e)/T; q^k) > 0$. The required result holds if $p(M/(T \cup e); q^k) > 0$. Therefore we may assume that $p(M/(T \cup e); q^k) = 0$. Now $(M - e)/T = (M/T) - e$ and therefore $p((M/T) - e; q^k) > 0$. Hence e is not a loop or an isthmus of M/T. Thus

$$p(M/T; q^k) = p((M/T) - e; q^k) - p(M/(T \cup e); q^k) > 0.$$

We conclude that $c(M/T; q) \leqslant k$ and, since $T \subseteq S - e$, the lemma follows. □

6.4.7. Proposition. *Let* N *be a matroid and* λ *be a real number such that* $p(N'; \lambda) \geqslant 0$ *for all minors* N' *of* N. *Suppose that* T *is a subset of* $E(N)$ *for which* $p(N/T; \lambda) > 0$. *Then* $p(N - T; \lambda) > 0$.

Proof. This follows by a similar induction argument to that given in the last proof and is left as an exercise for the reader.   □

6.4.8. Corollary. (*Oxley, 1978a*) *If* T *is a subset of* $E(M)$, *then*

$$c(M - T; q) \leqslant c(M/T; q).$$

Proof. This follows by taking λ equal to each of $q, q^2, q^3, \ldots$ in 6.4.7.   □

We are now ready to prove Proposition 6.4.5.

Proof of Proposition 6.4.5. We shall show the equivalence of 6.4.5(i), 6.4.5(ii), and the following statement, which is easily seen to be equivalent to 6.4.5(iii).

(iii') S is minimal with the property that, for some subset T of S, $c((M - T)/(S - T); q) \leqslant k$.

We begin by showing that (i) implies (iii'). Let S be a minimal set for which $c(M - S; q) \leqslant k$. Then $c((M - S)/\emptyset; q) \leqslant k$. Suppose that S' and T are sets for which $T \subseteq S' \subsetneqq S$ and $c((M - T)/(S' - T); q) \leqslant k$. Then, by Corollary 6.4.8, $c((M - T) - (S' - T); q) \leqslant k$, that is, $c(M - S'; q) \leqslant k$. This contradicts the choice of S.

To show that (iii') implies (ii), suppose that S satisfies (iii'). Then, as $(M - T)/(S - T) = (M/(S - T)) - T$, the latter has critical exponent not exceeding k. Hence, by Lemma 6.4.6, there is a subset T' of T such that $c((M/(S - T))/(T - T'); q) \leqslant k$, that is, $c(M/(S - T'); q) \leqslant k$, or equivalently, $c((M - \emptyset)/(S - T'); q) \leqslant k$. By the choice of S, it follows that $T' = \emptyset$. Hence $c(M/S; q) \leqslant k$. Moreover, if $S' \subsetneqq S$ and $c(M/S'; q) \leqslant k$, then $c((M - \emptyset)/S'; q) \leqslant k$, contrary to the choice of S.

A similar argument shows that (ii) implies (i) and this completes the proof of Proposition 6.4.5. $\qquad\qquad\square$

A matroid M coordinatizable over $GF(q)$ is called *affine* if $c(M; q) = 1$. To justify this terminology, note that, from Corollary 6.4.3, M is affine if and only if the simplification of M is a subgeometry of the affine space $AG(r, q)$ for some r. Recall here that $AG(r, q)$ is obtained from the projective space $PG(r, q)$ by deleting the points of a hyperplane.

The next two observations come from combining Proposition 6.4.5 with Corollary 6.4.3 and Lemma 6.4.6.

$$c(M; q) = \min\left\{n \in \mathbb{N}: E(M) = \bigcup_{i=1}^{n} S_i \text{ and } M(S_i) \text{ is affine for all } i\right\}. \qquad (6.50)$$

$$c(M; q) = \min\left\{n \in \mathbb{N}: E(M) = \bigcup_{i=1}^{n} S_i \text{ and } M/(E - S_i) \text{ is affine for all } i\right\}. \qquad (6.51)$$

For a loopless graph Γ, the chromatic number $\chi(\Gamma)$ satisfies

$$\chi(\Gamma) = \min\{n \in \mathbb{Z}^+: p(M_\Gamma; n) > 0\}. \qquad (6.52)$$

It turns out to be quite fruitful to exploit the similarity between this and the definition of the critical exponent (6.48). Many bounds on the chromatic number of a graph are expressed in terms of vertex degrees. By analogy with this, the next result bounds $c(M; q)$ in terms of the sizes of its bonds. For a matroid N, we denote the set of simple submatroids of N by $\mathbb{R}(N)$. The set of bonds of N will be denoted by $\mathscr{C}^*(N)$.

6.4.9. Proposition. *If M is a loopless matroid coordinatizable over $GF(q)$, then*

$$c(M; q) \leqslant \left\lceil \log_q \left(1 + \max_{N \in \mathscr{R}(M)} \left(\min_{C^* \in \mathscr{C}^*(N)} |C^*| \right) \right) \right\rceil.$$

The proof of this result depends upon the following useful lemma for the characteristic polynomial.

6.4.10. Lemma. *(Oxley, 1978a) Let $\{e_1, e_2, ..., e_k\}$ be a bond C^* of a matroid M. Then*

$$p(M; \lambda) = (\lambda - k)p(M - C^*; \lambda)$$

$$+ \sum_{j=2}^{k} \sum_{i=1}^{j-1} p(M - \{e_1, ..., e_{i-1}, e_{i+1}, ..., e_{j-1}\}/\{e_i, e_j\}; \lambda).$$

Proof. We argue by induction on k. If $k = 1$, then e_1 is an isthmus of M and the result is immediate. Assume the result holds for $k < n$ and let $k = n \geqslant 2$. Then

$$p(M; \lambda) = p(M - e_1; \lambda) - p(M/e_1; \lambda). \tag{6.53}$$

If e_2 is not a loop of M/e_1, then

$$p(M/e_1; \lambda) = p(M/e_1 - e_2; \lambda) - p(M/e_1/e_2; \lambda). \tag{6.54}$$

But, if e_2 is a loop of M/e_1, then $p(M/e_1; \lambda) = 0$ and $M/e_1 - e_2 \cong M/\{e_1, e_2\}$, hence, (6.54) also holds in this case. On substituting (6.54) into (6.53), we obtain

$$p(M; \lambda) = p(M - e_1; \lambda) - p(M/e_1 - e_2; \lambda) + p(M/\{e_1, e_2\}; \lambda). \tag{6.55}$$

If $k = 2$, then $M - e_1$ and $M - e_2$ have e_2 and e_1 respectively as isthmuses. Thus $p(M - e_1; \lambda) = p(M - e_2; \lambda) = (\lambda - 1)p(M - C^*; \lambda)$, and $M/e_1 - e_2 \cong M - e_2/e_1 \cong M - e_2 - e_1 \cong M - C^*$. On combining these observations with (6.55), we deduce that the required result holds for $k = 2$. We may therefore suppose that $k > 2$. Then, as $M/e_1 - e_2 = M - e_2/e_1$, and e_1 is neither a loop nor an isthmus of $M - e_2$, we have

$$p(M/e_1 - e_2; \lambda) = p(M - e_2 - e_1; \lambda) - p(M - e_2; \lambda). \tag{6.56}$$

On substituting (6.56) into (6.55), we get that

$$p(M; \lambda) = p(M - e_1; \lambda) + p(M - e_2; \lambda) - p(M - \{e_1, e_2\}; \lambda) + p(M/\{e_1, e_2\}; \lambda).$$

As $C^* - A$ is a bond of $M - A$ for every proper subset A of C^*, we may now apply the induction assumption to each of the matroids $M - e_1$, $M - e_2$, and $M - \{e_1, e_2\}$ to get the required result. The straightforward details are omitted here. $\square$

Proof of Proposition 6.4.9. We argue by induction on $|E(M)|$. The result is true for $|E(M)| = 1$. Assume it to be true for all matroids on sets with fewer than n elements and suppose that $|E(M)| = n$. If M has an element e that is in a 2-element circuit, then $\mathscr{R}(M - e) = \mathscr{R}(M)$, $c(M - e; q) = c(M; q)$ and we can deduce the result by applying the induction assumption to $M - e$. We may now suppose that M is simple. Then, for a bond D^* of M of minimum size, we have, by (6.47) and Lemma 6.4.10, that $c(M; q) \leqslant \max\{\lceil \log_q(|D^*| + 1)\rceil, c(M - D^*; q)\}$. But

$$|D^*| = \min_{C^* \in \mathscr{C}^*(M)} |C^*| \leqslant \max_{N \in \mathscr{R}(M)} \left(\min_{C^* \in \mathscr{C}^*(N)} |C^*| \right).$$

Moreover, by the induction assumption,

$$c(M - D^*; q) \leqslant \left\lceil \log_q \left(1 + \max_{N \in \mathscr{R}(M - D^*)} \left(\min_{C^* \in \mathscr{C}^*(N)} |C^*| \right) \right) \right\rceil$$

$$\leqslant \left\lceil \log_q \left(1 + \max_{N \in \mathscr{R}(M)} \left(\min_{C^* \in \mathscr{C}^*(N)} |C^*| \right) \right) \right\rceil.$$

The required result now follows by induction.   □

The proof of the following consequence of Proposition 6.4.9 is left as an exercise for the reader.

6.4.11. Corollary. *Suppose that M is coordinatizable over $GF(q)$. If there is a covering of $E(M)$ with bonds each having fewer than q^k elements, then $c(M; q) \leqslant k$.*

If $E(M)$ can be covered by *disjoint* bonds, then we have the following:

6.4.12. Proposition. *(Oxley, 1978a) Suppose that M is coordinatizable over $GF(q)$ and $E(M)$ is a disjoint union of bonds. Then M is affine, that is, $c(M; q) = 1$.*

For $q = 2$, the converse of the last proposition holds (Brylawski, 1972b) (see Exercise 6.50). To see that the converse does not hold for $q > 2$, consider the affine plane $AG(2, q)$.

6.4.B. Minimal and Tangential Blocks

If M is a loopless matroid coordinatizable over $GF(q)$, then M and its simplification have the same characteristic polynomial and therefore have the same critical exponent. Thus, for the moment, we shall suppose that M is simple. Then M can be embedded as a submatroid of $PG(n - 1, q)$ for some n. In general, several different embeddings are possible. However, using the fact that $PG(n - 1, q)$ is isomorphic to the simple matroid associated with $V(n, q)$, we deduce from Corollary 6.4.3 that the value of $c(M; q)$ does not depend on the embedding.

6.4.13. Proposition. *If M is isomorphic to the restriction of $PG(n - 1, q)$ to the set E, then*

$$c(M; q) = \min\left\{ j \in \mathbb{N}: PG(n-1, q) \text{ has hyperplanes } H_1, H_2, ..., H_j \text{ such that} \right.$$

$$\left. \left(\bigcap_{i=1}^{j} H_i\right) \cap E = \varnothing \right\}$$

$$= \min\left\{ j \in \mathbb{N}: PG(n-1, q) \text{ has a flat of rank } n-j \text{ having empty} \right.$$

$$\left. \text{intersection with } E \right\}.$$

For any positive integer k, we shall call a simple matroid M a *k-block over* $GF(q)$ if $c(M; q) > k$. M is a *minimal k-block over* $GF(q)$ if M is a k-block over $GF(q)$ but no proper submatroid of M is. It follows easily from Corollary 6.4.3 that M is a minimal k-block over $GF(q)$ if and only if $c(M; q) = k + 1$ and, for all proper submatroids N of M, $c(N; q) \leqslant k$.

An elementary geometric argument shows that $PG(k, q)$ is a minimal k-block. Moreover, one can easily show using the characteristic polynomial that if Γ is a graph that is edge-minimal with the property of being properly $(q^k + 1)$-colorable, then its polygon matroid $M(\Gamma)$ is a minimal k-block. One important such graph is $M(K_{q^k+1})$. An infinite family of minimal k-blocks can be constructed from these examples by using the fact that if M and N are these minimal k-blocks over $GF(q)$, so is their series connection, $S(M, N)$ (Oxley, 1980) (Exercise 6.60). In view of this observation, it seems natural to consider a strengthened notion of minimality for k-blocks. A subclass of the class of minimal k-blocks that has received considerable attention is the class of *tangential k-blocks*. A simple matroid M that is coordinatizable over $GF(q)$ is a *tangential k-block over* $GF(q)$ if M is a k-block over $GF(q)$ but no simple proper minor of M is. It is not difficult to check that both $PG(k, q)$ and $M(K_{q^k+1})$ are tangential k-blocks. Moreover, since M and N are both minors of $S(M, N)$ (Brylawski, 1971), one cannot create new tangential k-blocks simply by taking series connections of these blocks.

The straightforward proof of the next result is left to the reader (Exercise 6.61).

6.4.14. Proposition. *The following statements are equivalent for a simple matroid M that is coordinatizable over $GF(q)$.*

(i) *M is a tangential k-block over $GF(q)$;*

(ii) *$c(M; q) > k$ and $c(N; q) \leqslant k$ for all loopless proper minors N of M;*

(iii) *$c(M; q) = k + 1$ and $c(M/F; q) \leqslant k$ for all non-empty flats F of M.*

Tangential blocks were studied originally by Tutte (1966a). He concentrated on tangential 1- and 2-blocks over $GF(2)$ and began by showing that there is only one tangential 1-block over $GF(2)$. Recall that a matroid M coordinatizable over $GF(q)$ is affine if $c(M; q) = 1$.

6.4.15. Proposition. (*Tutte, 1966a*) *The unique tangential 1-block over* $GF(2)$ *is* $M(K_3)$.

Proof. This follows immediately from the fact that a binary matroid is affine if and only if it has no odd circuits. The proof of this is left to the reader (Exercise 6.50).   □

We have already noted that $M(K_5)$ and $PG(2, 2)$ are tangential 2-blocks over $GF(2)$. Moreover, as the Petersen graph P_{10} has no 4-flow, $M^*(P_{10})$ is a 2-block over $GF(2)$. Indeed, it is not difficult to check that this 2-block is tangential.

The next theorem is the main result of Tutte's paper (1966a). F_7 denotes the Fano matroid, $PG(2, 2)$.

6.4.16. Theorem. *The only tangential 2-blocks over* $GF(2)$ *of rank at most 6 are* F_7, $M(K_5)$, *and* $M^*(P_{10})$.

Tutte also conjectured that the restriction on the rank in this theorem could be dropped:

6.4.17. Conjecture. *The only tangential 2-blocks over* $GF(2)$ *are* F_7, $M(K_5)$, *and* $M^*(P_{10})$.

Using geometric methods, Datta (1976b; 1981) proved that there are no tangential 2-blocks over $GF(2)$ of rank 7 or 8. Conjecture 6.4.17 remains one of the most important unsolved problems in this area of combinatorics. The most significant advance toward its solution was made by Seymour (1981b) who proved the following result.

6.4.18. Theorem. *Let* M *be a tangential 2-block over* $GF(2)$ *and suppose that* M *is not isomorphic to* F_7 *or* $M(K_5)$. *Then* M *is cographic.*

The proof of this theorem uses a number of very powerful results including the Four Color theorem (6.3.2) and Seymour's decomposition theorem for regular matroids (1980). We omit the details and refer the reader to Welsh (1982) for an outline of the proof.

An interesting consequence of Theorem 6.4.18 is that Conjecture 6.4.17 is equivalent to Tutte's 4-flow conjecture (6.3.11). The proof of this equivalence is straightforward and is based on the observation that a graph Γ without isthmuses has a nowhere-zero 4-flow if and only if $M^*(\Gamma)$ is a 2-block over $GF(2)$.

There are a number of results for tangential blocks over fields other than $GF(2)$. These results indicate that the binary case is certainly the nicest. By arguing in terms of the characteristic polynomial it is straightforward to prove the following result.

6.4.19. Proposition. *Suppose that M is coordinatizable over GF(q), and j and k are positive integers such that j divides k. Then M is a tangential k-block over GF(q) if and only if M is a tangential j-block over $GF(q^{k/j})$.*

We observe here that the assumption that M is coordinatizable over $GF(q)$ is redundant above if M is a tangential k-block over $GF(q)$, but may be needed if M is a tangential j-block over $GF(q^{k/j})$, since such a matroid need not be coordinatizable over $GF(q)$. The next result was proved by Walton & Welsh (1982).

6.4.20. Proposition. *The only tangential 1-blocks over GF(3) are $M(K_4)$ and $U_{2,4}$.*

The proof of this will use the following result of Brylawski (1971).

6.4.21. Proposition. *Let Γ be a loopless series–parallel network. Then Γ has a proper 3-coloring.*

Proof. We argue by induction on $|E(\Gamma)|$ to show that $p(M(\Gamma); 3) > 0$. If Γ is a forest having m edges, then $p(M(\Gamma); \lambda) = (\lambda - 1)^m$, hence $p(M(\Gamma); 3) > 0$. Thus the proposition is true in this case. Assume it to be true for $|E(\Gamma)| < n$ and let $|E(\Gamma)| = n$. We may suppose that Γ is not a forest. Then $E(\Gamma)$ has a subset $\{e_1, e_2\}$ that is either a circuit or a bond of $M(\Gamma)$. In the first case, $p(M(\Gamma); \lambda) = p(M(\Gamma - e_2); \lambda)$ and the result follows by the induction assumption. In the second case, by Lemma 6.4.10,

$$p(M(\Gamma); \lambda) = (\lambda - 2)p(M(\Gamma) - \{e_1, e_2\}; \lambda) + p(M(\Gamma)/\{e_1, e_2\}; \lambda).$$

By the induction assumption, when $\lambda = 3$, the first term on the right-hand side is positive. Since the second term is non-negative, the result follows. □

Proof of Proposition 6.4.20. $M(K_4) = M(K_{3^1+1})$ and $U_{2,4} \cong PG(1, 3)$, hence both $M(K_4)$ and $U_{2,4}$ are tangential 1-blocks over $GF(3)$. If M is a tangential 1-block having no minor isomorphic to $M(K_4)$ or $U_{2,4}$, then, by Table 7.1 (p. 146) of White (1986), $M \cong M(\Gamma)$ where Γ is a series–parallel network. By Proposition 6.4.21, $p(M(\Gamma); 3) > 0$. Hence $c(M; 3) = 1$; a contradiction. □

Proposition 6.4.20 and Conjecture 6.4.17 suggest that tangential blocks are relatively scarce. Indeed, Welsh (1980) made several conjectures to this effect. Subsequently, he and Seymour (Walton, 1981) and, independently, Whittle (1987) gave a number of examples to disprove these conjectures, thereby showing that there are many more tangential blocks than had previously been thought.

6.4.22. Example. (Seymour & Welsh, in Walton, 1981) Let q be a prime power exceeding two and let x_1, x_2, and x_3 be three non-collinear points of $PG(2, q)$. Let S_1 be the set of points of $PG(2, q)$ that lie on one of the lines spanned by $\{x_1, x_2\}$, $\{x_2, x_3\}$, and $\{x_3, x_1\}$. Let x be an arbitrary point that is on the line spanned by $\{x_3, x_1\}$ but different from x_1 and x_3. Let S_2 be the set of points of $PG(2, q)$ that lie on one of the lines spanned by $\{x_1, x_2\}$ and $\{x_2, x_3\}$, together with the point x. The geometries N_q and M_q are obtained from $PG\{2, q)$ by restricting to the sets $S_1 - \{x_1, x_2, x_3\}$ and $S_2 - \{x_1, x_3\}$, respectively. Affine representations for these geometries are shown in Figure 6.7, where we note that a number of lines have been left out to avoid cluttering the diagrams. We remark, without proof, that $N_4 \cong AG(2, 3)$ (Exercise 6.62).

Figure 6.7.

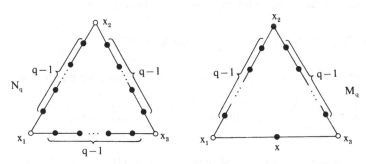

6.4.23. Proposition. M_q and N_q are tangential 1-blocks over $GF(q)$.

Proof. In view of Proposition 6.4.14, to establish that a member M of $\{M_q, N_q\}$ is a minimal 1-block over $GF(q)$, we need to check that the following hold. We leave these checks to the reader.

(1) The ground set E of M intersects every line of $PG(2, q)$.

(2) For every point p of M, there is a line of $PG(2, q)$ that meets E in p only.

Evidently, for any point e of M, the simplification of M/e is not $PG(1, q)$. Thus no minor of M of lesser rank is also a 1-block. It follows from this that M is indeed a tangential 1-block. $\qquad\square$

6.4.24. Example. (Whittle, 1989a) Let A be a subgroup of order m of $GF(q)^*$, the multiplicative group of $GF(q)$. Let $\{v_1, v_2, ..., v_r\}$ be a basis B for $V(r, q)$ and let D be $\{v_i + (-1)^{i-j+1}av_j : 1 \leqslant i < j \leqslant r$ and $a \in A\}$. The matroids $Q_r(A)$ and $Q'_r(A)$ are obtained by restricting $V(r, q)$ to the sets $B \cup D$ and D, respectively. Thus, for example, $Q'_2(GF(q)^*)$ is precisely the matroid N_q in Figure 6.7, while $Q_2(GF(q)^*)$ is obtained from N_q by adjoining the points x_1,

x_2, and x_3 to N_q. If A is the trivial group, then it is straightforward to show that $Q_r(A) \cong M(K_{r+1})$. More generally, it can be shown that $Q_r(A)$ depends only on r and the group A and not on the prime power q.

The matroid $Q_r(A)$ was introduced by Dowling (1973a, b) and is now known as the *rank r Dowling geometry based on the group A*. In fact, Dowling defined such matroids when A is an arbitrary finite group. Our main interest here will be in the special case defined above, although we remark that Whittle (1989a) has described an interesting extension of the critical problem to Dowling geometries in general.

Since $Q_2'(GF(q)^*) \cong N_q$, Proposition 6.4.23 implies that $Q_2'(GF(q)^*)$ is a tangential 1-block over $GF(q)$. Moreover, when A is the trivial group and r is q^k, $Q_r(A)$ is the tangential k-block $M(K_{q^k+1})$ over $GF(q)$. These observations are special cases of the following result of Whittle (1989a).

6.4.25. Proposition. *Let A be a subgroup of $GF(q)^*$. Then*

(i) *for $r = \dfrac{q^k - 1}{|A|} + 1$, $Q_r(A)$ is a tangential k-block over $GF(q)$; and*

(ii) *for $r = \dfrac{q^k - 1}{|A|} + 2$, $Q_r'(A)$ is a tangential k-block over $GF(q)$.*

In an important sequence of papers Whittle (1987, 1988, 1989b) noted several general constructions for using known tangential blocks to find others. Next we describe the simplest of these constructions, which, curiously, was the last to be found. Suppose that M is a rank r geometry that is coordinatizable over $GF(q)$. Let E be a subset of $PG(r, q)$ such that the restriction of $PG(r, q)$ to this set is isomorphic to M. Clearly $\mathrm{cl}_P(E)$ is a hyperplane of $PG(r, q)$ where cl_P denotes the closure operator of $PG(r, q)$. Now take a point p of $PG(r, q)$ that is not in $\mathrm{cl}_P(E)$. Let E' be the set of all points of $PG(r, q)$ lying on some line that contains p and some point of E, that is, $E' = \bigcup_{x \in E} \mathrm{cl}_P(\{x, p\})$. We call the restriction of $PG(r, q)$ to E' a *q-lift* of M. Thus, for example, $PG(2, 2)$ is a 2-lift of the 3-point line $U_{2,3}$, while a 3-lift of $U_{2,3}$ is the complement in $PG(2, 3)$ of $U_{2,3}$.

6.4.26. Proposition. *(Whittle, 1989b). If M is a tangential k-block over $GF(q)$ and M' is a q-lift of M, then M' is a tangential $(k + 1)$-block over $GF(q)$.*

The last construction produces tangential $(k + 1)$-blocks from tangential k-blocks. Next we shall describe two special cases of a quotient construction of Whittle (1988) that produces tangential k-blocks from tangential k-blocks.

A description of the general quotient construction can be found in Exercise 6.67. A rank r matroid M is *supersolvable* if there is a set $\{F_0, F_1, F_2, ..., F_r\}$ of modular flats of M with $r(F_i) = i$ for $0 \leqslant i \leqslant r$ and $F_i \supseteq F_{i-1}$ for $1 \leqslant i \leqslant r$. We call the set $\{F_0, F_1, F_2, ..., F_r\}$ a *maximal chain of modular flats*.

Now suppose that M is a supersolvable rank r tangential k-block over $GF(q)$ with $r > k + 1$. Then we can embed M in $PG(r-1, q)$. Let $\{F_0, F_1, F_2, ..., F_r\}$ be a maximal chain of modular flats in M. Then $F_0 = \varnothing$ and $F_r = E(M)$. Because $r > k + 1$, $M \not\cong PG(k, q)$. Let m be the least element of the set $\{i: 2 \leqslant i \leqslant r$ and $M(F_i) \not\cong PG(i-1, q)\}$. Since M does not have $PG(k, q)$ as a minor, $m \leqslant k + 1$. But $r > k + 1$ and so F_m is a proper flat of M. As $M(F_m) \not\cong PG(m-1, q)$, the closure of F_m in $PG(r-1, q)$ contains an element x that is not in F_m. Let M'' be the elementary quotient of M by the element x, that is, M'' is formed by first extending M by adding x and then contracting x.

6.4.27. Theorem. (*Whittle, 1987*) *The simplification of M'' is a rank $(r-1)$ tangential k-block over $GF(q)$.*

We shall not prove this result here but instead we note the following important consequence of it.

6.4.28. Corollary. (*Whittle, 1987*) *For all r such that $k + 1 \leqslant r \leqslant q^k$, there is a rank r tangential k-block over $GF(q)$.*

Proof. $M(K_{q^k+1})$ is a supersolvable tangential k-block over $GF(q)$. By repeatedly applying the above construction, the corollary follows. $\square$

The second special case of Whittle's quotient construction that we shall consider involves the complete principal truncation $\bar{T}_F(M)$ of the matroid M with respect to the flat F (see White, 1986, p. 149). If $r(F) = j > 0$, we recall that $\bar{T}_F(M)$ is formed from M by putting a set P of $j - 1$ independent points freely on F and then contracting P. The bases of $\bar{T}_F(M)$ are the subsets of $E(M)$ of the form B or $B' \cup x$ where x is a non-loop element of F, and B and B' are subsets of $E - F$ that are independent in M such that $|B| = r - j + 1$, $|B'| = r - j$, and $r(B \cup F) = r(B' \cup F) = r$.

6.4.29. Theorem. (*Whittle, 1987*) *Let M be a tangential k-block over $GF(q)$ and F be a proper non-empty modular flat of M. If $\bar{T}_F(M)$ is coordinatizable over $GF(q)$, then the simple matroid associated with $\bar{T}_F(M)$ is a tangential k-block over $GF(q)$.*

From this, Whittle deduced the following.

6.4.30. Corollary. *For all m with $2 \leqslant m \leqslant q$, the simplification of $\bar{T}_{M(K_m)}(M(K_{q^k+1}))$ is a tangential k-block over $GF(q)$ with rank $q^k - m + 2$ and with $\frac{1}{2}(q^k + 1)q^k - \frac{1}{2}m(m+1) + 1$ elements.*

It is not difficult to check that, for $n = q^k$, the simplification of $\bar{T}_{M(K_{n-1})}(M(K_{n+1}))$ is isomorphic to the matroid M_n in Example 6.4.22.

We conclude this discussion of tangential blocks with some diagrams of such matroids from Whittle (1985) and with some unsolved problems. The matroid shown in Figure 6.8a is the simplification of $\bar{T}_{M(K_3)}(M(K_6))$ and is a tangential 1-block over $GF(5)$. The restriction of this matroid to the

Figure 6.8.

(a)

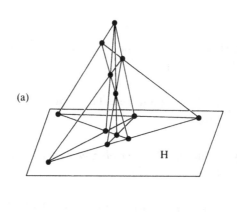

H

(b)

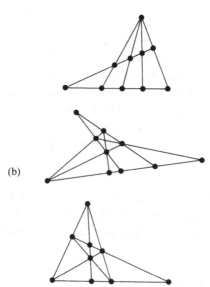

hyperplane H is the simplification of $\bar{T}_{M(K_3)}(M(K_5))$ and is a tangential 1-block over $GF(4)$.

The three matroids shown in Figure 6.8b are all examples of tangential 1-blocks over $GF(5)$. The first is the simplification of $\bar{T}_{M(K_4)}(M(K_6))$. We note that both the first and second have characteristic polynomial equal to $(\lambda - 1)(\lambda - 4)(\lambda - 5)$, while the third has characteristic polynomial $(\lambda - 1)(\lambda - 3)(\lambda - 5)$. Each of these three matroids is the simplification of some quotient of $M(K_6)$.

In connection with his constructions, Whittle raised several questions. Call a tangential k-block over $GF(q)$ *normal* if it is the simplification of a quotient of $M(K_{q^k+1})$. Not every tangential k-block is normal; for example, $M^*(P_{10})$ is a non-normal 2-block over $GF(2)$.

6.4.31. Problem ·

(i) Are there any non-normal supersolvable tangential k-blocks?
Less strongly:
(ii) Are there any non-normal tangential k-blocks with modular hyperplanes?
Conversely:
(iii) Does every normal tangential k-block have a modular hyperplane?
More strongly:
(iv) Is every normal tangential k-block supersolvable?

Another interesting unsolved problem raised by Whittle (1985) is the following.

6.4.32. Problem. Do all tangential k-blocks contain a spanning bond?

Given that tangential blocks are much more abundant than was once thought, the approach to classifying such objects has somewhat changed. Whittle (1987) showed that one group of well-behaved tangential blocks is those with modular hyperplanes. He also showed (Whittle, 1989a, b) that if $|A| \geqslant 2$, then $Q'_r(A)$ has no modular hyperplanes unless $r = 3$ and $|A| = 2$; that $M^*(P_{10})$ has no modular hyperplanes; and that a q-lift of a matroid M has no modular hyperplanes if and only if M has no modular hyperplanes. On combining these observations with Propositions 6.4.15 and 6.4.20, we deduce that there are tangential k-blocks over $GF(q)$ with no modular hyperplanes for all prime powers q and all positive integers k except when k is 1 and q is 2 or 3. Interestingly, it is precisely in the exceptional cases just noted that the problem of finding all tangential k-blocks over $GF(q)$ has been solved. Indeed, Whittle (1989b) asserts that the existence of tangential blocks without modular hyperplanes lies at the heart of the problem of determining all tangential blocks.

6.4.C. Bounding the Critical Exponent for Classes of Matroids

In this section we survey a number of results and conjectures concerned with determining an upper bound on the critical exponent of a matroid M when M is in some class of matroids characterized by excluded minors. Most of this work has appeared since 1980 and is related to the following conjecture of Brylawski (1975c).

6.4.33. Conjecture. *If M is a loopless matroid coordinatizable over $GF(q)$ and M has no minor isomorphic to $M(K_4)$, then $c(M; q) \leqslant 2$.*

Restated in terms of k-blocks this conjecture asserts that every tangential 2-block over $GF(q)$ has a minor isomorphic to $M(K_4)$.

The following very general extension of this conjecture was proposed by Whittle (1985).

6.4.34. Conjecture. *Every tangential k-block over $GF(q)$ has a minor isomorphic to $M(K_{k+2})$.*

This conjecture is easily seen to be true for $k = 1$. In support of the general case of the conjecture, Whittle (1987) has proved the following result.

6.4.35. Theorem. *A tangential k-block over $GF(q)$ that has a modular hyperplane has $M(K_{k+2})$ as a submatroid.*

Return now to Brylawski's conjecture. It is certainly true for $q = 2$. To see this, recall from the proof of Proposition 6.4.20 that a loopless matroid with no minor isomorphic to $U_{2,4}$ or $M(K_4)$ is isomorphic to $M(\Gamma)$ for some series–parallel network Γ. By Proposition 6.4.21, Γ is 3-colorable. Hence Γ is certainly 4-colorable, so $p(M(\Gamma); 4) > 0$ and hence $c(M(\Gamma); 2) \leqslant 2$.

For larger values of q, the conjecture is much more difficult. The following resolution of the conjecture for the case $q = 3$ and partial result for the case $q = 4$ were obtained by Oxley (1987b) as consequences of non-trivial structure theorems for the classes of matroids involved. The matroid $\mathcal{W}^3$ is the rank 3 whirl; an affine representation for it is shown in Figure 6.9.

Figure 6.9.

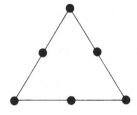

6.4.36. Proposition. *Let M be a loopless ternary matroid having no minor isomorphic to $M(K_4)$. Then $c(M; 3) \leqslant 2$.*

6.4.37. Proposition. *Let M be a loopless matroid that is coordinatizable over GF(4). Suppose that M has no minor isomorphic to $M(K_4)$ or $\mathcal{W}^3$. Then $c(M; 4) \leqslant 2$.*

The graph K_4 is isomorphic to the 3-spoked wheel graph $\mathcal{W}_3$ where $\mathcal{W}_r$ is shown in Figure 6.10. The polygon matroids of the wheel graphs are of fundamental structural importance in the class of matroids (see, for example, Tutte, 1966b), and the following conjecture of Oxley is an alternative strengthening of Brylawski's conjecture in the case $q = 2$.

Figure 6.10.

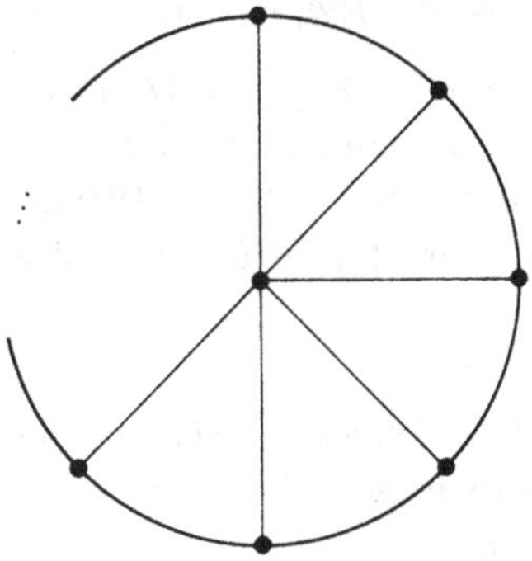

6.4.38. Conjecture. *Let M be a loopless binary matroid having no minor isomorphic to $M(\mathcal{W}_r)$. Then $c(M; 2) \leqslant r - 1$.*

The truth of Conjecture 6.4.34 would imply the truth of this conjecture. However, both conjectures seem very difficult. Conjecture 6.4.38 holds when $r = 3$ since it is equivalent to the case $q = 2$ of Conjecture 6.4.33, and we showed above that the latter is true. Moreover, Oxley proved Conjecture 6.4.38 in general when $r = 4$ (Oxley, 1987a) and for regular matroids when $r = 5$ (Oxley, 1989a). In the latter special case, the stronger bound $c(M; 2) \leqslant r - 2$ holds. Again both these results were derived from results on the structure of the relevant classes of matroids.

Prior to Whittle's advancing Conjecture 6.4.34, Walton & Welsh (1980) had proposed the following:

6.4.39. Conjecture. *If M is a loopless binary matroid having no minor isomorphic to $M(K_5)$, then $c(M; 2) \leqslant 3$.*

Welsh (1979) also offered the weaker conjecture that there is a fixed positive integer k so that, for all loopless binary matroids M having no minor isomorphic to $M(K_5)$, $c(M; 2) \leqslant k$. The corresponding weakenings of each of Conjectures 6.4.34 and 6.4.38 are both open and would certainly be sensible starting points for the conjectures themselves. We shall describe next some results of Kung that resolve Welsh's conjecture as well as the weakened form of Conjecture 6.4.33 that seeks only a fixed bound on $c(M; q)$ for all loopless matroids M that are coordinatizable over $GF(q)$ and have no minor isomorphic to $M(K_4)$.

Let $\mathscr{G}$ be a class of geometries that is closed under deletion. Kung defines its *size function*, $h(\mathscr{G}, r)$, to be the function with domain $D = \{r \in \mathbb{N} : \mathscr{G}$ contains a rank r geometry$\}$ for which $h(\mathscr{G}, r) = \max\{|E(M)| : M \in \mathscr{G}, r(M) = r\}$.

The *growth rate*, $g(\mathscr{G}, r)$, of $\mathscr{G}$ is defined, for all positive integers r in D, by

$$g(\mathscr{G}, r) = h(\mathscr{G}, r) - h(\mathscr{G}, r - 1).$$

The *maximum growth rate* $g(\mathscr{G})$ of $\mathscr{G}$ is $\max\{g(\mathscr{G}, r) : r \in D - \{0\}\}$, provided this maximum exists, and is infinite otherwise.

The following conjecture is due to Kung (1986a). A class of geometries is *minor-closed* if every geometry that is a minor of a member of the class is also in the class.

6.4.40. Conjecture. *Let $\mathscr{G}$ be a minor-closed class of geometries coordinatizable over $GF(q)$. Then the maximum growth rate of $\mathscr{G}$ is finite if and only if $\max\{c(M; q) : M \in \mathscr{G}\}$ is finite.*

In one direction this conjecture is proved by the following result (Kung, 1986a). The other direction remains open.

6.4.41. Proposition. *Let $\mathscr{G}$ be a class of geometries coordinatizable over $GF(q)$ and closed under deletion. Suppose that the size function of $\mathscr{G}$ satisfies $h(\mathscr{G}, r) \leqslant cr$ for some integer c. Then $c(M; q) \leqslant c$ for all M in $\mathscr{G}$.*

Proof. Let $M(E)$ be a member of $\mathscr{G}$. As $\mathscr{G}$ is closed under deletion, for all subsets E' of E, $|E'| \leqslant cr(E')$. Thus, by Edmonds' covering theorem (1965b), E can be partitioned into c independent sets. Since independent sets are affine, it follows by (6.50) that $c(M; q) \leqslant c$. $\square$

Kung (1986a, 1987) has proved a number of results on growth rates of various classes of geometries and from these has deduced, using the last result, bounds on the critical exponents of members of these classes. The next two propositions are examples of such results. The first proves the conjecture of Welsh stated after Conjecture 6.4.39. In each, M is a loopless binary matroid.

6.4.42. Proposition. *If M has no minor isomorphic to $M(K_5)$, then $c(M; 2) \leqslant 8$.*

6.4.43. Proposition. *If M has no minor isomorphic to $M(K_{3,3})$, then $c(M; 2) \leqslant 10$.*

The proofs of these results are long and involve operations on the bond graphs of binary geometries. For a binary geometry M having a hyperplane H, the *bond graph* $\Gamma(H, M)$ is the labelled graph defined as follows. The vertex set of $\Gamma(H, M)$ is the set of points of M not in H. Two vertices a and b are joined by an edge ab if there is a third point c on the line of M spanned by a and b. As $\{a, b, c\}$ is a circuit of M having an odd number of elements and M is binary, $\{a, b, c\}$ cannot be contained in the bond $E–H$ of M. Thus $c \in H$. The edge ab of $\Gamma(H, M)$ is labelled by c.

An approach to Conjecture 6.4.39 that provides an alternative to that offered by Proposition 6.4.42 is to try to retain the original bound, but to prove the result for a subclass of the original class. The next two results are of this form. The first is due to Walton & Welsh (1980), the second to Kung (1986a). As above, M is a loopless binary matroid.

6.4.44. Proposition. *Suppose that M has no minor isomorphic to $M(K_5)$ or F_7^*. Then $c(M; 2) \leqslant 3$.*

For an outline of the proof of this proposition, see Exercise 6.55.

6.4.45. Proposition. *Suppose that M has no minor isomorphic to $M(K_5)$ or F_7. Then $c(M; 2) \leqslant 3$.*

The following result of Kung (1988) employs a modification of the technique used to prove Propositions 6.4.42, 6.4.43, and 6.4.45 to obtain a partial result toward Conjecture 6.4.33.

6.4.46. Proposition. *Let M be a loopless matroid coordinatizable over $GF(q)$ and having no $M(K_4)$-minor. Then $c(M; q) \leqslant 6q^3$.*

We conclude this section with a solution to the critical problem for the class of transversal matroids. Brylawski (1975c) proved that, for a loopless principal transversal matroid M that is coordinatizable over $GF(q)$, $c(M; q) \leqslant 2$. He also proposed the following extension of that result. Recall that a gammoid is a minor of a transversal matroid.

6.4.47. Conjecture. *Let M be a loopless gammoid coordinatizable over $GF(q)$. Then $c(M; q) \leqslant 2$.*

As $M(K_4)$ is not a gammoid, this conjecture is weaker than Conjecture 6.4.33. Moreover, since, the latter holds for $q = 2$ and $q = 3$, so does the former. The following result of Whittle (1984) verifies Brylawski's conjecture for transversal matroids.

6.4.48. Proposition. *Let M be a loopless transversal matroid coordinatizable over $GF(q)$. Then $c(M; q) \leqslant 2$.*

To prove this proposition we shall need to recall some basic facts about transversal matroids. Let $M(E)$ be such a matroid and $(A_1, A_2, ..., A_n)$ be a presentation of M, that is, the independent sets of M are the partial transversals of this family of sets. A *cyclic flat* of M is a flat that is a union of circuits. From Corollary 5.1.3 of White (1987), we deduce that if F is a proper cyclic flat of M, then

$$F = \cap (E - A_i), \qquad (6.57)$$

where the intersection is taken over all i for which $F \cap A_i \neq \varnothing$.

The proof of Proposition 6.4.48 will use the next result of Bondy & Welsh (1972) and a lemma.

6.4.49. Proposition. *If $M(E)$ is a rank r transversal matroid, then M has a presentation $(A_1, A_2, ..., A_r)$ such that each A_i is a bond of M.*

6.4.50. Lemma. *Let $M(E)$ be a simple matroid coordinatizable over $GF(q)$. If E' is a subset of E that intersects all cyclic flats of M, then $c(M(E); q) \leqslant c(M(E'); q) + 1$.*

Proof. We identify $M(E)$ with a submatroid of $PG(r - 1, q)$ to which it is isomorphic. Let $c(M(E'); q) = k$. Then, by Corollary 6.4.3, there are hyperplanes, $H_1, H_2, ..., H_k$, of $PG(r - 1, q)$ such that $\left(\bigcap_{i=1}^{k} H_i \right) \cap E' = \varnothing$. If $\left(\bigcap_{i=1}^{k} H_i \right) \cap E$ contains a circuit C, then, since $\left(\bigcap_{i=1}^{k} H_i \right) \cap E$ is a flat of M, it contains $\mathrm{cl}_M(C)$, the closure in M of C. This set contains a point of E', hence so must $\left(\bigcap_{i=1}^{k} H_i \right) \cap E'$; a contradiction. We conclude that $\left(\bigcap_{i=1}^{k} H_i \right) \cap E$ is independent in M. Therefore the restriction of M to this set is affine, that is, $c(M(E - E'); q) = 1$. But, by Proposition 6.4.4, $c(M; q) \leqslant c(M(E'); q) + c(M(E - E'); q)$. Hence, $c(M; q) \leqslant c(M(E'); q) + 1$, as required. $\qquad \square$

Proof of Proposition 6.4.48. We may assume that M is simple and that $r(M) = r$. Thus we can identify $M(E)$ with a submatroid of $PG(r - 1, q)$ to which it is isomorphic. By Proposition 6.4.49, M has a presentation $(A_1, A_2, ..., A_r)$ such that each A_i is a bond of M. Let $I = \{1, 2, ..., r\}$ and, for all i in I, let H_i be the hyperplane of $PG(r - 1, q)$ that is spanned by $E - A_i$.

We suppose first that $\bigcap_{i \in I} H_i = \emptyset$. For all j in I, consider $\bigcap_{i \in I - j} H_i$. As $\bigcap_{i \in I} H_i = \emptyset$ and H_j is a modular flat of $PG(r - 1, q)$, $\bigcap_{i \in I - j} H_i$ has rank one and so contains a single point, say x_j. It follows, without difficulty, that $X = \{x_1, x_2, ..., x_r\}$ is independent in $PG(r - 1, q)$. Hence X is a basis of $PG(r - 1, q)$. Let N be the submatroid of $PG(r - 1, q)$ on $E \cup X$. We shall show next that X meets every cyclic flat of N. Assume that F is such a flat and that $F \cap X = \emptyset$. Then $r(F) \leqslant r - 1$, so F is a proper cyclic flat of M. Hence by (6.57), for some subset J of I,

$$F = \bigcap_{j \in J} (E - A_j) = E \cap \left(\bigcap_{j \in J} H_j \right).$$

But $\mathrm{cl}_N(F) = (E \cup X) \cap \mathrm{cl}_M(F) = (E \cup X) \cap \left(E \cap \left(\bigcap_{j \in J} H_j \right) \right) = (E \cup X) \cap \left(\bigcap_{j \in J} H_j \right)$.

Thus $\mathrm{cl}_N(F)$ contains $\{x_i : i \notin J\}$, that is, $\mathrm{cl}_N(F) \cap X \neq \emptyset$. Since $\mathrm{cl}_N(F) = F$, this is a contradiction. We conclude that X does indeed meet all cyclic flats of N. Thus, by Lemma 6.4.50, $c(N; q) \leqslant c(N(X); q) + 1$. But X is independent, so $N(X)$ is affine. Hence $c(N; q) \leqslant 2$, and so, by Proposition 6.4.4, $c(M; q) \leqslant 2$. This completes the proof in the case that $\bigcap_{i \in I} H_i = \emptyset$.

Now suppose that $\bigcap_{i \in I} H_i \neq \emptyset$ and choose x from $\bigcap_{i \in I} H_i$. This time we let N be the submatroid of $PG(r - 1, q)$ on $E \cup x$. A similar argument to the above again shows that $c(N; q) \leqslant 2$ and hence that $c(M; q) \leqslant 2$. $\square$

6.5. Linear Codes

In this section, we consider the work of Greene, Dowling, Jaeger, and Rosenstiehl & Read that applies Tutte–Grothendieck techniques to various problems related to linear codes.

To begin, recall that an $[n, r]$ *linear code* C over $GF(q)$ is an r-dimensional subspace of the n-dimensional vector space $V(n, q)$ over $GF(q)$. We call r and n, respectively, the *dimension* and *length* of C. If U is an $r \times n$ matrix over $GF(q)$, the rows of which form a basis for C, then U is called a *generator matrix* for C. It is straightforward to check that, for such a matrix U, the matroid on the columns of U depends only on C and not on U. We denote this matroid by $M(C)$. The *dual code* C^* of C is defined by $C^* = \{v \in V(n, q) : v \cdot w = 0 \text{ for all } w \text{ in } C\}$. Evidently C^* is an $[n, n - r]$ linear code. Moreover, it follows by Proposition 5.4.1 of White (1986) that $M(C^*)$ is isomorphic to $M^*(C)$, the dual of $M(C)$.

The members of a linear code C are called *codewords*. If $\mathbf{v}$ is the codeword $(v_1, v_2, \ldots, v_n)$, its *weight* $w(\mathbf{v})$ is the cardinality of its support, that is, $w(\mathbf{v}) = |\{v_i: v_i \neq 0\}|$. Dimension and length are two of the three fundamental parameters associated with a linear code C. The third of these parameters is the *distance* d of C. This is defined to be $\min\{w(\mathbf{v}): \mathbf{v} \in C - \mathbf{0}\}$. Evidently d is the size of a smallest bond in $M(C)$, or equivalently, the size of a smallest circuit in $M^*(C)$. It is therefore easily determined from $t(M(C))$ using the formula

$$d = n - r + 1 - \max\{j: b_{ij} > 0, \text{ for some } i > 0\}. \tag{6.58}$$

As a much deeper application of T–G techniques, we next present Greene's (1976) result that the distribution of codeweights in a linear code is a generalized T–G invariant. For a linear code C, the *codeweight polynomial* $A(C; q, z)$ of C is defined by

$$A(C; q, z) = \sum_{\mathbf{v} \in C} z^{w(\mathbf{v})}.$$

Thus, if a_i is the number of codewords $\mathbf{v}$ in C having weight i, then

$$A(C; q, z) = \sum_{i=0}^{n} a_i z^i.$$

The proofs below of Propositions 6.5.1 and 6.5.4 will use the following notation. If W is a subspace of $V(n, q)$, then W_0 will denote the subspace consisting of those vectors in W whose first entry is zero; $\hat{W}$ will denote the vector space obtained from W by removing the first entry of every vector. By convention, $\hat{W}_0 = \hat{W}$ where $W = W_0$.

6.5.1. Proposition.

$$A(C; q, z) = (1 - z)^r z^{n-r} t\left(M(C); \frac{1 + (q-1)z}{1 - z}, \frac{1}{z}\right).$$

Proof. Let $f(M(C)) = A(C; q, z)$. We shall show that f is well defined and that it is a generalized T–G invariant for which $\sigma = z$ and $\tau = 1 - z$. We begin by noting that f is well defined if C has length 1 for, in that case,

$$f(M(C)) = \begin{cases} 1, & \text{if } M(C) \text{ is a loop,} \\ 1 + (q-1)z, & \text{if } M(C) \text{ is an isthmus.} \end{cases}$$

Assume that f is well defined if C has length less than m, and suppose that C has length m where $m \geq 2$.

Let U be a generator matrix for C and suppose that the element e of $M(C)$ is neither a loop nor an isthmus. Without loss of generality, we may assume that e corresponds to the first column of U. Let U' be the matrix obtained by row reducing U so that the first entry in the first column is 1 and all other entries in this column are 0. Clearly U' is also a generator matrix for C. From considering the matrix U', we easily deduce that

$$M(\hat{C}_0) = M(C)/e \quad \text{and} \quad M(\hat{C}) = M(C) - e. \tag{6.59}$$

Now consider the map $g: C - C_0 \rightarrow \hat{C} - \hat{C}_0$ that removes the first entry of each codeword, that is, $g((v_1, \mathbf{v}')) = \mathbf{v}'$. We show next that

$$g \text{ is a bijection.} \tag{6.60}$$

First observe that if $(v_1, \mathbf{v}')$ and $(u_1, \mathbf{v}')$ are both in C for some distinct v_1 and u_1, then $(v_1 - u_1, \mathbf{0}) \in C$. But this implies the contradiction that e is an isthmus of $M(C)$. Hence the image of g is indeed $\hat{C} - \hat{C}_0$ and (6.60) holds. By definition, $A(C) = \sum_{\mathbf{v} \in C} z^{w(\mathbf{v})}$. Thus

$$A(C) = \sum_{(v_1, \mathbf{v}') \in C_0} z^{w((v_1, \mathbf{v}'))} + \sum_{(v_1, \mathbf{v}') \in C - C_0} z^{w((v_1, \mathbf{v}'))} = \sum_{\mathbf{v}' \in \hat{C}_0} z^{w(\mathbf{v}')} + z \sum_{(v_1, \mathbf{v}') \in C - C_0} z^{w(\mathbf{v}')}.$$

Therefore, by (6.60),

$$A(C) = \sum_{\mathbf{v}' \in \hat{C}_0} z^{w(\mathbf{v}')} + z \sum_{\mathbf{v}' \in \hat{C} - \hat{C}_0} z^{w(\mathbf{v}')} = z \sum_{\mathbf{v}' \in \hat{C}} z^{w(\mathbf{v}')} + (1 - z) \sum_{\mathbf{v}' \in \hat{C}_0} z^{w(\mathbf{v}')}$$

$$= zA(\hat{C}) + (1 - z)A(\hat{C}_0).$$

Hence, by (6.59) and the induction assumption, if e is neither a loop nor an isthmus of $M(C)$, then

$$A(C) = zf(M(C) - e) + (1 - z)f(M(C)/e). \tag{6.61}$$

Now suppose that e is a loop of $M(C)$. Then $A(C) = A(\hat{C}_0)$, so, by the induction assumption,

$$A(C) = f(L)f(M(C)/e). \tag{6.62}$$

Finally, if e is an isthmus of $M(C)$, then C is the direct sum of $\hat{C}$ with a one-dimensional space. Hence $A(C) = (1 + (q - 1)z)A(\hat{C})$ so, by the induction assumption,

$$A(C) = f(I)f(M(C) - e). \tag{6.63}$$

On combining (6.61)–(6.63), we conclude by induction that f is well defined. Moreover, the same equations imply that f is a generalized T–G invariant. The proposition now follows easily by Corollary 6.2.6. $\square$

We now sketch an alternative derivation of Proposition 6.5.1 that mimics the treatment of two-variable coloring in section 6.3.F. As there, T denotes the $(n + 1) \times (n + 1)$ matrix with $T(i, j) = \binom{j}{i}$ where i and j both range over the set $\{0, 1, 2, ..., n\}$; and M_{KC} is the $(n + 1) \times (r + 1)$ matrix with rows and columns indexed by $\{0, 1, 2, ..., n\}$ and $\{0, 1, 2, ..., r\}$, respectively, such that

$M_{KC}(i, j)$ is the number of i-element subsets of $M(C)$ having corank j, that is, rank $r - j$. Now, recall that $A(C) = \sum_{i=0}^{r} a_i z^i$. Then, as in (6.43), we have the matrix equation

$$M_{KC} \cdot \mathbf{q} = T \cdot \mathbf{c}_q \qquad (6.64)$$

where $\mathbf{q} = (1, q, q^2, ..., q^r)^t$ and $\mathbf{c}_q = (a_n, a_{n-1}, ..., a_0)^t$. Both sides of (6.64) are column vectors with rows indexed by i where $0 \leqslant i \leqslant n$. To verify (6.64), we observe that, for fixed i, the corresponding entries in these column vectors count the number of pairs $(P, \mathbf{w})$ where P is an i-element subset of the set of columns of the generator matrix U of C and $\mathbf{w}$ is a codeword that has entry 0 in every column corresponding to P. The left-hand side of (6.64) chooses P first according to its corank, while the right-hand side chooses $\mathbf{w}$ first according to its number of zero entries.

Proposition 6.5.1 can be derived from (6.64) by arguing as in section 6.3.F. We leave the details of this to the reader as an exercise. Again, as for two-variable coloring, we can invert Proposition 6.5.12 to obtain

$$t(M(C); x, y) = \frac{y^n}{(y-1)^r} A\left(C; (x-1)(y-1), \frac{1}{y}\right). \qquad (6.65)$$

Hence we can recover the Tutte polynomial of a vector matroid M if we know its associated linear code over sufficiently many finite fields. For example, if M is coordinatizable over $GF(q)$, we could use the fields $GF(q^i)$ for $1 \leqslant i \leqslant r$.

The next result is the celebrated MacWilliams duality formula for linear codes (MacWilliams, 1963; Greene, 1976). It can be proved analogously to the proof of Proposition 6.3.27. We leave the details to the reader.

6.5.2. Proposition.

$$A(C^*; q, z) = \frac{(1 + (q-1)z)^n}{q^r} A\left(C; q, \frac{1-z}{1 + (q-1)z}\right).$$

Apart from Greene's work just described, the other pioneering work in the matroid invariant theory of linear codes was by Dowling (1971). He showed that a fundamental problem of coding theory, that of finding the maximum possible dimension r for a linear code over $GF(q)$ having length n and distance at least d, is a special case of the critical problem for matroids.

The precise statement of Dowling's result will require another definition. The *punctured Hamming ball*, $H_q(n, d - 1)$, consists of all non-zero vectors of $V(n, q)$ having fewer than d non-zero coordinates. Evidently C is a maximum-dimension length n linear code having distance at least d if and only if C is a maximum-dimension subspace of $V(n, q)$ containing no member of $H_q(n, d-1)$. Thus the problem of maximizing the dimension of a code of distance at least d is equivalent to the problem of finding the critical exponent of the vector matroid on $H_q(n, d-1)$. In fact:

6.5.3. Proposition. (*Dowling, 1971*) *If r is the maximum dimension of a linear code over $GF(q)$ having length n and distance at least d, and c is the critical exponent of the vector matroid on $H_q(n, d-1)$, then $r = n - c$.*

Proof. This follows directly from the definition of the critical exponent. $\square$

Let $G_q(n, d-1)$ be the simplification of the vector matroid on $H_q(n, d-1)$. Then $G_q(n, 2)$ is $Q_n(GF(q)^*)$, the rank n Dowling geometry based on the multiplicative group of $GF(q)$ (see section 6.4.B). When $q = 2$, $G_q(n, 2)$ is isomorphic to the polygon matroid of K_{n+1}. In general, the calculation of the characteristic polynomial of $G_q(n, m)$ is difficult except in the cases where m is 1, 2, $n-1$, or n, when the geometry is supersolvable (see Exercise 6.78).

Next we turn our attention to an interesting T–G invariant for *binary codes*, that is, linear codes over $GF(2)$. This invariant was discovered for graphs by Rosenstiehl & Read (1978), and Jaeger (1989b) noted that their result could be extended to binary matroids. The dimension of a vector space V will be denoted by dim V.

6.5.4. Proposition. *Let C be a binary code of length n. Then*

(i) $$t(M(C); -1, -1) = (-1)^n 2^{\dim(C \cap C^*)}.$$

Hence

(ii) $$|t(M(C); -1, -1)| = |C \cap C^*|.$$

Proof. The proof in Rosenstiehl & Read (1978) was in graph-theoretic terms. We generalize these ideas to binary vector spaces. Let $h(M(C)) = (-1)^{n(C)} 2^{\dim(C \cap C^*)}$ where $n(C)$ is the length of C. We shall show that h is a well defined T–G invariant. First note that h is well defined if $n(C) = 1$ for, in that case, $C \cap C^* = \{0\}$ and so $h(M(C)) = -1$. Assume that h is well defined if $n(C) < m$ and let $n(C) = m \geqslant 2$.

Let $B = B(C) = C \cap C^*$ and suppose that the element e of $M(C)$ is neither a loop nor an isthmus. We may assume that e corresponds to the first column of a generator matrix U for C. If $x \in \hat{C}$, then either $(1, x)$ or $(0, x)$ is in C, but not both, otherwise $(1, 0) \in C$, and e is an isthmus of $M(C)$.

Now observe that

either $B = B_0$, or, for some vector x having first entry 1, $B = B_0 + \langle x \rangle$.

(6.66)

In view of (6.59), we shall write $C - e$ for $\hat{C}$, and C/e for $\hat{C}_0$. One easily checks that $(C - e)^*$ is C^*/e, that is,

$$(C - e)^* = (\widehat{C^*})_0.$$

Thus if $x \in B(C - e) = (C - e) \cap (C - e)^*$, then $(0, x) \in C^*$, while either $(0, x) \in C$ or $(1, x) \in C$, but not both. Dually, if $y \in B(C/e)$, then $(0, y) \in C$, while either $(0, y) \in C^*$ or $(1, y) \in C^*$, but not both. The proof of the next result is straightforward and is left as an exercise.   □

6.5.5. Lemma. *Either*

(i) *for some* x *in* $B(C - e)$, $(1, x) \in C$ *and so* $B(C - e) = \hat{B}_0 + \langle x \rangle$;

or

(ii) *for all* x *in* $B(C - e)$, $(0, x) \in C$ *and* $B(C - e) = \hat{B}_0$.

Now suppose that dim $B = k$. We shall show that one of the following three possibilities must occur.

$$\dim B(C - e) = k - 1 = \dim B(C/e). \tag{6.67}$$

$$\dim B(C - e) = k + 1, \dim B(C/e) = k. \tag{6.68}$$

$$\dim B(C - e) = k, \dim B(C/e) = k + 1. \tag{6.68*}$$

First we note that each statement in the following is equivalent to its successor.

(1) $B = B_0$.

(2) $(1, 0) \in B^*$.

(3) $(1, 0) \in (C \cap C^*)^* = C + C^*$.

(4) For some z,

    (a) $(1, z) \in C$ and $(0, z) \in C^*$; or

    (b) $(1, z) \in C^*$ and $(0, z) \in C$.

Suppose that $B \neq B_0$. Then (i) cannot occur otherwise $(1, x) \in C$ and $(0, x) \in C^*$ so $(1, 0) \in C + C^*$; a contradiction. Thus $B(C - e) = \hat{B}_0$ and, dually, $B(C/e) = \hat{B}_0$. Hence dim $B(C - e) = \dim \hat{B}_0 = \dim B(C/e)$. But, by (6.66), as $B \neq B_0$, dim $B = \dim B_0 + 1 = \dim \hat{B}_0 + 1$. Thus if $B \neq B_0$, then (6.67) occurs.

We may now assume that $B = B_0$. Then, from above, (4a) or (4b) occurs. In the former case, for some z_1, $(1, z_1) \in C$ and $(0, z_1) \in C^*$, so $z_1 \in B(C - e)$. In the latter case, for some z_2, $(0, z_2) \in C$ and $(1, z_2) \in C^*$ so $z_2 \in B(C/e)$. Moreover, exactly one of (4a) and (4b) occurs; otherwise, for some z_1 and z_2, both $(1, z_1)$ and $(0, z_2)$ are in C and $(0, z_1)$ and $(1, z_2)$ are in C^*. Hence $(1, z_1) \cdot (1, z_2) = 0$ and $(0, z_1) \cdot (0, z_2) = 0$, so $1 = 0$; a contradiction.

Suppose that (4a) occurs. Then, as $z_1 \in B(C - e)$, we have, by (i), that dim $B(C - e) = \dim \hat{B}_0 + 1 = \dim B_0 + 1 = \dim B + 1$. Moreover, as (4b) does not occur, the dual of Lemma 6.5.5 implies that $B(C/e) = \hat{B}_0$, so dim $B(C/e) = \dim \hat{B}_0 = \dim B_0 = \dim B$. Hence, if (4a) occurs, then so does (6.68) and, by duality, if (4b) occurs, so does (6.68*). We conclude that one of (6.67), (6.68), and (6.68*) must occur. It is routine to check that, in each case,

$$(-1)^n 2^{\dim B} = (-1)^{n(C-e)} 2^{\dim B(C-e)} + (-1)^{n(C/e)} 2^{\dim B(C/e)}.$$

Hence, by (6.59) and the induction assumption, if e is neither a loop nor an isthmus of $M(C)$, then

$$(-1)^{n(C)}2^{\dim B} = h(M(C) - e) + h(M(C)/e).$$

It is also easy to check that

$$(-1)^{n(C)}2^{\dim B} = \begin{cases} h(I)h(M(C) - e) & \text{if } e \text{ is an isthmus,} \\ h(L)h(M(C) - e) & \text{if } e \text{ is a loop.} \end{cases}$$

From the last two equations, we deduce by induction that h is well defined and, moreover, that h is a T–G invariant. As $h(I) = h(L) = -1$, it follows by Theorem 6.2.2 that $h(M(C)) = t(M(C); -1, -1)$. □

6.5.6. Corollary. *Let C be a binary code. Then $C \cap C^*$ is trivial if and only if $M(C)$ has an odd number of bases.*

Proof. For an arbitrary matroid M, if both $t(M; 1, 1)$ and $t(M; -1, -1)$ are evaluated modulo 2, we obtain the same result. But, by 6.2.11(i), $t(M; 1, 1)$ is the number of bases of M. Using these observations, the result follows easily from Proposition 6.5.4. □

On putting $(q, z) = (4, -1)$ in Proposition 6.5.1 and using Proposition 6.5.4, we get that, for a binary code C,

$$2^{k+r} = |c_e - c_o| \tag{6.69}$$

where $k = \dim(C \cap C^*)$ and c_e and c_o are the number of even- and odd-weight codewords, respectively, in the linear code over $GF(4)$ that is generated by C.

A consequence of (6.69) is that $C \subseteq C^*$ if and only if, in the code over $GF(4)$ that is generated by C, all codewords have even weight. We leave the verification of this to the reader noting that one can derive this directly from the fact that $C \subseteq C^*$ if and only if, in a binary generator matrix U for C, any two rows are orthogonal over $GF(2)$.

In contrast to the binary case, if C is a linear code over $GF(q)$ for $q \geq 4$ then $\dim(C \cap C^*)$ is not, in general, a matroid invariant. For example, consider the following representation of the 4-point line over $GF(13)$ where $a \notin \{0, 1\}$:

$$\begin{bmatrix} 1 & 0 & 1 & 1 \\ 0 & 1 & 1 & a \end{bmatrix}.$$

We leave the reader to check that

$$\dim(C \cap C^*) = \begin{cases} 1 & \text{if } a \in \{6, 8\}, \\ 0 & \text{otherwise.} \end{cases}$$

When $q = 3$, Jaeger (1989c) showed that $(\sqrt{3})^{\dim(C \cap C^*)}$ is the modulus of the complex number $t(M(C); j, j^2)$ where $j = e^{2\pi i/3}$.

6.6. Other Tutte Invariants

The bibliography at the end of this chapter indicates how widespread the use of T–G techniques has been in combinatorics. We have tried to provide some indication of this in this chapter. However, the large number of diverse applications, together with our desire to avoid providing a superficial skim through the theory, has meant that some important topics have been omitted. To compensate partially for this we shall, in this short section, briefly survey some of the other applications. Still more applications are touched on in the exercises and we hope that the extensive bibliography will provide the reader with adequate opportunities for further reading on matroid invariant theory. One particularly exciting recent development has seen the application of T–G techniques in knot theory where a family of new polynomial invariants has been discovered. We shall not discuss this rapidly growing area here, but instead, refer the interested reader to Kauffman's survey (1988) of the area and to the following papers in the bibliography: Jaeger (1988c), Jaeger, Vertigan & Welsh (1990), Jones (1985), Kauffman (1987, 1988), Lickorish (1988), Lipson (1986), Thistlethwaite (1987, 1988a, b), Traldi (1989), and Vertigan (1990).

6.6.A. Basis Activities

We consider here what is essentially the most basic group invariant. It was introduced for graphs by Tutte in his founding work (1954) and was later extended to matroids by Crapo (1969). For a matroid M, recall that $t(M; 1, 1)$ enumerates the bases of M. Thus, it is clear that we may partition $\mathscr{B}(M)$, the set of bases of M, into blocks $\mathscr{B}_{ij}$ where $|\mathscr{B}_{ij}| = b_{ij}$, the coefficient of $x^i y^i$ in $t(M; x, y)$. We now describe one way to obtain such a partition.

First we linearly order the ground set E of M by relabeling the elements of E as $1, 2, ..., n$. For a basis B of M, the *internal activity*, $\iota(B)$, of B is equal to the number of elements e of B for which e is the least element in the unique bond contained in $(E - B) \cup e$. Similarly, the *external activity*, $\varepsilon(B)$, of B is the number of elements e of $E - B$ for which e is the least element in the unique circuit contained in $B \cup e$. On letting $\mathscr{B}_{ij}$ be the number of bases of M of internal activity i and external activity j, we obtain the desired partition of $\mathscr{B}(M)$, that is,

$$b_{ij} = |\{B \in \mathscr{B}(M): \iota(B) = i, \varepsilon(B) = j\}|. \tag{6.70}$$

To verify this, one shows that if $f(M) = |\mathscr{B}_{ij}|$, then provided the greatest element n is neither a loop nor an isthmus,

$$f(M) = f(M - n) + f(M/n). \tag{6.71}$$

One striking consequence of (6.70) is the fact that $|\{B \in \mathcal{B}(M): \iota(B) = i, \varepsilon(B) = j\}|$ does not depend upon the particular ordering chosen for E.

Whitney (1932) introduced another concept that has been fruitfully developed in this context. A *broken circuit* is an independent set that is obtained from a circuit by deleting its least element. Evidently a basis has external activity equal to zero if and only if it contains no broken circuits. Now define $p^+(M; \lambda)$ by

$$p^+(M; \lambda) = t(M; \lambda + 1, 0). \tag{6.72}$$

By (6.20), if w_i is the coefficient of λ^{r-i} in the characteristic polynomial $p(M; \lambda)$ of M, then

$$p^+(M; \lambda) = \sum_{i=0}^{r} |w_i| \lambda^{r-i}.$$

The numbers w_i, $0 \leqslant i \leqslant r$, are called the *Whitney numbers of the first kind*. A detailed discussion of their properties can be found in Aigner (1987). It is not difficult to show, by verifying (6.71) for $f(M) = |w_i|$, that

$$|w_i| = |\{I \subseteq E: |I| = i \text{ and } I \text{ contains no broken circuits}\}|. \tag{6.73}$$

Hence, by (6.72),

$t(M; 2, 0)$ is $\sum_{i=0}^{r} |w_i|$, the number of subsets of E that contain no broken circuits.

$$\tag{6.74}$$

A bijective proof of the relationship between (6.70) and (6.73) is given in Brylawski (1977c), and other properties of the associated invariants may be found in Beissinger (1982), Björner (1980, 1982), Brylawski (1977b, 1982), Brylawski & Oxley (1980, 1981), Wilf (1976), and Zaslavsky (1983). Among these are toplogical properties of the broken-circuit complex, the simplicial complex whose simplices are the subsets that contain no broken circuits. Berman (1977) gives an activity-theoretic interpretation of b_{ij} for acyclic orientations. A generalization of his result to a three-variable polynomial has been given by Las Vergnas (1978; 1984).

6.6.B. Hyperplane Arrangements

A finite set of hyperplanes in Euclidean d-space $\mathbb{E}^d$ is called an *arrangement of hyperplanes*. Such an arrangement decomposes $\mathbb{E}^d$, and various counting problems associated with this decomposition have been extensively studied. See Brylawski (1976, 1985), Cordovil (1980, 1982, 1985), Cordovil & Silva (1985, 1987), Cordovil, Las Vergnas & Mandel (1982), Greene (1977), Greene & Zaslavsky (1983), Las Vergnas (1977), Winder (1966), Zaslavsky (1975a, 1976, 1977, 1979, 1981a, b, 1983), Buck (1943), Orlik (1989), Schläfi (1950), and Stanley (1980). Moreover, Zaslavsky (1975b) has given a comprehensive treatment of such problems. In this section we briefly survey some of his results which can be derived using T–G techniques.

Let $\{H_1, H_2, ..., H_n\}$ be a set of hyperplanes in $\mathbb{E}^d$ and consider the set of intersections $\left\{\bigcap_{i \in I} H_i : I \subseteq \{1, 2, ..., n\}\right\}$. This set is partially ordered by reverse inclusion. In general, this poset P need not be a geometric lattice, although it will be if no intersection of hyperplanes is parallel to another hyperplane, or, more precisely, if the following condition holds (Exercise 6.87).

Whenever $J \subseteq \{1, 2, ..., n\}$ and $\bigcap_{i \in J} H_i$ contains a line, $\bigcap_{i \in J} H_i$ meets H_j

for all j in $\{1, 2, ..., n\} - J$.   (6.75)

In the results that follow we assume that P is a geometric lattice and we let M denote the simple matroid for which P is the lattice of flats. For the generalizations of these results to arbitrary arrangements of hyperplanes in projective as well as Euclidean space, we refer the reader to Zaslavsky's paper (1975b).

When the hyperplanes $H_1, H_2, ..., H_n$ are removed from $\mathbb{E}^d$, the remainder of the space falls into components, each a d-dimensional open polyhedron. We call these polyhedra *regions* of the arrangement. Such regions may be bounded or unbounded. An arrangement of hyperplanes is called *central* if the hyperplanes have non-empty common intersection.

6.6.1. Proposition. *The total number of regions of a non-central arrangement is* $t(M; 2, 0) - t(M; 1, 0) = t(M; 2, 0) - |\mu(M)|$.

6.6.2. Proposition. *The number of bounded regions of a non-central arrangement is* $t(M; 1, 0) = |\mu(M)|$.

For central arrangements, the situation is a little different. In particular, (6.75) always holds for such an arrangement so the poset of hyperplane intersections will certainly be a geometric lattice in this case. It is not difficult to show that a central arrangement has no bounded regions. On the other hand:

6.6.3. Proposition. *The number of (unbounded) regions in a central arrangement is* $t(M; 2, 0)$.

Given a central arrangement $\{H_1, H_2, ..., H_n\}$ having associated matroid M, suppose we perturb one of the hyperplanes, say H_i, by translation from its initial position. Let H_i' be the perturbation of H_i.

6.6.4. Proposition. *The numbers of bounded regions of the arrangements* $\{H_1, H_2, ..., H_n\} \cup \{H_i'\}$ *and* $(\{H_1, H_2, ..., H_n\} \cup \{H_i'\}) - \{H_i\}$ *are the same and are equal to* $\beta(M)$.

As a further development of these ideas, Greene & Zaslavsky (1983) have given an interpretation in terms of arrangements for the coefficients of the characteristic polynomial. Moreover, many of the above results can be generalized to oriented matroids (see, for example, Cordovil, Las Vergnas & Mandel, 1982; Las Vergnas, 1975a; 1984). Finally, we note that Cordovil (1980) showed that a conjecture of Grünbaum on the minimum number of regions of a pseudoline arrangement in the real projective plane can be deduced from certain general inequalities for the Whitney numbers.

6.6.C. Separation of Points by Hyperplanes

We now consider various results that are obtained by projectively dualizing the results of the previous subsection. When this is done, the image of a hyperplane is a point and the image of a region is a topologically connected family of hyperplanes. Suppose E is a finite subset of $\mathbb{E}^d$ and M is the affine matroid induced on E. We consider those subsets E' of E that can be separated from their complements $E - E'$ by some hyperplane of $\mathbb{E}^d$. The number of such *hyperplane-separable subsets* is a T–G invariant. In fact:

6.6.5. Proposition. *The number of hyperplane-separable subsets of E equals* $t(M; 2, 0)$.

To verify the fundamental T–G recursion in this case, one considers an extreme point e of E. Then every separation of a subset E' of $E - e$ gives a separation of either E' or $E' \cup e$. Both E' and $E' \cup e$ are hyperplane-separable if and only if E' can be separated from $E - (E' \cup e)$ by a hyperplane through e. But separations of the latter type are in one-to-one correspondence with separations in M/e, where here one projects from e onto a hyperplane H which is in general position further from e than any point of $E - e$.

Similar arguments to the above can be used to establish the following results.

6.6.6. Proposition. *Let C be the convex hull of a basis B of M and suppose that $C \cap E = B$. Then the number of subsets of E that can be separated from their complements by a hyperplane intersecting C is given by* $t(M; 2, 0) - 2t(M; 1, 0) = t(M; 2, 0) - 2|\mu(M)|$.

6.6.7. Proposition. *Let ε be sufficiently small and B_ε be an epsilon ball in $\mathbb{E}^d$ centered at an element e of M that is neither a loop nor an isthmus. Then the* number of subsets of E that can be separated by a hyperplane that passes through B_ε is given by $t(M; 2, 0) - 2b_{10} = t(M; 2, 0) - 2\beta(M)$.

To conclude this section, we note that, apart from their close links with hyperplane dissections, hyperplane separations are also closely related to acyclic orientations of graphs. Details of this relationship, including combinatorial correspondences between the objects involved, can be found in Brylawski (1985), Greene (1977), and Greene & Zaslavsky (1983).

6.6.D. Intersection Theory

We saw in sections 6.3.F and 6.5 that the coboundary polynomial $\bar{\chi}(M; \lambda, v)$ enumerates generalized colorings as well as codeweights. We now put these two facts into a general framework: the combinatorial structure of the way an embedded matroid intersects the flats of its ambient geometry. More details of this *intersection theory* can be found in Brylawski (1979b, 1981b).

A matroid $M(E)$ is said to be *embedded into a geometry* $G(T)$ if there is a mapping $f: E \to T \dot\cup \hat{0}$ such that $r_M(E') = r_G(f(E'))$ for all subsets E' of E. Equivalently, the simplification of M is a subgeometry of G. The element $\hat{0}$ here serves merely as the image of any loops in M.

A rank r geometry G is called *upper combinatorially uniform* if it has the same number $W_2(i, j)$ of flats of corank j in every upper interval of rank i. The numbers $W_2(i, j)$ are called the (doubly indexed) *Whitney numbers* of G *of the second kind*. The equation $W_2(r, r - k) = W_k$ relates these numbers to the (singly indexed) Whitney numbers W_k of the second kind discussed by Aigner (1987). The reader should note that these doubly indexed Whitney numbers of the second kind differ from their namesakes W_{ij}, which were studied by Greene & Zaslavsky (1983). Examples of upper combinatorially uniform geometries include finite affine and projective geometries, where the Whitney numbers are Gaussian coefficients; Boolean algebras, where the Whitney numbers are binomial coefficients; polygon matroids of complete graphs, where the Whitney numbers are Stirling numbers of the second kind; and perfect matroid designs, that is, matroids in which flats of the same rank have equal cardinalities. In the last case, a formula for $W_2(i, j)$ in terms of the sizes of the flats can be found in Brini (1980), Brylawski (1979b, 1982), and Young, Murty & Edmonds (1970).

The *intersection matrix* $I_G(M)$ of the embedding of $M(E)$ into the upper combinatorially uniform geometry G is defined by $I_G(M; i, j) = |\{F: F$ is a corank j flat of G with $|f^{-1}(F)| = i\}|$, that is, $I_G(M; i, j)$ counts the flats of G of corank j that contain i points of E. The *intersection polynomial* $i_G(M; u, v)$ is then given by

$$i_G(M; u, v) = \sum_i \sum_j I_G(M; i, j) u^i v^j.$$

The principal idea of intersection theory is that the numbers $I_G(M; i, j)$ do not depend upon the embedding but only on the Whitney numbers of G and the Tutte polynomial of M. Indeed, if W_2 denotes the matrix of Whitney numbers $W_2(i, j)$ of G, then we have the matrix equation

$$T^{-1} \cdot M_{\mathrm{KC}} \cdot W_2 = I_G(M) \tag{6.76}$$

where $T^{-1}(i, j) = (-1)^{i+j} \binom{j}{i}$ for all i, j in $\{0, 1, ..., |E|\}$ and M_{KC} is the cardinality-corank matrix of M (see section 6.5). After seeing similar arguments in sections 6.3.F and 6.5, the reader will not be surprised that (6.76) is derived by counting in two different ways: it is not difficult to check that

$$(T \cdot I_G)(i, j) = (M_{\mathrm{KC}} \cdot W_2)(i, j) = N(i, j) \tag{6.77}$$

where $N(i, j) = |\{(F, E'): F \text{ is a corank } j \text{ flat of } G, E' \subseteq E, |E'| = i$, and $E' \subseteq f^{-1}(F)\}|$. Hence $N(i, j)$ is the number of ordered pairs consisting of a flat of corank j and a subset of size i embedded in the flat.

The intersection polynomial $i_G(M; u, v)$ is not a T–G invariant. However, it does satisfy the same recursion as the coboundary polynomial $\bar{\chi}$:

$$i_G(M) = i_G(M - e) + (u - 1)i_{\overline{G/f(e)}}(M/e) \tag{6.78}$$

where e is neither a loop nor an isthmus of M, and M/e is embedded into $\overline{G/f(e)}$, the simplification of $G/f(e)$. The formula (6.78) is derived from the following recursion:

$$I_G(M; i, j) = I_G(M - e; i, j) + I_{\overline{G/f(e)}}(M/e; i - 1, j) - I_{\overline{G/f(e)}}(M/e; i, j). \tag{6.79}$$

To verify (6.79), observe that $I_G(M - e; i, j)$ counts the i-point flats of M that do not contain e, together with the $(i + 1)$-point flats of M containing e. But there are precisely $I_{\overline{G/f(e)}}(M/e; i, j)$ flats of the latter type. Since $I_{\overline{G/f(e)}}(M/e; i - 1, j)$ counts those i-point flats that do contain e, (6.79) follows.

On combining (6.76) and the matrix-theoretic proof of 6.3.26, we get the polynomial equation

$$i_G(M; u, v) = (u - 1)^r t\left(M; \frac{u + \lambda - 1}{u - 1}, u \right)\Bigg|_{\lambda^i \mapsto \sum_j W_2(i, j)v^j}. \tag{6.80}$$

Here we use the same evaluation of the Tutte polynomial as in 6.3.26 and then replace λ^i in the resulting polynomial for $\bar{\chi}(M)$ by $\sum_j W_2(i, j)v^j$. To justify this, compare (6.43) and (6.77).

For any upper combinatorially uniform geometry, G, every rank k upper interval has the same characteristic polynomial. Denoting this polynomial by $p_k(G; \lambda)$, we see that $p_r(G; \lambda) = p(G; \lambda)$. Now let $W_1(i, j)$ denote the coefficient of λ^j in $p_i(G; \lambda)$, where both i and j are chosen from the set $\{0, 1, 2, ..., r\}$. We call the numbers $W_1(i, j)$ the (doubly indexed) *Whitney numbers of the first kind*. The equation $W_1(r, r - k) = w_k$ relates these numbers to the (singly indexed) Whitney numbers w_k of the first kind which are discussed in section 6.6.A and in more detail by Aigner (1987). As with the Whitney numbers of the second kind, the numbers $W_1(i, j)$ differ from the numbers $w_{i,j}$, which Greene & Zaslavsky (1983) call 'doubly indexed Whitney numbers of the first kind'.

One can show that W_1 and W_2 are inverse matrices (Exercise 6.92b). When G is the polygon matroid of a complete graph, this inverse relationship is precisely the relationship between the Stirling numbers of the first and second kinds. When G is a Boolean algebra B, the inverse relationship is that exploited earlier between the binomial coefficient matrix T and the signed binomial coefficient matrix T^{-1}. Indeed, one easily checks that $p_i(B; \lambda) = (\lambda - 1)^i$, so that, in this case, $W_1(i, j) = (-1)^{i+j}\binom{j}{i}$, that is, $W_1(i, j) = T^{-1}(i, j)$. Dowling (1973b) was the first to prove the formula for evaluating $p_i(G; \lambda)$ from an inverse matrix. This formula also appears in a slightly more general form in Brylawski (1979b).

Using the fact that W_1 is the inverse of W_2, we get immediately from (6.76) that

$$M_{KC} = T \cdot I_G(M) \cdot W_1. \tag{6.81}$$

Moreover, by inverting (6.80) and using (6.45), it can be shown that

$$\bar{\chi}(M; u, \lambda) = i_G(M; u, v)|_{v^i \mapsto p_i(M; \lambda)}. \tag{6.82}$$

The above theory can be used in a straightforward manner to compute, for example $i_{G'}(M)$ from $i_G(M)$ where G' is another upper combinatorially uniform geometry in which M is embedded (Brylawski, 1979b). Hence, the intersection numbers for an embedding of an n-vertex graph Γ into K_n yield those for a linear representation of M_Γ. Further, $i_G(M)$ can be computed from $i_G(M^*)$, this result being the intersection analog of the MacWilliams duality formula. Also, if G is a perfect matroid design and G' is a subgeometry of G, then one can compute $t(G')$ from $t(G - G')$. To see this, we note that if a_j is the size of each corank j flat of G, then

$$I_G(G'; i, j) = I_G(G - G'; a_j - i, j).$$

Some examples of the above calculations are given in the exercises.

Among the other applications of intersection theory are various reconstruction results for the Tutte polynomial. In particular, Proposition 5.1 of Brylawski (1982) enables one to reconstruct $t(M)$ from the Tutte polynomials of its single-element contractions. With $S_{KC}(M; x, y)$ equal to the cardinality-corank polynomial of $M(E)$, we have the straightforward formula

$$S_{KC}(M; x, y) = \int \left(\sum_{e \in E} S_{KC}(M/e; x, y) \right) dx + y^{r(M)} \tag{6.83}$$

where $\int dx$ is the formal integral operator: $\int x^k y^m dx = \dfrac{x^{k+1} y^m}{k+1}$.

Brylawski (1981c) reconstructed $t(M)$ from the multiset of isomorphism classes of hyperplanes of M; in the graphic case (1981b), he reconstructed $t(M_\Gamma)$ from the multiset of isomorphism classes of single-vertex deletions of Γ. To make the latter calculation and to find the intersection numbers for graph complements, a finer graph invariant called the *polychromate* was introduced. For a graph Γ having m vertices and n edges, this is defined by

$$\hat{\chi}(\Gamma; y, z_1, z_2, ..., z_m) = \sum_i y^i \sum_\pi M(\Gamma, i, \pi) \mathbf{z}^\pi. \qquad (6.84)$$

The second sum here is over all integer partitions π of m, while if $\pi = 1^{a_1} 2^{a_2} ... m^{a_m}$, then $\mathbf{z}^\pi = z_1^{a_1} z_2^{a_2} ... z_m^{a_m}$. Further, $M(\Gamma, i, \pi)$ denotes the number of vertex partitions of type π in which there are exactly i edges of Γ joining two vertices in the same class.

It is an easy matter to reconstruct the polychromate $\hat{\chi}$ from vertex or edge deletions, as well as to determine its behavior under the addition of isolated vertices or the taking of complements. Moreover,

$$i_{K_m}(\Gamma; u, v) = \hat{\chi}(\Gamma; u, v, v, ..., v). \qquad (6.85)$$

One can use these ideas to construct non-isomorphic graphic matroids of arbitrarily high connectivity having the same polychromate, and hence the same Tutte polynomial (see Exercise 6.19).

As a final application of intersection theory, we sketch suggestively similar statements and proofs of two extremal theorems, one due to Bose & Burton (1966), the other to Turán (see, for example, Erdös, 1967). Another proof of the former is outlined in Exercise 6.65. In both these theorems, the ambient geometry is a supersolvable upper combinatorially uniform geometry G. Each theorem has two parts: the first part asserts that a smallest subset of G that meets every modular flat of rank $c+1$ has the same size as a smallest flat of corank c; the second part asserts that every such smallest subset of G is a smallest flat of corank c. Our proof will be only of the second part of these theorems, the characterization of the extremal subsets. Since any flat of corank c meets all modular flats of rank $c+1$, to prove this second part it suffices to show that every extremal subset is a flat.

In the Bose–Burton theorem, the ambient geometry is $PG(d, q)$ and the assertion of the second part of the theorem is that if a subset A of the points of this projective space has the same size as a subspace F of rank $c+1$ and A meets every subspace of corank c, then $A \cong F \cong PG(c, q)$.

In Turán's theorem, the ambient geometry is $M(K_n)$. To state the theorem we shall need some more notation. For natural numbers p and n, the *Turán graph* $T_p(n)$ is the unique n-vertex p-partite graph for which every vertex class has either $\left\lfloor \dfrac{n}{p} \right\rfloor$ or $\left\lceil \dfrac{n}{p} \right\rceil$ members. It is not difficult to check that $T_p(n)$ has the largest number of edges among all complete p-partite graphs on n vertices (Exercise 6.91). Hence a smallest corank c flat F in $M(K_n)$ contains $\dbinom{n}{2} - |E(T_{c+1}(n))|$ edges. The second part of Turán's theorem asserts that if a subset A of the edges of K_n has this minimum number $|F|$ of edges and A meets every $(c+2)$-clique, then A is isomorphic to the complement in K_n of $T_{c+1}(n)$.

Both of these theorems have equivalent formulations in terms of a subset of the ambient geometry that is of maximum size with respect to the property of not containing any modular flat of corank $c + 1$. Indeed, Turán's theorem is probably more commonly stated in this way.

For fixed c, the common proof of the second part of these two theorems is by induction on $r(G) - (c + 1)$, each result being trivial when this quantity is 0. Let F be as in the theorems and X be a subset of G of size $|G - F|$ which, like $G - F$, does not contain any rank $(c + 1)$ modular flats. Then all the single-element contractions X/e of X are essentially extremal. It then follows by induction that X/e is isomorphic to $(G - F)/e$. Thus we may employ (6.83) to show that X and $G - F$ have the same Tutte polynomial and the same intersection matrix. The only flat of G of corank c that avoids $G - F$ is F itself, that is, $I_G(G - F; 0, c) = 1$. Hence $I_G(X; 0, c)$ is also 1 so X must avoid, and therefore be complementary to, a flat of the same corank and size as F.

We shall illustrate this idea further in the graphical case by looking in more detail at what happens when $c = 1$. We leave the task of completing the details in the projective case as an exercise. Evidently $|E(T_2(n))| = \left\lfloor \dfrac{n^2}{4} \right\rfloor$. Now take a subset X of $E(K_n)$ that has $\left\lfloor \dfrac{n^2}{4} \right\rfloor$ edges and no triangles. We shall also let X denote the subgraph of K_n induced by this set of edges. Let e be an edge of X with endpoints u and v. Then

$$E(X/e) = \left\lfloor \frac{n^2}{4} \right\rfloor - 1 = \left\lfloor \frac{(n-2)^2}{4} \right\rfloor + n - 2,$$

and none of the edges of X/e is multiple. If $\bar{v}$ is the vertex of X/e formed by identifying u and v, then

$$|E(X/e - \bar{v})| = |E(X - \{u, v\})| = \left\lfloor \frac{(n-2)^2}{4} \right\rfloor + n - 2 - \deg \bar{v}.$$

But $X - \{u, v\}$ contains no triangles and therefore has at most $\left\lfloor \dfrac{(n-2)^2}{4} \right\rfloor$ edges. Hence $\deg \bar{v} = n - 2$ and $|E(X/e - \bar{v})| = \left\lfloor \dfrac{(n-2)^2}{4} \right\rfloor$. Thus, by the induction assumption, $X/e - \bar{v}$ is isomorphic to $T_2(n-2)$ and $\bar{v}$ is adjacent to every vertex of this graph. Thus the isomorphism type and intersection matrix of X/e are determined for all e in X. Hence X itself is determined and is isomorphic to $T_2(n)$.

Exercises

Those with one asterisk are more difficult; those with two asterisks are unsolved.

Section 6.2

6.1.   (a) Determine the Tutte polynomials of the three matroids M_1, M_2, and M_3 for which affine representations are shown below.

Figure 6.11.

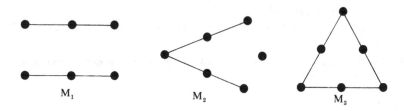

M_1 M_2 M_3

   (b) (Brylawski, 1972b) Show that M_1 and M_2 are the unique smallest pair of non-isomorphic matroids with the same Tutte polynomial.

* (c) (Brylawski, 1972b) Determine when two rank 3 geometries have the same Tutte polynomial.

6.2.   Define a function t^* on the class $\mathcal{M}$ of all matroids by $t^*(M) = t(M^*; x, y)$. Show that t^* is a T–G invariant and use this to give an alternative proof of the fact that $t(M^*; x, y) = t(M; y, x)$.

6.3.   (a) Show that, for $m \geqslant 2$,
$$t(U_{1,m}; x, y) = x + y + y^2 + \ldots + y^{m-1}.$$

   (b) Determine $t(U_{r,m}; x, y)$.

6.4.   If a_{ij} is the coefficient of $x^i y^j$ in $S(M; x, y)$, prove that it is a group invariant and determine its value in terms of the coefficients of $t(M; x, y)$.

6.5.   Let F be a field and $\sigma, \tau, \gamma,$ and δ be non-zero elements of F. Suppose that f is a function from $\mathcal{M}$ into $F[x, y]$ having the following properties.

   (1) $f(I; x, y) = x$ and $f(L; x, y) = y$.

   If e is an element of the matroid M, then

   (2) $f(M; x, y) = \sigma f(M - e; x, y) + \tau f(M/e; x, y)$ if e is neither an isthmus nor a loop,

   (3) $f(M; x, y) = \gamma f(M(e); e, y) f(M - e; x, y)$ if e is an isthmus, and

   (4) $f(M; x, y) = \delta f(M(e); x, y) f(M - e; x, y)$ if e is a loop.

   (a) Find an example to show that, in order for f to be well defined, we must have $\gamma = \delta$.

   (b) Show that if $\gamma = \delta$, then

$$f(M; x, y) = \gamma^{-1}\sigma^{|E|-r(E)}\tau^{r(E)}t\left(M; \frac{x\gamma}{\tau}, \frac{y\gamma}{\sigma}\right).$$

(c) If $\gamma = \delta$, show that

(5) $f(M_1 \oplus M_2; x, y) = \gamma f(M_1; x, y)f(M_2; x, y)$

and that (b) still holds if this condition replaces (3) and (4).

6.6. Use Theorem 6.2.2 to prove 6.2.11(ii).

6.7. Prove Proposition 6.2.20.

6.8. For each of the following determine whether there is a matroid having the specified polynomial as its Tutte polynomial. If there is such a matroid, determine whether or not it is unique.

(a) $x^2 + x + 3y + y^2$.

(b) $x^3 + 2x^2 + x + x^2y + 3xy + xy^2 + y + 2y^2 + y^3$.

(c) $x^4 + 4x^3 + 7x^2 + 3x^2y + 7xy + 6xy^2 + 3xy^3 + xy^4$.

(d) $x^3 + 3x^2 + 2xy + y^2$.

(e) $x^3 + 3x^2y + 2x^2y^2 + x^2y^3 + 3xy^2 + 4xy^3 + 3xy^4 + xy^5 + y^3 + 2y^4 + 2y^5 + y^6$.

6.9. Let $S(M; x, y) = \sum_i \sum_j a_{ij}x^iy^j$ where $S(M; x, y)$ is the rank generating polynomial of $M(E)$.

(a) Show that

$$\max\{i: a_{ij} > 0 \text{ for some } j\} = r(M),$$
$$\max\{j: a_{ij} > 0 \text{ for some } i\} = n(M)$$

and $0 \leqslant a_{ij} \leqslant \binom{m}{\lfloor m/2 \rfloor}$ for all i and j where $m = |E|$.

(b) Use (a) to deduce that, corresponding to matroids on m elements there are at most $2^{(m+1)^3/4}$ distinct rank generating polynomials.

(c) If $g(m)$ denotes the number of non-isomorphic geometries on a set of cardinality m, then Knuth (1974) has shown that

$$g(M) \geqslant \frac{1}{m!} 2^{\left\lceil\binom{m}{\lfloor m/2 \rfloor}/2m\right\rceil}.$$

Use this result with (b) to deduce that, for any number N, there are at least N non-isomorphic geometries having the same Tutte polynomial.

6.10. Let H be a circuit and a hyperplane in the matroid $M(E)$. Let $\mathcal{B}'$ consist of the set of bases of M together with the set H.

(a) Show that $\mathcal{B}'$ is the set of bases of a matroid M' on E.

(b) Use the rank generating polynomial to show that $t(M'; x, y) = t(M; x, y) - xy + x + y$.

6.11. Show that

$$t(M(K_4); x, y) = x^3 + 3x^2 + 2x + 4xy + 2y + 3y^2 + y^3,$$
$$t(M(K_5); x, y) = x^4 + 6x^3 + 11x^2 + 6x + 10x^2y$$
$$+ 20xy + 15xy^2 + 5xy^3 + 6y + 15y^2$$
$$+ 15y^3 + 10y^4 + 4y^5 + y^6,$$

and

$$t(M(K_{3,3}); x, y) = x^5 + 4x^4 + 10x^3 + 11x^2 + 5x$$
$$+ 9x^2y + 15xy + 6xy^2$$
$$+ 5y + 9y^2 + 5y^3 + y^4.$$

6.12. (Brylawski, 1982) Let $\mathscr{G}$ be the set of isomorphism classes of geometries and R be a commutative ring. If M is a matroid, $\bar{M}$ denotes its simplification. A *geometric T–G invariant* is a function f from $\mathscr{G}$ into R such that if e is an element of a geometry G, then

(1) $f(G) = f(G(e))f(G - e)$ if e is an isthmus, and
(2) $f(G) = f(G - e) + f(\overline{G/e})$ otherwise.

(a) Give a non-trivial example of a geometric T–G invariant.
(b) Show that if G is a geometry, then $f(G) = t(G; f(I), 0)$.

6.13. For a matroid M having Tutte polynomial $t(M; x, y) = \sum\limits_{i=0}^{r(M)} \sum\limits_{j=0}^{n(M)} b_{ij}x^iy^j$, show that $\sum\limits_{j=0}^{n(M)} b_{r(M)-1,j} = n(M)$ and $\sum\limits_{i=0}^{r(M)} b_{i,n(M)-1} = r(M)$.

6.14. (Oxley, 1983a) Let f be a generalized T–G invariant and suppose that $f(I) = x$ and $f(L) = \sigma + \tau$ (see 6.2.6(ii)).

(a) Prove that if H is a hyperplane of $M(E)$ and $E - H = \{x_1, x_2, ..., x_k\}$, then
$$f(M) = \sigma^{k-1}(x + (k-1)\tau)f(M(H))$$
$$+ \tau^2 \sum_{j=2}^{k} \sum_{i=1}^{j-1} \sigma^{j-2}f((M - x_1, x_2, ..., x_{i-1}, x_{i+1}, ..., x_{j-1})/x_i, x_j).$$

(b) State the corollary of this result obtained by taking $f(M) = p(M; \lambda)$.
(c) Interpret (a) when x, σ, and τ are all equal to one.
(d) Use (b) to obtain an identity relating $\beta(M)$ and $\beta(M(H))$.

6.15. A matroid M_1 is a *series–parallel extension* of a matroid M_2 if M_1 can be obtained from M_2 by repeated application of the operations of series and parallel extension.

(a) Show that if M_1 is a series–parallel extension of the loopless matroid M_2, then $\beta(M_1) = \beta(M_2)$.

(b) (Crapo, 1967) If $M(\mathscr{W}_r)$ is the polygon matroid of the r-spoked wheel and $\mathscr{W}^r$ is the rank r whirl, show that $\beta(M(\mathscr{W}_r)) = r - 1$ and $\beta(\mathscr{W}^r) = r$.

A matroid $M(E)$ is *3-connected* if $M(E)$ is connected and there is no partition $\{X, Y\}$ of S such that $|X|, |Y| \geqslant 2$ and $r(X) + r(Y) - r(M) = 1$.

(c) (Seymour, 1980) Prove that a connected matroid M is not 3-connected if and only if M is the 2-sum of two matroids on at least three elements.
(d) Use Exercise 6.1 to show that 3-connectedness is not a Tutte invariant.
(e) (Oxley, 1982a) Let M be a matroid with $\beta(M) = k > 1$. Prove that either
    (1) M is a series–parallel extension of a 3-connected matroid N such that $\beta(N) = k$, or
    (2) M is the 2-sum of two matroids each having $\beta < k$.
* (f) (Oxley, 1982a) Let N be a minor of the matroid M and suppose that $\beta(N) = \beta(M) > 0$. Prove that if N is 3-connected, then M is a series–parallel extension of N.

(g) Prove that $\beta(M) = 2$ if and only if M is a series–parallel extension of the 4-point line, $U_{2,4}$, or $M(K_4)$.

(h) Prove that $\beta(M) = 3$ if and only if M is a series–parallel extension of $U_{2,5}$, $U_{3,5}$, F_7, F_7^*, $M(\mathcal{W}_4)$, or $\mathcal{W}^3$.

* (i) Determine all matroids M for which $\beta(M) = 4$.

6.16. Find two geometries G and G' with the same Tutte polynomial where G is isomorphic to its dual but G' is not. (Hint: Form two distinct matroids from $AG(3, 2)$ each retaining seven circuit-hyperplanes, and use Exercise 6.10.)

Figure 6.12.

6.17. (a) Show that for M_1 and M_2, shown in Figure 6.12, $t(M_1^*) = t(M_2^*)$ but that M_1^* is Hamiltonian, that is, has a spanning circuit, whereas M_2^* is not.

(b) Show that the size of a smallest circuit in a matroid is a Tutte invariant.

** (c) Is the size of a largest circuit in a graph a Tutte invariant? (Recently Schwärzler, 1991, has answerede this question in the negative.)

6.18. (Brylawski, 1975a) Prove that if $p(M; \lambda) = \sum_{k=0}^{r} c_k \lambda^k$ and $T(M)$ is the truncation of M, then

$$p(T(M); \lambda) = \sum_{k=1}^{r} c_k \lambda^{k-1} + c_0.$$

Section 6.3.A

6.19. (a) Find two simple graphs Γ and Δ such that $\chi_\Gamma(M) = \chi_\Delta(M)$, but M_Γ and M_Δ have different Tutte polynomials.

* (b) (Tutte, 1974) Now find two simple graphs Γ and Δ such that $t(M_\Gamma; x, y) = t(M_\Delta; x, y)$ but M_Γ and M_Δ are not isomorphic.

* (c) (Brylawski, 1981b) Show that Γ and Δ can be chosen in (b) to have arbitrarily high connectivity.

Figure 6.13.

6.20. For $\lambda = 1, 2, 3, 4$, and 5, find the number of proper λ-colorings of the graph Γ shown in Figure 6.13, and use these numbers to calculate $\chi_\Gamma(\lambda)$ for arbitrary λ.

6.21. If $\mathscr{W}_n$ is the n-spoked wheel, show that its chromatic polynomial is $\lambda[(\lambda - 2)^n + (-1)^n(\lambda - 2)]$.

6.22. (a) Let $K_n - e$ be the graph obtained from K_n by deleting an edge. Show that its chromatic polynomial is $\lambda(\lambda - 1)(\lambda - 2) \dots (\lambda - n + 3)(\lambda - n + 2)^2$.

   (b) If e and f are adjacent edges of K_n, find the chromatic polynomial of $K_n - e - f$.

   (c) Show that the chromatic polynomials of $K_{3,3}$ and $K_{3,4}$ are

$$\lambda(\lambda - 1)(\lambda^4 - 8\lambda^3 + 28\lambda^2 - 47\lambda + 31)$$

   and

$$\lambda(\lambda - 1)(\lambda^5 - 11\lambda^4 + 55\lambda^3 - 147\lambda^2 + 204\lambda - 115),$$

   respectively.

Section 6.3.B

6.23. If Γ is a graph, find the degree of its flow polynomial.

6.24. Show that the flow polynomials of $K_{3,3}$ and K_5 are

$$(\lambda - 1)(\lambda - 2)(\lambda^2 - 6\lambda + 10)$$

   and

$$(\lambda - 1)(\lambda^2 - 4\lambda + 5)(\lambda^3 - 5\lambda^2 + 11\lambda - 9),$$

   respectively.

6.25. If Γ is a connected planar graph and Γ^* is a geometric dual of Γ, give a one-to-λ correspondence between the set of nowhere-zero λ-flows of Γ and the set of proper λ-colorings of Γ^*.

6.26. Let Γ be a cubic graph. Show that the number of proper edge 3-colorings of Γ equals $(-1)^{r(M_\Gamma^*)}t(M_\Gamma; 0, -3)$. (Hint: Relate the coloring to a nowhere-zero flow over an appropriate group.)

6.27. (Negami, 1987) The two 3-variable polynomials $f(\Gamma; t, x, y)$ and $f^*(\Gamma; t, x, y)$ are recursively defined for graphs as follows:

   (1) $f(\bar{K}_n) = t^n$ for all $n \geqslant 1$ where $\bar{K}_n$ is the complement of K_n;

   (2) $f(\Gamma) = yf(\Gamma - e) + xf(\Gamma/e)$ for all e in $E(\Gamma)$;

   and

   (3) $f^*(\bar{K}_n) = t^n$ for all $n \geqslant 1$;

   (4) $f^*(\Gamma) = xf^*(\Gamma - e) + yf^*(\Gamma/e)$ for all edges e of Γ that are not loops or isthmuses;

   (5) $f^*(\Gamma) = (x + ty)f^*(\Gamma - e)$ if e is a loop;

   (6) $f^*(\Gamma) = (x + y)f^*(\Gamma/e)$ if e is an isthmus.

   (a) (Negami, 1987) Prove that if Γ has $k(\Gamma)$ components, then

$$f(\Gamma; (x - 1)(y - 1), 1, y - 1) = (y - 1)^{|V(\Gamma)|}(x - 1)^{k(\Gamma)}t(M_\Gamma; x, y).$$

   (b) (Oxley, 1989a) Prove that

$$f(\Gamma; t, x, y) = \left(\frac{ty}{x}\right)^{k(\Gamma)}\left(\frac{x}{y}\right)^{|V(\Gamma)|}y^{|E(\Gamma)|}t\left(M_\Gamma; 1 + \frac{ty}{x}, 1 + \frac{x}{y}\right)$$

   and

$$f^*(\Gamma; t, x, y) = \left(\frac{tx}{y}\right)^{k(\Gamma)} \left(\frac{y}{x}\right)^{|V(\Gamma)|} x^{|E(\Gamma)|} t\left(M_\Gamma; 1 + \frac{x}{y}, 1 + \frac{ty}{x}\right).$$

(c) Prove that

$$f^*(\Gamma; t, x, y) = t^{k(\Gamma) - |V(\Gamma)|} f(\Gamma; t, ty, x).$$

(d) Show that

$$f(\Gamma; \lambda, -1, 1) = \chi_\Gamma(\lambda)$$

and

$$\lambda^{-|V(\Gamma)|} f(\Gamma; \lambda, \lambda, -1) = \chi_\Gamma^*(\lambda).$$

(e) Let Γ be a plane graph and Γ^* be a geometric dual of Γ. Show that

$$f^*(\Gamma) = f(\Gamma^*).$$

Section 6.3.C

6.28. Prove that if a graph Γ has a nowhere-zero n-flow, then Γ has a nowhere-zero $(n + 1)$-flow.

6.29. (Jaeger, 1976b) Let the complement of a spanning tree in a connected graph Γ be a *cotree*.

(a) Prove that if $E(\Gamma)$ is the union of m cotrees, then Γ has a nowhere-zero $\mathbb{Z}_2^m$-flow.

(b) Use Edmonds' covering theorem for matroids (Edmonds, 1965b) to prove that every 3-edge connected graph is the union of three cotrees.

(c) Use (a) and (b) (and not Theorem 6.3.10) to prove that every bridgeless graph has a nowhere-zero 8-flow.

(d) Deduce that every bridgeless graph can be covered by three Eulerian

Figure 6.14.

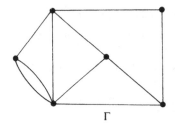   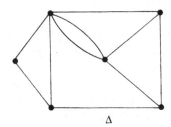

Γ Δ

subgraphs, where the latter is a subgraph whose edge set can be partitioned into circuits.

(e) Use the same technique that was used to prove (c) to show that every 4-edge connected graph has a nowhere-zero 4-flow.

6.30. (a) (Tutte, 1974) Consider the two graphs shown in Figure 6.14. Show that the Tutte polynomials of their polygon matroids are equal but that these polygon matroids are non-isomorphic.

(b) Show that any orientation of Γ has 48 nowhere-zero $\mathbb{Z}$-flows taking values in $[-2, 2]$, but any orientation of Δ has 52 such flows.

6.31. Prove Lemma 6.3.12.

6.32. Prove that S-closure is a closure operator.

6.33. (Walton, 1981) Let the graph Γ be formed from a set of 2-connected planar graphs by taking 3-sums. Prove that Γ has a nowhere-zero 4-flow.

Section 6.3.D

6.34. (Greene & Zaslavsky, 1983) A *totally cyclic orientation* of a graph Γ is an orientation θ of Γ such that every edge of Γ_θ is in some directed cycle. If Γ is planar and Γ^* is a geometric dual of Γ, find a combinatorial correspondence between the set of acyclic orientations of Γ and the set of totally cyclic orientations of Γ^*.

6.35. (Greene & Zaslavsky, 1983) If uv is an edge of a graph Γ and $N(\Gamma)$ denotes the number of acyclic orientations of Γ having u as the unique source and v as the unique sink, prove that

$$N(\Gamma) = \beta(M_\Gamma).$$

Section 6.3.E

6.36. Let Γ be the graph shown in Figure 6.15.
If edge e_i has retention probability p_i, and $p_1 = \frac{3}{4}$, $p_2 = \frac{5}{16}$, $p_3 = \frac{1}{4}$, $p_4 = \frac{3}{8}$, $p_5 = \frac{11}{16}$, and $p_6 = \frac{15}{16}$, construct a graph $\check{\Gamma}$ from Γ so that every edge of $\check{\Gamma}$ has retention probability $\frac{1}{2}$ and the probability that there is an (s, t)-path in Γ is equal to the probability that there is an (s, t)-path in $\check{\Gamma}$.

Figure 6.15.

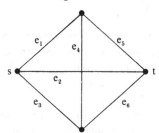

6.37. Supply the argument that proves 6.3.21(ii).

*6.38. If $\Pr(\mathscr{D}(M)) = \sum\limits_{i=1}^{|M|} b_i p^i$, determine the coefficients b_i.

6.39. In $M = M_d(E \cup d)$, suppose the retention probability p_i of every point e_i of E is equal to the finite binary decimal p. Derive the formula for $\Pr(\mathscr{D}(M))$ from the corresponding formula when $p = \frac{1}{2}$. By taking limits give another proof of 6.3.21(i).

6.40. (a) Complete the proof of Lemma 6.3.24.
  (b) Let $M_{(k)}$ be the matroid that is obtained from M by replacing every non-isthmus of M by k elements in series and replacing every isthmus by k isthmuses. Find $t(M_{(k)}; x, y)$ in terms of $t(M; x, y)$.
  * (c) (Brylawski, 1982, Proposition 4.10) Let M_d be a pointed matroid in which d is neither a loop nor an isthmus. Let M be a matroid on the set $\{p_1, p_2, \ldots, p_k\}$. The *tensor product* $M \otimes M_d$ is the matroid N_k formed from M as follows: let $N_0 = M$ and, for $i = 1, 2, \ldots, k$, let N_i be the 2-sum of the basepointed matroids (N_{i-1}, p_i) and (M_d, d), this 2-sum being $(N_{i-1}/p_i) \oplus (M_d/d)$ if p_i is a loop, and $(N_{i-1} \setminus p_i) \oplus (M_d \setminus d)$ if p_i is an isthmus.

If $t_P(M_d) = x'f(x, y) + y'g(x, y)$, show that

$$t(M \otimes M_d; x, y) =$$

$$(f(x, y))^{|M| - r(M)}(g(x, y))^{r(M)}t\left(M; \frac{(x-1)f+g}{g}, \frac{f+(y-1)g}{f}\right).$$

(d) Deduce Lemma 6.3.24 and part (b) from part (c).

Section 6.3.F

6.41. Complete the details of each of the proofs of Proposition 6.3.26.

6.42. Verify that $T^{-1}(i, j) = (-1)^{i+j}\binom{j}{i}$ where T^{-1} is the inverse of the matrix T for which $T(i, j) = \binom{j}{i}$ for all i, j in $\{0, 1, 2, ..., n\}$.

6.43. Argue directly from Theorem 6.2.2 to show that

$$t(M; x, y) = (y-1)^{-r}\bar{\chi}(M; (x-1)(y-1), y).$$

6.44. Show that if M has rank r,
   (a) $\bar{\chi}(M; 0, \lambda) = p_M(\lambda)$;
   (b) $\bar{\chi}(M; \lambda, v) = S_{KC}(M; \lambda - 1, v)$; and
   (c) $S_{KC}(M; u, v) = u^r S\left(M; \frac{v}{u}, u\right)$.

Section 6.3.G

6.45. Use deletion–contraction arguments to prove 6.3.28 and 6.3.29.

6.46. (a) Find the medial graphs of each of K_2, K_3, and a planar embedding of K_4.
   (b) Show that the medial graph of an n-cycle C_n is $C_n^{(2)}$, the graph obtained from C_n by doubling every edge.
   (c) Find the medial graph of $C_n^{(2)}$.
   (d) If Δ is the graph of the octahedron, find a graph Γ such that $\Gamma_m = \Delta$.

6.47. Prove that if Γ is a plane graph and Γ^* is its geometric dual, then $\Gamma_m = (\Gamma^*)_m$.

6.48. Show how deletion and contraction of an edge e in a plane graph Γ correspond to decomposing Γ_m into two smaller 4-regular graphs.

Section 6.4.A

6.49. Prove Corollary 6.4.11.

6.50. Let $M(E)$ be a rank r binary matroid.
   (a) Prove that the following statements are equivalent.
      (1) M is affine.
      (2) If ϕ is a coordinatization of M in $V(r, 2)$, then there is a linear functional f on $V(r, 2)$ that distinguishes $\phi(E)$.
      (3) All circuits of M have even cardinality.
      (4) All hyperplane complements of M^* have even cardinality.
      (5) There is a partition of E into bonds of M.
   (b) Assume that $M(E) \cong M_\Gamma$ for a graph Γ. Prove that (1)–(5) are equivalent to (6) Γ is 2-colorable.
   (c) Assume that $M(E) \cong M_\Gamma^*$ for a connected graph Γ. Prove that (1)–(5) are equivalent to (7) Γ is Eulerian.

*6.51. Prove Proposition 6.4.12.

6.52. Show that if $c(M; q) = k$, but $c(M - e; q) < k$ for all e, then $c(M/e; q) < k$ for all e.

6.53. For a loopless matroid M, there are several ways one may attempt to define the chromatic number. Two possibilities are

$$\chi(M) = \min\{j \in \mathbb{Z}^+: p(M; j) > 0\}$$

and

$$\pi(M) = \min\{j \in \mathbb{Z}^+: p(M; j + k) > 0 \text{ for } k = 0, 1, 2, \ldots\}.$$

(a) Evidently $\chi(M) \leq \pi(M)$. Give examples to show that $\pi(M) - \chi(M)$ can become arbitrarily large.

(b) Show that if $T \subseteq E(M)$, then $\chi(M(T))$ can exceed $\chi(M)$ and $\pi(M(T))$ can exceed $\pi(M)$.

(c) Let M be a geometry and $\mathscr{C}^*(M)$ be its set of bonds. Prove that

$$\pi(M) \leq 1 + \max_{C^* \in \mathscr{C}^*(M)} |C^*|.$$

(d) (Heron, 1972b) Prove that, for a matroid $M(E)$, $\pi(M) \leq |E| - r + 2$.

(e) (Lindström, 1978) Prove that if M is a regular matroid, then $\chi(M) = \pi(M)$.

*(f) (Walton, 1981) Extend (e) to show that if M is binary and has no minor isomorphic to F_7, then $\chi(M) = \pi(M)$.

(g) (Oxley, 1978a) Prove that for a regular matroid M having $\mathscr{R}(M)$ as its set of simple submatroids,

$$\pi(M) \leq 1 + \max_{N \in \mathscr{R}(M)} \left(\min_{C^* \in \mathscr{C}^*(N)} |C^*| \right).$$

(h) Deduce from (g) the following result of Lindström (1978). If a loopless regular matroid can be covered by bonds of size less than n, then $\pi(M) \leq n$.

*(i) (Oxley, 1978a) Use (g) to prove that if M is a connected regular geometry, then

$$\pi(M) \leq \max_{C^* \in \mathscr{C}^*(M)} |C^*|$$

unless M is an odd circuit or an isthmus.

*6.54. (Walton, 1981) With $\pi(M)$ as defined in the previous exercise, prove that if M is loopless and representable over $GF(q)$, then, provided M has no minor isomorphic to any of the four matroids shown in Figure 6.16,

$$\pi(M) \leq q + 1.$$

Figure 6.16.

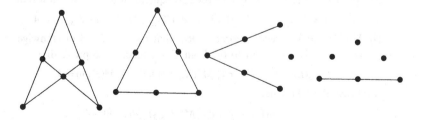

6.55.* (a) (Brylawski, 1975a) Let M be the generalized parallel connection of the matroids M_1 and M_2 across the modular flat X. Prove that

$$p(M; \lambda) = p(M_1; \lambda)p(M_2; \lambda)[p(M_1(X); \lambda)]^{-1}.$$

(b) (Walton & Welsh, 1980) Let M be the 2-sum of matroids M_1 and M_2 with basepoint z. Show that

$$p(M; \lambda) = (\lambda - 1)^{-1} p(M_1; \lambda) p(M_2; \lambda) + p(M_1/z; \lambda) p(M_2/z; \lambda).$$

(c) (Walton & Welsh, 1980) Let M_1 and M_2 be binary matroids and $X = \{a, b, c\}$ be a 3-circuit of M_1 and M_2. Let M be the generalized parallel connection of M_1 and M_2 across X, and N be the 3-sum of M_1 and M_2, that is, $N = M - X$. Show that

$$p(N; \lambda) = \frac{p(M_1; \lambda) p(M_2; \lambda)}{(\lambda - 1)(\lambda - 2)}$$
$$+ p((M - \{a, b\})/c; \lambda) + p((M - a)/b; \lambda) + p(M/a; \lambda).$$

Let $\mathscr{F} = EX(M_1, M_2, ..., M_n)$ be the class of matroids having no minor isomorphic to any one of $M_1, M_2, ..., M_n$. A *splitter* N for $\mathscr{F}$ is a 3-connected member of $\mathscr{F}$ such that if $M \in \mathscr{F}$ and M has a minor isomorphic to N, then $M \cong N$.

(d) Let $\mathscr{F}$ be a class of matroids closed under isomorphism and the taking of minors. Prove that if N is a splitter for $\mathscr{F}$, then N^* is a splitter for $\mathscr{F}^* = \{M : M^* \in \mathscr{F}\}$.

(e) (Walton & Welsh, 1982) Let $\mathscr{A}$ be a class of matroids and $\pi(\mathscr{A}) = \max\{\pi(M): M \text{ is a loopless member of } \mathscr{A}\}$. Prove that if $\pi(EX(N, M_1, M_2, ..., M_n)) = k$ and N is a splitter for $EX(M_1, M_2, ..., M_n)$, then $\pi(EX(M_1, M_2, ..., M_n)) \leqslant \max\{k, \pi(N)\}$.

(f) Show that Seymour's 6-flow theorem (6.3.10) is equivalent to the assertion that, for a loopless member M of $EX(F_7, F_7^*, M(K_5), M(K_{3,3}))$, $\pi(M) \leqslant 6$.

*(g) (Seymour, 1980) Prove that F_7 is a splitter for $EX(U_{2,4}, F_7^*)$, that $M(K_5)$ is a splitter for $EX(U_{2,4}, F_7, F_7^*, M(K_{3,3}))$, and that R_{10} is a splitter for $EX(U_{2,4}, F_7^*)$.

*(h) (Walton & Welsh, 1980) Prove the following:

$$\pi(EX(U_{2,4}, F_7, M(K_5))) \leqslant 6;$$
$$\pi(EX(U_{2,4}, F_7^*, M(K_5))) \leqslant 6;$$
$$\pi(EX(U_{2,4}, F_7, M(K_{3,3}))) \leqslant 6;$$
$$\pi(EX(U_{2,4}, F_7^*, M(K_{3,3}))) \leqslant 6.$$

(i) Show that the Four Color theorem is equivalent to the assertion that

$$\pi(EX(U_{2,4}, F_7, F_7^*, M^*(K_5), M^*(K_{3,3}), M(K_5), M(K_{3,3}))) = 4.$$

(j) Use (i) and Wagner's theorem (1964) that the case $k = 4$ of Hadwiger's conjecture is equivalent to the Four Color theorem to prove that

$$\pi(EX(U_{2,4}, F_7, F_7^*, M^*(K_5), M^*(K_{3,3}), M(K_5))) = 4.$$

(k) Prove the following:

$$\pi(EX(U_{2,4}, F_7, M^*(K_{3,3}), M(K_5))) = 4;$$
$$\pi(EX(U_{2,4}, F_7^*, M^*(K_{3,3}), M(K_5))) = 5;$$
$$\pi(EX(U_{2,4}, F_7, M^*(K_{3,3}), M(K_{3,3}))) = 5;$$
$$\pi(EX(U_{2,4}, F_7^*, M^*(K_{3,3}), M(K_{3,3}))) = 5.$$

(l) Deduce from (k) that a loopless graph having no subgraph contractible to $K_{3,3}$ is 5-colorable.

(m) Deduce from (k) that a graph without isthmuses having no subgraph contractible to $K_{3,3}$ has a nowhere-zero 4-flow.

(n) Use (m) to show that a cubic graph without isthmuses having no subgraph contractible to $K_{3,3}$ has edge-chromatic number 3.

*6.56. (Walton, 1981) Use Seymour's decomposition result for regular matroids to prove that if $M \in EX(U_{2,4}, F_7)$, then $p(M; \lambda) \geqslant 0$ for all λ in $\mathbb{Z}^+$.

6.57. (a) Show that $p(PG(n, q); \lambda) = \sum_{i=0}^{n+1} (\lambda - q^i)$.

(b) Show that if $M = AG(n, q)$, then

$$p(M; \lambda) = (\lambda - 1)[\lambda^n - (q^n - 1)\lambda^{n-1} + (q^n - 1)(q^{n-1} - 1)\lambda^{n-2} + \cdots$$
$$+ (-1)^k(q^n - 1)(q^{n-1} - 1) \cdots (q^{n-k+1} - 1)\lambda^{n-k} + \cdots$$
$$+ (-1)^n(q^n - 1)(q^{n-1} - 1) \cdots (q - 1)].$$

6.58. Find three non-isomorphic matroids each having characteristic polynomial equal to $(\lambda - 1)(\lambda - 3)^2$.

Section 6.4.B

6.59. (a) Find all minimal 1-blocks over $GF(2)$.

** (b) Find all minimal 1-blocks over $GF(3)$.

6.60. Prove that if M and N are minimal k-blocks over $GF(q)$, then so is their series connection.

6.61. Prove Proposition 6.4.14.

6.62. Check that the matroid N_4 in Example 6.4.22 is isomorphic to $AG(3,2)$.

6.63. (a) (Oxley, 1980) Let M be coordinatizable over $GF(q)$ and C^* be a bond of M such that $c(M; q) - 1 = c(M - C^*; q) = k$. Prove that $|C^*| \geqslant q^k$.

(b) Show that if M is a tangential k-block and C^* is a bond of M having exactly q^k elements, then $E - C^*$ is a modular hyperplane of M.

6.64. (a) (Walton, 1981) Prove that a tangential k-block over $GF(q)$ is 3-connected.

(b) Give an example of a 3-connected minimal t-block that is not a tangential t-block.

(c) (Walton, 1981) Prove that a tangential k-block over $GF(2)$ is not a 3-sum.

6.65. (Mullin & Stanton, 1979) A $(q, m. k)$-matroid is a submatroid of $PG(k - 1, q)$ having rank k and critical exponent greater than m. A minimal (q, m, k)-matroid is a (q, m, k)-matroid for which no submatroid is also a (q, m, k)-matroid. Let $\eta(q, m, k)$ denote the least number of elements in a (q, m, k)-matroid.

* (a) (Oxley, 1979a) Prove that $\eta(q, m, k) = \dfrac{q^{m+1} - 1}{q - 1} + k - m - 1$.

(b) (Brylawski, 1975c) Show that if $r, q > 2$ and $U_{r,n}$ is coordinatizable over $GF(q)$, then $c(U_{r,n}; q) = 1$.

* (c) Prove that a (q, m, k)-matroid having $\eta(q, m, k)$ elements is isomorphic to $PG(m, q) \oplus U_{k-m-1,k-m-1}$ for $q = 2$ and $m \geqslant 2$, and for $q > 2$ and $m \geqslant 1$.

(d) Deduce from (c) the following result of Bose & Burton (1966). Let M be a loopless matroid coordinatizable over $GF(q)$ and suppose M has critical exponent greater than m. Then M has at least $\dfrac{q^{m+1}-1}{q-1}$ elements. Moreover,

if M has exactly $\dfrac{q^{m+1}-1}{q-1}$ elements, then $M \cong PG(m, q)$.

6.66. Use an argument involving coloring and the critical exponent to prove that $AG(3, 2)$ has no minor isomorphic to $M(K_4)$.

6.67. (Whittle, 1985) Recall that if M and N are matroids having a common ground set E, then N is a *quotient* of M if every flat of N is also a flat of M. Let M be a tangential k-block over $GF(q)$ and N be a loopless quotient of M that is coordinatizable over $GF(q)$. Let $r(M) - r(N) = m$. Assume that N has a proper non-empty flat F such that
   (1) F is a modular flat of M;
   (2) $r_M(F) - r_N(F) = m$; and
   (3) for every proper flat F' of $N(F)$, $p(N(F)/F'; q^k) > 0$.
 * (a) Prove that the simplification of N is a tangential k-block over $GF(q)$.
   (b) Deduce Theorem 6.4.27 from (a).
   (c) Deduce Theorem 6.4.29 from (a).

6.68. Suppose that the matroid M is represented over $GF(q)$ by a matrix A. We call M' a $GF(q)$-*vector quotient* of M (see, for example, White, 1986, 7.4.8) if M' can be obtained from M by adjoining a linearly independent set of columns to A and then contracting those elements of the resulting matroid that correspond to the newly adjoined columns.
   (a) Show that if M' is a $GF(q)$-vector quotient of M, then $c(M'; q) \geqslant c(M; q)$.
 * (b) (Jaeger, 1981) Prove the following analog of Hajós' theorem (1961) characterizing all graphs of chromatic number at least k. If M is coordinatizable over $GF(q)$, then $c(M; q) \geqslant k$ if and only if M has as a restriction a matroid that can be constructed from copies of $PG(k-1, q)$ by a sequence of series connections and $GF(q)$-vector quotients.

Section 6.4.C

6.69. Both K_4 and K_5 have two different types of hyperplanes. Find the four bond graphs that arise from K_4 and K_5.

6.70. Add the argument omitted in the last paragraph of the proof of Proposition 6.4.48.

6.71. Give an example of a transversal matroid coordinatizable over $GF(q)$ having critical exponent 2.

6.72.**(a) (Jaeger, 1982) Let I_1 and I_2 be independent sets in $PG(r-1, q)$ and M be the submatroid on $I_1 \cup I_2$. For which values of q is $c(M; q) = 1$?
   **(b) If $I_1, I_2, ..., I_k$ are independent sets in $PG(r-1, q)$, when is $c(I_1 \cup I_2 \cup ... \cup I_k; q) \leqslant k$? When is k best-possible here?

Section 6.5

6.73. (a) Show that $M(C)$ depends only on C and not on the generator matrix U for C.
   (b) Show that $M(C^*) \cong M^*(C)$.

6.74. Verify (6.58).

6.75. Complete the derivation of Proposition 6.5.1 from (6.64).

6.76. Prove Proposition 6.5.2.

6.77. Let C be a binary code. Use Corollary 6.5.6 to verify that, in the code over $GF(4)$ that is generated by C, all codewords have even weight.

6.78. (a) Calculate $p_M(\lambda)$ when M is $G_q(n, 2)$ and when M is $G_q(n, n-1)$.

    (b) Calculate $p_M(\lambda)$ when M is $G_2(5, 3)$.

    (c) Can the critical exponent of $G_2(n+1, 2k+1)$ be computed directly from that of $G_2(n, 2k)$? (They are the same.)

6.79. Calculate the codeweight polynomial $A(C)$ for

    (a) the projective code C_p where $M(C_p) = PG(r-1, q)$;

    (b) the dual C_p^* of C_p (This is the *Hamming code*.);

    (c) the optimal code C where $M(C) = U_{r,n}$.

6.80. (a) Prove that if M is coordinatizable over $GF(q)$, then, provided q is sufficiently large, $C(M)$ has codewords of every possible weight.

    (b) Prove that, for $q = 2$, $A(C(M))$ always has a_n or a_{n-1} equal to zero unless M has an isthmus.

6.81. With $A(C; q, z) = \sum_{i=0}^{n} a_i z^i$, define $T_j = \dfrac{\sum_i a_{n-i} \binom{i}{j}}{q^{r-j}}$. Use (6.64) to prove (a)–(c).

    (a) T_j is always an integer.

    (b) $T_0 = 1$.

    (c) T_1 determines the number of loops in $M(C)$.

    If C^* has distance d, calculate T_i for $i \leqslant d$.

6.82. (Asano, Nishizeki, Saito & Oxley, 1984) Let U be an $r \times n$ matrix over $GF(q)$. The *chain-group N generated* by U is the linear code C generated by U, a *chain* being a codeword in C. The support $\sigma(f)$ of a chain f is the support of the corresponding codeword. Let $M = M(C)$ and E denote the set of elements of M.

    (a) Prove that $c(M; q) \leqslant k$ if and only if N contains k chains $f_1, f_2, \ldots, f_k$ such that $E = \sum_{i=1}^{k} \sigma(f_i)$.

        Suppose that $S \subseteq T \subseteq E(M)$. Prove that

    (b) $c(M(S); q) \leqslant k$ if and only if N contains k chains $f_1, f_2, \ldots, f_k$ such that $S \subseteq \bigcup_{i=1}^{k} \sigma(f_i)$;

    (c) $c(M/(E-S); q) \leqslant k$ if and only if N contains k chains $f_1, f_2, \ldots, f_k$ such that $S = \bigcup_{i=1}^{k} \sigma(f_i)$;

    (d) $c(M(T)/(T-S); q) \leqslant k$ if and only if $N(T)$, the set of restrictions to T of chains in N, contains k chains $f_1', f_2', \ldots, f_k'$ such that $S = \bigcup_{i=1}^{k} \sigma(f_i')$.

    (e) Use (b)–(d) to give another proof of Proposition 6.4.5.

6.83. (Jaeger, 1989a) Let C be a binary code and E be the ground set of $M(C)$. Let U^* be a generator matrix for C^*. If $e \in E$, then to *double e in series*, one replaces the column of U^* corresponding to e by two copies of this column. If $F \subseteq E$, let $U^*:F$ be the matrix obtained from U^* by doubling every element of F in series. Let $C^*:F$ be the code generated by $U^*:F$ and let $C:F$ be $(C^*:F)^*$.

(a) Show that, for e in E, $M(C:\{e\})$ is obtained from $M(C)$ by adding an element in series with e unless e is an isthmus of $M(C)$; in the exceptional case, $M(C:\{e\})$ is obtained from $M(C)$ by adjoining another isthmus.

Let N be the number of ordered pairs (X_1, X_2) of subsets of E for which $X_1 \cup X_2 = E$ and each X_i is a disjoint union of circuits of $M(C)$.

(b) Show that N equals the number of ordered pairs of codewords of C^*, the union of whose supports is E.

* (c) Prove that

$$N = (-\tfrac{1}{2})^r \sum_{F \subseteq E} (-1)^{|F|}(-2)^{\dim B(C:F)}$$

where $r = r(M(C))$.

* (d) Prove that

$$N = (-1)^{r(M^*(C))} t(M(C); 0, -3).$$

(e) Use (c), (d), and Proposition 6.5.4 to prove that, for a binary matroid M having ground set E and rank r,

$$t(M; 0, -3) = (\tfrac{1}{2})^r \sum_{F \subseteq E} t(M:F; -1, -1).$$

Here $M:F$ is obtained from M by adjoining an element in series to each non-isthmus element of F, and adjoining an isthmus to M for each isthmus of M in F.

(f) Let M be a rank r matroid on E and α, β, and γ be numbers with $1 + \gamma \neq 0$ and $1 + \gamma(\alpha + 1) \neq 0$. Using induction, prove that

$$t\left(M; \frac{\alpha + \gamma\alpha^2}{1 + \gamma}, \frac{\beta + \gamma(\alpha + \beta)}{1 + \gamma(\alpha + 1)}\right)$$
$$= \left(\frac{1}{1 + \gamma}\right)^r \left(\frac{1}{1 + \gamma(\alpha + 1)}\right)^{|E| - r} \sum_{F \subseteq E} \gamma^{|F|} t(M:F; \alpha, \beta).$$

(g) If C is a binary code, show that

$$|F| + \dim(B(C:F)) \geqslant \dim B(C).$$

(h) Use (f), (g) and Proposition 6.5.4 to prove that, if M is a binary matroid, then $t(M; 3, 3) = kt(M; -1, -1)$ for some odd integer k.

(i) Show that the Four Color theorem is equivalent to the statement that if M is the polygon matroid of a planar loopless graph, then $t(M; -3, 0)$ is non-zero with the sign of $(-1)^{r(M)}$.

(j) Use (f) and (i) to prove that if M is the polygon matroid of a planar loopless graph, then , for all γ in $(-1, 0]$, $t\left(M; \dfrac{-3 + 9\gamma}{1 + \gamma}, \dfrac{3\gamma}{2\gamma - 1}\right)$ is non-zero with the sign of $(-1)^{r(M)}$.

Section 6.6.A

6.84. Prove that if $f(M) = |\mathscr{B}_{ij}(M)|$ and the greatest element n of M is neither a loop nor an isthmus, then $f(M) = f(M - n) + f(M/n)$.

6.85. Prove 6.73.

6.86. (Brylawski, 1977c) Let M be a matroid of rank r on $\{1, 2, ..., n\}$. Show that the number of ways to color the elements of M with $\{1', 2', ..., n'\}$ so that no broken circuit is colored entirely with $1'$ is equal to $\lambda^{n-r}p^+(M; \lambda)$.

Section 6.6.B

6.87. Show that if (6.75) holds, then the poset of hyperplane intersections is a geometric lattice.

6.88. (Buck, 1943) Suppose that the hyperplanes $H_1, H_2, ..., H_n$ are in general position in $\mathbb{E}^d$. Show that

(a) the number of regions of this arrangement is $\binom{n}{0} + \binom{n}{1} + ... + \binom{n}{d}$;

(b) the number of bounded regions of this arrangement is $\binom{n-1}{d}$.

6.89. Let Γ be a loopless graph having vertex set $\{1, 2, ..., d\}$. For each edge $e = ij$ of Γ, let H_e be the hyperplane $\{(x_1, x_2, ..., x_n): x_i = x_j\}$ of $\mathbb{E}^d$. Show that the following hold.

(a) For the arrangement $\{H_e: e \in E(\Gamma)\}$, the poset of hyperplane intersections is isomorphic to the lattice of flats of M_Γ.

(b) There is a bijection between the set of acyclic orientations of Γ and the regions of the arrangement $\{H_e: e \in E(\Gamma)\}$ determined as follows: the region corresponding to the acyclic orientation α of Γ is $\{(x_1, x_2, ..., x_d): x_i < x_j$ if α directs the edge ij of Γ from i to $j\}$.

Section 6.6.C

6.90. Let E be a set of n points in general position in $\mathbb{E}^d$. Show that the number of hyperplane-separable subsets of E is $\left[2\binom{n-1}{0} + \binom{n-1}{1} + ... + \binom{n-1}{d} \right]$.

(Zaslavsky, 1975b, p. 72, discusses the history of this result.)

Section 6.6.D

6.91. Show that the Turán graph $T_p(n)$ has the largest number of edges among all complete p-partite graphs on n vertices.

6.92. (a) For G equal to $PG(r-1, q)$ and $AG(r-1, q)$, find the matrices W_1 and W_2 (see Exercise 6.57).

* (b) (Brylawski, 1979b) Prove that, for any upper combinatorially uniform geometry G, the matrices W_1 and W_2 are inverses of each other.

6.93. (Brylawski, 1979b) Show that if M_{KC} is the cardinality-corank matrix of the matroid M, then the corresponding matrix M^*_{KC} for M^* satisfies

$$M^*_{KC}(i, j) = M_{KC}(n - i, i + j + r - n)$$

for all i in $\{0, 1, ..., n\}$ and all j in $\{0, 1, ..., n - r\}$.

6.94. (Brylawski, 1981b) Let Γ and Δ be the graphs shown in Figure 6.14.

(a) Show that, for each of the matroids M_Γ and M_Δ, the matrix M_{KC} is the following.

$$
\begin{array}{c}
\begin{array}{cccccc} 0 & 1 & 2 & 3 & 4 & 5 \end{array} \\
\begin{array}{c} 0 \\ 1 \\ 2 \\ 3 \\ 4 \\ 5 \\ 6 \\ 7 \\ 8 \\ 9 \\ 10 \end{array}
\begin{bmatrix}
0 & 0 & 0 & 0 & 0 & 1 \\
0 & 0 & 0 & 0 & 10 & 0 \\
0 & 0 & 0 & 44 & 1 & 0 \\
0 & 0 & 108 & 12 & 0 & 0 \\
0 & 151 & 58 & 1 & 0 & 0 \\
98 & 142 & 12 & 0 & 0 & 0 \\
151 & 58 & 1 & 0 & 0 & 0 \\
108 & 12 & 0 & 0 & 0 & 0 \\
44 & 1 & 0 & 0 & 0 & 0 \\
10 & 0 & 0 & 0 & 0 & 0 \\
1 & 0 & 0 & 0 & 0 & 0
\end{bmatrix}
\end{array}
$$

(b) Show that for $M(K_6)$ the matrix W_2 is the following.

$$
\begin{bmatrix}
1 & 0 & 0 & 0 & 0 & 0 \\
1 & 1 & 0 & 0 & 0 & 0 \\
1 & 3 & 1 & 0 & 0 & 0 \\
1 & 7 & 6 & 1 & 0 & 0 \\
1 & 15 & 25 & 10 & 1 & 0 \\
1 & 31 & 90 & 65 & 15 & 1
\end{bmatrix}
$$

(c) Find W_1 for $M(K_6)$.

(d) Use (6.76) to show that, for M in $\{M_\Gamma, M_\Delta\}$, $I_{M(K_6)}(M)$ is the following matrix.

$$
\begin{bmatrix}
0 & 0 & 2 & 8 & 6 & 1 \\
0 & 0 & 10 & 24 & 8 & 0 \\
0 & 2 & 28 & 24 & 1 & 0 \\
0 & 4 & 24 & 8 & 0 & 0 \\
0 & 7 & 19 & 1 & 0 & 0 \\
0 & 8 & 6 & 0 & 0 & 0 \\
0 & 5 & 1 & 0 & 0 & 0 \\
0 & 4 & 0 & 0 & 0 & 0 \\
0 & 1 & 0 & 0 & 0 & 0 \\
0 & 0 & 0 & 0 & 0 & 0 \\
1 & 0 & 0 & 0 & 0 & 0
\end{bmatrix}
$$

6.95. (Brylawski, 1979b) Let G and G' be upper combinatorially uniform geometries into which the matroid M is embedded. Show that

$$
I_{G'}(M) = I_G(M) \cdot W_1 \cdot W_2'
$$

where W_1 is the matrix of doubly indexed Whitney numbers of the first kind of G, and W_2' is the matrix of doubly indexed Whitney numbers of the second kind of G'.

6.96. (Brylawski, 1979b) Let M_1 be the matroid that is represented over any field by the following matrix:

$$
\begin{array}{cccccccc}
 & a & b & c & d & e & f & g_1 \\
\left[\begin{array}{ccccccc}
1 & 0 & 0 & 0 & 1 & 0 & -1 \\
0 & 1 & 0 & 0 & 1 & 1 & 0 \\
0 & 0 & 1 & 0 & 0 & 1 & 1 \\
0 & 0 & 0 & 1 & 0 & 1 & 1
\end{array}\right]
\end{array}
$$

Let M_2 be the matroid that is represented over any field except F_2 by the following matrix, where $\alpha \neq 0, 1$.

$$
\begin{array}{cccccccc}
 & a & b & c & d & e & f & g_2 \\
\left[\begin{array}{ccccccc}
1 & 0 & 0 & 0 & 1 & 0 & 0 \\
0 & 1 & 0 & 0 & 1 & 1 & 1 \\
0 & 0 & 1 & 0 & 0 & 1 & 1 \\
0 & 0 & 0 & 1 & 0 & 1 & \alpha
\end{array}\right]
\end{array}
$$

(a) Show that M_1 and M_2 have the affine embeddings shown in Figure 6.17.

Figure 6.17.

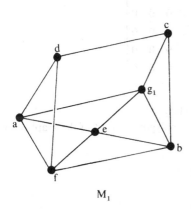

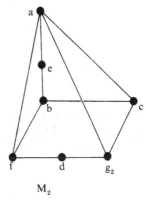

M_1

M_2

(b) Show that, for both M_1 and M_2, the matrix M_{KC} is the following.

	0	1	2	3	4
0	0	0	0	0	1
1	0	0	0	7	0
2	0	0	21	0	0
3	0	33	2	0	0
4	24	11	0	0	0
5	20	1	0	0	0
6	7	0	0	0	0
7	1	0	0	0	0

(c) Show that $t(M_1; x, y) = t(M_2; x, y)$.

(d) Show that if $M \in \{M_1, M_2\}$, then $I_{PG(3,3)}(M)$ is the following matrix.

$$\begin{bmatrix} 0 & 2 & 58 & 33 & 1 \\ 0 & 7 & 55 & 7 & 0 \\ 0 & 17 & 15 & 0 & 0 \\ 0 & 7 & 2 & 0 & 0 \\ 0 & 6 & 0 & 0 & 0 \\ 0 & 1 & 0 & 0 & 0 \\ 0 & 0 & 0 & 0 & 0 \\ 1 & 0 & 0 & 0 & 0 \end{bmatrix}$$

(e) Show by using (6.76) and also by using the result of Exercise 6.95 that $I_{PG(3,2)}(M_1)$ is the following matrix.

$$\begin{bmatrix} 0 & 0 & 5 & 8 & 1 \\ 0 & 1 & 13 & 7 & 0 \\ 0 & 2 & 15 & 0 & 0 \\ 0 & 5 & 2 & 0 & 0 \\ 0 & 6 & 0 & 0 & 0 \\ 0 & 1 & 0 & 0 & 0 \\ 0 & 0 & 0 & 0 & 0 \\ 1 & 0 & 0 & 0 & 0 \end{bmatrix}$$

(f) Why is the matrix $I_{PG(3,2)}(M_2)$ undefined?

(g) Show that M_1^* is isomorphic to the polygon matroid of the graph shown in Figure 6.18.

Figure 6.18.

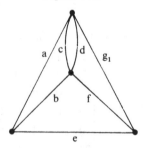

(h) Use Exercise 6.93 to find M_{KC} for M_1^* and then (6.76) to show that $I_{M(K_4)}(M_1^*)$ is the following matrix. Check your calculations by finding $I_{M(K_4)}(M_1^*)$ directly from the definition.

$$\begin{bmatrix} 0 & 0 & 0 & 1 \\ 0 & 0 & 5 & 0 \\ 0 & 2 & 1 & 0 \\ 0 & 3 & 0 & 0 \\ 0 & 2 & 0 & 0 \\ 0 & 0 & 0 & 0 \\ 0 & 0 & 0 & 0 \\ 1 & 0 & 0 & 0 \end{bmatrix}$$

(i) Find M_{KC} for M_1^c, where M_1^c is the matroid $PG(3, 2) - X$ where X is the set of columns of the matrix representing M_1.

6.97. (Brylawski, 1981b) Show that the graphs Γ and Δ in Figure 6.14, which have the same Tutte polynomial, have different polychromates.

References

Aigner, M. (1987). Whitney numbers, in N. White (ed.). *Combinatorial Geometries*, Encyclopedia of Mathematics and Its Applications 29, pp. 139–60. Cambridge University Press.

Appel, K. & Haken, W. (1976). Every planar map is four-colorable, *Bull. Amer. Math. Soc.* **82**, 711–12.

Arrowsmith, D. K. & Jaeger, F. (1982). On the enumeration of chains in regular chain-groups, *J. Comb. Theory Ser. B* **32**, 75–89.

Asano, T., Nishizeki, T., Saito, N. & Oxley, J. (1984). A note on the critical problem for matroids, *Europ. J. Comb.* **5**, 93–7.

Baclawski, K. (1975). Whitney numbers of geometric lattices, *Adv. Math.* **16**, 125–38.

Baclawski, K. (1979). The Möbius algebra as a Grothendieck ring, *J. Algebra* **57**, 167–79.

Bari, R. A. & Hall, D. W. (1977). Chromatic polynomials and Whitney's broken circuits, *J. Graph Theory* **1**, 269–75.

Barlotti, A. (1965). Some topics in finite geometrical structures, Institute of Statistics Mimeograph Series 439, Department of Statistics, University of North Carolina, Chapel Hill, North Carolina.

Barlotti, A. (1966). Bounds for k-caps in $PG(r, q)$ useful in the theory of error-correcting codes, Institute of Statistics Mimeograph Series 484.2, Department of Statistics, University of North Carolina, Chapel Hill, North Carolina.

Barlotti, A. (1980). Results and problems in Galois geometry, in J. Srivastava (ed.), *Combinatorial Mathematics, Optimal Designs and their Applications*, Ann. Discrete Math. 6, pp. 1–5. North-Holland, Amsterdam.

Beissinger, J. S. (1982). On external activity and inversion in trees, *J. Comb. Theory Ser. B* **33**, 87–92.

Bender, E., Viennot, G. & Williamson, S. G. (1984). Global analysis of the delete–contract recursion for graphs and matroids, *Linear and Multilinear Algebra* **15**, 133–60.

Berman, G. (1977). The dichromate and orientations of a graph, *Can. J. Math.* **29**, 947–56.

Berman, G. (1978). Decomposition of graph functions, *J. Comb. Theory Ser. B* **25**, 151–65.

Biggs, N. (1974). *Algebraic Graph Theory*, Cambridge University Press.

Biggs, N. (1979). Resonance and reconstruction, in B. Bollobás (ed.), *Surveys in Combinatorics, Proceedings of the Seventh British Combinatorial Conference*, pp. 1–21. Cambridge University Press.

Biggs, N. L., Lloyd, E. K. & Wilson, R. J. (1976). *Graph Theory: 1736–1936*, Oxford University Press.

Birkhoff, G. D. (1912–13). A determinant formula for the number of ways of coloring a map. *Ann. Math.* (2) **14**, 42–46.

Birkhoff, G. D. (1913). The reducibility of maps, *Amer. J. Math.* **35**, 115–28.

Birkhoff, G. D. (1930). On the number of ways of coloring a map, *Proc. Edinburgh Math. Soc.* (2) **2**, 83–91.

Birkhoff, G. D. & Lewis, D. C. (1946). Chromatic polynomials, *Trans. Amer. Math. Soc.* **60**, 355–451.

Bixby, R. E. (1975). A composition for matroids, *J. Comb. Theory Ser. B* **18**, 59–73.

Bixby, R. E. (1977). Kuratowski's and Wagner's theorems for matroids, *J. Comb. Theory Ser. B* **22**, 31–53.

Björner, A. (1980). Some matroid inequalities, *Discrete Math.* **31**, 101–3.

Björner, A. (1982). On the homology of geometric lattices, *Algebra Universalis* **14**, 107–28.

Blake, I. F. & Mullin, R. C. (1976). *An Introduction to Algebraic and Combinatorial Coding Theory*, Academic Press, New York.

Bland, R. G. & Las Vergnas, M. (1978). Orientability of matroids, *J. Comb. Theory Ser. B* **24**, 94–123.

Bollobás, B. (1976). *Extremal Graph Theory*, Academic Press, London.

Bondy, J. A. & Hemminger, R. L. (1977). Graph reconstruction – a survey, *J. Graph Theory* **1**, 227–68.

Bondy, J. A. & Murty, U. S. R. (1976). *Graph Theory with Applications*, Macmillan, London; American Elsevier, New York.

Bondy, J. A. & Welsh, D. J. A. (1972). Some results on transversal matroids and constructions for identically self-dual matroids, *Quart. J. Math. Oxford* (2) **23**, 435–51.

Bose, R. C. & Burton, R. B. (1966). A characterization of flat spaces in a finite geometry and the uniqueness of the Hamming and MacDonald codes, *J. Comb. Theory* **1**, 96–104.

Bosman, O. (1982). Nowhere-zero flows in graphs, Honours Thesis, Australian National University.

Brini, A. (1980). A class of rank-invariants for perfect matroid designs, *Europ. J. Comb.* **1**, 33–8.

Broadbent, S. R. & Hammersley, J. M. (1957). Percolation processes I. Crystals and mazes, *Proc. Camb. Phil. Soc.* **53**, 629–41.

Brooks, R. L. (1941). On colouring the nodes of a network, *Proc. Camb. Phil. Soc.* **37**, 194–7.

Brouwer, A. E. & Schrijver, A. (1978). The blocking number of an affine space, *J. Comb. Theory Ser. A* **24**, 251–3.

Bruen, A. A. & de Resmini, M. (1983). Blocking sets in affine planes, *Combinatorics '81* (Rome, 1981), North-Holland Mathematics Studies 78, pp. 169–75. North-Holland, Amsterdam.

Bruen, A. A. & Thas, J. A. (1977). Blocking sets, *Geom. Dedicata* **6**, 193–203.

Brylawski, T. (1971). A combinatorial model for series–parallel networks, *Trans. Amer. Math. Soc.* **154**, 1–22.

Brylawski, T. (1972a). The Tutte–Grothendieck ring, *Algebra Universalis* **2**, 375–88.

Brylawski, T. (1972b). A decomposition for combinatorial geometries, *Trans. Amer. Math. Soc.* **171**, 235–82.

Brylawski, T. (1974). Reconstructing combinatorial geometries, in R. A. Bari & F. Harary (eds), *Graphs and Combinatorics*, Lecture Notes in Mathematics 406, pp. 226–35. Springer-Verlag, Berlin.

Brylawski, T. (1975a). Modular constructions for combinatorial geometries, *Trans. Amer. Math. Soc.* **203**, 1–44.

Brylawski, T. (1975b). On the nonreconstructibility of combinatorial geometries, *J. Comb. Theory Ser. B* **19**, 72–6.

Brylawski, T. (1975c). An affine representation for transversal geometries, *Stud. Appl. Math.* **54**, 143–60.

Brylawski, T. (1976). A combinatorial perspective on the Radon convexity theorem, *Geom. Dedicata* **54**, 459–66.

Brylawski, T. (1977a). A determinantal identity for resistive networks, *SIAM J. Appl. Math.* **32**, 443–9.

Brylawski, T. (1977b). Connected matroids with the smallest Whitney numbers, *Discrete Math.* **18**, 243–52.

Brylawski, T. (1977c). The broken-circuit complex, *Trans. Amer. Math. Soc.* **234**, 417–33.

Brylawski, T. (1977d). Geometrie combinatorie e loro applicazioni, University of Rome lecture series (unpublished).

Brylawski, T. (1977e). Funzioni di Möbius, University of Rome lecture series (unpublished).

Brylawski, T. (1979a). Teoria dei codici e matroidi, University of Rome lecture series (unpublished).

Brylawski, T. (1979b). Intersection theory for embeddings of matroids into uniform geometries, *Stud. Appl. Math.* **61**, 211–44.

Brylawski, T. (1980). The affine dimension of the space of intersection matrices, *Rend. Mat.* (6) **13**, 59–68.

Brylawski, T. (1981a). Matroidi coordinabili, University of Rome lecture series (unpublished).

Brylawski, T. (1981b). Intersection theory for graphs, *J. Comb. Theory Ser. B* **30**, 233–46.

Brylawski, T. (1981c). Hyperplane reconstruction of the Tutte polynomial of a geometric lattice, *Discrete Math.* **35**, 25–38.

Brylawski, T. (1982). The Tutte polynomial. Part 1: General theory, in A. Barlotti (ed.), *Matroid Theory and Its Applications, Proceedings of the Third Mathematics Summer Center* (C.I.M.E., 1980), pp. 125–275. Liguori, Naples.

Brylawski, T. (1985). Coordinatizing the Dilworth truncation, in L. Lovász & A. Recski (eds), *Matroid Theory*, Colloq. Math. Soc. János Bolyai 40, pp. 61–95. North-Holland, New York.

Brylawski, T. (1986). Blocking sets and the Möbius function, in *Combinatorica*, Symposia Mathematica 28, pp. 231–49. Academic Press, New York.

Brylawski, T. & Kelly, D. G. (1978). Matroids and combinatorial geometries, in G.-C. Rota (ed.), *Studies in Combinatorics*, pp. 179–217. Mathematical Association of America, Washington, D.C.

Brylawski, T. & Kelly, D. G. (1980). *Matroids and Combinatorial Geometries*, Carolina Lecture Series 8, Department of Mathematics, University of North Carolina, Chapel Hill, North Carolina.

Brylawski, T. & Lucas, T. D. (1976). Uniquely representable combinatorial geometries, in *Colloquio Internazionale Sulle Teorie Combinatorie* (Roma, 1973), Atti dei Convegni Lincei 17, Tomo I, pp. 83–104. Accad. Naz. Lincei, Rome.

Brylawski, T. & Oxley, J. G. (1980). Several identities for the characteristic polynomial of a combinatorial geometry, *Discrete Math.* **31**, 161–70.

Brylawski, T. & Oxley, J. G. (1981). The broken-circuit complex: its structure and factorizations, *Europ. J. Comb.* **2**, 107–21.

Brylawski, T., Lo Re, P. M., Mazzocca, F. & Olanda, D. (1980). Alcune applicazioni della teoria dell'intersezione alle geometrie di Galois, *Richerche Mat.* **29**, 65–84.

Buck, R. C. (1943). Partition of space, *Amer. Math. Monthly* **50**, 541–4.

Cardy, S. (1973). The proof of and generalisations to a conjecture by Baker and Essam, *Discrete Math.* **4**, 101–22.

Cartier, P. (1981). Les arrangements d'hyperplans: Un chapitre de géométrie combinatoire, in *Séminaire Bourbaki, Vol. 1980/81 Exposés 561–578*, Lecture Notes in Mathematics, 901, pp. 1–22. Springer-Verlag, Berlin.

Chaiken, S. (1989). The Tutte polynomial of a ported matroid, *J. Comb. Theory Ser. B* **46**, 96–117.

Cordovil, R. (1979). Contributions à la théorie des géométries combinatoires, Thesis, l'Université Pierre et Marie Curie, Paris.

Cordovil, R. (1980). Sur l'évaluation $t(M; 2, 0)$ du polynôme de Tutte d'un matroïde et une conjecture de B. Grünbaum relative aux arrangements de droites du plan. *Europ. J. Comb.* **1**, 317–22.

Cordovil, R. (1982). Sur les matroïdes orientés de rang 3 et les arrangements du pseudodroites dans le plan projectif réel, *Europ. J. Comb.* **3**, 307–18.

Cordovil, R. (1985). A combinatorial perspective on the non-Radon partitions, *J. Comb. Theory Ser. A* **38**, 38–47; erratum **40**, 194.

Cordovil, R. & Silva, I. P. (1985). A problem of McMullen on the projective equivalences of polytopes, *Europ. J. Comb.* **6**, 157–61.

Cordovil, R. & Silva, I. P. (1987). Determining a matroid polytope by non-Radon partitions, *Linear Algebra Appl.* **94**, 55–60.

Cordovil, R., Las Vergnas, M. & Mandel, A. (1982). Euler's relation, Möbius functions, and matroid identities, *Geom. Dedicata* **12**, 147–62.

Cossu, A. (1961). Su alcune proprietà dei (k, n)-archi di un piano proiettivo sopra un corpo finito, *Rend. Mat.* (5) **20**, 271–77.

Crapo, H. H. (1966). The Möbius function of a lattice, *J. Comb. Theory* **1**, 126–31.

Crapo, H. H. (1967). A higher invariant for matroids, *J. Comb. Theory* **2**, 406–17.

Crapo, H. H. (1968a). Möbius inversions in lattices, *Arch. Math. (Basel)* **19**, 595–607.

Crapo, H. H. (1968b). The joining of exchange geometries, *J. Math. Mech.* **17**, 837–52.

Crapo, H. H. (1969). The Tutte polynomial, *Aequationes Math.* **3**, 211–29.

Crapo, H. H. (1970). Chromatic polynomials for a join of graphs, in P. Erdös, A. Rényi &

V. Sós (eds), *Combinatorial Theory and its Applications*, Colloq. Math. Soc. János Bolyai 4, pp. 239–45. North-Holland, Amsterdam.

Crapo, H. H. (1971). Constructions in combinatorial geometries, Notes, National Science Foundation Advanced Science Seminar in Combinatorial Theory, Bowdoin College, Maine (unpublished).

Crapo, H. H. & Rota, G.-C. (1970). *On the Foundations of Combinatorial Theory: Combinatorial Geometries*, preliminary edition, M.I.T. Press, Cambridge, Mass.

Cunningham, W. H. & Edmonds, J. (1980). A combinatorial decomposition theory, *Can. J. Math.* **32**, 734–65.

d'Antona, O. & Kung, J. P. S. (1980). Coherent orientations and series–parallel networks, *Discrete Math.* **32**, 95–8.

Datta, B. T. (1976a). On tangential 2-blocks, *Discrete Math.* **21**, 1–22.

Datta, B. T. (1976b). Nonexistence of six-dimensional tangential 2-blocks, *J. Comb. Theory Ser. B* **21**, 171–93.

Datta, B. T. (1979). On Tutte's conjecture for tangential 2-blocks, in J. A. Bondy & U. S. R. Murty (eds), *Graph Theory and Related Topics*, pp. 121–32. Academic Press, New York.

Datta, B. T. (1981). Nonexistence of seven-dimensional tangential 2-blocks, *Discrete Math.* **36**, 1–32.

Dawson, J. E. (1984). A collection of sets related to the Tutte polynomial of a matroid, in K. M. Koh & H. P. Yap (eds), *Graph Theory*, Lecture Notes in Mathematics 1073, pp. 193–204. Springer-Verlag, Berlin.

Deza, M. (1977). On perfect matroid designs, in *Construction and Analysis of Designs* (Japanese), Proceedings of a Symposium, pp. 98–108. Research Institute for Mathematical Sciences, Kyoto University, Kyoto.

Deza, M. (1984). *Perfect matroid designs*, Reports of the Department of Mathematics of the University of Stockholm, Sweden, No. 8.

Deza, M. & Singhi, N. M. (1980). Some properties of perfect matroid designs, in J. Srivastava (ed.), *Combinatorial Mathematics, Optimal Designs and their Applications*, Ann. Discrete Math. 6, pp. 57–76. North-Holland, Amsterdam.

Dirac, G. A. (1952). A property of 4-chromatic graphs and some remarks on critical graphs, *J. London Math. Soc.* **27**, 85–92.

Dirac, G. A. (1957). A theorem of R. L. Brooks and a conjecture of H. Hadwiger, *Proc. London Math. Soc.* (3) **7**, 161–95.

Dirac, G. A. (1964). Generalisations of the five colour theorem, *Theory of Graphs and Its Applications* (Smolenice, 1963), pp. 21–7. Academic Press, New York.

Dowling, T. A. (1971). Codes, packing, and the critical problem, *Atti del Convegno di Geometria Combinatoria e sue Applicazioni*, pp. 209–24. Institute of Mathematics, University of Perugia, Perugia.

Dowling, T. A. (1973a). A class of geometric lattices based on finite groups, *J. Comb. Theory* **13**, 61–87.

Dowling, T. A. (1973b). A q-analog of the partition lattice, in J. N. Srivastava *et al.* (eds), *A Survey of Combinatorial Theory*, pp. 101–15. North-Holland, Amsterdam.

Dowling, T. A. & Wilson, R. M. (1974). The slimmest geometric lattices, *Trans. Amer. Math. Soc.* **196**, 203–15.

Edelman, P. (1984a). A partial order on the regions of $\mathbb{R}^n$ dissected by hyperplanes, *Trans. Amer. Math. Soc.* **283**, 617–32.

Edelman, P. H. (1984b). The acyclic sets of an oriented matroid, *J. Comb. Theory Ser. B* **36**, 26–31.

Edmonds, J. (1965a). Lehman's switching game and a theorem of Tutte and Nash-Williams, *J. Res. Nat. Bur. Stand. Section B* **69**, 73–7.

Edmonds, J. (1965b). Minimum partition of a matroid into independent sets, *J. Res. Nat. Bur. Stand. Section B* **69**, 67–72.

Erdös, P. (1967). Extremal problems in graph theory, in F. Harary (ed.), *A Seminar on Graph Theory*, pp. 54–9. Holt, Rinehart & Winston, New York.

Essam, J. W. (1971). Graph theory and statistical physics, *Discrete Math.* **1**, 83–112.

Farrell, E. J. (1979). On a general class of graph polynomials, *J. Comb. Theory Ser. B* **26**, 111–22.

Goldman, J. & Rota, G.-C. (1969). The number of subspaces of a vector space, in W. T. Tutte (ed.), *Recent Progress in Combinatorics*, pp. 75–83. Academic Press, New York.

Good, I. J. & Tideman, T. N. (1977). Stirling numbers and a geometric structure from voting theory, *J. Comb. Theory Ser. A* **23**, 34–45.

Greene, C. (1973). On the Möbius algebra of a partially ordered set, *Adv. Math.* **10**, 177–87.

Greene, C. (1975). An inequality for the Möbius function of a geometric lattice, *Stud. Appl. Math.* **54**, 71–4.

Greene, C. (1976). Weight enumeration and the geometry of linear codes, *Stud. Appl. Math.* **55**, 119–28.

Greene, C. (1977). Acyclic orientations (notes from the talk), in M. Aigner (ed.), *Higher Combinatorics*, pp. 65–8. Reidel, Dordrecht.

Greene, C. & Zaslavsky, T. (1983). On the interpretation of Whitney numbers through arrangements of hyperplanes, zonotopes, non-Radon partitions and orientations of graphs, *Trans. Amer. Math. Soc.* **280**, 97–126.

Greenwell, D. L. & Hemminger, R. L. (1969). Reconstructing graphs, in G. Chartrand & S. F. Kapoor (eds), *The Many Facets of Graph Theory*, Lecture Notes in Mathematics 110, pp. 91–114. Springer-Verlag, Berlin.

Hadwiger, H. (1943). Über eine Klassifikation der Streckenkomplexe, *Vierteljschr. Naturforsch. Ges. Zürich* **88**, 133–42.

Hajós, G. (1961). Über eine Konstruktion nicht färbbärer Graphen, *Wiss. Z. Martin-Luther-Univ. Halle-Wittenberg A* **10**, 116–17.

Hall, D. W., Siry, J. W. & Vanderslice, B. R. (1959). The chromatic polynomial of the truncated icosahedron, *Proc. Amer. Math. Soc.* **66**, 405–7.

Heron, A. P. (1972a). Matroid polynomials, in D. J. A. Welsh & D. R. Woodall (eds), *Combinatorics*, pp. 164–203. Institute of Mathematics and Its Applications, Southend-on-Sea, U.K.

Heron, A. P. (1972b). Some topics in matroid theory, D. Phil. Thesis, Oxford.

Huseby, A. B. (1989). Domination theory and the Crapo β-invariant, *Networks* **19**, 135–49.

Jaeger, F. (1976a). Balanced valuations and flows in multigraphs, *Proc. Amer. Math. Soc.* **55**, 237–42.

Jaeger, F. (1976b). On nowhere-zero flows in multigraphs, in C. Nash-Williams & J. Sheehan (eds), *Proceedings of the Fifth British Combinatorial Conference*, Congressus Numerantium 15, pp. 373–9. Utilitas Mathematica, Winnipeg.

Jaeger, F. (1979). Flows and generalized coloring theorems in graphs, *J. Comb. Theory Ser. B* **26**, 205–16.

Jaeger, F. (1981). A constructive approach to the critical problem for matroids, *Europ. J. Comb.* **2**, 137–44.

Jaeger, F. (1982). Problem, Matroid Theory (Szeged, 1982), Colloq. Math. Soc. János Bolyai 40 (unpublished).

Jaeger, F. (1983). Geometrical aspects of Tutte's 5-flow conjecture, in *Graphs and Other Combinatorial Topics*, pp. 124–33. Teubner, Leipzig.

Jaeger, F. (1985). On five-edge-colorings of cubic graphs and nowhere-zero flow problems, *Ars Comb.* **20**, 229–44.

Jaeger, F. (1988a). On Tutte polynomials and cycles of plane graphs, *J. Comb. Theory Ser. B* **44**, 127–46.

Jaeger, F. (1988b). Nowhere-zero flow problems, in L. W. Beineke & R. J. Wilson (eds), *Selected Topics in Graph Theory. 3*, pp. 71–95. Academic Press, London.

Jaeger, F. (1988c). On Tutte polynomials and link polynomials, *Proc. Amer. Math. Soc.* **103**, 647–54.

Jaeger, F. (1989a). On edge-colorings of cubic graphs and a formula of Roger Penrose, in L. D. Andersen (ed.), *Graph Theory in Memory of G. A. Dirac*, Ann. Discrete Math. 41, pp. 267–80. North-Holland, Amsterdam.

Jaeger, F. (1989b). On Tutte polynomials of matroids representable over GF(q), *Europ. J. Comb.* **10**, 247–55.

Jaeger, F. (1989c). Tutte polynomials and bicycle dimension of ternary matroids, *Proc. Amer. Math. Soc.* **107**, 17–25.

Jaeger, F., Vertigan, D. L. & Welsh, D. J. A. (1990). On the computational complexity of the Jones and Tutte polynomials, *Math. Proc. Camb. Phil. Soc.* **108**, 35–53.

Jambu, M. & Terao, H. (1984). Free arrangements of hyperplanes and supersolvable lattices, *Adv. Math.* **52**, 248–58.

Joyce, D. (1984). Generalized chromatic polynomials, *Discrete Math.* **50**, 51–62.

Jones, V. F. R. (1985). A polynomial invariant for knots via von Neumann algebras,

Bull. Amer. Math. Soc. **12**, 103–11.

Kahn, J. & Kung, J. P. S. (1980). Varieties and universal models in the theory of combinatorial geometries, *Bull. Amer. Math. Soc.* **3**, 857–8.

Kászonyi, L. (1978a). An example for geometries having property $K(4)$ and being not four-colourable, in A. Hajnal & V. Sós (eds), *Combinatorics*, Colloq. Math. Soc. János Bolyai 18, pp. 635–8. North-Holland, New York.

Kászonyi, L. (1978b). On half-planar geometries, in A. Hajnal & V. Sós (eds), *Combinatorics*, Colloq. Math. Soc. János Bolyai 18, pp. 639–51. North-Holland, New York.

Kauffman, L. H. (1987). *On Knots*, Annals of Mathematical Studies 115, Princeton University Press.

Kauffman, L. H. (1988). New invariants in the theory of knots, *Amer. Math. Monthly* **95**, 195–242.

Kelly, D. G. & Rota, G.-C. (1973). Some problems in combinatorial geometry, in J. N. Srivastava *et al.* (eds), *A Survey of Combinatorial Theory*, pp. 309–13. North-Holland, Amsterdam.

Knuth, D. E. (1974). The asymptotic number of geometries, *J. Comb. Theory Ser. A* **17**, 398–401.

Kung, J. P. S. (1980). The Rédei function of a relation, *J. Comb. Theory Ser. A* **29**, 287–96.

Kung, J. P. S. (1986a). Growth rates and critical exponents of classes of binary combinatorial geometries, *Trans. Amer. Math. Soc.* **293**, 837–59.

Kung, J. P. S. (1986b). *A Source Book in Matroid Theory*, Birkhäuser, Boston.

Kung, J. P. S. (1987). Excluding the cycle geometries of the Kuratowski graphs from binary geometries, *Proc. London Math. Soc.* (3) **55**, 209–42.

Kung, J. P. S. (1988). The long-line graph of a combinatorial geometry: 1. Excluding $M(K_4)$ and the $(q + 2)$-point line as minors, *Quart. J. Math. Oxford* (2) **39**, 223–34.

Kung, J. P. S., Murty, M. R. & Rota, G.-C. (1980). On the Rédei zeta function, *J. Number Theory* **12**, 421–36.

Las Vergnas, M. (1975a). Matroïdes orientables, *C.R. Acad. Sci. Paris Sér. A* **280**, 61–4.

Las Vergnas, M. (1975b). Extensions normales d'une matroïde, polynôme de Tutte d'un morphisme, *C.R. Acad. Sci. Paris Sér. A* **280**, 1479–82.

Las Vergnas, M. (1977). Acyclic and totally cyclic orientations of combinatorial geometries, *Discrete Math.* **20**, 51–61.

Las Vergnas, M. (1978). Sur les activités des orientations d'une géométrie combinatoire, *Colloque Mathématiques Discrètes: Codes et Hypergraphs* (Brussels 1978), Cahiers Centre Études Recherche Opérations 20, pp. 293–300. Brussels.

Las Vergnas, M. (1979). On Eulerian partitions of graphs, in R. J. Wilson (ed.), *Graph Theory and Combinatorics*, Research Notes in Mathematics 34, pp. 62–75. Pitman, San Francisco.

Las Vergnas, M. (1980). On the Tutte polynomial of a morphism of matroids, in M. Deza & I. G. Rosenberg (eds), *Combinatorics 79*, Part 1, Ann. Discrete Math. 8, pp. 7–20. North-Holland, Amsterdam.

Las Vergnas, M. (1981). Eulerian circuits of 4-valent graphs imbedded in surfaces, in L. Lovász & V. Sós (eds), *Algebraic Methods in Graph Theory*, Colloq. Math. Soc. János Bolyai 25, pp. 451–78. North-Holland, New York.

Las Vergnas, M. (1983). Le polynôme de Martin d'un graphe Eulérien, in C. Berge *et al.* (eds), *Combinatorial Mathematics*, Ann. Discrete Math. 17, pp. 397–411. North-Holland, Amsterdam.

Las Vergnas, M. (1984). The Tutte polynomial of a morphism of matroids, II: Activities of orientations, in J. A. Bondy & U. S. R. Murty (eds), *Progress in Graph Theory*, pp. 367–80. Academic Press, New York.

Las Vergnas, M. (1988). On the evaluation at (3, 3) of the Tutte polynomial of a graph, *J. Comb. Theory Ser. B* **44**, 367–72.

Lee, L. A. (1975). On chromatically equivalent graphs, Thesis, George Washington University.

Lickorish, W. B. R. (1988). Polynomials for links, *Bull. London Math. Soc.* **20**, 558–88.

Lindner, C. C. & Rosa, A. (1978). Steiner quadruple systems – a survey, *Discrete Math.* **22**,

147–81.

Lindström, B. (1978). On the chromatic number of regular matroids, *J. Comb. Theory Ser. B* **24**, 367–9.

Lipson, A. S. (1986). An evaluation of a link polynomial, *Math. Proc. Camb. Phil. Soc.* **100**, 361–4.

Lucas, T. D. (1974). Properties of rank-preserving weak maps, *Bull. Amer. Math. Soc.* **80**, 127–31.

Lucas, T. D. (1975). Weak maps of combinatorial geometries, *Trans. Amer. Math. Soc.* **206**, 247–79.

MacWilliams, F. J. (1963). A theorem on the distribution of weights in a systematic code, *Bell. Sys. Tech. J.* **42**, 79–94.

Martin, P. (1977). Enumérations eulériennes dans les multigraphes et invariants de Tutte–Grothendieck, Thesis, Grenoble.

Martin, P. (1978). Remarkable valuation of the dichromatic polynomial of planar multigraphs. *J. Comb. Theory Ser. B* **24**, 318–24.

Mason, J. (1972). Matroids: unimodal conjectures and Motzkin's theorem, in D. J. A. Welsh & D. R. Woodall (eds), *Combinatorics*, pp. 207–21. Institute of Mathematics and Its Applications, Southend-on-Sea, U.K.

Mason, J. (1977). Matroids as the study of geometrical configurations, in M. Aigner (ed.), *Higher Combinatorics*, pp. 133–76. Reidel, Dordrecht.

Matthews, K. R. (1977). An example from power residues of the critical problem of Crapo and Rota, *J. Number Theory* **9**, 203–8.

Minty, G. J. (1966). On the axiomatic foundations of the theories of directed linear graphs, electrical networks, and network programming, *J. Math. Mech.* **15**, 485–520.

Minty, G. J. (1967). A theorem on three-coloring the edges of a trivalent graph, *J. Comb. Theory* **2**, 164–7.

Mullin, R. C. & Stanton, R. G. (1979). A covering problem in binary spaces of finite dimension, in J. A. Bondy & U. S. R. Murty (eds), *Graph Theory and Related Topics*, pp. 315–27. Academic Press, New York.

Murty, U. S. R. (1971). Equicardinal matroids, *J. Comb. Theory* **11**, 120–6.

Nash-Williams, C. St. J. A. (1966). An application of matroids to graph theory, in *Theory of Graphs*, International Symposium (Rome), pp. 263–5. Dunod, Paris.

Negami, S. (1987). Polynomial invariants of graphs, *Trans. Amer. Math. Soc.* **299**, 601–22.

Ore, O. (1967). *The Four-Color Problem*, Academic Press, New York.

Orlik, P. (1989). *Introduction to Arrangements*, C.B.M.S. Regional Conference Series 72, American Mathematical Society, Providence, Rhode Island.

Orlik, P. & Solomon, L. (1980). Combinatorics and topology of complements of hyperplanes, *Invent. Math.* **56**, 167–89.

Oxley, J. G. (1978a). Colouring, packing, and the critical problem, *Quart. J. Math. Oxford* (2) **29**, 11–22.

Oxley, J. G. (1978b). Cocircuit coverings and packings for binary matroids, *Math. Proc. Camb. Phil. Soc.* **83**, 347–51.

Oxley, J. G. (1979a). A generalization of a problem of Mullin and Stanton for matroids, in A. F. Horadam & W. D. Wallis (eds), *Combinatorial Mathematics VI*, Lecture Notes in Mathematics 748, pp. 92–7. Springer-Verlag, Berlin.

Oxley, J. G. (1979b). On cographic regular matroids, *Discrete Math.* **25**, 89–90.

Oxley, J. G. (1980). On a covering problem of Mullin and Stanton for binary matroids, *Aequationes Math.* **20**, 104–12.

Oxley, J. G. (1982a). On Crapo's beta invariant for matroids, *Stud. Appl. Math.* **66**, 267–77.

Oxley, J. G. (1982b). A note on half-planar geometries, *Period. Math. Hungar.* **13**, 137–9.

Oxley, J. G. (1983a). On a matroid identity, *Discrete Math.* **44**, 55–60.

Oxley, J. G. (1983b). On the numbers of bases and circuits in simple binary matroids, *Europ. J. Comb.* **4**, 169–78.

Oxley, J. G. (1987a). The binary matroids with no 4-wheel minor, *Trans. Amer. Math. Soc.* **301**,

63–75.

Oxley, J. G. (1987b). A characterization of the ternary matroids with no $M(K_4)$-minor, *J. Comb. Theory Ser. B* **42**, 212–49.

Oxley, J. G. (1989a). The regular matroids with no 5-wheel minor, *J. Comb. Theory Ser. B* **46**, 292–305.

Oxley, J. G. (1989b). A characterization of certain excluded-minor classes of matroids, *Europ. J. Comb.* **10**, 275–9.

Oxley, J. G. (1989c). A note on Negami's polynomial invariants for graphs, *Discrete Math.* **76**, 279–81.

Oxley, J. G. (1990). On an excluded-minor class of matroids, *Discrete Math.* **82**, 35–52.

Oxley, J. G. & Welsh, D. J. A. (1979a). On some percolation results of J. M. Hammersley, *J. Appl. Prob.* **16**, 526–40.

Oxley, J. G. & Welsh, D. J. A. (1979b). The Tutte polynomial and percolation, in J. A. Bondy & U. S. R. Murty (eds), *Graph Theory and Related Topics*, pp. 329–39. Academic Press, New York.

Oxley, J. G., Prendergast, K. & Row, D. H. (1982). Matroids whose ground sets are domains of functions, *J. Austral. Math. Soc. Ser. A* **32**, 380–7.

Penrose, R. (1971). Applications of negative dimensional tensors, in D. Welsh (ed.), *Combinatorial Mathematics and Its Applications*, pp. 221–44. Academic Press, London.

Pezzoli, L. (1984). On *D*-complementation, *Adv. Math.* **51**, 226–39.

Provan, J. S. & Ball, M. O. (1984). Computing network reliability in time polynomial in the number of cuts, *Oper. Res.* **32**, 516–26.

Purdy, G. B. (1979). Triangles in arrangements of lines, *Discrete Math.* **25**, 157–63.

Purdy, G. B. (1980). On the number of regions determined by n lines in the projective plane, *Geom. Dedicata* **9**, 107–9.

Rado, R. (1978). Monochromatic paths in graphs, in B. Bollobás (ed.), *Advances in Graph Theory*, Ann. Discrete Math. 3, pp. 191–4. North-Holland, Amsterdam.

Read, R. C. (1968). An introduction to chromatic polynomials, *J. Comb. Theory* **4**, 52–71.

Rosenstiehl, P. (1975). Bicycles et diagonales des graphes planaires, in *Colloques sur la Théorie des Graphes* (Paris, 1974), Cahiers Centre Études Recherche Opérations 17, pp. 365–83. Brussels.

Rosenstiehl, P. & Read, R. C. (1978). On the principal edge tripartition of a graph, in B. Bollobás (ed.), *Advances in Graph Theory*, Ann. Discrete Math. 3, pp. 195–226. North-Holland, Amsterdam.

Rota, G.-C. (1964). On the foundations of combinatorial theory I: Theory of Möbius functions, *Z. Wahrsch. verw. Gebiete* **2**, 340–68.

Rota, G.-C. (1967). Combinatorial analysis as a theory, Hedrick Lectures, Mathematical Association of America summer meeting (Toronto, 1967) (unpublished).

Rota, G.-C. (1971). Combinatorial theory, old and new, in *Proceedings of the International Mathematics Congress* (Nice, 1970), 3, pp. 229–33. Gauthier-Villars, Paris.

Satyanarayana, A. & Tindell, R. (1987). Chromatic polynomials and network reliability, *Discrete Math.* **67**, 57–79.

Scafati Tallini, M. (1966). (k, n)-archi di un piano grafico finito, con particolare riguardo a quelli condue caratteri, Nota I, II, *Rend. Acc. Naz. Lincei* (8) **40**, 812–18, 1020–5.

Scafati Tallini, M. (1973). Calotte di tipo (m, n) in uno spazia di Galois $S(r, q)$, *Rend Acc. Naz. Lincei* (8) **53**, 71–81.

Schläfi, L. (1950). *Gesammelte mathematische Abhandlungen*, Band I, Birkhäuser, Basel.

Schwärzler, W. (1991). Being Hamiltonian is not a Tutte invariant, *Discrete Math.* (to appear).

Segre, B. (1961). *Lectures on Modern Geometry*, Edizioni Cremonese, Rome.

Seymour, P. D. (1979). On multi-colourings of cubic graphs, and conjectures of Fulkerson and Tutte, *Proc. London Math. Soc.* (3) **38**, 423–60.

Seymour, P. D. (1980). Decomposition of regular matroids, *J. Comb. Theory Ser. B* **28**, 305–59.

Seymour, P. D. (1981a). Nowhere-zero 6-flows, *J. Comb. Theory Ser. B* **30**, 130–5.

Seymour, P. D. (1981b). On Tutte's extension of the four-colour problem, *J. Comb. Theory*

Ser. B **31**, 82–94.

Seymour, P. D. (1981c). Some applications of matroid decomposition, in L. Lovász &
V. Sós (eds), *Algebraic Methods in Graph Theory*, Colloq. Math. Soc. János Bolyai 25, pp.
713–26. North-Holland, New York.

Seymour, P. D. (1991). Matroid structure, in *Handbook of Combinatorics*, R. Graham,
M. Grötschel & L. Lovász (eds) (to appear).

Seymour, P. D. & Welsh, D. J. A. (1975). Combinatorial applications of an inequality
from statistical mechanics, *Math. Proc. Camb. Phil. Soc.* **77**, 485–97.

Shank, H. (1975). The theory of left–right paths, in A. P. Street & W. D. Wallis (eds),
Combinatorial Mathematics III, Lecture Notes in Mathematics 452, pp. 42–54.
Springer-Verlag, Berlin.

Shepherd, G. C. (1974). Combinatorial properties of associated zonotopes, *Can. J. Math.* **26**,
302–21.

Smith, C. A. B. (1969). Map colourings and linear mappings, in D. J. A. Welsh (ed.),
Combinatorial Mathematics and Its Applications, pp. 259–83. Academic Press, London.

Smith, C. A. B. (1972). Electric currents in regular matroids, in D. J. A. Welsh & D. R. Woodall
(eds), *Combinatorics*, pp. 262–85. Institute of Mathematics and Its Applications,
Southend-on-Sea, U.K.

Smith, C. A. B. (1974). Patroids, *J. Comb. Theory* **16**, 64–76.

Smith, C. A. B. (1978). On Tutte's dichromatic polynomial, in B. Bollobás (ed.), *Advances
in Graph Theory*, Ann. Discrete Math. 3, pp. 247–57. North-Holland, Amsterdam.

Stanley, R. P. (1971a). Modular elements of geometric lattices, *Algebra Universalis* **1**, 214–17.

Stanley, R. P. (1971b). Supersolvable semimodular lattices, in *Proceedings of the Conference
on Möbius Algebras*, pp. 80–142. University of Waterloo, Ontario.

Stanley, R. P. (1972). Supersolvable lattices, *Algebra Universalis* **2**, 197–217.

Stanley, R. P. (1973a). A Brylawski decomposition for finite ordered sets, *Discrete Math.* **4**, 77–82.

Stanley, R. P. (1973b). Acyclic orientations of graphs, *Discrete Math.* **5**, 171–8.

Stanley, R. P. (1980). Decompositions of rational convex polytopes, in J. Srivastava (ed.),
Combinatorial Mathematics, Optimal Designs and Their Applications, Ann. Discrete Math. 6,
pp. 333–42. North-Holland, Amsterdam.

Stanley, R. P. (1986). *Enumerative Combinatorics, Vol. I*, Wadsworth & Brooks/Cole,
Monterey, California.

Szekeres, G. & Wilf, H. (1968). An inequality for the chromatic number of a graph, *J.
Comb. Theory* **4**, 1–3.

Terao, H. (1980). Arrangement of hyperplanes and their freeness, I and II, *J. Faculty of
Sci., Univ. Tokyo, Sci. IA*, **27**, 293–312, 313–20.

Terao, H. (1981). Generalized exponents of a free arrangement of hyperplanes and
Shepard–Todd–Brieskorn formula, *Invent. Math.* **63**, 159–79.

Thistlethwaite, M. B. (1987). A spanning tree expansion of the Jones polynomial, *Topology*
26, 297–309.

Thistlethwaite, M. B. (1988a). Kauffman's polynomial and altering links, *Topology* **27**, 311–18.

Thistlethwaite, M. B. (1988b). On the Kauffman polynomial of an adequate link, *Invent.
Math.* **93**, 285–96.

Traldi, L. (1989). A dichromatic polynomial for weighted graphs and link polynomials,
Proc. Amer. Math. Soc. **106**, 279–86.

Tutte, W. T. (1947). A ring in graph theory, *Proc. Camb. Phil. Soc.* **43**, 26–40.

Tutte, W. T. (1954). A contribution to the theory of chromatic polynomials, *Can. J. Math.* **6**,
80–91.

Tutte, W. T. (1956). A class of abelian groups, *Can. J. Math.* **8**, 13–28.

Tutte, W. T. (1958). A homotopy theorem for matroids I, II, *Trans. Amer. Math. Soc.* **88**,
144–60, 161–74.

Tutte, W. T. (1959). Matroids and graphs, *Trans. Amer. Math. Soc.* **90**, 527–52.

Tutte, W. T. (1965). Lectures on matroids, *J. Res. Nat. Bur. Stand. Section B* **69**, 1–47.

Tutte, W. T. (1966a). On the algebraic theory of graph colorings, *J. Comb. Theory* **1**, 15–50.

Tutte, W. T. (1966b). Connectivity in matroids, *Can. J. Math.* **18**, 1301–24.

Tutte, W. T. (1967). On dichromatic polynomials, *J. Comb. Theory* **2**, 301–20.

Tutte, W. T. (1969a). Projective geometry and the 4-color problem, in W. Tutte (ed.), *Recent Progress in Combinatorics*, pp. 199–207. Academic Press, New York.

Tutte, W. T. (1969b). A geometrical version of the four color problem, in R. C. Bose & T. A. Dowling (eds), *Combinatorial Mathematics and its Applications*, pp. 553–61. University of North Carolina Press, Chapel Hill, North Carolina.

Tutte, W. T. (1974). Codichromatic graphs, *J. Comb. Theory* **16**, 168–75.

Tutte, W. T. (1976). The dichromatic polynomial, in C. Nash-Williams & J. Sheehan (eds), *Proceedings of the Fifth British Combinatorial Conference*, Congressus Numerantium 15, pp. 605–35. Utilitas Mathematics, Winnipeg.

Tutte, W. T. (1979). All the king's men (a guide to reconstruction), in J. A. Bondy & U. S. R. Murty (eds), *Graph Theory and Related Topics*, pp. 15–33. Academic Press, New York.

Tutte, W. T. (1980a). 1-factors and polynomials, *Europ. J. Comb.* **1**, 77–87.

Tutte, W. T. (1980b). Rotors in graph theory, in J. Srivastava (ed.), *Combinatorial Mathematics, Optimal Designs and Their Applications*, Ann. Discrete Math. 6, pp. 343–7. North-Holland, Amsterdam.

Tutte, W. T. (1984). *Graph Theory*, Encyclopedia of Mathematics and Its Applications 21, Cambridge University Press.

Van Lint, J. H. (1971). *Coding Theory*, Lecture Notes in Mathematics 201, Springer-Verlag, Berlin.

Veblen, O. (1912). An application of modular equations in Analysis Situs, *Ann. Math.* (2) **14**, 86–94.

Vertigan, D. L. (1991). The computational complexity of Tutte invariants for planar graphs (to appear).

Wagner, K. (1964). Beweis einer Abschwächung der Hadwiger-Vermutung, *Math. Ann.* **153**, 139–41.

Walton, P. N. (1981). Some topics in combinatorial theory, D. Phil. Thesis, Oxford.

Walton, P. N. & Welsh, D. J. A. (1980). On the chromatic number of binary matroids, *Mathematika* **27**, 1–9.

Walton, P. N. & Welsh, D. J. A. (1982). Tangential 1-blocks over GF(3), *Discrete Math.* **40**, 319–20.

Welsh, D. J. A. (1969). Euler and bipartite matroids, *J. Comb. Theory* **6**, 375–77.

Welsh, D. J. A. (1971). Combinatorial problems in matroid theory, in D. J. A. Welsh (ed.), *Combinatorial Mathematics and Its Applications*, pp. 291–307. Academic Press, London.

Welsh, D. J. A. (1976). *Matroid Theory*, Academic Press, London.

Welsh, D. J. A. (1977). Percolation and related topics, *Science Prog.* **64**, 65–83.

Welsh, D. J. A. (1979). Colouring problems and matroids, in B. Bollobás (ed.), *Surveys in Combinatorics, Proceedings of the Seventh British Combinatorial Conference*, pp. 229–57. Cambridge University Press.

Welsh, D. J. A. (1980). Colourings, flows, and projective geometry, *Nieuw Archief Wiskunde* (3) **28**, 159–76.

Welsh, D. J. A. (1982). Matroids and combinatorial optimisation, in A. Barlotti (ed.), *Matroid Theory and Its Applications, Proceedings of the Third Mathematics Summer Center* (C.I.M.E., 1980), pp. 323–416. Liguori, Naples.

Welsh, D. J. A. (1988). Matroids and their applications, in L. W. Beineke & R. J. Wilson (eds), *Selected Topics in Graph Theory*, 3, pp. 43–70. Academic Press, London.

White, N. (1972). The critical problem and coding theory, research paper, SPS-66 Vol. III, section 331, Jet Propulsion Lab., Pasadena, California.

White, N. (ed.) (1986). *Theory of Matroids*, Encyclopedia of Mathematics and Its Applications 26, Cambridge University Press.

White, N. (ed.) (1987). *Combinatorial Geometries*, Encyclopedia of Mathematics and Its Applications 29, Cambridge University Press.

Whitney, H. (1932). A logical expansion in mathematics, *Bull. Amer. Math. Soc.* **38**, 572–9.

Whitney, H. (1933a). The coloring of graphs. *Ann. Math.* (2) **33**, 688–718.

Whitney, H. (1933b). 2-isomorphic graphs, *Amer. J. Math.* **55**, 245–54.

Whitney, H. (1933c). A set of topological invariants for graphs, *Amer. J. Math.* **55**, 231–5.

Whitney, H. (1935). On the abstract properties of linear dependence, *Amer. J. Math.* **57**, 509–33.

Whittle, G. P. (1984). On the critical exponent of transversal matroids, *J. Comb. Theory Ser. B* **37**, 94–5.

Whittle, G. P. (1985). Some aspects of the critical problem for matroids, Ph.D. Thesis, University of Tasmania.

Whittle, G. P. (1987). Modularity in tangential *k*-blocks, *J. Comb. Theory Ser. B* **42**, 24–35.

Whittle, G. P. (1988). Quotients of tangential *k*-blocks, *Proc. Amer. Math. Soc.* **102**, 1088–98.

Whittle, G. P. (1989a). Dowling group geometries and the critical problem, *J. Comb. Theory Ser. B* **47**, 80–92.

Whittle, G. P. (1989b). *q*-lifts of tangential *k*-blocks, *J. London Math. Soc.* (2) **39**, 9–15.

Wilf, H. S. (1976). Which polynomials are chromatic? in *Colloquio Internazionale sulle Teorie Combinatorie* (Roma, 1973), Atti dei Convegni Lincei 17, Tomo I, pp. 247–56. Accad. Naz. Lincei, Rome.

Winder, R. O. (1966). Partitions of *N*-space by hyperplanes, *SIAM J. Appl. Math.* **14**, 811–18.

Woodall, D. R. & Wilson, R. J. (1978). The Appel–Haken proof of the four-color theorem, in L. W. Beineke & R. J. Wilson (eds), *Selected Topics in Graph Theory*, pp. 83–101. Academic Press, London.

Young, P. O. & Edmonds, J. (1972). Matroid designs, *J. Res. Nat. Bur. Stand. Section B* **72**, 15–44.

Young, P., Murty, U. S. R. & Edmonds, J. (1970). Equicardinal matroids and matroid designs, in *Combinatorial Mathematics and Its Applications*, pp. 498–542. University of North Carolina, Chapel Hill, North Carolina.

Zaslavsky, T. (1975a). Counting the faces of cut-up spaces, *Bull. Amer. Math. Soc.* **81**, 916–18.

Zaslavsky, T. (1975b). Facing up to arrangements: Face-count formulas for partitions of space by hyperplanes, *Mem. Amer. Math. Soc.*, No. 154.

Zaslavsky, T. (1976). Maximal dissections of a simplex, *J. Comb. Theory Ser. A* **20**, 244–57.

Zaslavsky, T. (1977). A combinatorial analysis of topological dissections, *Adv. Math.* **25**, 267–85.

Zaslavsky, T. (1979). Arrangements of hyperplanes; matroids and graphs, in *Proceedings of the Tenth Southeastern Conference on Combinatorics, Graph Theory and Computing*, Congressus Numerantium 24, pp. 895–911. Utilitas Mathematica, Winnipeg.

Zaslavsky, T. (1981a). The geometry of root systems and signed graphs, *Amer. Math. Monthly* **88**, 88–105.

Zaslavsky, T. (1981b). The slimmest arrangements of hyperplanes: II. Basepointed geometric lattices and Euclidean arrangements, *Mathematika* **28**, 169–90.

Zaslavsky, T. (1982a). Signed graphs, *Discrete Appl. Math.* **4**, 47–74.

Zaslavsky, T. (1982b). Signed graph coloring, *Discrete Math.* **39**, 215–28.

Zaslavsky, T. (1982c). Chromatic invariants of signed graphs, *Discrete Math.* **42**, 287–312.

Zaslavsky, T. (1982d). Bicircular geometry and the lattice of forests of a graph, *Quart. J. Math. Oxford* (2) **33**, 493–511.

Zaslavsky, T. (1983). The slimmest arrangements of hyperplanes: I. Geometric lattices and projective arrangements, *Geom. Dedicata* **14**, 243–59.

Zaslavsky, T. (1985a). Geometric lattices of structured partitions I: Gain-graphic matroids and group-valued partitions, manuscript.

Zaslavsky, T. (1985b). Geometric lattices of structured partitions II: Lattices of group-valued partitions based on graphs and sets, manuscript.

Zaslavsky, T. (1987). The Möbius function and the characteristic polynomial, in N. White (ed.), *Combinatorial Geometries*, Encyclopedia of Mathematics and Its Applications 29, pp. 114–38. Cambridge University Press.

Zaslavsky, T. (1991a). Biased graphs IV: Geometrical realizations, *J. Comb. Theory Ser. B*, (to appear).

Zaslavsky, T. (1991b). Biased graphs III: Chromatic and dichromatic invariants, *J. Comb. Theory Ser. B* (to appear).

7

The Homology and Shellability of Matroids and Geometric Lattices

ANDERS BJÖRNER

7.1. Introduction

With a finite matroid M are associated several simplicial complexes that are interrelated in an appealing way. They carry some of the significant invariants of M as face numbers and Betti numbers, and give rise to useful algebraic structures. In this chapter we shall study three such complexes: (1) the matroid complex $IN(M)$ of independent subsets, (2) the broken circuit complex $BC_\omega(M)$ relative to an ordering ω of the ground set, and (3) the order complex $\Delta(\bar{L})$ of chains in the associated geometric lattice L.

To systematize our approach to the combinatorial and homological properties of these complexes we utilize the notion of shellability. A complex is said to be shellable if its maximal faces are equicardinal and can be arranged in a certain order that is favorable for induction arguments. Shellability was established for matroid and broken circuit complexes by Provan (1977) and for order complexes of geometric lattices by Björner (1980a). One key property of a shellable complex Δ that we bring into play is the existence of a polynomial $h_\Delta(x)$ with non-negative integer coefficients that encodes the basic combinatorial invariants of Δ. The coefficients of $h_\Delta(1 + \lambda)$ are the face numbers of Δ and $h_\Delta(0)$ is the top Betti number of Δ, all other Betti numbers being zero. Since each coefficient in $h_\Delta(x)$ has an interpretation as counting certain of the maximal faces of Δ, the determination of the homology of a shellable complex becomes a purely combinatorial task once the basic theory of such complexes has been established.

The simplicial complex that most naturally comes to mind in connection with a matroid M is the collection $IN(M)$ of independent sets in M. While exploring the shellability of such complexes we are naturally led to the concepts of internal and external activity in a basis of M, and from there to the consideration of a two-variable generating function $T_M(x, y)$, the Tutte polynomial, such that $T_M(x, 1)$ and $T_M(1, y)$ are the shelling polynomials of

IN(M) and *IN(M*)* respectively (*M** is the orthogonal matroid). As an application, several matroid inequalities are derived.

The broken circuit complex $BC_\omega(M)$ of $M = M(S)$ relative to an ordering ω of S is the collection of those subsets of S that do not contain any broken circuit, that is, a circuit with deleted first element. This notion was developed by Whitney (1932), Rota (1964), Wilf (1976), and Brylawski (1977a), originally for enumerative purposes. The broken circuit complex carries the 'chromatic' properties of M: the shelling polynomial of $BC_\omega(M)$ equals $T_M(x, 0)$, the face numbers are the Whitney numbers of the first kind, and $BC_\omega(M)$ is a cone over a related complex whose top Betti number is $\beta(M)$, the beta invariant of M.

The homology of geometric lattice complexes $\Delta(\bar{L})$ was determined in the pioneering work of Folkman (1966), and has had a significant role since then. On the one hand, Folkman's vanishing theorem for homology made geometric lattices one of the motivating examples for the theory of Cohen–Macaulay posets (Stanley, 1977; Baclawski, 1980; see Björner, Garsia & Stanley, 1982). On the other hand, Orlik & Solomon (1980) showed that the singular cohomology ring of the complement of a complex arrangement of hyperplanes can be described entirely in terms of the order homology of the geometric lattice of intersections. Hence, in these connections (and others, such as in Gel'fand & Zelevinsky, 1986) geometric lattice homology is related to interesting applications of matroids within mathematics. Folkman's theorem is here deduced as a simple consequence of the shellability of $\Delta(\bar{L})$.

In connection with the homology of matroid complexes *IN(M)* and geometric lattice complexes $\Delta(\bar{L})$, an interesting role is played by broken circuit complexes. Namely, $BC_\omega(M^*)$ induces cycles that form a characteristic-free basis for the homology of *IN(M)*, and $BC_\omega(M)$ similarly determines a basis for the homology of $\Delta(\bar{L})$. These, together with the Orlik–Solomon algebra, are examples of a certain universality of the broken circuit idea for constructing bases for algebraic objects associated with matroids and geometric lattices.

This chapter aims to give a unified and concise, yet gentle, introduction to the topics that have been outlined. A minimum of prerequisites will be assumed. Sections 7.2–7.6 are entirely combinatorial; all algebraic aspects have been deferred to the last four sections. A simple presentation of the relevant parts of simplicial homology in section 7.7 makes the chapter essentially self-contained. Only the most basic ideas are developed in the text; additional results and ramifications appear among the exercises. The last section of the chapter contains all references to original sources and related comments.

7.2. Shellable Complexes

We begin by recalling the fundamental definitions. A *simplicial complex* (or just *complex*) Δ is a collection of subsets of a finite set V such that (1) if $F \in \Delta$ and $G \subseteq F$ then $G \in \Delta$ and (2) if $v \in V$ then $\{v\} \in \Delta$. Throughout this chapter we will assume that complexes are non-void. Note that $\Delta \neq \emptyset$ implies $\emptyset \in \Delta$. The elements of V are called *vertices* and the members of Δ are called *simplices* or *faces*. A face that is not properly contained in any other face is called a *facet*. The *dimension* of a face $F \in \Delta$ is one less than its cardinality, and the *dimension* of the complex is the maximal dimension of a face. That is, $\dim F = |F| - 1$ and $\dim \Delta = \max\{\dim F | F \in \Delta\}$. A complex is said to be *pure* if all its facets are equicardinal.

For a simplicial complex Δ let f_k denote the number of faces of cardinality k. Thus $f_0 = 1$, $f_1 = |V|$ and $f_k = 0$ for $k > r = \dim \Delta + 1$.

The convention in the literature is to let f_k denote the number of faces of dimension k, but from a combinatorial point of view, and particularly for the purposes of this chapter, our definition has definite advantages.

It is convenient to express the *face numbers* f_k by their generating function, the *face enumerator*

$$f_\Delta(\lambda) = \lambda^r + f_1 \lambda^{r-1} + \ldots + f_r = \sum_{i=0}^{r} f_i \lambda^{r-i}. \tag{7.1}$$

The *Euler characteristic* of Δ is $\chi(\Delta) = -1 + f_1 - f_2 + \ldots = (-1)^{r-1} f_\Delta(-1)$. (A topologist would call this the 'reduced' Euler characteristic; it is one less than the usual topological Euler characteristic.) A complex Δ for which every facet contains a certain vertex v is called a *cone* with *apex* v. Since the number of even faces must equal the number of odd faces (there is a pairing with respect to containment of v), it follows that

$$\chi(\Delta) = 0, \text{ if } \Delta \text{ is a cone.} \tag{7.2}$$

7.2.1. Example. Let Δ be the two-dimensional simplicial complex of Figure 7.1, having facets $A = \{a, b, c\}$, $B = \{a, b, d\}$, $C = \{a, c, d\}$, $D = \{b, d, e\}$, $E = \{c, d, e\}$, and $F = \{b, c, d\}$. The face numbers are $f_0 = 1$, $f_1 = 5$, $f_2 = 9$,

Figure 7.1.

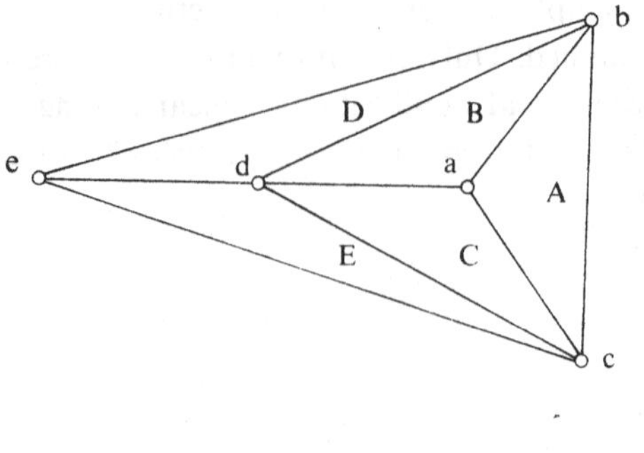

and $f_3 = 6$, hence $\chi(\Delta) = -1 + 5 - 9 + 6 = 1$.

Let Δ be a pure simplicial complex. A *shelling* of Δ is a linear order of the facets of Δ such that each facet meets the complex generated by its predecessors in a non-void union of maximal proper faces. In other words, the linear order $F_1, F_2, \ldots, F_t$ of the facets of Δ is a shelling if and only if

> *for each pair F_i, F_j of facets such that $1 \leqslant i < j \leqslant t$ there is a facet F_k satisfying*
> $1 \leqslant k < j$ *and an element* $x \in F_j$ *such that* $F_i \cap F_j \subseteq F_k \cap F_j = F_j - x$.   (7.3)

A complex is said to be *shellable* if it is pure and admits a shelling. It is easy to see that every zero-dimensional complex is shellable, while a one-dimensional complex (a simple graph) is shellable if and only if it is connected. The intuitive idea with a shelling is that of building the pure d-dimensional complex Δ stepwise by introducing one facet at a time and attaching it onto the complex already constructed in such a way that the intersection is topologically a $(d-1)$-ball or a $(d-1)$-sphere.

As an example, consider the complex Δ in Figure 7.1. It is easy to verify that an arbitrary permutation of A, B, and C followed by an arbitrary permutation of D, E, and F gives a shelling of Δ, whereas linear orders of the facets that begin with A, B, E, ..., or with D, E, C, A, ... are not shellings.

For the remainder of this section let us keep the following notation fixed. Δ is an $(r-1)$-dimensional shellable complex, and $F_1, F_2, \ldots, F_t$ are the facets of Δ listed in a shelling order. For $i = 1, 2, \ldots, t$ let $\Delta_i = \{G \in \Delta \mid G \subseteq F_k$ for some $k \leqslant i\}$, that is, Δ_i is the subcomplex of Δ generated by the i first facets. Also, for $i = 1, 2, \ldots, t$ let $\mathscr{R}(F_i) = \{x \in F_i : F_i - x \in \Delta_{i-1}\}$; $\mathscr{R}(F_i)$ is called the *restriction* of F_i induced by the shelling. Thus $\mathscr{R}(F_i) = \varnothing$ if and only if $i = 1$ and $\mathscr{R}(F_i) = F_i$ if and only if all proper subsets of F_i are contained in Δ_{i-1}. In the following proposition we consider Δ to be partially ordered by set inclusion of the faces, so that we may speak of Boolean intervals $[G_1, G_2] = \{G \in \Delta : G_1 \subseteq G \subseteq G_2\}$ as subsets of Δ.

7.2.2. Proposition. *The intervals* $[\mathscr{R}(F_i), F_i]$, $i = 1, 2, \ldots, t$, *partition the shellable complex* Δ.

Proof. The sequence $F_1, F_2, \ldots, F_i$ of facets of Δ_i is a shelling of Δ_i, so inductively it will suffice to show that

(1) $\Delta_{t-1} \cup [\mathscr{R}(F_t), F_t] = \Delta$ and

(2) $\Delta_{t-1} \cap [\mathscr{R}(F_t), F_t] = \varnothing$.

Let $G \in \Delta - \Delta_{t-1}$. Then $G \subseteq F_t$. If $x \notin G$ for some $x \in \mathscr{R}(F_t)$ then $G \subseteq F_t - x \in \Delta_{t-1}$, which contradicts $G \notin \Delta_{t-1}$. Hence, $\mathscr{R}(F_t) \subseteq G$, and (1) is proved. (2) is equivalent to $\mathscr{R}(F_t) \notin \Delta_{t-1}$. If $\mathscr{R}(F_t) \in \Delta_{t-1}$, then $\mathscr{R}(F_t) \subseteq F_i \cap F_t$ for some i, $1 \leqslant i < t$, and so by definition (7.3) $\mathscr{R}(F_t) \subseteq F_i \cap F_t \subseteq F_k \cap F_t = F_t - x$ for some k, $1 \leqslant k < t$, and $x \in F_t$. But $F_t - x \subseteq F_k \in \Delta_{t-1}$ entails that $x \in \mathscr{R}(F_t)$, which contradicts $\mathscr{R}(F_t) \subseteq F_t - x$. So $\mathscr{R}(F_t) \notin \Delta_{t-1}$.   □

The preceding result shows that when the facet F_i (with all its subfaces) is added to the complex Δ_{i-1} during the shelling process, then $\mathscr{R}(F_i)$ is the *unique* minimal face of F_i which is 'new' in Δ_i, that is, which lies in $\Delta_i - \Delta_{i-1}$.

With the shellable complex Δ we shall associate the *shelling polynomial* $h_\Delta(x)$, defined by

$$h_\Delta(x) = \sum_{i=1}^{t} x^{|F_i - \mathscr{R}(F_i)|}. \tag{7.4}$$

7.2.3. Proposition. *Let Δ be a shellable complex with shelling polynomial $h_\Delta(x)$ and face enumerator $f_\Delta(\lambda)$. Then*

$$h_\Delta(1 + \lambda) = f_\Delta(\lambda).$$

Hence, the polynomial $h_\Delta(x)$ is independent of shelling order.

Proof.

$$h_\Delta(1 + \lambda) = \sum_{i=1}^{t} (1 + \lambda)^{|F_i - \mathscr{R}(F_i)|} = \sum_{i=1}^{t} \sum_{k=0}^{r} \binom{|F_i - \mathscr{R}(F_i)|}{k} \lambda^k$$

$$= \sum_{k=0}^{r} \left(\sum_{i=1}^{t} \binom{|F_i - \mathscr{R}(F_i)|}{k} \right) \lambda^k,$$

and, by Proposition 7.2.2,

$$f_{r-k} = \sum_{i=1}^{t} \binom{|F_i - \mathscr{R}(F_i)|}{k}.   □$$

We see from the above that $h_\Delta(1)$ equals the number of facets and $h_\Delta(2)$ equals the number of faces of Δ. More important is that $h_\Delta(0) = f_\Delta(-1) = (-1)^{r-1}\chi(\Delta)$. Directly from definition (7.4) we get, however, that $h_\Delta(0)$ equals the number of facets F such that $\mathscr{R}(F) = F$.

7.2.4. Corollary. $(-1)^{r-1}\chi(\Delta)$ *equals the number of facets F such that $\mathscr{R}(F) = F$.*

To illustrate these ideas, let us choose the shelling A, B, C, D, E, F for the complex Δ of Example 7.2.1. Then $\mathscr{R}(A) = \varnothing$, $\mathscr{R}(B) = \{d\}$, $\mathscr{R}(C) = \{c, d\}$, $\mathscr{R}(D) = \{e\}$, $\mathscr{R}(E) = \{c, e\}$, $\mathscr{R}(F) = F$. Hence, the shelling polynomial is

$$h_\Delta(x) = x^3 + x^2 + x + x^2 + x + 1 = x^3 + 2x^2 + 2x + 1.$$

We can now check that $h_\Delta(2) = 21$ is the total number of faces, $h_\Delta(1) = 6$ is the number of facets, and $h_\Delta(0) = 1$ equals the Euler characteristic $\chi(\Delta)$. In fact, of course, $h_\Delta(1 + \lambda) = \lambda^3 + 5\lambda^2 + 9\lambda + 6 = f_\Delta(\lambda)$.

Let $f_\Delta(\lambda) = f_0 \lambda^r + f_1 \lambda^{r-1} + \dots + f_r$ and $h_\Delta(x) = h_0 x^r + h_1 x^{r-1} + \dots + h_r$ be the face enumerator and the shelling polynomial of Δ. The two number sequences $(f_0, f_1, \dots, f_r)$ and $(h_0, h_1, \dots, h_r)$, called respectively the *f-vector* and the *h-vector* of Δ, are intimately related.

By comparing coefficients in the relation $f_\Delta(\lambda) = h_\Delta(1 + \lambda)$ (Proposition 7.2.3) we get

$$f_k = \sum_{i=0}^{r} h_i \binom{r-i}{k-i}, \ k = 0, 1, ..., r. \tag{7.5}$$

Similarly, the relation $h_\Delta(x) = f_\Delta(x - 1)$ implies the inverse formula

$$h_k = \sum_{i=0}^{r} (-1)^{i+k} f_i \binom{r-i}{k-i}, \ k = 0, 1, ..., r. \tag{7.6}$$

We deduce that $h_0 = f_0 = 1$ and $h_1 = f_1 - r$. This also follows directly from the definition (7.4) of $h_\Delta(x)$, which can be restated:

$$h_k = \text{card}\{\text{facets } F \text{ such that } |\mathcal{R}(F)| = k\}, \ 0 \leqslant k \leqslant r. \tag{7.7}$$

From this we have that $h_k \geqslant 0$ for $0 \leqslant k \leqslant r$.

We shall now use the correlation between *f*-vectors and *h*-vectors to uncover some enumerative facts about shellable complexes. These will be used only in section 7.5.

7.2.5. Proposition. *Let* Δ *be an* $(r-1)$*-dimensional shellable complex on the vertex set* V, $|V| = v > r$. *Then*

(i) $f_k < f_j$, *for all* $0 \leqslant k < j \leqslant r - k$,

(ii) $f_k \leqslant f_{r-k+1}$, *if* $1 \leqslant k < (r+1)/2$ *and* $h_{r-k+1} \geqslant 1$, *with equality if and only if in addition* $h_2 = 1$,

(iii) *in the case where all e-element subsets of* V *belong to* Δ, *for* $e < r$, *then*

$$f_k \geqslant \sum_{i=0}^{e} \binom{v-r+i-1}{i} \binom{r-i}{k-i}$$

for $k = 0, 1, ..., r$, *and the following conditions are equivalent:*

(α) *equality holds for some* $k > e$

(β) *equality holds for all* k

(γ) $f_r = \binom{v-r+e}{e}$.

Proof. Part (i) follows directly from (7.5) and the non-negativity (7.7) of h_k, together with elementary properties of binomial coefficients. For part (ii) a little more of the structure of *h*-vectors must be used; see Exercise 7.2.

Now, suppose as in (iii) that all *e*-element subsets of V are in Δ, that is,

$$f_k = \binom{v}{k}$$ for $k = 0, 1, ..., e$. Using standard combinatorial identities such as

$$(-1)^y \binom{-x}{y} = \binom{x+y-1}{y}$$ and the Vandermonde convolution formula

$$\binom{x+y}{k} = \sum_{i=0}^{k} \binom{x}{i}\binom{y}{k-i},$$ (7.6) can then be developed as follows for $k \leqslant e$:

$$h_k = \sum_{i=0}^{k} (-1)^{i+k} \binom{r-i}{k-i}\binom{v}{i} = \sum_{i=0}^{k} \binom{k-r-1}{k-i}\binom{v}{i} = \binom{v-r+k-1}{k}. \quad (7.8)$$

Inserting these values into (7.5) and using that $h_k \geqslant 0$ for $k > e$ we see that the inequality in (iii) arises from (7.5) by singling out the first $e+1$ terms. It is also evident by this argument that if

(δ) $h_{e+1} = h_{e+2} = \ldots = h_r = 0,$

then equality in (iii) holds for all k. Thus, $(\delta) \Rightarrow (\beta) \Rightarrow (\gamma) \Rightarrow (\alpha)$, since (γ) means that equality holds for $k = r$. If equality holds in (iii) for some $k > e$ then (7.5) shows that $h_{e+1} = h_{e+2} = \ldots = h_k = 0$. It is a consequence of the following lemma, in view of (7.7), that then $h_{k+1} = h_{k+2} = \ldots = h_r = 0$ also, so (α) implies (δ).   □

7.2.6. Lemma. *Let $\mathscr{F} = \{\mathscr{R}(F_i) : 1 \leqslant i \leqslant t\}$, where $F_1, F_2, \ldots, F_t$ is a shelling of Δ. Then given $A \in \mathscr{F} - \{\varnothing\}$ there exists $x \in A$ such that $A - x \in \mathscr{F}$.*

Proof. Suppose that $1 < g \leqslant t$ and let $X_1, X_2, \ldots, X_s$ be the maximal proper subsets of $\mathscr{R}(F_g)$. They all belong to the subcomplex Δ_{g-1} since $\mathscr{R}(F_g)$ is the unique minimal face in $\Delta_g - \Delta_{g-1}$. Hence by Proposition 7.2.2 there are unique facets $F_{i_1}, F_{i_2}, \ldots, F_{i_s}$ satisfying $\mathscr{R}(F_{i_j}) \subseteq X_j \subseteq F_{i_j}$ and $i_j < g$ for $j = 1, 2, \ldots, s$. We have that $F_{i_j} \neq F_{i_k}$ when $j \neq k$, since if $X_j \subseteq F_{i_k}$ and $X_k \subseteq F_{i_j}$ then $\mathscr{R}(F_g) = X_j \cup X_k \subseteq F_{i_j}$ which contradicts 7.2.2. Assume that indices have been chosen so that $i_1 < i_2 < \ldots < i_s$. Then every proper subset of X_s belongs to Δ_{i_s-1}, so $\mathscr{R}(F_{i_s})$ cannot be strictly contained in X_s. Hence, $\mathscr{R}(F_{i_s}) = X_s \subset \mathscr{R}(F_g)$ and $|\mathscr{R}(F_g)| = |\mathscr{R}(F_{i_s})| + 1$.   □

7.3. Matroid Complexes

If $M = M(S)$ is a matroid of rank r on the finite set S, the family of all independent sets in M forms an $(r-1)$-dimensional simplicial complex, which we denote by $IN(M)$. Complexes of this kind are called *matroid complexes*. It is one of the first facts of matroid theory that a matroid complex is pure, and we will soon see that it is also shellable. In fact, matroid complexes can be characterized both in terms of purity and in terms of shellability (cf. Exercise 7.4 and Theorem 7.3.4).

A number of remarkable properties of matroids are revealed by, but not dependent on, assigning a linear order to the underlying point set. Our approach to these results will be aided by the following definitions and conventions. By an *ordered matroid* $M(S, \omega)$ we will mean a matroid $M(S)$ together with a linear ordering ω of the underlying point set S. Let us agree to write a k-subset $A = \{x_1, x_2, ..., x_k\} \subseteq S$ as an ordered k-tuple $[x_1, x_2, ..., x_k]$ if and only if $x_1 < x_2 < ... < x_k$ under the order ω. The k-subsets of S are linearly ordered by the *lexicographic order* defined as follows: $[x_1, x_2, ..., x_k]$ precedes $[y_1, y_2, ..., y_k]$ if and only if $x_i = y_i$ for $i < e$ and $x_e < y_e$ for some position e. In what follows, when we compare bases of an ordered matroid it is always with respect to this induced linear order.

Recall that if B is a basis of $M(S)$ and $p \in S - B$, then there is a unique circuit $\operatorname{ci}(B, p)$ contained in $B \cup p$. Dually, if $b \in B$ there is a unique bond $\operatorname{bo}(B, b)$ contained in $(S - B) \cup b$. The *basic circuit* and *basic bond* are characterized as follows.

7.3.1. Lemma. $b \in \operatorname{ci}(B, p) \Leftrightarrow (B - b) \cup p$ *is a basis* $\Leftrightarrow p \in \operatorname{bo}(B, b)$.

Let C be a circuit of an ordered matroid $M = M(S, \omega)$ and c the least element in C. Then the set $C - c$ is called a *broken circuit*. A basis of M that contains no broken circuits will be referred to as an *nbc-basis*. This concept will figure prominently in later sections. For now it will be used for future reference in the following technical lemma, which provides the key to the results of this and the following section.

7.3.2. Lemma. *Let $M(S, \omega)$ be an ordered matroid. Assume that the basis B precedes the basis C in the induced lexicographic order. Then $B \cap C \subseteq A \cap C$ for some basis A that also precedes C, and such that $|A \cap C| = |C| - 1$. Further, if C is an nbc-basis then A can be chosen to be an nbc-basis.*

Proof. Let $B = [b_1, b_2, ..., b_r]$, $C = [c_1, c_2, ..., c_r]$ and assume that $b_i = c_i$ for $i = 1, 2, ..., e - 1$, and $b_e \neq c_e$. Then $b_e < c_i$ for $i = e, e + 1, ..., r$. By the basis exchange property there is an element $y \in C - B$ such that $A_1 = (C - y) \cup b_e$ is a basis. For the first claim we can let $A = A_1$ be this basis.

For the second claim, assume that C is an nbc-basis. If A_1 is not an nbc-basis then there is an element $a_1 \in S - A_1$ such that a_1 is the least element of the basic circuit $\operatorname{ci}(A_1, a_1)$. Furthermore, b_e must belong to $\operatorname{ci}(A_1, a_1)$, since otherwise C would contain the broken circuit $\operatorname{ci}(A_1, a_1) - a_1$. Hence, $A_2 = (A_1 - b_e) \cup a_1$ is a basis, A_2 precedes A_1, $B \cap C \subseteq A_2 \cap C$, and $|A_2 \cap C| = r - 1$. If A_2 is not an nbc-basis the argument can be repeated until after a finite number of steps we reach a basis $A = A_k$ with the required properties. $\square$

Disregarding the last sentence (about nbc-bases), Lemma 7.3.2 is equivalent to the following (cf. (7.3)):

7.3.3. Theorem. *Let $M(S, \omega)$ be an ordered matroid. Then the ω-lexicographic order of bases of M is a shelling of $IN(M)$. In particular, all matroid complexes are shellable.*

Thus every linear ordering of the ground set induces a shelling, and this rich supply of shellings in fact characterizes matroids.

7.3.4. Theorem. *A simplicial complex Δ is a matroid complex if and only if Δ is pure and every ordering of the vertices induces a shelling.*

Proof. In one direction this is Theorem 7.3.3.

For the converse, suppose that a pure complex Δ is not a matroid complex. If V is the vertex set of Δ, then for some subset $U \subset V$ the induced subcomplex $\Delta_U = \{F \in \Delta : F \subseteq U\}$ is not pure (cf. Exercise 7.4). Let F be a facet of Δ_U of minimal dimension, and among the facets of Δ_U of dimension greater than $\dim F$ choose G such that $|F \cap G|$ is as large as possible. Now, order the vertex set V in such a way that the elements of $F - G$ come first, then the elements of G, and the remaining elements last. Let $\tilde{G}$ be the first facet of Δ that contains G, and let $\tilde{F}$ be an arbitrarily chosen facet of Δ that contains F. Clearly, $\tilde{F} \cap U = F$ and $\tilde{G} \cap U = G$, since F and G are maximal in Δ_U. The chosen ordering of the vertices ensures that a facet H of Δ precedes $\tilde{G}$ if and only if $H \cap (F - G) \neq \varnothing$. In particular, $\tilde{F}$ precedes $\tilde{G}$. F and G are facets of Δ_U and $|F| < |G|$, so $|F \cap G| \leqslant |G| - 2$. Hence, $|\tilde{F} \cap \tilde{G}| \leqslant |\tilde{G}| - 2$. Assume that $\tilde{F} \cap \tilde{G} \subseteq H \cap \tilde{G} = \tilde{G} - g$, for some facet H of Δ that precedes $\tilde{G}$ and $g \in \tilde{G}$. Then $g \in \tilde{G} - \tilde{F}$, and $H - \tilde{G} = h \in F - G$. If $g \in \tilde{G} - G$, then $G \cup h \subseteq H \cap U$, hence $G \cup h \in \Delta_U$, which contradicts the maximality of G in Δ_U. If $g \in G - F$, then $G' = (G - g) \cup h$ is a facet of Δ_U satisfying $\dim G' = \dim G > \dim F$ and $|F \cap G'| > |F \cap G|$, which contradicts the choice of G. Hence, such a facet H cannot exist, and the induced order is not a shelling. $\qquad\square$

Having established the shellability of matroid complexes we shall now find their shelling polynomials. Let B be a basis of an ordered matroid $M = M(S, \omega)$. An element $p \in S - B$ is said to be *externally active* in B if p is the least element in the basic circuit $\mathrm{ci}(B, p)$. Otherwise p is *externally passive* in B. Dually, an element $p \in B$ is said to be *internally active* in B if p is the least element in the basic bond $\mathrm{bo}(B, p)$. Otherwise p is *internally passive* in B. Denote by $EA(B)$, $EP(B)$, $IA(B)$, and $IP(B)$, respectively, the sets of externally active, externally passive, internally active, and internally passive elements in B. We will call the number $i(B) = |IA(B)|$ the *internal activity* of B, and $e(B) = |EA(B)|$ the *external activity* of B. It is obvious that p is internally active in B if and only if p is externally active in the basis $S - B$ of the orthogonal ordered matroid $M^* = M^*(S, \omega)$.

7.3.5. Example. Let M be the matroid defined either by points and lines in affine space as in Figure 7.2a, or by the graph of Figure 7.2b. We will list the 8 bases of M in lexicographic order, and indicate for each basis which elements are active in it.

Figure 7.2.

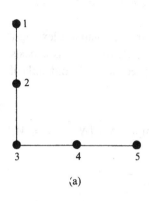

(a)

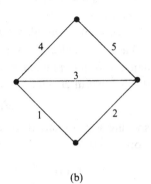

(b)

Basis	Internally active	Externally active
$B_1 = [1, 2, 4]$	1, 2, 4	—
$B_2 = [1, 2, 5]$	1, 2	—
$B_3 = [1, 3, 4]$	1, 4	—
$B_4 = [1, 3, 5]$	1	—
$B_5 = [1, 4, 5]$	1	3
$B_6 = [2, 3, 4]$	4	1
$B_7 = [2, 3, 5]$	—	1
$B_8 = [2, 4, 5]$	—	1, 3

A picture of the complex $IN(M)$ appears in Figure 7.3. It takes the form of the surface of a triangular bipyramid together with the two interior triangles

Figure 7.3.

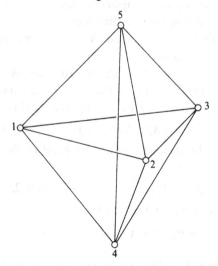

145 and 245. It is instructive to observe how the sequence of bases $B_1, \ldots, B_8$ gives a shelling of $IN(M)$.

Let $M = M(S, \omega)$ be an ordered matroid and consider again the lexicographic order of bases. It is clear, in view of Lemma 7.3.1, that if B is a basis and $b \in B$ then $B - b$ is contained in a basis that precedes B if and only if b is internally passive in B. Thus

$$\mathcal{R}(B) = IP(B). \tag{7.9}$$

The shelling polynomial $h_\Delta(x)$ of a matroid complex $\Delta = IN(M)$, $M = M(S, \omega)$ is therefore equal to

$$h_\Delta(x) = \sum_B x^{|B - \mathcal{R}(B)|} = \sum_B x^{|IA(B)|} = \sum_B x^{i(B)}. \tag{7.10}$$

Dually, the shelling polynomial $h_{\Delta^*}(y)$ of the orthogonal matroid complex $\Delta^* = IN(M^*)$, $M^* = M^*(S, \omega)$ is equal to

$$h_{\Delta^*}(y) = \sum_B y^{e(B)}.$$

In both cases the summation is over all bases B of M. It thus seems tempting to combine these two shelling polynomials associated with M into one polynomial of two variables

$$T_M(x, y) = \sum_B x^{i(B)} y^{e(B)}. \tag{7.11}$$

However, there is a complication. Whereas we know that the evaluations

$$T_M(x, 1) = h_{IN(M)}(x) \quad \text{and} \quad T_M(1, y) = h_{IN(M^*)}(y) \tag{7.12}$$

are independent of the order ω (Proposition 7.2.3), it is not immediately clear that $T_M(x, y)$ itself is independent of the ordering of S. We will soon see that this is the case, so that $T_M(x, y)$, called the *Tutte polynomial* of M, depends only on the matroid structure of M.

Before proving independence of the ordering, let us again look at the matroid M of Example 7.3.5. From the table of internally and externally active elements we conclude that its Tutte polynomial is

$$T_M(x, y) = x^3 + x^2 + x^2 + x + xy + xy + y + y^2 = x^3 + 2x^2 + y^2 + 2xy + x + y.$$

This implies that the shelling polynomial $h_\Delta(x)$ of the matroid complex $\Delta = IN(M)$ equals

$$h_\Delta(x) = T_M(x, 1) = x^3 + 2x^2 + 3x + 2,$$

and for the orthogonal matroid $\Delta^* = IN(M^*)$

$$h_{\Delta^*}(y) = T_M(1, y) = y^2 + 3y + 4.$$

7.3.6. Proposition. *Let $M = M(S, \omega)$ be an ordered matroid. Then the family of intervals $[IP(B), S - EP(B)]$, one for each basis B of M, partitions the Boolean algebra of subsets of S.*

Proof. We must prove for every subset $A \subseteq S$ that there is a basis B_A such that $IP(B_A) \subseteq A \subseteq S - EP(B_A)$, and that such a basis B_A is unique.

Let $A \subseteq S$. Then let X_A be the lexicographically greatest basis in the submatroid M_A that is induced on the subset A by M. The set X_A is independent in M and therefore there is a unique basis B_A such that $IP(B_A) \subseteq X_A \subseteq B_A$, by Proposition 7.2.2 and (7.9). Let $a \in A - B_A = A - X_A$, and assume that a is externally passive in B_A. This means that there is an element $b \in B_A$ such that $b < a$ and $(B_A - b) \cup a$ is a basis. In the case $b \notin A$ this implies that $X_A \cup a$ is independent, which is impossible since X_A is a basis in M_A and $a \notin X_A$. If $b \in A$, then $b \in B_A \cap A = X_A$ and $(X_A - b) \cup a$ would be a basis in M_A strictly preceded by X_A, which contradicts the choice of X_A. Hence, all elements of $A - B_A$ are externally active in B_A. We have shown that $IP(B_A) \subseteq X_A \subseteq A \subseteq S - EP(B_A)$.

Next, suppose that $IP(B) \subseteq A \subseteq S - EP(B)$ for some basis B of M. Observe that

$$EA(B) \text{ is contained in the closure of } IP(B). \tag{7.13}$$

To see this, suppose that $p \in S - B$ is externally active in B and let C be the basic circuit $ci(B, p)$ and C' the corresponding broken circuit $ci(B, p) - p$. If $q \in C'$ then $p < q$ and $(B - q) \cup p$ is a basis, so q is internally passive in B. Since p lies in the closure of C' and $C' \subseteq IP(B)$, statement (7.13) follows. This fact, (7.13), implies that $X = B \cap A$ is a basis of M_A. Suppose that $X \neq X_A$, where X_A is defined as above. Let $X = \{x_1, x_2, \ldots, x_a\}$, $x_1 < x_2 < \ldots < x_a$, and $X_A = \{y_1, y_2, \ldots, y_a\}$, $y_1 < y_2 < \ldots < y_a$. Since X_A is the greatest basis of M_A there is an index e such that $x_i = y_i$ for $i = 1, 2, \ldots, e - 1$ and $x_e < y_i$ for $i = e, e + 1, \ldots, a$. By the basis exchange axiom $(X - x_e) \cup y_j$ is a basis for some $j \geq e$. But then y_j is externally passive in X, and hence also in B, contradicting the assumption $A \subseteq S - EP(B)$. Hence, $X = X_A$ and, since $IP(B) \subseteq X \subseteq B$, also $B = B_A$. $\qquad\square$

7.3.7. Theorem. *Let $M = M(S, \omega)$ be an ordered matroid with Tutte polynomial $T_M(x, y)$. Then*

$$T_M(1 + \xi, 1 + \eta) = \sum_{A \subseteq S} \xi^{r(S) - r(A)} \eta^{r^*(S) - r^*(S - A)},$$

where r and r^* denote the rank functions of M and the orthogonal matroid M^* respectively.

In particular, the Tutte polynomial does not depend on the ordering ω of S.

Proof. Let B be a basis of M and assume that $IP(B) \subseteq A \subseteq S - EP(B)$. The observation (7.13) that $EA(B)$ is contained in the closure of $IP(B)$ implies that $r(A) = |A \cap B|$, so $r(S) - r(A) = |B - A|$. Dually, $r^*(S) - r^*(S - A) = |(S - B) - (S - A)| = |A - B|$. Thus, using Proposition 7.3.6 we get

$$\sum_{A \subseteq S} \xi^{r(S)-r(A)} \eta^{r^*(S)-r^*(S-A)} = \sum_{B \text{ basis}} \sum_{IP(B) \subseteq A \subseteq S - EP(B)} \xi^{|B-A|} \eta^{|A-B|}$$

$$= \sum_{B \text{ basis}} \sum_{j,k=0}^{\infty} \binom{i(B)}{j} \binom{e(B)}{k} \xi^j \eta^k$$

$$= \sum_{B \text{ basis}} (1 + \xi)^{i(B)} (1 + \eta)^{e(B)}$$

$$= T_M(1 + \xi, 1 + \eta). \qquad \square$$

The Tutte polynomial is treated in great detail in Chapter 6 of this volume, to which the reader is referred for more information.

Before ending this section, let us mention that (7.12) and Corollary 7.2.4 imply for the Euler characteristic of $IN(M)$ that $\chi(IN(M)) = (-1)^{r-1} T_M(0, 1)$. Another expression will be given in Proposition 7.4.7.

7.4. Broken Circuit Complexes

Let $M = M(S, \omega)$ be an ordered matroid. Recall that when the least element of a circuit is deleted we call the remaining set a *broken circuit*. The family of all subsets of S that contain no broken circuit forms a simplicial complex, which we denote $BC_\omega(M)$ and call the *broken circuit complex* of M. Note that $BC_\omega(M)$ is defined if and only if M is loopless, since if M has a loop then every subset of S contains the broken circuit $\varnothing$.

When discussing broken circuit complexes we may, whenever convenient, exclude the existence not only of loops but also of parallel elements in a matroid. Two elements $x, y \in S$ in a loopless matroid $M(S)$ are said to be *parallel* if $\text{rank}_M(\{x, y\}) = 1$. Parallelism is an equivalence relation. Define a rank function on the set $\bar{S}$ of equivalence classes by $\text{rank}_{\bar{M}}(\{X_1, X_2, ..., X_n\}) = \text{rank}_M(X_1 \cup X_2 \cup ... \cup X_n)$. This determines a matroid $\bar{M}(\bar{S})$ without loops or parallel elements, the *simplification* of M. If $M(S, \omega)$ is ordered let $\bar{M}(\bar{S}, \bar{\omega})$ be ordered by the first elements in the respective parallelism classes. The following observation is straightforward.

7.4.1. Proposition. *The two broken circuit complexes $BC_\omega(M)$ and $BC_{\bar{\omega}}(\bar{M})$ are isomorphic.*

The role that is played by the ordering ω in the construction of broken circuit complexes should perhaps be elucidated. Different orderings of the point set of a given matroid may yield non-isomorphic broken circuit complexes, as illustrated in Example 7.4.4 below. However, we will find that the important invariants of such complexes are independent of order.

Let $M(S, \omega)$ be an ordered loopless matroid with first element e. The family of all subsets of $S - e$ that contain no broken circuit will be called the *reduced broken circuit complex* and denoted $\overline{BC}_\omega(M)$. Let us gather some initial observations.

7.4.2. Proposition. *Let $M = M(S, \omega)$ be an ordered loopless matroid of rank r. Then*

(i) $\overline{BC}_\omega(M) \subseteq BC_\omega(M) \subseteq IN(M)$,

(ii) $BC_\omega(M)$ *is a pure $(r-1)$-dimensional complex whose facets are the nbc-bases of M,*

(iii) $BC_\omega(M)$ *is a cone over $\overline{BC}_\omega(M)$ with apex e,*

(iv) $\overline{BC}_\omega(M)$ *is a pure $(r-2)$-dimensional complex.*

Proof. Suppose that $X \subseteq S$ contains no broken circuit. Then, *a fortiori*, X contains no circuit and hence $X \in IN(M)$. Being independent, X is included in some basis of M. Let B be the lexicographically first such basis. Suppose that B contains a broken circuit C. In this situation it would be possible to find elements y and z in S such that $y \notin B$, $C \cup y = \text{ci}(B, y)$, $z \in C - X$, and $y < z$. Then $(B - z) \cup y$ would be a basis which contains X and precedes B, which contradicts the choice of B. Thus B cannot contain any broken circuit, that is, B is an nbc-basis. We have shown parts (i) and (ii).

If a basis B of M does not contain the first element e of S then B includes the broken circuit $\text{ci}(B, e) - e$. Thus all nbc-bases contain e so that $BC_\omega(M)$ is a cone with apex e. Parts (iii) and (iv) now follow. □

If B is an nbc-basis, let us call $B - e$ a *reduced nbc-basis*. The following result is basic to this section.

7.4.3. Theorem. *Let $M(S, \omega)$ be an ordered loopless matroid. Then $BC_\omega(M)$ and $\overline{BC}_\omega(M)$ are shellable. In both cases the ω-lexicographic ordering of facets (nbc-bases and reduced nbc-bases respectively) is a shelling.*

Proof. For $BC_\omega(M)$ the result was proved in Lemma 7.3.2. By Proposition 7.4.2(iii) and Exercise 7.1 the result follows also for $\overline{BC}_\omega(M)$. □

7.4.4. Example. Let $M = M(S)$ be the matroid on the set $S = \{1, 2, 3, 4, 5\}$ of Example 7.3.5. Under the natural ordering ω of S the broken circuits are $\{2, 3\}, \{4, 5\}$, and $\{2, 4, 5\}$ and the nbc-bases are $B_1 = \{1, 2, 4\}, B_2 = \{1, 2, 5\}$, $B_3 = \{1, 3, 4\}$, and $B_4 = \{1, 3, 5\}$. Under the ordering $\omega': 1 < 2 < 4 < 3 < 5$ of S the broken circuits are $\{2, 3\}, \{3, 5\}$, and $\{2, 4, 5\}$ and the nbc-bases are $B_1 = \{1, 2, 4\}, B_2 = \{1, 2, 5\}, B_3 = \{1, 3, 4\}$, and $B_5 = \{1, 4, 5\}$. Thus, $BC_\omega(M)$ and $BC_{\omega'}(M)$ are non-isomorphic; the corresponding reduced complexes are illustrated in Figure 7.4. Observe here also how in both cases the respective lexicographic ordering of edges gives a shelling.

Anders Björner

Figure 7.4. (a) $\overline{BC}_\omega(M)$; (b) $\overline{BC}_{\omega'}(M)$.

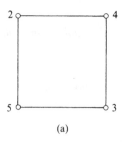

(a)

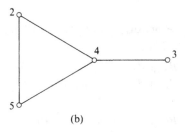

(b)

We will now turn to the shelling polynomials of (reduced) broken circuit complexes and the related enumerative aspects. Suppose that B is an nbc-basis in an ordered loopless matroid $M(S, \omega)$ and $b \in B$. If $B - b$ is contained in a lexicographically smaller nbc-basis then, as in (7.9), $b \in IP(B)$. If on the other hand b is internally passive in B, then $B - b$ is contained in some basis A that precedes B and, as was shown in the proof of Proposition 7.4.2, the earliest basis that contains $B - b$ is an nbc-basis. Thus we have shown that in the ω-lexicographic shelling of $BC_\omega(M)$

$$\mathcal{R}(B) = IP(B). \tag{7.14}$$

Consequently, the shelling polynomial $h_\Delta(x)$ of $\Delta = BC_\omega(M)$ equals $\sum_B x^{i(B)}$ with summation over all nbc-bases B. But B contains no broken circuit if and only if $e(B) = 0$. Hence,

$$h_{BC_\omega(M)}(x) = T_M(x, 0). \tag{7.15}$$

By similar reasoning it is straightforward to show directly, or it can be deduced from (7.15) via Proposition 7.4.2(iii) and Exercise 7.1, that the shelling polynomial of the reduced broken circuit complex equals

$$h_{\overline{BC}_\omega(M)}(x) = \frac{1}{x} T_M(x, 0). \tag{7.16}$$

Formula (7.15) implies that the face numbers of broken circuit complexes are independent of the ordering ω. See Example 7.4.4 for an illustration of this fact. For both orderings considered there the (non-isomorphic) broken circuit complexes have f-vector $(1, 5, 8, 4)$.

The face numbers of broken circuit complexes and their face enumerator $T_M(1 + \lambda, 0)$ are among the most interesting numerical invariants in matroid theory. In the following we shall assume familiarity with the *Möbius function* $\mu(x, y)$ and the *characteristic polynomial* $p(L; \lambda) = \sum_{x \in L} \mu(\hat{0}, x)\lambda^{r-r(x)} = \sum_{k=0}^{r} w_k \lambda^{r-k}$ of a rank r geometric lattice L. For this, see e.g. Chapters 7 and 8 in White

(1987), or Stanley (1986). The non-negative integers $\tilde{w}_k = (-1)^k w_k$ are called the *(unsigned) Whitney numbers of the first kind*.

7.4.5. Proposition. *Let $M(S, \omega)$ be an ordered loopless matroid and L the corresponding geometric lattice. For $x \in L$, put $\mathbf{nbc}(x) = \{A \in BC_\omega(M): \bar{A} = x\}$. Then*

$$|\mathbf{nbc}(x)| = (-1)^{r(x)} \mu(\hat{0}, x).$$

Proof. We will show that $(-1)^{r(y)} |\mathbf{nbc}(y)|$ satisfies the same recursion as $\mu(\hat{0}, y)$. Since only the empty set spans the empty flat we have that $|\mathbf{nbc}(\hat{0})| = 1$, as required.

Assume that $x \neq \hat{0}$, let S' be the set points on the flat x and let $M'(S', \omega')$ be the restriction of the ordered matroid $M(S, \omega)$ to S'. It is straightforward to check that a subset of S' is a broken circuit in M' if and only if it is a broken circuit in M. Hence, $BC_{\omega'}(M') = \cup_{y \leq x} \mathbf{nbc}(y)$, and the union is disjoint. Since all members of $\mathbf{nbc}(y)$, being independent and spanning y, have cardinality $r(y)$ and since $BC_{\omega'}(M')$ is a cone, it follows from (7.2) that

$$\sum_{\hat{0} \leq y \leq x} (-1)^{r(y)} |\mathbf{nbc}(y)| = -\chi(BC_{\omega'}(M')) = 0.$$

The proof is complete. □

Summing the left-hand side in Proposition 7.4.5 over all flats x of rank k gives the total number of broken-circuit-free subsets of size k. The same summation for the right-hand side gives the Whitney number $\tilde{\omega}_k$. Consequently we have proved:

7.4.6. Theorem. *For a loopless matroid, the Whitney numbers of the first kind coincide with the face numbers of the broken circuit complex (induced by any order).*

In view of (7.15) and Proposition 7.2.3 this result can be expressed as

$$p(M; \lambda) = (-1)^r h_{BC_\omega(M)}(1 - \lambda) = (-1)^r T_M(1 - \lambda, 0). \tag{7.17}$$

To illustrate this result, consider once again the matroid M of Example 7.3.5. Its Tutte polynomial $T_M(x, y) = x^3 + 2x^2 + y^2 + 2xy + x + y$ was computed following (7.12), and we get $T_M(1 - \lambda, 0) = 4 - 8\lambda + 5\lambda^2 - \lambda^3 = -p(M; \lambda)$. This should be compared with the *f*-vector $(1, 5, 8, 4)$ of M's broken circuit complexes; see Example 7.4.4.

We end this section with the determination of the Euler characteristics of matroid and (reduced) broken circuit complexes. For this we will need the *beta invariant*

$$\beta(M) = (-1)^{r(S)} \sum_{A \subseteq S} (-1)^{|A|} r(A), \tag{7.18}$$

of a matroid $M = M(S)$, discussed in Chapter 7 of White (1987). Also, we define the *Möbius invariant* $\tilde{\mu}(M)$ by

$$\tilde{\mu}(M) = \begin{cases} |\mu_L(\hat{0}, \hat{1})| & \text{if } M \text{ is loopless,} \\ 0 & \text{if } M \text{ has loops,} \end{cases} \tag{7.19}$$

where in the first case L is the lattice of flats of M. Clearly, $\tilde{\mu}(M) = \tilde{w}_r(L)$.

7.4.7. Proposition. *Let M be a rank r matroid. Then*

(i) $\chi(IN(M)) = (-1)^{r-1}\tilde{\mu}(M^*)$,

(ii) $\chi(BC_\omega(M)) = 0$,

(iii) $\chi(\overline{BC}_\omega(M)) = (-1)^r\beta(M)$,

where in (ii) *and* (iii) *M is presumed to be loopless and ordered by ω.*

Proof. Part (ii) is immediately clear, since $BC_\omega(M)$ is a cone. To prove (i) and (iii) we use the relation $\chi(\Delta) = (-1)^{r-1}h_\Delta(0)$ from Corollary 7.2.4.

Proof of (i): relations (7.12) and (7.17) give

$$h_{IN(M)}(0) = T_M(0, 1) = T_{M^*}(1, 0) = (-1)^{n-r}p(M^*; 0) = \tilde{\mu}(M^*)$$

if M^* is loopless. If M^* has a loop then $IN(M)$ is a cone and both sides equal zero.

Proof of (iii): relation (7.16) and Theorem 7.3.7 give

$$h_{\overline{BC}_\omega(M)}(0) = \frac{\partial T_M}{\partial x}(0, 0)$$

$$= \sum_{A \subseteq S} (r(S) - r(A))(-1)^{r(S)-r(A)-1+r^*(S)-r^*(S-A)}$$

$$= (-1)^{r(S)} \sum_{A \subseteq S} (r(A) - r(S))(-1)^{|A|} = \beta(M). \qquad \square$$

It is a consequence of the preceding result that $\beta(M) \geqslant 0$ for all matroids M. For future reference we state the following properties of the beta invariant; see Chapter 7 of White (1987) for proofs.

7.4.8. Proposition. *For any matroid $M = M(S)$ with $|S| \geqslant 2$:*

(i) $\beta(M) = \beta(M^*)$, *and*

(ii) $\beta(M) > 0$ *if and only if M is connected.*

7.5. Application to Matroid Inequalities

The material developed in the preceding sections provides a good framework for dealing with certain inequalities for independence numbers and for Whitney numbers of the first kind. This section, which can be skipped without loss of continuity, is devoted to this application. Underlying the inequalities considered lies the deeper and largely unsolved problem of understanding the h-vectors of matroid and broken circuit complexes.

Throughout this section r and n will always denote the rank and cardinality of the matroid M under consideration, and c_M (the *girth* of M) will be the smallest size of a circuit, that is, $c_M = \min\{|C| : C$ is a circuit in $M\}$.

We begin with the face numbers of matroid complexes, often called *independence numbers*. For a given matroid $M = M(S)$ let I_k denote the number of k-element independent subsets of S, $k = 0, 1, \ldots, r$. The number $b(M) = I_r$ of bases of M is of particular interest.

7.5.1. Proposition. *Let M be a loopless matroid with $n > r$. Then*

(i) $I_k < I_j$, *for all* $0 \leqslant k < j \leqslant r - k$,
(ii) $I_k \leqslant I_{r-k+1}$, *if* $1 \leqslant k < (r + 1)/2$ *and M has fewer than k isthmuses.*

For the proof of this proposition see Proposition 7.2.5 and Exercise 7.19.

7.5.2. Proposition. *Let M be a loopless matroid and $c = c_M$. Then*

$$I_k \geqslant \sum_{i=0}^{c-1} \binom{n-r+i-1}{i}\binom{r-i}{k-i}$$

for $k = 0, 1, \ldots, r$, and equality holds for some $k \geqslant c$, or equivalently for all k, if and only if M is isomorphic to the direct product of the free matroid B_{r-c+1} and the $(c-1)$-uniform matroid $U_{c-1,n-r+c-1}$.

Proof. The inequalities result from letting $\Delta = IN(M)$ in Proposition 7.2.5.

Suppose that M is such that equality holds for all k. Then $b(M) = \binom{n-r+c-1}{c-1}$, and in Proposition 7.2.5 we proved that the h-vector of $IN(M)$ must equal $\left(1, n-r, \ldots, \binom{n-r+c-2}{c-1}, 0, 0, \ldots, 0\right)$. The form of the Tutte polynomial $T_M(x, 1) = x^r + (n-r)x^{r-1} + \ldots + \binom{n-r+c-2}{c-1}x^{r-c+1}$ reveals that M has exactly $r - c + 1$ isthmuses (cf. Exercise 7.18) which together form a free submatroid B_{r-c+1}. Thus, $M = A \oplus B_{r-c+1}$, where A is a matroid of rank $c - 1$ on $n - r + c - 1$ points. Since $b(A) = b(M) = \binom{n-r+c-1}{c-1}$, we are forced to conclude that $A = U_{c-1,n-r+c-1}$. $\quad\square$

The preceding result shows in the particular case $k = r$ that for any loopless matroid M

$$b(M) \geqslant \binom{n-r+c-1}{c-1}. \tag{7.20}$$

Let M be a loopless matroid and let $(h_0, h_1, \ldots, h_r)$ be the h-vector of $IN(M)$. We know from (7.12) that $T_M(x, 1) = \sum_{i=0}^{r} h_i x^{r-i}$, and from (7.6) and Proposition 7.4.7 that $h_0 = 1$, $h_1 = n - r$ and $h_r = \tilde{\mu}(M^*)$. In the case where M is connected more can be said about the h-vector.

7.5.3. Proposition. *If M has no isthmuses then $h_i \geq n - r$ for $i = 2, 3, \ldots, r - 1$, and if in addition M is connected or else no connected component of M is a circuit then also $h_r \geq n - r$.*

Proof. Let us first assume that $M = M(S)$ is connected. The only connected matroid on two elements is the 2-point circuit for which $\mathbf{h} = (1, 1)$, so the initial case is appropriate for an induction argument on n. Let $e \in S$. Since M is connected we know that either the contraction M/e or the restriction $M - e$ is connected (cf. White, 1986, p. 181). Let $(h_0^c, h_1^c, \ldots, h_{r-1}^c)$ and $(h_0^r, h_1^r, \ldots, h_r^r)$ be the h-vectors of $IN(M/e)$ and $IN(M - e)$, respectively. The Tutte polynomial identity $T_M(x, y) = T_{M/e}(x, y) + T_{M-e}(x, y)$ evaluated at $y = 1$ shows that $h_i = h_{i-1}^c + h_i^r$ for $i = 1, 2, \ldots, r$.

Suppose first that M/e is connected. Since M/e is of rank $(r - 1)$ on $(n - 1)$ elements the induction assumption at once gives $h_i^c \geq n - r$ for $i = 1, 2, \ldots, r - 1$. Next, suppose that $M - e$ is connected. Then $h_i^r \geq n - r - 1$ for $i = 1, 2, \ldots, r$, since $M - e$ is of rank r on $(n - 1)$ elements. The orthogonal of the contraction $(M/e)^*$, being the restriction $M^* - e$ of the connected matroid M^*, cannot contain a loop. Hence, $\tilde{\mu}((M/e)^*) > 0$, that is, $h_{r-1}^c \geq 1$. But that forces $h_i^c \geq 1$ also, for $i = 1, 2, \ldots, r - 2$, as a consequence of (7.7) and Lemma 7.2.6.

Thus, in either case it follows that $h_i \geq n - r$ for $i = 2, 3, \ldots, r$, and the connected case is settled.

Assume now that it has been shown for loopless matroids without isthmuses having $p - 1$ connected components that $h_i \geq n - r$ for $i = 1, 2, \ldots, r - 1$ and $h_r \geq 1$. Suppose that our matroid M has p components, one of which is M_1. Then $M = M_1 \oplus M_2$ where M_2 has $p - 1$ components. If r_j, n_j, and $(h_0^j, h_1^j, \ldots, h_{r_j}^j)$ denote the rank, cardinality, and h-vector of $M_j, j = 1, 2$, then we know that $r_1 + r_2 = r$, $n_1 + n_2 = n$, $h_i^1 \geq n_1 - r_1 \geq 1$ for $i = 1, 2, \ldots, r_1$, and $h_i^2 \geq n_2 - r_2$ for $i = 1, 2, \ldots, r_2 - 1$, $h_{r_2}^2 \geq 1$. A comparison of coefficients in the Tutte polynomial formula $T_M(x, 1) = T_{M_1}(x, 1) T_{M_2}(x, 1)$ shows that $h_i = \sum_{i=j+k} h_j^1 h_k^2$. Thus we arrive at the desired conclusion: $h_i \geq n_1 - r_1 + n_2 - r_2 = n - r$ for $i = 1, 2, \ldots, r - 1$.

In the case where no connected component of M is a circuit then $n_1 - r_1 \geq 2$, and we can assume that $h_{r_2}^2 \geq n_2 - r_2 \geq 2$ in the induction assumption of the preceding paragraph. Thus $h_r = h_{r_1}^1 h_{r_2}^2 \geq (n_1 - r_1)(n_2 - r_2) \geq n_1 - r_1 + n_2 - r_2 = n - r$. $\square$

7.5.4. Proposition. *Let* M *be a matroid without loops or isthmuses and* $c = c_M$. *Then*

$$I_k \geqslant \sum_{i=0}^{c-1} \binom{n-r+i-1}{i}\binom{r-i}{k-i} + (n-r)\binom{r-c+1}{k-c}, \quad \text{for } k = 0, 1, \ldots, r-1.$$

If M *is connected or else no connected component of* M *is a circuit then the formula holds also for* $k = r$, *that is,*

$$b(M) \geqslant \binom{n-r+c-1}{c-1} + (n-r)(r-c+1).$$

Proof. We know that $h_i = \binom{n-r+i-1}{i}$ for $i = 0, 1, \ldots, c-1$ by (7.8), and $h_i \geqslant n-r$ for $i = c, c+1, \ldots, r-1$, (r) by 7.5.3, so the result follows from formula (7.5).   □

Taking $c = 2$ in the above formulas we get

$$I_k \geqslant \binom{r}{k} + (n-r)\binom{r}{k-1}, \quad k = 0, 1, \ldots, r-1, \tag{7.21}$$

and

$$b(M) \geqslant 1 + r(n-r), \tag{7.22}$$

which hold for all matroids M without loops and isthmuses except that (7.22) fails by a trifle for a few products involving circuits (cf. Exercise 7.24). For *connected* matroids it can be shown that equality holds for all $k \leqslant r$ in (7.21), or equivalently in (7.22), if and only if M is isomorphic to the parallel connection of an $(r+1)$-point circuit and an $(n-r)$-point atom.

Taking $c = 3$ in Proposition 7.5.4 we get formulas that are valid for all connected simple matroids.

We will now turn our attention to the face numbers of broken circuit complexes. Recall from Theorem 7.4.6 that these are the Whitney numbers of the first kind $\tilde{w}_0, \tilde{w}_1, \ldots, \tilde{w}_r$. In particular, $\tilde{w}_r = \bar{\mu}(M)$, the Möbius invariant (7.19). It will be assumed that all matroids are *simple*, i.e. lack loops and parallel elements. This results in no lack of generality (cf. Proposition 7.4.1).

7.5.5. Proposition. *Let* M *be a simple matroid with* $n > r$. *Then*

(i) $\tilde{w}_k < \tilde{w}_j$, *for all* $0 \leqslant k < j \leqslant r-k$,
(ii) $\tilde{w}_k \leqslant \tilde{w}_{r-k+1}$, *if* $2 \leqslant k < (r+1)/2$ *and* M *has fewer than* k *connected components, or if* $k = 1$, $4 \leqslant r \leqslant n-2$ *and* M *is connected.*

7.5.6. Proposition. *Let M be a simple matroid and $c = c_M$. Then*

$$\tilde{w}_k \geq \sum_{i=0}^{c-2} \binom{n-r+i-1}{i}\binom{r-i}{k-i}$$

for $k = 0, 1, \ldots, r$, and equality holds for some $k \geq c - 1$, or equivalently for all k, if and only if M is isomorphic to the direct product of the free matroid B_{r-c+1} and the $(c-1)$-uniform matroid $U_{c-1,n-r+c-1}$.

Taking $k = r$ in Proposition 7.5.6 we obtain

$$\tilde{\mu}(M) \geq \binom{n-r+c-2}{c-2}. \tag{7.23}$$

It also follows (since $c \geq 3$ for simple matroids) that

$$\tilde{w}_k \geq \binom{r}{k} + (n-r)\binom{r-1}{k-1} \tag{7.24}$$

for $k = 0, 1, \ldots, r$, with equality exclusively characterizing the direct products of free matroids and lines. It is interesting that the lower bounds 7.5.2 and 7.5.6 are attained by precisely the same class of matroids.

Proof. These results arise from applying Proposition 7.2.5 to a broken circuit complex of M. For 7.5.5 see Exercise 7.20.

We need here only verify the characterization of equivalence in 7.5.6. If $M = B \oplus U$ where $B = B_{r-c+1}$ and $U = U_{c-1,n-r+c-1}$ then $\tilde{w}_r = \tilde{\mu}(M) = \tilde{\mu}(B)\tilde{\mu}(U) = \tilde{\mu}(U) = \binom{n-r+c-2}{c-2}$, which according to 7.2.5 implies equivalence in 7.5.6 for all k.

Suppose now that M is such that equivalence holds in 7.5.6 for all k; then, in particular, $\tilde{\mu}(M) = \binom{n-r+c-2}{c-2}$. While proving Proposition 7.2.5 we showed that this implies that the Tutte polynomial $T_M(x, 0)$, being the shelling polynomial of the broken circuit complex, equals $T_M(x, 0) = x^r + (n-r)x^{r-1} + \ldots + \binom{n-r+c-3}{c-2}x^{r-c+2}$. From the form of $T_M(x, 0)$ we can deduce that M is the direct product of $r - c + 2$ connected simple matroids $M_1, M_2, \ldots, M_{r-c+2}$ (cf. Exercise 7.18). Assume that M_i is of rank r_i and cardinality n_i for $i = 1, 2, \ldots, r - c + 2$. Clearly, $\tilde{\mu}(M_i) \leq \binom{n_i - 1}{r_i - 1}$ since there are $\tilde{\mu}(M_i)$ nbc-bases in M_i all of which contain the first element under some ordering. From the same viewpoint it is apparent that $\tilde{\mu}(M_i) = \binom{n_i - 1}{r_i - 1}$ if and only if M_i is r_i-uniform. We get that

$$\tilde{\mu}(M) = \prod_{i=1}^{r-c+2} \tilde{\mu}(M_i) \leqslant \prod_{i=1}^{r-c+2} \binom{n_i-1}{r_i-1} \leqslant \binom{n-r+c-2}{c-2} = \tilde{\mu}(M).$$

But $\binom{a_1}{b_1}\binom{a_2}{b_2} \leqslant \binom{a_1+a_2}{b_1+b_2}$ with equality if and only if $a_1 = b_1 = 0$ or $a_2 = b_2 = 0$. Thus $n_i = r_i = 1$ for all i except, say, $i = 1$, for which $n_1 = n-r+c-1$, $r_1 = c-1$ and $\tilde{\mu}(M_1) = \binom{n_1-1}{r_1-1}$. So $M_1 = U_{c-1,n-r+c-1}$ and $M_2 \oplus \cdots \oplus M_{r-c+2} = B_{r-c+1}$. $\qquad \square$

Let $(h_0, h_1, \ldots, h_r)$ be the h-vector of a broken circuit complex of a simple matroid M. Thus $T_M(x, 0) = h_0 x^r + h_1 x^{r-1} + \cdots + h_r$. We know from (7.6), Proposition 7.4.7, and Exercise 7.1 that in general $h_0 = 1$, $h_1 = n-r$, $h_{r-1} = \beta(M)$, and $h_r = 0$. However, as was the case for the matroid complex, when M is connected more can be said.

7.5.7. Proposition. *If M is connected then $h_i \geqslant n-r$ for $i = 2, 3, \ldots, r-2$, and $h_{r-1} \geqslant 1$.*

Proof. $\beta(M) \geqslant 1$ is equivalent to M being connected (Proposition 7.4.8). For the other inequalities see Brylawski (1977b), Theorem 3.1.2 $\qquad \square$

7.5.8. Proposition. *Let M be a connected simple matroid and $c = c_M$. Then*

$$\tilde{w}_k \geqslant \sum_{i=0}^{c-2} \binom{n-r+i-1}{i}\binom{r-i}{k-i} + (n-r)\binom{r-c+2}{k-c+1} \text{ for } k = 0, 1, \ldots, r-2,$$

$$\tilde{w}_{r-1} \geqslant \sum_{i=0}^{c-2} (r-i)\binom{n-r+i-1}{i} + (n-r)\left[\binom{r-c+2}{2} - 1\right] + \beta(M),$$

and

$$\tilde{\mu}(M) \geqslant \binom{n-r+c-2}{c-2} + (n-r)(r-c) + \beta(M).$$

Proof. The inequalities arise from (7.5), that is,

$$\tilde{w}_k = \sum_{i=0}^{r} h_i \binom{r-i}{k-i}, \quad k = 0, 1, \ldots, r,$$

since $h_i = \binom{n-r+i-1}{i}$ for $i = 0, 1, \ldots, c-2$ by (7.8), and $h_i \geqslant n-r$ for $i = c-1, c, \ldots, r-2$, by 7.5.7. (The last two inequalities, the $k = r-1$ and $k = r$ cases, actually require that $c \leqslant r$. If $c > r$ then M is uniform and $\tilde{w}_{r-1} = \binom{n}{r-1}$, $\tilde{\mu}(M) = \binom{n-1}{r-1}$.) $\qquad \square$

By taking $c = 3$ in the above formulas we get inequalities that hold for all connected simple matroids, namely

$$\tilde{w}_k \geqslant \binom{r}{k} + (n - r)\binom{r}{k-1}, \text{ for } k = 0, 1, ..., r - 2,$$

$$\tilde{w}_{r-1} \geqslant r + (n - r)\left[\binom{r}{2} - 1\right] + \beta(M), \text{ and} \qquad (7.25)$$

$$\tilde{\mu}(M) \geqslant 1 + (n - r)(r - 2) + \beta(M).$$

To eliminate $\beta(M)$, the last two of these formulas can be replaced by

$$\tilde{w}_{r-1} \geqslant r + (n - r)\binom{r}{2} - 1, \text{ and} \qquad (7.26)$$

$$\tilde{\mu}(M) \geqslant (n - r)(r - 1),$$

which hold for all connected simple matroids with the exception of the parallel connection of three 3-point lines, for which $r = 4$, $n = 7$, $\tilde{\mu} = 8$, and $\tilde{w}_3 = 20$; see Brylawski (1977b). The connected simple matroids that achieve equality in (7.25) and (7.26) have not been characterized; some examples are parallel connections of circuits and lines, the Fano projective plane, and the complete graph K_4.

7.6. Order Complexes of Geometric Lattices

Let L be a finite geometric lattice with proper part $\bar{L} = L - \{\hat{0}, \hat{1}\}$. The chains $x_0 < x_1 < ... < x_k$ of $\bar{L}$ are the simplices of a simplicial complex $\Delta(\bar{L})$, called the *order complex* of L. If rank $(L) = r$, then $\Delta(\bar{L})$ is a pure $(r - 2)$-dimensional complex.

We assume familiarity with the cryptomorphic correspondence between simple matroids $M = M(S)$ and geometric lattices L; see section 3.4 of White (1986). Under this correspondence the ground set S is identified with the set L^1 of atoms of L. For $0 \leqslant k \leqslant r = \text{rank}(L)$ let $L^k = \{x \in L : r(x) = k\}$, and for $x \in L^k$ let $\square x$ denote the corresponding flat $\square x = \{p \in L^1 : p \leqslant x\}$ in M. To ordered matroids $M = M(S, \omega)$ correspond *atom-ordered geometric lattices* (L, ω), i.e. ω is a linear ordering of L^1.

Let (L, ω) be an atom-ordered geometric lattice. Denote by $\text{Cov}(L) \subseteq L \times L$ the set of *coverings*, i.e., pairs (x, y) such that if $x \in L^i$ then $y > x$ and $y \in L^{i+1}$. An *edge labeling* $\lambda = \lambda_\omega : \text{Cov}(L) \to L^1$ is defined by the rule

$$\lambda(x, y) = \min_\omega(\square y - \square x). \qquad (7.27)$$

Note that the label $\lambda(x, y)$ is well defined since the indicated set of atoms must be non-empty. Also, the labeling $\lambda = \lambda_\omega$ depends, of course, on ω.

7.6.1. Example. Figure 7.5(a) shows the geometric lattice of flats of the matroid from Example 7.3.5. Figure 7.5(b) shows the edge labeling induced by the natural ordering $1 < 2 < 3 < 4 < 5$ of its atoms.

Figure 7.5.

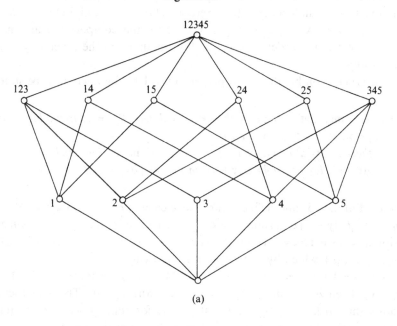

(a)

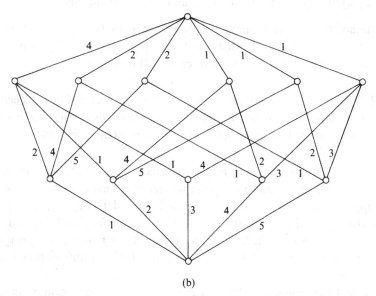

(b)

The edge labeling (7.27) induces a labeling of unrefinable chains. If the chain $\mathbf{c}: x_0 < x_1 < \ldots < x_k$ is unrefinable, meaning that (x_{i-1}, x_i) is a covering for $1 \leqslant i \leqslant k$, let

$$\lambda(\mathbf{c}) = (\lambda(x_0, x_1), \lambda(x_1, x_2), \ldots, \lambda(x_{k-1}, x_k)). \qquad (7.28)$$

This label $\lambda(\mathbf{c})$ is an ordered k-tuple of atoms. As an (unordered) k-subset of atoms, $\lambda(\mathbf{c})$ is independent (cf. Exercise 7.28). Let us call $\lambda(\mathbf{c})$ *increasing* if $\lambda(x_0, x_1) < \lambda(x_1, x_2) < \ldots < \lambda(x_{k-1}, x_k)$, where the comparisons are made with respect to the order ω of L^1. If all inequalities go the other way $\lambda(\mathbf{c})$ is *decreasing*.

The crucial combinatorial properties of this labeling λ will now be stated.

7.6.2. Lemma. *Let* x, $y \in L$ *with* $x < y$, *and let* $\mathcal{M}_{x,y} = \{$*unrefinable chains* $x = x_0 < x_1 < \ldots < x_k = y\}$. *Then*

(i) *there exists a unique chain* $\mathbf{c}_{x,y} \in \mathcal{M}_{x,y}$ *with increasing label,*

(ii) $\lambda(\mathbf{c}_{x,y})$ *is lexicographically least in the set* $\{\lambda(\mathbf{c}) : \mathbf{c} \in \mathcal{M}_{x,y}\}$.

Proof. Put $x_0 = x$, and define recursively elements $p_i \in L^1$ and $x_i \in L^{r(x_0)+i}$ by $p_i = \min_\omega(\square y - \square x_{i-1})$ and $x_i = x_{i-1} \vee p_i$. This recursive definition will end with $x_k = y$, where $k = r(y) - r(x)$. Let $\mathbf{c}_{x,y} : x_0 < x_1 < \ldots < x_k$. Then $\lambda(\mathbf{c}_{x,y}) = (p_1, p_2, \ldots, p_k)$, which by construction is increasing.

Suppose that $\mathbf{c} : x = y_0 < y_1 < \ldots < y_k = y$, with $y_i = x_i$ for $0 \leqslant i \leqslant e-1$ and $y_e \neq x_e$. Then $\lambda(\mathbf{c}) = \{p_1, \ldots, p_{e-1}, q_e, \ldots, q_k\}$ with $q_e \neq p_e$. The construction shows that in fact $p_e < q_e$, and that $p_e = q_f$ for some $f > e$. Hence $\lambda(\mathbf{c})$ is lexicographically greater than $\lambda(\mathbf{c}_{x,y})$, and $\lambda(\mathbf{c})$ is not increasing. $\square$

The facets of the order complex $\Delta(\bar{L})$ are the maximal chains in $\bar{L}$. Extending these by $\hat{0}$ and $\hat{1}$ we get an identification with the set $\mathcal{M} = \mathcal{M}_{\hat{0},\hat{1}}$ of maximal chains in L. This leads to the main result of this section.

7.6.3. Theorem. *Let* (L, ω) *be an atom-ordered geometric lattice. Then the* ω-*lexicographic order of labels* $\lambda(\mathbf{c})$ *determines a shelling order of the set* $\mathcal{M}$ *of maximal chains (facets of* $\Delta(\bar{L})$). *In particular,* $\Delta(\bar{L})$ *is shellable.*

Proof. By (7.3) the following must be verified: If $\mathbf{c}$, $\mathbf{d} \in \mathcal{M}$ with $\lambda(\mathbf{d}) < \lambda(\mathbf{c})$, then there exist $\mathbf{e} \in \mathcal{M}$ and $x \in \mathbf{c}$ such that $\lambda(\mathbf{e}) < \lambda(\mathbf{c})$ and $\mathbf{c} \cap \mathbf{d} \subseteq \mathbf{c} \cap \mathbf{e} = \mathbf{c} - x$.

Suppose that $\mathbf{c} : \hat{0} = x_0 < x_1 < \ldots x_r = \hat{1}$, and $\mathbf{d} : \hat{0} = y_0 < y_1 < \ldots y_r = \hat{1}$, and $\lambda(\mathbf{d}) < \lambda(\mathbf{c})$. Suppose furthermore that $x_i = y_i$ for $0 \leqslant i \leqslant f$, and that $x_{f+1} \neq y_{f+1}$. Let g be the least integer greater than f such that $x_g = y_g$ (g is well defined since $x_r = y_r$). Then $g - f \geqslant 2$ and $f < i < g$ implies that $x_i \neq y_i$. See Figure 7.6.

The fact that $\lambda(\mathbf{d}) < \lambda(\mathbf{c})$ shows that $x_f < x_{f+1} < \ldots < x_g$ cannot have the lexicographically least label in the set $\mathcal{M}_{x_f, x_g}$. Hence by Lemma 7.6.2 there exists $f < i < g$ such that $\lambda(x_{i-1}, x_i) > \lambda(x_i, x_{i+1})$. Again by the lemma there exists a unique $z_i \in L^i$ such that $x_{i-1} < z_i < x_{i+1}$ and $\lambda(x_{i-1}, z_i) < \lambda(z_i, x_{i+1})$. Replace x_i in $\mathbf{c}$ by z_i. This gives a new maximal chain $\mathbf{e}$ such that $\mathbf{c} \cap \mathbf{d} \subseteq \mathbf{c} \cap \mathbf{e} = \mathbf{c} - x_i$, and (by the lemma) $\lambda(\mathbf{e}) < \lambda(\mathbf{c})$. $\square$

Figure 7.6.

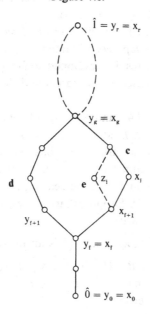

Let $c : x_1 < x_2 < \ldots < x_{r-1}$ be a maximal chain in $\bar{L}$. As before, c may tacitly be identified with its extension by $x_0 = \hat{0}$ and $x_r = \hat{1}$ to a maximal chain in L. The shellability proof (and in essence Lemma 7.6.2) shows that there exists a maximal chain e such that $\lambda(e) < \lambda(c)$ and $e \cap c = c - x_i$ if and only if $\lambda(x_{i-1}, x_i) > \lambda(x_i, x_{i+1})$. Hence, the restriction operator induced by the shelling in Theorem 7.6.3 is given by

$$\mathscr{R}(c) = \{ x_i \in c : \lambda(x_{i-1}, x_i) > \lambda(x_i, x_{i+1}) \}. \tag{7.29}$$

Let us write simply $\mu(L)$ for the Möbius function value $\mu(\hat{0}, \hat{1})$, computed over the geometric lattice L. A well known theorem of P. Hall gives that $\mu(L)$ is equal to the number of odd cardinality chains in $\bar{L}$ minus the number of even cardinality chains in $\bar{L}$ (including the empty chain); see Rota (1964) or Stanley (1986). Part (i) of the following proposition is a restatement of Hall's theorem for L. Part (ii) is then implied via (7.29) and Corollary 7.2.4.

7.6.4. Proposition. *Let L be a geometric lattice of rank r. Then*

(i) $\chi(\Delta(\bar{L})) = \mu(L)$,

(ii) $\mu(L) = (-1)^r |\{ c \in \mathscr{M} : \lambda(c) \text{ is decreasing}\}|$, *for any labeling $\lambda = \lambda_\omega$ induced by an atom ordering ω as in (7.27) and (7.28).*

For an illustration of this result, take a look at the rank 3 geometric lattice in Figure 7.5. Direct computation of the Möbius function, using its recursive definition, gives that $\mu(L) = -4$. The four decreasing labels of maximal chains are $(4, 2, 1)$, $(4, 3, 1)$, $(5, 2, 1)$, and $(5, 3, 1)$.

The combinatorics of the edge labelings (7.27) of geometric lattices L extends to so-called *rank-selected subposets*, obtained by deleting an arbitrary set of rank levels L^k from L. The results developed so far in this section have more general versions for rank-selected subposets and their order complexes; see Exercise 7.30.

Combining Proposition 7.4.5 and part (ii) of Proposition 7.6.4 we find that the number of nbc-bases in L equals the number of maximal chains with decreasing label. This fact can also be established by an explicit bijection.

For an atom-ordered geometric lattice (L, ω) of rank r, let $\mathcal{DM}$ denote the set of maximal chains in L with decreasing labels, and let $\mathbf{nbc} = \mathbf{nbc}(\hat{1})$ denote the set of nbc-bases of L. Suppose that $\mathbf{c} \in \mathcal{DM}$, with $\lambda(\mathbf{c}) = (\lambda_1, \lambda_2, ..., \lambda_r)$, $\lambda_1 > \lambda_2 > ... > \lambda_r$. Then clearly

$$\phi(\mathbf{c}) = \{\lambda_r, \lambda_{r-1}, ..., \lambda_1\} \text{ is an nbc-basis.} \tag{7.30}$$

Conversely, suppose that $B = \{b_1, b_2, ..., b_r\} \in \mathbf{nbc}$, $b_1 < b_2 < ... < b_r$, and construct the maximal chain $\psi(B) : \hat{0} < b_r < (b_r \vee b_{r-1}) < (b_r \vee b_{r-1} \vee b_{r-2}) < ... < (b_r \vee ... \vee b_1) = \hat{1}$. Then

$$\lambda(\psi(B)) = (b_r, b_{r-1}, ..., b_1), \text{ hence } \psi(B) \in \mathcal{DM}. \tag{7.31}$$

Since $\phi \circ \psi(B) = B$ for all $B \in \mathbf{nbc}$, and similarly for the opposite composition, we conclude that ϕ and ψ are bijections $\mathcal{DM} \leftrightarrow \mathbf{nbc}$.

Statement (7.31) is a consequence of

$$\text{if } A \in BC_\omega(L), \text{ then } \min_\omega \bar{A} \in A, \tag{7.32}$$

that is, the least element in A is also the least element in the flat spanned by A, which in turn follows directly from the definitions. Statement (7.32) also implies that if $B = \{b_1, b_2, ..., b_r\} \in \mathbf{nbc}$, $b_1 < b_2 < ... < b_r$, and if $\pi \in S_r$ is a permutation, then

$$\psi_\pi(B) \in \mathcal{DM} \text{ if and only if } \pi = \text{id}, \tag{7.33}$$

where $\psi_\pi(B) : \hat{0} < b_{\pi(r)} < (b_{\pi(r)} \vee b_{\pi(r-1)}) < ... < (b_{\pi(r)} \vee ... \vee b_{\pi(1)}) = \hat{1}$.

7.7. Homology of Shellable Complexes

This section will review the construction of simplicial homology and the basic facts about the homology of a shellable complex. The presentation is essentially self-contained and should be accessible to readers without a background in algebraic topology. Readers having such a background can proceed directly to Theorem 7.7.2.

Let Δ be a simplicial complex on vertex set V. We will temporarily assume that V is linearly ordered, but the particular order chosen is of no significance for the end product. Let us agree to permit a non-void face $F = \{v_0, v_1, ..., v_k\}$

of Δ to be written $F = [v_0, v_1, ..., v_k]$ if and only if $v_0 < v_1 < ... < v_k$ in the given order of V. Then let $C_k(\Delta)$ denote the free Abelian group generated by the symbols $[v_0, v_1, ..., v_k]$, i.e., by the set of k-dimensional faces of Δ written in canonical form. The elements of $C_k(\Delta)$ are formal linear combinations with integer coefficients of k-dimensional faces. Thus, $C_{-1}(\Delta) \cong \mathbb{Z}$, $C_0(\Delta) \cong \mathbb{Z}^{|V|}$, the direct sum of $|V|$ copies of $\mathbb{Z}$, and $C_k(\Delta) = 0$ for $k < -1$ and $k > d = \dim \Delta$. Let group homomorphisms

$$\partial_k : C_k(\Delta) \rightarrow C_{k-1}(\Delta)$$

be defined on the basis elements by

$$\partial_k[v_0, v_1, ..., v_k] = \sum_{i=0}^{k} (-1)^i [v_0, v_1, ..., \hat{v}_i, ..., v_k], \qquad (7.34)$$

and extend linearly to all of $C_k(\Delta)$, for $k = 0, 1, ..., d$. Put $\partial_k = 0$ otherwise. Here the circumflex has the meaning that what stands underneath should be deleted. Thus, for instance $\partial_0[v_0] = \varnothing$ and $\partial_2[v_0, v_1, v_2] = [v_1, v_2] - [v_0, v_2] + [v_0, v_1]$.

The elements ρ of $C_k(\Delta)$ such that $\partial_k(\rho) = 0$ are called k-dimensional *cycles*; they form a subgroup denoted by $Z_k(\Delta)$. The elements σ of $C_k(\Delta)$ that satisfy $\sigma = \partial_{k+1}(\tau)$ for some $\tau \in C_{k+1}(\Delta)$ are called k-dimensional *boundaries* and form a subgroup $B_k(\Delta)$. It is easy to verify that $\partial_k \circ \partial_{k+1} = 0$ for all $k \in \mathbb{Z}$. Thus, $B_k(\Delta) \subseteq Z_k(\Delta)$ for all $k \in \mathbb{Z}$. The quotient group

$$H_k(\Delta) = Z_k(\Delta)/B_k(\Delta) \qquad (7.35)$$

is the k-dimensional *homology group* of Δ with integer coefficients.

The homology groups we have defined are referred to in topology as 'reduced' homology and denoted $\tilde{H}_k(\Delta)$. Since we have no occasion to consider any other variation of homology, we adhere to the simpler terminology. A technical feature of this definition is that $H_{-1}(\Delta) \cong \mathbb{Z}$ for the 'empty complex' $\Delta = \{\varnothing\}$, while $H_{-1}(\Delta) = 0$ as soon as there are vertices (non-degenerate complexes). Also, the rank of $H_0(\Delta)$ is one less than the number of connected components of Δ.

The rank of the Abelian group $H_k(\Delta)$ is called the k-th *Betti number* of the complex Δ. By the *rank* of a finitely generated Abelian group we mean the maximum number of linearly independent elements of infinite order.

A complex Δ is called *acyclic* (*over* $\mathbb{Z}$) if $H_k(\Delta) = 0$ (i.e., $H_k(\Delta)$ is isomorphic to the trivial group) for all $k \in \mathbb{Z}$. The following two facts will be needed:

$$\text{a cone is acyclic, and} \qquad (7.36)$$

if Δ_1, Δ_2 and $\Delta_1 \cap \Delta_2$ are acyclic complexes then $\Delta_1 \cup \Delta_2$ is also acyclic.

$$(7.37)$$

Both facts are completely elementary given some knowledge of topology. For instance, (7.37) follows from the Mayer–Vietoris long exact sequence.

They are also elementary in the sense that straightforward proofs directly from the definitions are easy.

7.7.1. Lemma. *Given a shelling of a complex* Δ, *let* $\Delta' = \Delta - \{facets\ F\ such that\ \mathscr{R}(F) = F\}$, *where* $\mathscr{R}(F)$ *is the induced restriction operator. Then* Δ' *is acyclic.*

Proof. The first observation to be made is that the ordering of facets of Δ, restricted to the facets in Δ', gives a shelling of Δ'. The second observation is that the restriction operator $\mathscr{R}(F)$ induced by this shelling of Δ' coincides with the original one and that in particular then $\mathscr{R}(F) \neq F$ for all facets $F \in \Delta'$. These observations are simple consequences of the definition (7.3).

We will prove that Δ' is acyclic inductively using its shelling $F_1, F_2, ..., F_t$. Let $\bar{F}_i = \{G \in \Delta' : G \subseteq F_i\}$, and $\Delta_i = \bar{F}_1 \cup \bar{F}_2 \cup ... \cup \bar{F}_i$, $1 \leqslant i \leqslant t$. We have that $\bar{F}_i$ is a cone with any of its vertices as apex, and $\Delta_{i-1} \cap \bar{F}_i$ is a cone with any vertex in $F_i - \mathscr{R}(F_i)$ as apex. Therefore by (7.36), $\bar{F}_i$ and $\Delta_{i-1} \cap \bar{F}_i$ are acyclic and, by (7.37), if Δ_{i-1} is acyclic then so is also Δ_i. Since $\Delta_1 = \bar{F}_1$ is acyclic, it follows by finite induction that so is $\Delta_t = \Delta'$. $\square$

We shall use the notation $\rho(F) = a$ to denote that the k-dimensional face F occurs with the coefficient $a \in \mathbb{Z}$ in the formal linear combination $\rho \in C_k(\Delta)$.

7.7.2. Theorem. *Let* Δ *be a shellable d-dimensional complex. Suppose furthermore that* $\{facets\ F\ such\ that\ \mathscr{R}(F) = F\} = \{F_1, F_2, ..., F_p\}$, *where* $\mathscr{R}(F)$ *is the restriction operator induced by some shelling. Then*

(i)
$$H_i(\Delta) \cong \begin{cases} \mathbb{Z}^p, & if\ i = d, \\ 0, & if\ i \neq d. \end{cases}$$

(ii) *There are cycles* $\rho_1, \rho_2, ..., \rho_p \in H_d(\Delta)$ *uniquely determined by*
$$\rho_k(F_j) = \begin{cases} 1, & if\ j = k, \\ 0, & if\ j \neq k. \end{cases}$$

(iii) $\{\rho_1, \rho_2, ..., \rho_p\}$ *is a basis of the free group* $H_d(\Delta)$.

Notice that for $d = 1$ the theorem states the familiar fact that the cycle space of a connected graph has a basis (unique up to signs) induced by the family of edges lying outside some fixed spanning tree. Notice also that rank $H_d(\Delta) = p = (-1)^d \chi(\Delta)$, by Corollary 7.2.4 (this is a special case of the Euler–Poincaré formula).

For illustration, consider the complex Δ of Example 7.2.1. Taking any shelling (for instance the one mentioned after 7.2.4) one concludes from the theorem that $H_2(\Delta) \cong \mathbb{Z}$ while $H_i(\Delta) = 0$ for all $i \neq 2$. This also follows from the fact (obvious upon inspection) that Δ is homotopy-equivalent to the 2-sphere.

Proof. The subcomplex $\Delta' = \Delta - \{F_1, F_2, ..., F_p\}$, which by Lemma 7.7.1 is acyclic, differs from Δ only in dimension d. So $C_k(\Delta') = C_k(\Delta)$ for all $k < d$, and consequently $H_k(\Delta) = H_k(\Delta') = 0$ for all $k < d - 1$. Also, $B_{d-1}(\Delta') \subseteq B_{d-1}(\Delta) \subseteq Z_{d-1}(\Delta) = Z_{d-1}(\Delta')$ and $H_{d-1}(\Delta') = 0$ imply that $H_{d-1}(\Delta) = 0$. Thus we have proved that $H_i(\Delta) = 0$ for all $i \neq d$.

For $1 \leqslant k \leqslant p$ we have that $\partial_d(F_k) \in B_{d-1}(\Delta) \subseteq Z_{d-1}(\Delta) = Z_{d-1}(\Delta')$, and since Δ' is acyclic there exists $\rho'_k \in C_d(\Delta')$ such that $\partial_d(\rho'_k) = \partial_d(F_k)$. Let $\rho_k = F_k - \rho'_k$. By construction $\partial_d(\rho_k) = 0$, so $\rho_k \in Z_d(\Delta) = H_d(\Delta)$, and also $\rho_k(F_j) = \delta_{jk}$ (Kronecker's delta), for $1 \leqslant j \leqslant p$. If $\sigma_k \in H_d(\Delta)$ satisfies $\sigma_k(F_j) = \delta_{jk}$, $1 \leqslant j \leqslant p$, then $\sigma_k - \rho_k \in H_d(\Delta') = 0$ and so $\sigma_k = \rho_k$. This proves part (ii).

We will now show that $\{\rho_1, \rho_2, ..., \rho_p\}$ is a basis of $H_d(\Delta)$. (Remark: one could argue that $H_d(\Delta) = Z_d(\Delta)$ must be free, since it is a subgroup of the free Abelian group $C_d(\Delta)$, but this will of course be a direct consequence of providing a basis.)

Linear independence: Let $\sigma = \sum_{k=1}^{p} a_k \rho_k$, with $a_k \in \mathbb{Z}$. Part (ii) shows that $\sigma(F_j) = a_j$, which means that $\sigma = 0$ only if $a_k = 0$ for all $1 \leqslant k \leqslant p$.

Generating property: Let $\sigma \in H_d(\Delta)$. Consider the cycle $\tau = \sigma - \sum_{k=1}^{p} \sigma(F_k)\rho_k$. Part (ii) shows that $\tau(F_j) = 0$ for all $1 \leqslant j \leqslant p$, which means that $\tau \in H_d(\Delta') = 0$. So

$$\sigma = \sum_{k=1}^{p} \sigma(F_k)\rho_k. \tag{7.38}$$

All claims about the structure of $H_d(\Delta)$ made in parts (i) and (iii) have now been established. $\qquad\qquad\qquad\qquad\qquad\qquad\qquad\qquad\qquad\qquad\qquad\qquad\qquad\square$

The cycles $\rho_1, ..., \rho_p$ induced by a shelling are sometimes (e.g. for matroid complexes and geometric lattices) the fundamental cycles of spherical subcomplexes. By this the following is meant. Suppose that a simplicial complex Δ is homeomorphic to the d-sphere. Then $H_d(\Delta) \cong \mathbb{Z}$ and the generator ρ of $H_d(\Delta)$, which is unique up to sign, is a linear combination in which every facet of Δ occurs with coefficient $+1$ or -1. This generator ρ is called the *fundamental cycle* of the spherical complex Δ.

7.8. Homology of Matroids

As a direct consequence of Theorem 7.7.2 and our work in sections 7.3 and 7.4 (see particularly 7.3.3, 7.4.3, and 7.4.7) we obtain the following two results.

7.8.1. Theorem. *Let $\Delta = IN(M)$ be the complex of independent sets in a matroid M of rank r. Then*

$$H_i(\Delta) \cong \begin{cases} \mathbb{Z}^{\tilde{\mu}(M^*)}, & \text{if } i = r - 1, \\ 0, & \text{if } i \neq r - 1. \end{cases}$$

7.8.2. Theorem. *Let* $\Delta = \overline{BC}_\omega(M)$ *be the reduced broken circuit complex of an ordered loopless matroid* $M = M(S, \omega)$ *of rank r. Then*

$$H_i(\Delta) \cong \begin{cases} \mathbb{Z}^{\beta(M)}, & \text{if } i = r - 2, \\ 0, & \text{if } i \neq r - 2. \end{cases}$$

For illustration, consider the rank 3 matroid M of Example 7.3.5. In Section 7.3 we computed its Tutte polynomial $T_M(x, y) = x^3 + 2x^2 + y^2 + 2xy + x + y$, whose values $\bar{\mu}(M^*) = T_M(0, 1) = 2$ and $\beta(M) = \dfrac{\partial T_M}{\partial x}(0, 0) = 1$ should be checked against the topology of the complexes $IN(M)$ and $\overline{BC}_\omega(M)$. These complexes are depicted in Figures 7.3 and 7.4.

Theorem 7.8.2 and Proposition 7.4.8 together imply a curious topological duality for reduced broken circuit complexes that seems to lack a systematic explanation. For any matroid $M = M(S)$ without loops or isthmuses, and for any orderings ω, ω' of its ground set S:

$$H_i(\overline{BC}_\omega(M)) \cong H_{|S|-i-4}(\overline{BC}_{\omega'}(M^*)), \tag{7.39}$$

for all $i \in \mathbb{Z}$.

We will now describe a basis for the homology of matroid complexes $IN(M)$. It follows from 7.3.3 and 7.7.2 that such a basis is implicitly determined by any ordering ω of the ground set. What we seek here is a simple explicit description of these bases directly in terms of matroid structure.

Let $M(S, \omega)$ be an ordered matroid of rank r with no isthmus. For each basis B of M construct a simplicial complex $\Sigma_{B,\omega}$ as follows:

For each $b \in B$ let $\phi(b) \notin B$ be the least element of the basic bond $\mathrm{bo}(B, b) - b$ (which is non-empty since b is not an isthmus), and define elements p_i by $\phi(B) = [p_1, p_2, ..., p_k]$ with $p_1 < ... < p_k$. Next, let $A_i = \{p_i\} \cup \phi^{-1}(p_i)$ for $1 \leqslant i \leqslant k$. The sets A_i are the blocks of a partition of the set $\phi(B) \cup B$. Finally, let $\Sigma_{B,\omega} = \{F \subseteq B \cup \phi(B) : A_i \nsubseteq F \text{ for all } 1 \leqslant i \leqslant k\}$.

7.8.3. Proposition.

(i) $B \in \Sigma_{B,\omega} \subseteq IN(M)$,

(ii) $\Sigma_{B,\omega}$ *is homeomorphic to the* $(r-1)$-*dimensional sphere.*

Proof. Veiwed as a simplicial complex $D(A) = \{E \subseteq A : E \neq A\}$, for $A \neq \varnothing$, is the boundary of an $(|A| - 1)$-simplex, and hence topologically an $(|A| - 2)$-sphere. We have that (writing $\Sigma_B = \Sigma_{B,\omega}$ for simplicity)

$$\Sigma_B = D(A_1) * D(A_2) * ... * D(A_k), \tag{7.40}$$

and $|A_i| \geqslant 2$ for $1 \leqslant i \leqslant k$, where the asterisk denotes simplicial join. (The *join* of two simplicial complexes Δ_1 and Δ_2 is defined by $\Delta_1 * \Delta_2 = \{F_1 \cup F_2 : F_1 \in \Delta_1$

and $F_2 \in \Delta_2$}.) From (7.40) it follows that $B \in \Sigma_B$ (delete p_i from A_i for all $1 \leqslant i \leqslant k$), and hence that dim $\Sigma_B = r - 1$. It is well known in topology that the join of two simplicial spheres is homeomorphic to a sphere (see e.g. Munkres, 1984, p. 370), so part (ii) follows.

The argument for part (i) hinges on the following technical observation:

$$1 \leqslant i < j \leqslant k \Rightarrow \operatorname{ci}(B, p_i) \cap A_j = \varnothing, \qquad (7.41)$$

which follows from the definition of A_j using Lemma 7.3.1.

We will prove by induction on $|F \cap \phi(B)|$ that every facet F of Σ_B is a basis of M. Clearly, $|F \cap \phi(B)| = 0$ if and only if $F = B$. Suppose that $F \cap \phi(B) \neq \varnothing$ and that i is *minimal* such that $p_i \in F \cap \phi(B)$. There is a unique $b_i \in A_i \cap B$ such that $b_i \notin F$. Let $F' = (F - p_i) \cup b_i$. Then F' is also a facet of Σ_B and $|F' \cap \phi(B)| = |F \cap \phi(B)| - 1$, so by the induction hypothesis F' is a basis. It follows from (7.41) that $\operatorname{ci}(B, p_i) = \operatorname{ci}(F', p_i)$, and from $\phi(b_i) = p_i$ that $b_i \in \operatorname{ci}(B, p_i)$. Hence, $b_i \in \operatorname{ci}(F', p_i)$ and we conclude using Lemma 7.3.1 that $F = (F' - b_i) \cup p_i$ is a basis. $\qquad \square$

For each basis B of M let $\sigma_{B,\omega}$ denote the fundamental cycle of the spherical complex $\Sigma_{B,\omega}$. There is also the explicit expression (cf. (7.40))

$$\sigma_{B,\omega} = \sum_{i_1 = 0}^{e_1} \cdots \sum_{i_k = 0}^{e_k} (-1)^{i_1 + \cdots + i_k} (A_1 - a_{i_1}^1) \cup \cdots \cup (A_k - a_{1_k}^k), \qquad (7.42)$$

where $A_j = [a_0^j, a_1^j, \ldots, a_{e_j}^j]$ is listed increasingly in the ω-ordering, for $1 \leqslant j \leqslant k$. Since $\Sigma_{B,\omega}$ is a full-dimensional subcomplex of $IN(M)$ we have that $\sigma_{B,\omega} \in H_{r-1}(IN(M))$. To remove the sign ambiguity we could demand that $\sigma_{B,\omega}(B) = 1$. (Remark: readers unhappy with the reliance on topology for the derivation of $\sigma_{B,\omega}$ can take (7.42) as its definition and check by direct computation that $\partial_{r-1}(\sigma_{B,\omega}) = 0$ for the simplicial boundary map (7.34).)

For the following result, recall that $i(B)$ denotes the internal activity in a basis B, defined in section 7.3.

7.8.4. Theorem. *Let $M = M(S)$ be a matroid of rank r with no isthmus. Then for every ordering ω of S the set of cycles $\{\sigma_{B,\omega} : i(B) = 0\}$ forms a basis for the free Abelian group $H_{r-1}(IN(M))$.*

Proof. We apply Theorem 7.7.2 to the ω-lexicographic shelling of $IN(M)$ constructed in section 7.3. In view of 7.7.2(ii) and (7.9) all that needs to be checked is that if $IP(B) = B$ (equivalently: $i(B) = 0$) and $F \in \Sigma_{B,\omega}$ for some basis $F \neq B$ then $IP(F) \neq F$.

Suppose then that $IP(B) = B$, $F \neq B$, and $F \in \Sigma_{B,\omega}$. Let $F = F_j = (B - \{b_1, \ldots, b_j\}) \cup \{\phi(b_1), \ldots, \phi(b_j)\}$, where $b_1, \ldots, b_j \in B$ and $\phi(b_j) < \ldots < \phi(b_1)$. We will prove by induction on $j \geqslant 1$ that $\phi(b_j)$ is internally active in F_j. For $j = 1$ this is clear, since $\phi(b_1)$ is the least element of $\operatorname{bo}(B, b_1) - b_1$ (by definition), $\phi(b_1) < b_1$ (since $b_1 \in IP(B)$), and $\operatorname{bo}(B, b_1) = \operatorname{bo}(F_1, \phi(b_1))$. For general $j > 1$, suppose that on the contrary there exists $q \in \mathrm{bo}(F_j, \phi(b_j))$ such that $q < \phi(b_j)$. One can then make the following observations:

(1) $q \notin B$ and $q \notin F_{j-1}$,
(2) $b_j \notin \mathrm{ci}(B, q)$,
(3) $b_j \in \mathrm{ci}(F_{j-1}, q)$,
(4) $\phi(b_i) \in \mathrm{ci}(F_{j-1}, q)$, for some $1 \leqslant i \leqslant j-1$.

For (1) use $q \notin F_j$ (by definition) and $q < \phi(b_j) \leqslant \phi(b_i) < b_i$ for $1 \leqslant i \leqslant j$, with the last inequality following from the fact that b_i is internally passive in B. Observation (2) is equivalent to $q \notin \mathrm{bo}(B, b_j)$, which is clear since by definition $\phi(b_j)$ is minimal in $\mathrm{bo}(B, b_j) - b_j$ and $q < \phi(b_j) < b_j$. Similarly, (3) is equivalent to $q \in \mathrm{bo}(F_{j-1}, b_j)$, which follows since $\mathrm{bo}(F_j, \phi(b_j)) = \mathrm{bo}(F_{j-1}, b_j)$. Finally, if (4) were false then $\mathrm{ci}(F_{j-1}, q) \subseteq B \cup q$, hence $\mathrm{ci}(F_{j-1}, q) = \mathrm{ci}(B, q)$, which would contradict (2) and (3).

Now, choose i *maximal* so that $\phi(b_i) \in \mathrm{ci}(F_{j-1}, q)$. Then $\mathrm{ci}(F_{j-1}, q) \subseteq F_i \cup q$, hence $\mathrm{ci}(F_{j-1}, q) = \mathrm{ci}(F_i, q)$, and we have that $\phi(b_i) \in \mathrm{ci}(F_i, q)$, or equivalently $q \in \mathrm{bo}(F_i, \phi(b_i))$. However, since $q < \phi(b_i)$ this contradicts the induction hypothesis, which says that $\phi(b_i)$ is internally active in F_i. We conclude that the existence of such an element q is impossible, and the induction step is complete.

We have shown that $\phi(b_j)$ is internally active in $F_j = F$, hence $IP(F) \neq F$. $\square$

7.8.5. Corollary. $IN(M) = \cup \Sigma_{B,\omega}$, with union over all bases B of zero internal activity.

Proof. Let B' be an arbitrary basis. From $\sigma_{B',\omega} = \sum\limits_{i(B)=0} n_B \sigma_{B,\omega}$ and $\sigma_{B',\omega}(B') = 1$ it follows that $\sigma_{B,\omega}(B') = \pm 1$, or equivalently $B' \in \Sigma_{B,\omega}$, for some B such that $i(B) = 0$. $\square$

In summary, we have shown that every isthmus-free matroid complex $IN(M)$ is the union of $\tilde{\mu}(M^*)$ spherical subcomplexes whose fundamental cycles give a basis for homology. Observe that a basis B of M satisfies $i(B) = 0$ if and only if $S - B$ is an nbc-basis of the orthogonal matroid M^*. Hence, the broken circuit complex $BC_\omega(M^*)$ plays a role for the homology of $IN(M)$ similar to that played by $BC_\omega(M)$ for $H_{r-2}(\bar{L})$, cf. Theorem 7.9.3.

7.8.6. Example. Let M be the matroid of Example 7.3.5. A picture of the complex $IN(M)$ appears in Figure 7.3. There are two bases of zero internal activity: 235 and 245. For 235 we find that $\phi(2) = \phi(3) = 1$ and $\phi(5) = 4$, so $\Sigma_{235} = D(123) * D(45)$ in the notation of (7.40). For 245 we have $\phi(2) = \phi(4) = \phi(5) = 1$, so $\Sigma_{245} = D(1245)$. These spherical complexes are shown in Figure 7.7, which should be compared with Figure 7.3.

Figure 7.7. (a) Σ_{235}; (b) Σ_{245}.

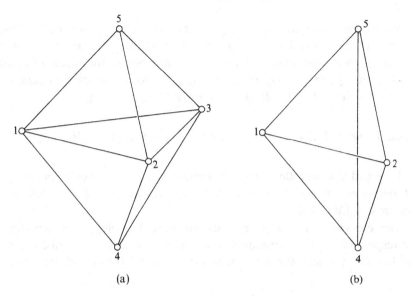

(a)   (b)

7.9. Homology of Geometric Lattices

The fact that there are interesting homological aspects of matroid theory was first made clear by the following result of Folkman (1966). Here $\tilde{\mu}(L) = (-1)^r \mu(L) > 0$ is the unsigned Möbius function value $\mu(L) = \mu(\hat{0}, \hat{1})$.

7.9.1. Theorem. *Let $\Delta = \Delta(\bar{L})$ be the order complex of a geometric lattice L of rank r. Then*

$$H_i(\Delta) \cong \begin{cases} \mathbb{Z}^{\tilde{\mu}(L)} & \text{if } i = r - 2, \\ 0 & \text{if } i \neq r - 2. \end{cases}$$

Proof. This follows directly from 7.6.3, 7.6.4, and 7.7.2.   □

This section will be devoted to a more detailed study of the homology of geometric lattices. First a basis for the non-vanishing order homology group $H_{r-2}(\Delta(\bar{L}))$ will be described, then a larger object, the *Whitney homology algebra* $H^W(L)$, will be constructed.

In the following, let (L, ω) be an atom-ordered geometric lattice of rank r. For each non-empty independent set of atoms $A = [a_1, a_2, \ldots, a_k]$, $a_1 < a_2 < \ldots < a_k$, and each permutation $\pi \in S_k$, let

$$c_{A,\pi} : a_{\pi(1)} < (a_{\pi(1)} \vee a_{\pi(2)}) < \ldots < (a_{\pi(1)} \vee a_{\pi(2)} \vee \ldots \vee a_{\pi(k-1)}). \quad (7.43)$$

Hence, $c_{A,\pi}$ is a maximal chain in the open interval $(\hat{0}, \bar{A})$ in L (if $k = 1$ this is the empty chain). Furthermore, let

$$\rho_A = \sum_{\pi \in S_k} \text{sign}(\pi) \mathbf{c}_{A,\pi}. \tag{7.44}$$

If one element, say $a_{\pi(1)} \vee \ldots \vee a_{\pi(j)}$, is removed from $\mathbf{c}_{A,\pi}$, then the resulting subchain is contained in exactly one other chain $\mathbf{c}_{A,\pi'}$, namely for the permutation π' that differs from π by transposition of the values of j and $j+1$. Then $\text{sign}(\pi') = -\text{sign}(\pi)$, so $\partial_{k-2}(\rho_A) = 0$ for the simplicial boundary operator ∂ defined by (7.34). We have proved the following.

7.9.2. Lemma. *If $A \neq \varnothing$ is independent and $\bar{A} = x$, then $\rho_A \in H_{r(x)-2}(\hat{0}, x)$.*

Here and in what follows we write simply $H_i(\hat{0}, x)$ for the order homology of the open interval $(\hat{0}, x) = \{z \in L : \hat{0} < z < x\}$, instead of the needlessly pedantic $H_i(\Delta((\hat{0}, x)))$.

The *elementary cycles* ρ_A provide non-zero homology representatives corresponding to all independent sets A. It is now easy to describe a basis for homology. Recall the notation $\mathbf{nbc}(x) = \{A \in BC_\omega(L) : \bar{A} = x\}$, for $x \in L$.

7.9.3. Theorem. *Let $x \in L - \{\hat{0}\}$. Then the elementary cycles $\{\rho_A : A \in \mathbf{nbc}(x)\}$ form a basis for the free Abelian group $H_{r(x)-2}(\hat{0}, x)$.*

Proof. We have earlier concluded that $\mathbf{nbc}(x)$ is the family of nbc-bases of the geometric lattice $[\hat{0}, x]$ with its induced atom ordering (see the proof of Proposition 7.4.5). Hence the present proof reduces to the case $x = \hat{1}$, i.e., we must show that the nbc-bases of L induce a basis of $H_{r-2}(\bar{L})$.

Using (7.29) and Theorem 7.7.2 all that needs to be checked is that if B is an nbc-basis then ρ_B contains exactly one chain $\mathbf{c}_{B,\pi}$ with decreasing label. But this was already shown in (7.33). □

It follows that for every geometric lattice L the order complex $\Delta(\bar{L})$ is the union of $\tilde{\mu}(L)$ spherical subcomplexes whose fundamental cycles give a basis for homology, cf. Exercise 7.48.

7.9.4. Example. Let L be the geometric lattice considered in Examples 7.3.5, 7.4.4, and 7.6.1, with the natural ordering $\mathbf{1} < \mathbf{2} < \mathbf{3} < \mathbf{4} < \mathbf{5}$ of its atoms, which we here print in boldface to distinguish them from integers.

It x is any atom of L, then $H_{-1}(\hat{0}, x) = H_{-1}(\{\varnothing\}) = \mathbb{Z}\varnothing$ and $\rho_{\{x\}} = \varnothing$.

If $x = \{\mathbf{1}, \mathbf{2}, \mathbf{3}\}$, then $H_0(\hat{0}, x) = \{c_1 \cdot \mathbf{1} + c_2 \cdot \mathbf{2} + c_3 \cdot \mathbf{3} : c_1 + c_2 + c_3 = 0\}$ and $\rho_{\{1,2\}} = \mathbf{1} - \mathbf{2}$, $\rho_{\{1,3\}} = \mathbf{1} - \mathbf{3}$.

Finally, let $x = \{\mathbf{1}, \mathbf{2}, \mathbf{3}, \mathbf{4}, \mathbf{5}\} = \hat{1}$, and let $B_1, B_2, \ldots, B_8$ be the bases of L indexed as in Example 7.3.5. The elementary cycles $\rho_i = \rho_{B_i}$ are better expressed by a picture than in algebraic notation; see for example ρ_8 in Figure 7.8.

Figure 7.8.

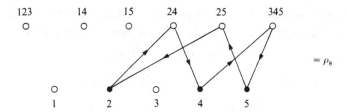

123   14   15   24   25   345

$= \rho_8$

The nbc-bases of L are B_1, B_2, B_3, and B_4, so $\{\rho_1, \rho_2, \rho_3, \rho_4\}$ is a basis of $H_1(\bar{L})$. To express another cycle in terms of this basis we need only check its coefficients on the chains with decreasing labels (by the general principle (7.38)). The four chains with decreasing labels in our example are shown in Figure 7.9, with the number of the corresponding nbc-basis indicated in parentheses.

From the two figures we immediately read off $\rho_8 = \rho_1 - \rho_2 - \rho_3 + \rho_4$, and similarly, $\rho_5 = -\rho_3 + \rho_4$, $\rho_6 = -\rho_1 + \rho_3$, and $\rho_7 = -\rho_2 + \rho_4$.

Figure 7.9.

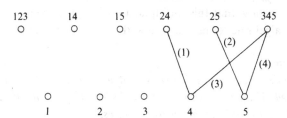

123   14   15   24   25   345

The following identity for elementary cycles will be of use later.

7.9.5. Lemma. *If $C = \{x_1, \ldots, x_k\}$ is a circuit, $x_1 < \ldots < x_k$, then*

$$\sum_{i=1}^{k} (-1)^i \rho_{C-x_i} = 0.$$

Proof. Each chain $x_{\pi(1)} < (x_{\pi(1)} \vee x_{\pi(2)}) < \ldots < (x_{\pi(1)} \vee \ldots \vee x_{\pi(k-2)})$, for given $\pi \in S_k$, appears twice in the expansion of the given sum; once in the term $\rho_{C-x_{\pi(k-1)}}$ and once, with the opposite sign, in the term $\rho_{C-x_{\pi(k)}}$. $\square$

Let L be a geometric lattice of rank r. Introduce a symbol $\rho_\varnothing$ (the elementary cycle corresponding to the empty set) and formally let $H_{-2}(\hat{0}, \hat{0}) = \mathbb{Z}\rho_\varnothing$. Then for each $x \in L$ we have a free Abelian group $H_{r(x)-2}(\hat{0}, x)$ of rank $(-1)^{r(x)}\mu(\hat{0}, x)$. We now combine all this homology into one global object:

$$H^W(L) = \bigoplus_{x \in L} H_{r(x)-2}(\hat{0}, x), \tag{7.45}$$

called the *Whitney homology* of L. This algebraic object can also be obtained as the homology of an algebraic chain complex (Exercise 7.53), and the name 'Whitney' stands to indicate that $H^W(L)$ is a direct sum of pieces

$$H_k^W = \bigoplus_{r(x)=k} H_{k-2}(\hat{0}, x),$$

whose ranks are the *Whitney numbers* of the first kind:

$$\text{rank } H_k^W = \sum_{r(x)=k} (-1)^k \mu(\hat{0}, x) = \tilde{w}_k. \tag{7.46}$$

From now on, fix an atom-ordering ω of L. The elementary cycles ρ_A of independent subsets $A \subseteq L^1$ are elements in $H^W(L)$, cf. Lemma 7.9.2, and we will now define a combinatorial multiplication rule for them. If $A, B \in IN(L)$, let

$$\rho_A \cdot \rho_B = \begin{cases} \pm\rho_{A \cup B} & \text{if } A \cap B = \varnothing \text{ and } A \cup B \in IN(L), \\ 0 & \text{otherwise.} \end{cases} \tag{7.47}$$

The sign of $\rho_{A \cup B}$ is positive if the elements of A followed by the elements of B give an even permutation of $A \cup B$ (all sets listed increasingly); it is otherwise negative. By counting inversions one finds that

$$\rho_A \rho_B = (-1)^{|A||B|} \rho_B \rho_A, \text{ for all } A, B \in IN(L). \tag{7.48}$$

Also, $\rho_A \rho_\varnothing = \rho_\varnothing \rho_A = \rho_A$ for all independent sets A.

We will now show that this multiplication uniquely extends to *all* cycles. The relevant algebraic terms are reviewed at the beginning of section 7.10.

7.9.6. Theorem.

   (i) *The Whitney homology $H^W(L)$ has a unique structure of an anticommutative L-graded $\mathbb{Z}$-algebra, with grading (7.45) and with multiplication that specializes to (7.47) on elementary cycles.*

   (ii) *A linear basis for $H^W(L)$ is given by $\{\rho_A : A \in BC_\omega(L)\}$.*

Proof. Part (ii) follows from Theorem 7.9.3. The work to be done for part (i) lies entirely in showing that the partial multiplication (7.47) extends so that $H_x \cdot H_y \subseteq H_{x \vee y}$, where $H_x = H_{r(x)-2}(\hat{0}, x)$. This global multiplication is then automatically anticommutative, since by (7.48) it is anticommutative on a basis.

The plan for the following proof is first to construct a more simple-minded algebra, whose multiplication is very manageable, and then show that $H^W(L)$ is a subalgebra.

Let $\mathscr{C}$ be the set of all lower chains in $L - \{\hat{0}\}$, i.e.,

$$\mathscr{C} = \{\varnothing\} \cup \{x_1 < x_2 < \dots < x_k : r(x_i) = i, 1 \leqslant i \leqslant k\}. \text{ Let } \mathbb{Z}\mathscr{C} = \left\{ \sum_{i=1}^t a_i c_i : a_i \in \mathbb{Z}, \right.$$

$$\left. c_i \in \mathscr{C} \right\} \text{ be the Abelian group freely generated by } \mathscr{C}. \text{ Define the product of}$$

basis elements $\mathbf{x} : x_1 < x_2 < \ldots < x_k$ and $\mathbf{y} : y_1 < y_2 < \ldots < y_l$ as follows:

$$\mathbf{x} \cdot \varnothing = \varnothing \cdot \mathbf{x} = \mathbf{x},$$

$$\mathbf{x} \cdot \mathbf{y} = \begin{cases} \sum_\sigma \text{sign}(\sigma) \cdot \mathbf{z}(\sigma, \mathbf{x}, \mathbf{y}) & \text{if } r(x_k \vee y_l) = k + l, \\ 0 & \text{otherwise.} \end{cases} \tag{7.49}$$

The summation in (7.49) is over all (k, l)-*shuffles* σ, i.e., all permutations σ of $\{1, 2, \ldots, k + l\}$ such that $\sigma^{-1}(1) < \ldots < \sigma^{-1}(k)$ and $\sigma^{-1}(k + 1) < \ldots < \sigma^{-1}(k + l)$; and $\mathbf{z}(\sigma, \mathbf{x}, \mathbf{y})$ denotes the lower chain $z_{\sigma(1)} < (z_{\sigma(1)} \vee z_{\sigma(2)}) < \ldots < (z_{\sigma(1)} \vee z_{\sigma(2)} \vee \ldots \vee z_{\sigma(k+l)})$, where $(z_1, z_2, \ldots, z_{k+l}) = (x_1, x_2, \ldots, x_k, y_1, y_2, \ldots, y_l)$.

The mapping $\mathscr{C} \times \mathscr{C} \to \mathbb{Z}\mathscr{C}$ thus defined extends by linearity to a multiplication $\mathbb{Z}\mathscr{C} \times \mathbb{Z}\mathscr{C} \to \mathbb{Z}\mathscr{C}$ (associativity requires checking), making $\mathbb{Z}\mathscr{C}$ into a ring with identity element $\varnothing$.

We have the following situation:

$$H^W(L) = \bigoplus_{x \in L} H_{r(x)-2}(\hat{0}, x) \overset{\phi}{\to} \bigoplus_{x \in L} C_{r(x)-2}(\hat{0}, x) \overset{\psi}{\to} \mathbb{Z}\mathscr{C}.$$

Here $C_{r(x)-2}(\hat{0}, x)$ is the chain group in the sense of section 7.7, i.e., the free Abelian group generated by all maximal chains in the open interval $(\hat{0}, x)$, and ϕ is the embedding of the subgroup of $(r(x) - 2)$-cycles, for all $x > \hat{0}$. Also, put $C_{-2}(\hat{0}, \hat{0}) = H_{-2}(\hat{0}, \hat{0}) = \mathbb{Z}\rho_\varnothing$. The mapping ψ is defined by sending the basis element $y_1 < y_2 < \ldots < y_{r(x)-1}$ in $C_{r(x)-2}(\hat{0}, x)$ to the corresponding basis element $y_1 < y_2 < \ldots < y_{r(x-1)} < x$ in $\mathbb{Z}\mathscr{C}$, with appropriate modification for the degenerate case $x = \hat{0}$. These mappings have the following key properties:

(1) ϕ is an injection of L-graded Abelian groups (obvious);

(2) ψ is an isomorphism of Abelian groups (obvious); and

(3) $\mathbb{Z}\mathscr{C}$ is an anticommutative L-graded algebra under the direct sum decomposition ψ^{-1} (follows from definition (7.49)).

We conclude that

$$\tau = \psi \circ \phi : H^W(L) \to \mathbb{Z}\mathscr{C}$$

is an L-graded embedding of $H^W(L)$ as a subgroup of the L-graded algebra $\mathbb{Z}\mathscr{C}$. To finish the proof we will show that

(4) $\qquad \tau(\rho_A) \cdot \tau(\rho_B) = \begin{cases} \pm \tau(\rho_{A \cup B}) & \text{if } A \cap B = \varnothing \text{ and } A \cup B \in IN(L) \\ 0 & \text{otherwise,} \end{cases}$

where $A, B \in IN(L)$ and the rule for signs is as in (7.47).

(5) $\qquad\qquad H^W(L) \cong \text{Im } \tau$ is multiplicatively closed in $\mathbb{Z}\mathscr{C}$.

If $A \cap B \neq \emptyset$ or if $A \cup B$ is dependent, then $r(\bar{A} \vee \bar{B}) < r(\bar{A}) + r(\bar{B})$, and if $A = [a_1, \ldots, a_k]$ and $B = [b_1, \ldots, b_l]$, rule (7.49) shows that $\tau(\rho_A) \cdot \tau(\rho_B) = 0$ in $\mathbb{Z}\mathscr{C}$. Otherwise $r(\bar{A} \vee \bar{B}) = r(\bar{A}) + r(\bar{B})$, and if $A = [a_1, \ldots, a_k]$ and $B = [b_1, \ldots, b_l]$, rule (7.49) shows that $\tau(\rho_A) \cdot \tau(\rho_B)$ is a linear combination of signed chains $z(\sigma, \tilde{c}_{A,\pi}, \tilde{c}_{B,\pi'})$ where $\sigma \in S_{k+l}/S_k \times S_l$ (the set of (k, l)-shuffles), $(\pi, \pi') \in S_k \times S_l$, and $\tilde{c}_{A,\pi}$ is the chain (7.43) augmented by $a_{\pi(1)} \vee a_{\pi(2)} \vee \ldots \vee a_{\pi(k)} = \bar{A}$ But via the natural bijection $(S_{k+l}/S_k \times S_l) \times (S_k \times S_l) \leftrightarrow S_{k+l}$ these chains are in sign-compatible bijection with the chains $\tilde{c}_{A \cup B, v}$, for $v \in S_{k+l}$, that occur in $\tau(\rho_{A \cup B})$. This proves (4).

We know that $\{\tau(\rho_A) : A \in BC_\omega(L)\}$ is a basis for Im τ, and (4) has shown that products of basis elements remain in Im τ. Hence, (5) follows. $\qquad \square$

7.10. The Orlik–Solomon Algebra

In this section a certain anticommutative L-graded algebra $\mathscr{A}(L)$ will be constructed for each geometric lattice L. The basic combinatorial properties of $\mathscr{A}(L)$ will be derived with emphasis on its intimate ties to the order homology of L.

We begin with a brief review of definitions and basic facts concerning graded algebras and exterior algebra. More details can be found in algebra books such as Bourbaki (1970).

Let M be a commutative monoid with identity element e, whose composition we write multiplicatively, and let R be a commutative ring. By an M-graded R-algebra A we mean a ring A together with a direct sum decomposition $A = \bigoplus_{x \in M} A_x$ (as an additive group) such that $A_e \cong R$ and $A_x \cdot A_y \subseteq A_{xy}$. In particular, via the identification $A_e = R$, each A_x is an R-module. An element $a \in A$ is called $homogeneous$ if $a \in A_x$ for some $x \in M$, and an ideal is $homogeneous$ if it is generated by homogeneous elements. Equivalently, an ideal $I \subseteq A$ is homogeneous if and only if $I = \bigoplus_{x \in M}(I \cap A_x)$. For any homogeneous ideal I there is a naturally induced structure of an M-graded R-algebra on its quotient: $A/I \cong \bigoplus_{x \in M}(A_x/I \cap A_x)$.

In most cases where graded algebras occur, $M = (\mathbb{N}^r, +)$ and R is a field. We shall mainly use $M = (L, \vee)$, where L is a geometric lattice, and $R = \mathbb{Z}$. The reason for using integer coefficients is to get sharper algebraic statements, in particular to show that no torsion arises. The following arguments would hold for an arbitrary ring R instead of $\mathbb{Z}$ if R-coefficients had been used in our previous work on homology also.

An L-graded algebra $A = \bigoplus_{x \in L} A_x$, where L is a geometric lattice, will be called $anticommutative$ if $a \cdot b = (-1)^{r(x) \| r(y)} b \cdot a$ for all $a \in A_x$, $b \in A_y$.

Let E be a free Abelian group with basis $\{e_1, e_2, \ldots, e_n\}$. The exterior algebra $\Lambda E = \bigoplus_{k=0}^{n} \Lambda^k E$ is an $\mathbb{N}$-graded $\mathbb{Z}$-algebra with main properties (multiplication denoted by $\wedge$):

(1) $x \wedge y = (-1)^{k \| l} y \wedge x$, for $x \in \Lambda^k E$ and $y \in \Lambda^l E$. In particular, for $x, y \in E = \Lambda^1 E : x \wedge y = -y \wedge x$ (so $x \wedge x = 0$).

(2) $\Lambda^k E$ is a free Abelian group with basis $\{e_A : \text{card } A = k\}$, where $A = [i_1, i_2, ..., i_k]$, $1 \leqslant i_1 < i_2 < ... < i_k \leqslant n$, and

$$e_A = e_{i_1} \wedge e_{i_2} \wedge ... \wedge e_{i_k}. \tag{7.50}$$

Consequently, rank $\Lambda^k E = \binom{n}{k}$.

(3) The multiplication of basis elements is, as follows from (1) and (2), given by

$$e_A \wedge e_B = \begin{cases} \pm e_{A \cup B} & \text{if } A \cap B = \varnothing, \\ 0 & \text{otherwise,} \end{cases}$$

with the sign determined by the parity of the permutation that brings A followed by B to $A \cup B$ (all sets listed in increasing order).

Now, let L be a geometric lattice and let $E(L)$ be the Abelian group freely generated by the set of its atoms $L^1 = \{e_1, e_2, ..., e_n\}$. Whenever convenient we may also assume as given an atom-ordering $\omega : e_1 < e_2 < ... < e_n$. For $A \subseteq L^1$ let $e_A \in \Lambda E(L)$ have the meaning (7.50). Then

$$\Lambda E(L) = \bigoplus_{x \in L} \Lambda_x, \text{ where } \Lambda_x = \bigoplus_{\bar{A} = x} \mathbb{Z} e_A,$$

and this direct sum decomposition gives $\Lambda E(L)$ the structure of an L-graded $\mathbb{Z}$-algebra, since $e_A \wedge e_B \in \mathbb{Z} e_{A \cup B} \subseteq \Lambda_{\bar{A} \vee \bar{B}}$.

For each circuit $C = \{x_1, x_2, ..., x_k\} \subseteq L^1$, $x_1 < ... < x_k$, let $\partial(e_C) = \sum_{i=1}^{k} (-1)^i e_{C - x_i}$, and let I be the ideal of $\Lambda E(L)$ generated by the elements $\partial(e_C)$ for all circuits C in L. The quotient

$$\mathscr{A}(L) = \Lambda E(L) / I \tag{7.51}$$

is the *Orlik–Solomon algebra* of L. Since I is a homogeneous ideal (the generators $\partial(e_C) \in \Lambda_C$ being homogeneous), $\mathscr{A}(L)$ inherits an L-grading from $\Lambda E(L)$:

$$\mathscr{A}(L) = \bigoplus_{x \in L} \mathscr{A}_x, \text{ where } \mathscr{A}_x = \Lambda_x / (I \cap \Lambda_x). \tag{7.52}$$

Denote by $\bar{e}_A$ the class of e_A in $\mathscr{A}(L)$, for subsets $A \subseteq L^1$.

7.10.1. Lemma.

(i) $\bar{e}_A \neq 0$ if and only if A is independent.

(ii) $\{\bar{e}_A : A \in BC_\omega(L)\}$ *linearly generates* $\mathscr{A}(L)$.

Proof. (i) Suppose A is dependent, and let $e_i \in C \subseteq A$ for some circuit C. Then $e_A = \pm e_i \wedge \partial(e_C) \wedge e_{A - C} \in I$. The converse, which will not be needed in what follows, is a consequence of Theorem 7.10.2.

(ii) Suppose that $\bar{e}_A \neq 0$, and that A contains the broken circuit $C - x_1$, where $C = \{x_1, ..., x_k\}$, $x_1 < ... < x_k$. Then A is independent by (i), so $x_1 \notin A$. The relation $\partial(e_C) \wedge e_{A-C} \in I$ takes the form

$$\bar{e}_A = \sum_{i=2}^{k} (-1)^i \bar{e}_{(A \cup x_1) - x_i},$$

expressing $\bar{e}_A$ as a linear combination of $\bar{e}_B$s with index set B lexicographically preceding A. If any such B contains a broken circuit, $\bar{e}_B$ can be similarly expanded in terms of lexicographically earlier elements, and so on. When the process eventually stabilizes, all index sets must be free of broken circuits. $\square$

We now come to the main result of this section, which relates the Orlik–Solomon algebra $\mathscr{A}(L)$ to the Whitney homology algebra $H^W(L)$ of section 7.9. For both objects a family of elements $\bar{e}_A$ and ρ_A, indexed by independent sets A, plays an important role.

7.10.2. Theorem.
  (i) *The correspondence $\bar{e}_A \leftrightarrow \rho_A$, for independent subsets $A \subseteq L^1$, extends to an isomorphism $\mathscr{A}(L) \cong H^W(L)$ of anticommutative L-graded algebras.*
  (ii) *$\{\bar{e}_A : A \in BC_\omega(L)\}$ is a linear basis for $\mathscr{A}(L)$.*

Proof. Define a surjective linear mapping

$$h : \Lambda E(L) \to H^W(L)$$

by letting $h(e_A) = \rho_A$, if A is independent, and $h(e_A) = 0$ otherwise. The rule (7.47) shows that h preserves multiplication, so h is actually a surjective ring-homomorphism. Furthermore, Lemma 7.9.5 shows that $h(\partial(e_C)) = 0$ for every circuit C, so $I \subseteq \text{Ker } h$. Therefore h induces a surjective ring-homomorphism

$$\bar{h} : \mathscr{A}(L) \to H^W(L),$$

such that $\bar{h}(\bar{e}_A) = \rho_A$, for all $A \in IN(L)$. Specializing to $A \in BC_\omega(L)$, and using Theorem 7.9.6(ii) and Lemma 7.10.1(ii), we then deduce the result from the following obvious lemma: If $f : G \to H$ is a surjective linear map of Abelian groups sending a generating set in G to a linearly independent set in H, then f is an isomorphism and both sets are bases. $\square$

For $0 \leqslant k \leqslant r = \text{rank}(L)$, let

$$\mathscr{A}^k(L) = \bigoplus_{r(x) = k} \mathscr{A}_x \cong \Lambda^k E(L)/(I \cap \Lambda^k E(L)). \qquad (7.53)$$

Then $\mathscr{A}(L) = \bigoplus_{k=0}^{r} \mathscr{A}^k(L)$ gives $\mathscr{A}(L)$ the structure of an $\mathbb{N}$-graded anticommutative algebra. An expression for the Hilbert–Poincaré polynomial of $\mathscr{A}(L)$

under this coarser grading follows directly from (7.46) and the isomorphism
$\mathcal{A}_x \cong H_{r(x)-2}(\hat{0}, x)$.

7.10.3. Corollary. $\displaystyle\sum_{k=0}^{r} (\text{rank } \mathcal{A}^k(L))t^k = \sum_{k=0}^{r} \tilde{w}_k t^k = (-t)^r p\left(L; -\frac{1}{t}\right),$
where $p(L; \lambda)$ is the characteristic polynomial of L (defined in section 7.4).

7.11. Notes and Comments

We end with references to original sources and some related remarks.
Additional results can be found among the exercises.

Section 7.1

There are several known topological aspects of matroid theory, in addition
to the topics treated here (which are mostly of an algebraic nature). I would
like to mention (1) the homotopy theorems of Tutte and Maurer, (2) the
topological theory of oriented matroids (particularly the Folkman–Lawrence
representation theorem), (3) simplicial matroids, and (4) matroid versions of
Sperner's lemma. Expository accounts with further references appear for (1)
and (2) in Björner (1991) and for (3) and (4) in White (1987), Chapter 6. For
a detailed treatment of (2), see Björner, Las Vergnas, Sturmfels, White &
Ziegler (1991).

Section 7.2

The notion of shellability originated in polytope theory in connection with
attempts from the mid-1800s onwards to prove the generalized Euler formula
by induction (Grünbaum, 1967). It has been most intensively studied for
polytopes and spheres (Danaraj & Klee, 1978), but recently also for many
other types of complexes (Björner, 1991). Many of the basic combinatorial
properties of a shellable complex, such as 7.2.2 and the role of the h-vector,
are due to McMullen (1970).

A numerical characterization of the h-vectors of shellable complexes was
given by Stanley (1977). The proper setting for this result is the theory of
Cohen–Macaulay complexes (Stanley, 1977; 1983). Proposition 7.2.5 depends
only on the most elementary properties of h-vectors (such as non-negativity),
and is valid for all Cohen–Macaulay complexes. Part (i) has been generalized
to all pure complexes by Stanley (1987), and further to all pure multicomplexes
by Hibi (1989).

Sections 7.3–7.4

The shellability of matroid and broken circuit complexes was first proven
by Provan (1977), see Provan & Billera (1980), using the recursive method
of 'vertex decomposability'. The lexicographic shelling method presented here
and its close connection to the concepts of internal/external activity and the

Tutte polynomial was discovered by Björner (1979). See section 8.6 for an extension to greedoids.

There are two complementary approaches to Tutte polynomials: the recursive approach based on contraction and deletion (see Chapter 6), and the constructive or generating function approach (of which a glimpse has been given here). The theory of Tutte polynomials was generalized from graphs to matroids by Crapo (1969), to whom 7.3.6 and 7.3.7 are due. A generalization of 7.3.6 appears in Dawson (1981).

The h-vectors and Stanley–Reisner rings of matroid complexes are discussed by Stanley (1977). He proves that such h-vectors are 'level sequences', but his conjecture that they are 'pure O-sequences' is still open. The result of Exercise 7.6 implies that $IN(M)$ is '$(n - m)$-Cohen–Macaulay connected', which Baclawski (1982) showed has interesting consequences for the Betti numbers and canonical module of the Stanley–Reisner ring of $IN(M)$. The h-vector of $IN(M)$ has also been studied by Dawson (1984).

Theorem 7.4.6, the key enumerative fact about broken-circuit-free sets, was discovered by Whitney (1932) for graphs and extended to matroids by Rota (1964). The proof given for 7.4.5 and 7.4.6 is from Björner & Ziegler (1991). The broken circuit complex was for the first time considered as a simplicial complex by Wilf (1976), for the purpose of studying Whitney numbers and chromatic polynomials of graphs. Many of the basic combinatorial properties of (reduced) broken circuit complexes, such as 7.4.7, are due to Brylawski (1977a). See also Brylawski & Oxley (1980, 1981) and Björner & Ziegler (1991).

Section 7.5

Inequalities for independence numbers and Whitney numbers of matroids have a sizeable literature. Much work in this area has been motivated by the still-open unimodality conjectures (Mason, 1972). For a survey of Whitney numbers see Chapter 8 of White (1987); for independence numbers see Welsh (1976), Dowling (1980), and Mahoney (1985). Applications of Whitney number inequalities to the enumeration of cells in hyperplane arrangements are discussed by Greene & Zaslavsky (1983) and Zaslavsky (1981, 1983).

Matroid inequalities reflect the deeper and more intrinsic question of characterizing the h-vectors of matroid and broken circuit complexes, about which little is known. The direct connection with characteristic polynomials (in the case of broken circuit complexes) indicates that these questions are likely to be very difficult.

The material in this section is from Björner (1979, 1980b). The $c = 2$ and $c = 3$ cases of 7.5.2 were independently found by Purdy (1982), and the $c = 3$ cases of 7.5.6 and 7.5.8, i.e., (7.24) and (7.25)–(7.26), are due to Dowling & Wilson (1974) and Brylawski (1977b) respectively. Heron (1972) had earlier found the inequalities 7.5.6, but did not characterize the case of equality. Formula (7.22) for connected matroids is due to G. Dinolt and U. Murty;

see Welsh (1976), p. 299.

Section 7.6

Theorem 7.6.3 is from Björner (1980a), and represents a special case of the method of lexicographic shellability for posets. The edge labelings (7.27) had earlier been used by Stanley (1974) to prove a more general version of 7.6.4(ii); see Exercise 7.30. The bijection (7.30), (7.31) also appears in that paper.

The lexicographic shellability of geometric lattices has been extended to some related classes of more general posets by Wachs & Walker (1986) and Laurent & Deza (1989). Also, the particular shellings constructed in Exercise 7.31(c) have been generalized from modular geometric lattices to all Tits buildings in Björner (1984).

The *facet graph* of a pure simplicial complex Δ has as its vertices the facets of Δ and as edges the pairs of facets that differ in only one element. Facet graphs of matroid complexes $IN(M)$ were studied by Maurer (1973), who obtained a characterization of this class of graphs. The facet graphs of geometric lattice complexes $\Delta(\bar{L})$ have been studied by Abels (1989, 1990). As discussed by Björner (1984) and Abels (1990) there are several structural similarities between geometric lattice complexes on the one hand and Tits buildings on the other. This is of course not surprising, since both structures generalize (in different directions) the properties of the subspace lattice of a finite-dimensional vector space. The similarities concern the role played by 'frames' (Boolean sublattices and apartments, respectively) and metric properties of the facet graphs.

Section 7.7

The material discussed is well known, except possibly for parts (ii) and (iii) of 7.7.2. Part (i) of 7.7.2 can be sharpened to the statement that Δ has the homotopy type of a wedge of p copies of the d-sphere; see Theorem 1.3 of Björner (1984) for an elementary proof.

For more about simplicial homology, and algebraic topology generally, see e.g. Munkres (1984). Björner (1991) surveys applications to combinatorics, including more details about shellable complexes.

Section 7.8

Theorems 7.8.1 and 7.8.2 appear to be due to Stanley; they are certainly implicit in Stanley (1977). The spheres $\Sigma_{B,\omega}$ were considered by Cordovil (1985) for other purposes (generalizations of Sperner's lemma). Proposition 7.8.3 is due to him. Theorem 7.8.4 appears to be new.

Section 7.9

Theorem 7.9.1 is due to Folkman (1966), and the basis results 7.9.3 and 7.9.6(ii) to Björner (1982). Whitney homology $H^W(L)$ was introduced via sheaf cohomology by Baclawski (1975). The exact relationship of Whitney homology to order homology, taken here as the definition of $H^W(L)$, was analyzed by Björner (1982) and Orlik & Solomon (1980); see Exercise 7.53. Our construction of the multiplicative structure on $H^W(L)$ is based on ideas of Orlik & Solomon (1980).

A surprising connection between the homology of finite partition lattices and free Lie algebras has been discovered by Barcelo (1990). She proves a direct correspondence between the homology basis of the broken-circuit complex and the free Lie algebra basis of Lyndon words.

Section 7.10

The definition of $\mathscr{A}(L)$ and the isomorphism 7.10.2(i) are due to Orlik & Solomon (1980). The basis result 7.10.2(ii) follows from Björner (1982) and Orlik & Solomon (1980), since it is immediately implied by 7.10.2(i) and 7.9.6(ii). It was independently discovered by Jambu & Leborgne (1986); see also Jambu & Terao (1989), Gel'fand & Zelevinsky (1986), and Zelevinsky (1990).

The algebra $\mathscr{A}(L)$ was introduced by Orlik and Solomon to give a combinatorial presentation of the cohomology ring of the complement in $\mathbb{C}^d$ to a finite union of central $(d-1)$-planes. See Orlik (1989) for an expository account and further references, and section 2.5 of Björner et al. (1991) for some matroid-theoretic aspects. This algebra has also found use in the work of Gel'fand, Zelevinskii, and coworkers, on hypergeometric functions.

A finite group acting on a geometric lattice L of rank r has an induced action on the homology $H_{r-2}(\bar{L})$. More generally, for each $J \subseteq \{1, 2, ..., r-1\}$ the group has a representation on the homology of the rank-selected subposet $L^J = \{x \in L : r(x) \in J\}$, and for each $0 \leqslant k \leqslant r$ a representation of degree $\tilde{w}_k$ on the graded component $\mathscr{A}^k(L)$ of the Orlik–Solomon algebra $\mathscr{A}(L)$. These two types of representation were first systematically studied by Stanley (1982) and Orlik & Solomon (1980) respectively. Later papers include Barcelo (1990), Barcelo & Bergeron (1990), Calderbank, Hanlon & Robinson (1986), Hanlon (1984, 1991), Lehrer (1987), Lehrer & Solomon (1986), and Rotman (1985).

Exercises

Problems whose solution is unknown to the author are denoted by an asterisk.

Section 7.2

7.1   Let Δ be a cone with apex $v \in V$, and let $\Delta' = \{F \in \Delta : v \notin F\}$. Show that Δ is shellable if and only if Δ' is shellable, and if so their shelling polynomials satisfy $h_\Delta(x) = x \cdot h_{\Delta'}(x)$.

7.2.   This exercise refers to Proposition 7.2.5.

(a) Prove part (i).

(b) Show that $h_i \geqslant 1$ implies $h_{i-1} \geqslant 1$ and $h_i \geqslant 2$ implies $h_{i-1} \geqslant 2$, for $1 \leqslant i \leqslant r$.

(c) Prove part (ii).

(d) Show that, for $r \geqslant 3$,

$$f_1 < f_2 < \cdots < f_{\lfloor r/2 \rfloor} \leqslant f_{\lfloor r/2 \rfloor + 1},$$

where if r is even the last inequality presupposes that $h_{\lfloor r/2 \rfloor + 1} \geqslant 1$. The last inequality is strict if r is odd or if $h_2 \geqslant 2$.

7.3. Say that a pure simplicial complex Δ is *strongly connected* if every pair of facets F and G are connected by a sequence of facets $F = F_0, F_1, ..., F_k = G$ such that $\text{codim}(F_{i-1} \cap F_i) = 1$, for $1 \leqslant i \leqslant k$. Show the following.

(a) Every shellable complex is strongly conected.

(b) If Δ is $(r-1)$-dimensional, strongly connected, and has v vertices, then

$$f_k \geqslant \binom{r}{k} + (v-r)\binom{r-1}{k-1},$$

for $0 \leqslant k \leqslant r$.

(c) Equality holds in (b) for all k (or, equivalently, for $k = r$) if and only if Δ is shellable and $h_2 = 0$.

Section 7.3

7.4. Show that a simplicial complex Δ on vertex set V is a matroid complex if and only if the induced subcomplex $\Delta_A = \{F \in \Delta : F \subseteq A\}$ is pure for all subsets $A \subseteq V$.

7.5. Let $M(S, \omega)$ be an ordered matroid, and let $\mathscr{F} = \{IP(B) : B \text{ a basis of } M\}$.

(a) (Dawson, 1984) Show that if $X, Y \in \mathscr{F}$ and $|X| > |Y|$, then there exists $x \in X - Y$ such that $Y \cup x \in \mathscr{F}$.

(b) Deduce using Lemma 7.2.6 and (7.9) that $\mathscr{F}$ is a greedoid. (See section 8.2 for the definition.)

(c) (Purtill, 1986) Show that $\mathscr{F}$ has the interval property.

* (d) Characterize those interval greedoids that arise from an ordered matroid in this way. What can be said about their f-vectors (i.e. the h-vectors of matroid complexes)?

7.6. Let $M(S)$ be a rank r matroid of size n, and define m to be the maximal size of any hyperplane. Show that for any subset $A \subseteq S$ such that $|A| < n - m$, the subcomplex of $IN(M)$ induced on $S - A$ is shellable and of dimension $r - 1$.

7.7. Consider the complete graph K_n on labeled vertices 1, 2, ..., n. Let c_k be the number of connected spanning subgraphs of K_n with k edges, and define $C_n(t) = \Sigma_k c_k t^k$.

(a) Show that $C_n(t) = t^{n-1} T_{K_n}(1, 1 + t)$. (Hint: Use (7.12) and Proposition 7.2.3.) Now, let $\mathscr{T}_n$ be the set of all spanning trees of K_n. Define an *inversion* of a tree $T \in \mathscr{T}_n$ to be an ordered pair (i, j) of vertices such that $i > j > 1$ and such that the unique path from 1 to j passes through i. Let $\text{inv}(T)$ be the number of inversions in T, and define $I_n(t) = \Sigma_{T \in T_n} t^{\text{inv}(T)}$.

(b) (Gessel & Wang, 1979) Show that $C_n(t) = t^{n-1} I_n(1 + t)$.

(c) Conclude that the number of trees in $\mathscr{T}_n$ with k inversions equals the number of trees in $\mathscr{T}_n$ with k externally active edges (with respect to a fixed arbitrary ordering of the edges of K_n).

(d) (Beissinger, 1982) Prove part (c) directly by constructing a bijection between the two classes of trees.

Section 7.4

7.8. (Brylawski, 1977a) Show that every matroid complex is a reduced broken circuit complex, i.e. given a matroid M construct an ordered matroid (M', ω) such that $IN(M) = \overline{BC}_\omega(M')$.

*7.9. (a) (Brylawski, 1977a) Characterize (reduced) broken circuit complexes.

(b) (Wilf, 1976, and others) Characterize the f-vectors (or h-vectors) of broken circuit complexes.

7.10. (Wilf, 1977; Brylawski, 1977a) Let $M(S)$ be simple matroid of rank r and size n with characteristic polynomial $p(M; \lambda)$. Let k be a positive integer and suppose that $k + 1$ colors are available, one of which is blue. Show that $k^{n-r}(-1)^r p(M; -k)$ equals the number of $(k + 1)$-colorings of S that have no blue broken circuit. (Broken circuits are defined with respect to some fixed but arbitrary ordering of S.)

7.11. (a) (Brylawski, 1977a) Let (L, ω) be an atom-ordered geometric lattice, and let A be an ω-initial segment of L^1. Show that $BC_\omega(L) = \Delta_A * \Delta_B$, where Δ_A and Δ_B are the subcomplexes induced on the complementary subsets A and $B = L^1 - A$, if and only if A is a modular flat. (The *join* of two simplicial complexes Δ_1 and Δ_2 is defined by $\Delta_1 * \Delta_2 = \{F_1 \cup F_2 : F_1 \in \Delta_1$ and $F_2 \in \Delta_2\}$.)

(b) (Stanley, 1971) Deduce using (a) that the characteristic polynomial $p([\hat{0}, x]; \lambda)$ of a modular flat $x \in L$ divides $p(L; \lambda)$.

7.12. Let L be a supersolvable geometric lattice with M-chain $\hat{0} = m_0 < m_1 < ... < m_r = \hat{1}$. For $1 \leqslant i \leqslant r$, let $A_i = \{p \in L^1 : p \leqslant m_i, p \nleqslant m_{i-1}\}$ and $a_i = |A_i|$. Show the following:

(a) if ω is an atom ordering such that all elements of A_i come before all of A_{i+1}, for $1 \leqslant i \leqslant r - 1$, then $BC_\omega(L) = \Delta_{A_1} * \Delta_{A_2} * ... * \Delta_{A_r}$;

(b) each subcomplex Δ_{A_i} in (a) is zero-dimensional;

(c) $p(L; \lambda) = (\lambda - a_1)(\lambda - a_2) ... (\lambda - a_r)$.

(A geometric lattice is called *supersolvable* if it has an M-chain, meaning a maximal chain of modular elements. This concept is due to Stanley (1972), who also proved (c). Parts (a) and (b) follow from Brylawski (1977a).)

7.13. (Björner & Ziegler, 1991) Show that the following conditions on a geometric lattice L of rank r are equivalent.

(a) L is supersolvable.

(b) For some atom ordering ω, the broken circuit complex $BC_\omega(L)$ is a multiple join of zero-dimensional subcomplexes.

(c) For some ω there exists a partition $L^1 = A_1 \cup A_2 \cup ... \cup A_r$, sujch that $|B \cap A_i| = 1$ for all nbc-bases B and all $1 \leqslant i \leqslant r$.

(d) For some ω, the 1-skeleton of $BC_\omega(L)$ is a complete r-partite graph.

(e) For some ω, the 1-skeleton of $BC_\omega(L)$ is an r-partite graph.

(f) For some ω, the inclusion-wise minimal broken circuits all have size 2.

(g) There exists a partition $L^1 = A_1 \cup A_2 \cup ... \cup A_r$ such that for any two distinct $x, y \in A_i$ there exists $z \in A_j$, for some $j < i$, such that $\{x, y, z\}$ is a circuit.

7.14. (Halsey, 1987; Björner & Ziegler, 1991) Show that a supersolvable geometric lattice is determined by its 3-truncation, i.e. it can be reconstructed from its point–line incidences.

7.15. (Björner, 1982) For every geometric lattice L, certain families of subsets of L^1 (the set of atoms), called *neat base-families*, are recursively defined as follows.

(1) If rank $L = 1$, then $\{\{\hat{1}\}\}$ is a neat base-family.

(2) If rank $L > 1$, then pick an arbitrary atom $p \in L^1$ and for each hyperplane $h \in L^{r-1}$ such that $h \not\geq p$ let $\mathcal{B}_h$ be a neat base-family in $[\hat{0}, h]$. Then $\mathcal{B} = \{A \cup p : A \in \mathcal{B}_h, h \not\geq p\}$ is a neat base-family in L.

Show the following.

(a) Every member of a neat base-family is a basis of L.

(b) Every neat base-family has $\tilde{\mu}(L)$ members.

(c) The facets of any broken circuit complex $BC_\omega(L)$ are a neat base-family, but the converse does not hold.

7.16. (Björner & Ziegler, 1991) Let L be a geometric lattice and $\pi : (L - \hat{0}) \to L^1$ a map such that $\pi(x) \leq x$ and $\pi(x) \leq y \leq x$ implies $\pi(y) = \pi(x)$ for all $x, y \in L - \hat{0}$. A subset $A \subseteq L^1$ is called *rooted* if $\pi(\bar{B}) \in B$ for all non-empty subsets $B \subseteq A$. The collection of rooted sets for L and π, denoted $RC_\pi(L)$, is called a *rooted complex*. Show the following.

(a) $RC_\pi(L)$ is a simplicial complex.

(b) $RC_\pi(L)$ is pure of dimension rank $L - 1$.

(c) All members of $RC_\pi(L)$ are independent. (That is, $RC_\pi(L)$ is a full-dimensional subcomplex of $IN(L)$.)

(d) $RC_\pi(L)$ is a cone with apex $\pi(\hat{1})$.

(e) Define π-*broken circuits* as subsets of the form $C - \pi(\bar{C})$, for circuits C such that $\pi(\bar{C}) \in C$. Then $RC_\pi(L)$ consists precisely of those subsets of L^1 that do not contain any π-broken circuit.

(f) For $x \in L - \hat{0}$, put $\mathbf{rc}(x) = \{A \in RC_\pi(L) : \bar{A} = x\}$. Then $\mathbf{rc}(x)$ is a neat base-family in $[\hat{0}, x]$.

(g) For all $x \in L$, $|\mathbf{rc}(x)| = (-1)^{r(x)}\mu(\hat{0}, x)$.

(h) The face numbers of $RC_\pi(L)$ coincide with the Whitney numbers of the first kind of L.

* (i) Is $RC_\pi(L)$ shellable?

7.17. (Björner & Ziegler, 1991) Show that every broken circuit complex $BC_\omega(M)$ is a rooted complex $RC_\pi(L)$, but the converse does not hold.

Section 7.5

7.18. For a loopless matroid M of rank r, show that

(a) $T_M(x, 0) = x^r + a_1 x^{r-1} + \ldots + a_k x^k$, with $a_k \neq 0$, if and only if M is the direct product of k connected simple matroids;

(b) $T_M(x, 1) = x^r + b_1 x^{r-1} + \ldots + b_k x^k$, with $b_k \neq 0$, if and only if M has exactly k isthmuses.

7.19. This exercise refers to Proposition 7.5.1.

(a) If M has exactly p isthmuses, show that $h_{r-p} \neq 0$ and $h_{r-p+1} = 0$ for the h-vector of $IN(M)$.

(b) Prove part (ii). Show that equality holds if and only if $n = r + 1$. (Hint: Proposition 7.5.3 is of use.)

(c) Show that, for $r \geq 3$,

$$I_1 < I_2 < \ldots < I_{[r/2]} \leq I_{[r/2]+1},$$

where if r is even the last inequality presupposes that M has fewer than

$r/2$ isthmuses. The last inequality is strict if r is odd or if $n > r + 1$. (Cf. Exercise 7.2.)

7.20. This exercise refers to Proposition 7.5.5.

(a) If M has exactly p connected components, show that $h_{r-p} \neq 0$ and $h_{r-p+1} = 0$, for the h-vector of $BC_\omega(M)$.

(b) Prove part (ii). Characterize the case of equality. (Hint: Proposition 7.5.7 is of use.)

(c) Show that, for $r \geq 3$,

$$\tilde{w}_1 < \tilde{w}_2 < \ldots < \tilde{w}_{[r/2]} \leq \tilde{w}_{[r/2]+1},$$

where if r is even the last inequality presupposes that M has fewer than $r/2$ connected components. The last inequality is strict if r is odd or if M is connected and $n > r + 1$. (Cf. Exercise 7.2.)

(d) Show that $\tilde{w}_{r-1} > \tilde{w}_r$. (Hint: here and for the $k = 1$ case in (b) use that $BC_\omega(M)$ is a cone.)

(e) Deduce that the sequence $(\tilde{w}_0, \tilde{w}_1, \ldots, \tilde{w}_r)$ is unimodal for $r \leq 6$.

7.21. (Dowling, 1980) Prove that the sequence $(I_0, I_1, \ldots, I_r)$ is unimodal for $r \leq 7$.

7.22. Suppose that a simple matroid M of rank r and size n has exactly q circuits of the minimal cardinality $c = c_M$. Show the following:

(a) $q \leq \dbinom{n-r+c-1}{c}$;

(b) $I_k \geq \displaystyle\sum_{i=0}^{c} \binom{n-r+i-1}{i}\binom{r-i}{k-i} - q\binom{r-c}{r-k}$, for $0 \leq k \leq r$;

(c) $\tilde{w}_k \geq \displaystyle\sum_{i=0}^{c-1} \binom{n-r+i-1}{i}\binom{r-i}{k-i} - q\binom{r-c+1}{r-k}$, for $0 \leq k \leq r$.

7.23. (Björner, 1982) Prove that $b(M) \geq (n/r)\tilde{\mu}(M)$, with equality if and only if M is an r-uniform matroid.

7.24. Show that inequality (7.22) is true for all matroids without loops and isthmuses with only the following exceptions: $C_r \oplus C_{n-r}^*$, for $2 \leq r \leq n - 2$, $C_2^2 \oplus C_3^*$, $C_2^2 \oplus C_3$, C_2^3, and C_2^4. Here C_i denotes the i-point circuit and C_2^k is the direct product of k copies of C_2. Show also that in these cases (7.22) fails by at most 2 units.

7.25. Let L be a geometric lattice of rank r.

(a) (Stanley, 1987) Linearly order the atoms of L and for each $x \in L$ let B_x be the lexicographically first basis of the interval $[\hat{0}, x]$. Show that $\{B_x : x \in L\}$ is a simplicial complex.

(b) (Wegner, 1984) Show that $(W_0, W_1, \ldots, W_r)$ is the f-vector of a simplicial complex, where $W_k = \text{card } L^k$ is a Whitney number of the second kind.

(c) Show that the complex in (a) is a subcomplex of the broken circuit complex.

Section 7.6

7.26. Deduce from Proposition 7.6.4 that $(-1)^{r(y)-r(x)}\mu(x, y) > 0$ for all $x \leq y$ in a geometric lattice L. (The inequality is due to Rota, 1964.)

7.27. Let L be a geometric lattice. Show that the order complex $\Delta(\bar{L} - A)$ is shellable for the following subsets $A \subseteq \bar{L}$.

(a) (Wachs & Walker, 1986) $A = [x, \hat{1}]$, for some $x \in L - \hat{0}$.

(b) (Baclawski, 1982) A is any subset containing no k-element antichain, supposing that every line of L has at least k points.

(c) (Björner, 1980a) A is the antichain of maximal complements of a fixed element $x \in \bar{L}$.

7.28. Show that the k-tuple $\lambda_\omega(\mathbf{c})$ of (7.28), considered as an unordered set of atoms, contains no broken circuit.

7.29. (Björner, Frankl & Stanley, 1987) For a geometric lattice L, let $\mathscr{F} = \{\mathscr{R}(\mathbf{c}):$ $\mathbf{c} \in \mathscr{M}\}$, where $\mathscr{M}$ is the set of maximal chains and $\mathscr{R}$ is the restriction operator (7.29).

(a) Show that $\mathscr{F}$ is a simplicial complex.

(b) Conclude that the h-vector of $\Delta(\bar{L})$ equals the f-vector of $\mathscr{F}$.

(c) Show that $\mathscr{F}$ is in general neither pure nor connected.

7.30. Let L be a geometric lattice of rank r, and for subsets $J \subseteq \{1, 2, ..., r-1\}$ define the *rank-selected* subposet $L^J = \{x \in L: r(x) \in J\} = \cup_{j \in J} L^j$. Let $\Delta(L^J)$ denote the order complex of L^J. Prove (a)–(c).

(a) $\Delta(L^J)$ is a pure $(|J| - 1)$-dimensional complex.

(b) $\mu(L^J) = \chi(\Delta(L^J)) = (-1)^{|J|-1} \cdot |\{\mathbf{c} \in \mathscr{M} : \mathscr{D}(\lambda(\mathbf{c})) = J\}|$.

(Here $\mu(L^J)$ denotes the Möbius function value $\mu(\hat{0}, \hat{1})$ computed on the poset $L^J \cup \{\hat{0}, \hat{1}\}$, $\lambda = \lambda_\omega$ is any labeling (7.28) of the set $\mathscr{M}$ of maximal chains in L, and $\mathscr{D}(\lambda_1, \lambda_2, ..., \lambda_r) = \{i : \lambda_i > \lambda_{i+1}\}$.)

(c) $\Delta(L^J)$ is shellable.

(d) Determine the homology of $\Delta(L^J)$.

* (e) Describe in matroid-theoretic terms a 'natural' basis for the homology group $H_{|J|-1}(\Delta(L^J))$ consisting of fundamental cycles of spherical subcomplexes.

(Part (b) is from Stanley, 1974, (c) and (d) from Björner, 1980a. See also Stanley, 1986.)

7.31. Let L be a supersolvable geometric lattice with M-chain $\hat{0} = m_0 < m_1 < ... < m_r = \hat{1}$ (for the definition see Exercise 7.12). Define an edge labeling $\lambda_M : \mathrm{cov}(L) \to \{1, 2, ..., r\}$ by the rule

$$\lambda_M(x, y) = \min\{i : m_i \vee x = m_i \vee y\},$$

and extend this to a labeling λ_M of unrefinable chains as in (7.28). Show the following.

(a) Lemma 7.6.2 holds for λ_M.

(b) Proposition 7.6.4, as well as its generalization in Exercise 7.30(b), holds for λ_M.

(c) Theorem 7.6.3 holds for λ_M and the natural lexicographic order of labels.

(The definitions of λ_M and part (b) are from Stanley, 1972; part (c) is from Björner, 1980a.)

7.32. Let L be a supersolvable geometric lattice as in the preceding exercise. For $x \in L$ define

$$\Lambda(x) = \{i : m_{i-1} \vee x = m_i \vee x\} \subseteq \{1, 2, ..., r\}$$

(a generalized Schubert symbol). Show that

(a) $|\Lambda(x)| = r(x)$, for all $x \in L$;

(b) $\{\lambda_M(x, y)\} = \Lambda(y) - \Lambda(x)$, for all $(x, y) \in \mathrm{Cov}(L)$.

Now, let P be any partial ordering of $\{1, 2, ..., r\}$ such that $i < j$ in P implies $i < j$ in $\mathbb{N}$. Define

$$L_P = \{x \in L : \Lambda(x) \text{ is an order ideal in } P\}.$$

Show the following.

(c) Every maximal chain in L_P has length r. Consequently, the order complex $\Delta(\bar{L}_P)$ is pure $(r-2)$-dimensional. $(\bar{L}_P = L_P - \{\hat{0}, \hat{1}\}.)$

(d) The labeling λ_M of maximal chains of L constructed in the preceding exercise restricts to a labeling of the maximal chains of L_P for which (the analogs of) Theorem 7.6.3 and Proposition 7.6.4 hold. In particular, $\Delta(\bar{L}_P)$ is shellable.

(e) If L is Boolean then L_P is a distributive lattice, and every finite distributive lattice arises this way.

(f) If L is the subspace lattice of a 4-dimensional vector space, and P is the ordering of $\{1, 2, 3, 4\}$ whose only comparability relation is $2 < 3$, then L_P is not a lattice.

(Björner & Stanley, 1985. See also Exercise 49b on p. 164 of Stanley, 1986. Part (e) is the fundamental theorem for finite distributive lattices, due to G. Birkhoff (Stanley, 1986, p. 106). When L is the subspace lattice of a vector space and P is a preordered linear forest, the poset L_P coincides with a quotient of a Tits building of type A, as studied by Wachs, 1986. Part (f) answers a question left open by Wachs.)

7.33. (Stanley, 1972; Björner & Stanley, 1985) Let $\Pi = \Pi_{r+1}$ be the lattice of partitions of the set $\{0, 1, ..., r\}$ ordered by refinement. A covering relation $x < y$ in Π corresponds to a merging of two distinct blocks B_1 and B_2 of x into one block $B_1 \cup B_2$ of y. Let

$$\lambda(x, y) = \max\{\min B_1, \min B_2\},$$

for all $(x, y) \in \mathrm{Cov}(\Pi)$. Here $\min B_i$ denotes the least element of B_i in the natural ordering of $\mathbb{N}$. Show the following:

(a) Π is a supersolvable lattice.

(b) The edge labeling λ of Π is a special case of the general construction in Exercise 7.31.

(c) There are exactly $r!$ maximal chains in Π with decreasing labels.

(d) $\mu(\Pi) = (-1)^r r!$

(e) $\mu(\Pi^J) = (-1)^{|J|-1} \Sigma_\sigma \sigma_1^* \, \sigma_2^* \, ... \, \sigma_r^*$, where Π^J is the rank-selected subposet defined as in Exercise 7.30, and the summation is over all permutations $\sigma \in S_r$ such that $\{i : \sigma_i > \sigma_{i+1}\} = J$, and $\sigma_k^* = \sigma_k - |\{i : i < k \text{ and } \sigma_i < \sigma_k\}|$. (E.g. if $\sigma = 42513$, then $\sigma_1^* \, ... \, \sigma_5^* = 4 \times 2 \times 3 \times 1 \times 1 = 24$.)

Now, let P be a partial ordering of $\{1, 2, ..., r\}$ such that $i < j$ in P implies $i < j$ in $\mathbb{N}$. Define

$$\Pi_P = \left\{(B_1, B_2, ..., B_k) \in \Pi : \bigcup_{i=1}^{k} (B_i - \{\min B_i\}) \text{ is an order ideal in } P\right\}.$$

(f) Show that the poset Π_P is a special case of the general construction in Exercise 7.32.

(g) Deduce that the order complex of every rank-selected subposet Π_P^J, for $J \subseteq \{1, 2, ..., r-1\}$, is shellable.

(h) Give a formula for $\mu(\Pi_P^J)$.

7.34. Show that the following subsets of the partition lattice Π_r, with induced ordering, have shellable order complexes.

(a) {partitions whose block sizes are $\equiv 0 \pmod{k}$}, if k divides r,

(b) {partitions whose block sizes are $\equiv 1 \pmod{k}$}, for any $k \geqslant 2$,

(c) {non-crossing partitions}, i.e. partitions such that for any blocks B_1 and B_2 the conditions $x_1, x_3 \in B_1$, $x_2, x_4 \in B_2$, and $x_1 < x_2 < x_3 < x_4$ imply $B_1 = B_2$.

(For (a) and (b) see Calderbank, Hanlon & Robinson, 1986; for (a) also Sagan, 1986. For (c) see Björner, 1980a. In each case the Möbius function has an interesting form; see the cited sources.)

7.35. Let L be a semimodular lattice of rank r. Say that two maximal chains in L are *adjacent* if they differ in exactly one element. Placing edges between adjacent pairs we get a graph $\mathcal{M}_L$ whose vertex set is the set of maximal chains in L. Let $d(\mathbf{c}, \mathbf{d})$ denote the usual graph distance (i.e. the length of a shortest connecting path) in $\mathcal{M}_L$.

Prove the following.

(a) The graph $\mathcal{M}_L$ is connected.

(b) (Abels, 1989) $d(\mathbf{c}, \mathbf{d}) = |\{c_i \vee d_j : 0 \leqslant i, j \leqslant r\}| - r - 1$, for any two chains $\mathbf{c} : c_0 < c_1 < \ldots < c_r$ and $\mathbf{d} : d_0 < d_1 < \ldots < d_r$.

(c) (Björner, 1980a) $\mathrm{diam}(\mathcal{M}_L) = \max_{\mathbf{c}, \mathbf{d}} d(\mathbf{c}, \mathbf{d}) \leqslant \binom{r}{2}$.

(d) If L is geometric, then $\mathrm{diam}(\mathcal{M}_L) = \binom{r}{2}$.

7.36. Let a *Boolean packing* of a rank r geometric lattice L mean a family of injective and cover-preserving mappings $\phi_i : B_i \to L$ from finite Boolean lattices B_i, $1 \leqslant i \leqslant t$, such that

(1) $r(\phi_i(\hat{0})) + r(\phi_i(\hat{1})) \geqslant r$, for all $1 \leqslant i \leqslant t$, and

(2) L is the disjoint union of the images $\phi_i(B_i)$.

(a) Show that every rank 3 geometric lattice has a Boolean packing.

* (b) Is the same true for rank $\geqslant 4$?

Section 7.7

7.37. The 3-element sets 123, 125, 126, 134, 136, 145, 234, 235, 246, 356, 456 are the facets of a pure two-dimensional simplicial complex Δ on the vertex set $\{1, \ldots, 6\}$. Show the following.

(a) Δ is shellable.

(b) $\Delta - \{123\}$ is not shellable. (Hint: This is a triangulation of a surface. Which one?)

(c) $H_2(\Delta) \cong \mathbb{Z}$, and in the generating 2-cycle (unique up to sign) the facet 123 has coefficient ± 2, while all other facets have coefficient ± 1.

*7.38. Find general conditions on a shellable complex Δ (or on a shelling) that guarantee that the basic cycles $\rho_1, \ldots, \rho_p$ of Theorem 7.7.2 are the fundamental cycles of subcomplexes of Δ homeomorphic to spheres, whose union is Δ. (This is the case if e.g. Δ is a pseudomanifold (Danaraj & Klee, 1974), for the lexicographic shelling of $\Delta(\bar{L})$ (Theorem 7.9.3 and Exercise 7.48), for the lexicographic shelling of $IN(M)$ (Theorem 7.8.4), and for the shelling of Tits buildings and their type-selected subcomplexes considered in Björner, 1984.)

Section 7.8

7.39. For any matroid M, show that the following hold.
   (a) $IN(M)$ is acyclic $\Leftrightarrow M$ has an isthmus $\Leftrightarrow IN(M)$ is a cone.
   (b) $NS(M)$ is acyclic $\Leftrightarrow M$ has a loop $\Leftrightarrow NS(M)$ is a cone.
   (c) $\overline{BC}_\omega(M)$ is acyclic $\Leftrightarrow M$ is not connected $\Leftrightarrow \overline{BC}_\omega(M)$ is a cone.
   Here $NS(M)$ is the complex of non-spanning subsets, and ω is an arbitrary ordering of the ground set. (For part (b), cf. Exercise 7.51.)

7.40. Show that if a cycle $\sigma_{B',\omega'}$, for arbitrary basis B' and ordering ω', is expressed in the basis $\{\sigma_{B,\omega} : i(B) = 0\}$ of Theorem 7.8.4 then all coefficients must equal $-1, 0$ or $+1$.

7.41. (a) Show that the mapping $\phi : B \rightarrow S - B$ of Proposition 7.8.3 is never injective, if M is a simple matroid. (Equivalently, the sphere $\Sigma_{B,\omega}$ is never a hyperoctahedron.)
   * (b) Is it true that for any two bases $B, B' \in IN(M)$ there exists some spherical subcomplex $\Sigma \subseteq IN(M)$ such that $B, B' \in \Sigma$?

7.42. (Björner & Ziegler, 1987) Compute the homology of reduced rooted complexes $\overline{RC}_\pi(L)$. These are defined by the property that $RC_\pi(L)$ is a cone over $\overline{RC}_\pi(L)$; see Exercise 7.16.

7.43. (a) For every simplicial complex Δ on a vertex set V such that $V \notin \Delta$, let $\Delta^* = \{A \subseteq V : V - A \notin \Delta\}$. Show that $\Delta^{**} = \Delta$.
   (b) Deduce from Alexander duality on the $(n-2)$-dimensional sphere, where $n = |V|$, that $H_i(\Delta) \cong H_{n-3-i}(\Delta^*)$, for all $i \in \mathbb{Z}$. (Here we use reduced homology with coefficients in a *field*.)
   (c) Let M be a matroid of cardinality n and rank ≥ 1. If $\Delta = NS(M)$, the complex of non-spanning subsets, then $\Delta^* = IN(M^*)$. Compare the duality $H_i(NS(M)) \cong H_{n-3-i}(IN(M^*))$ with the result of Exercise 7.51 in view of Theorems 7.8.1 and 7.9.1.
   (d) Let M be matroid of rank ≥ 2. Find a relationship between the homology of the complex of subsets contained in some cocircuit and the homology of the complex of subsets not containing any hyperplane.
   (e) Generalize parts (c) and (d) to greedoids (Chapter 8).
   (For Alexander duality see e.g. Munkres, 1984, p. 432. We require field coefficients here only to get a simpler statement avoiding cohomology.)

*7.44. Give an explicit combinatorial construction of a basis for the homology of
   (a) the reduced broken circuit complex $\overline{BC}_\omega(M)$ of a connected ordered matroid $M(S, \omega)$,
   (b) the dual complex G^* of a greedoid G (cf. section 8.6.C),
   (c) the order complex of a geometric semilattice, as defined by Wachs & Walker (1986).

7.45. Show that the following conditions are equivalent for a matroid M:
   (a) $IN(M)$ is homeomorphic to a sphere;
   (b) M is a direct product of circuits;
   (c) every independent set of corank 1 is contained in exactly two bases;

(d) $\tilde{\mu}(M^*) = 1$;

(e) $IN(M)$ has the homology of a sphere.

(For (c)$\Rightarrow$(b), see Provan & Billera, 1982.)

7.46. (a) Give an example of a matroid $M(S)$ and two orderings ω and ω' of S such that $\overline{BC}_\omega(M)$ is homomorphic to a sphere while $\overline{BC}_{\omega'}(M)$ is not.

*(b) Characterize those ordered matroids $M(S, \omega)$ for which $\overline{BC}_\omega(M)$ is homeomorphic to a sphere.

(It follows from Brylawski, 1971, that a necessary condition for (b) is that M is the matroid of a planar graph without a K_4 minor; see also Welsh, 1976, p. 237.)

Section 7.9

7.47. Show that the following conditions are equivalent for a semimodular lattice L:

(a) $\Delta(\bar{L})$ is homeomorphic to a sphere;

(b) L is Boolean;

(c) every interval in L of length 2 has cardinality 4;

(d) $\mu(x, y) = (-1)^{r(y) - r(x)}$, for all $x \le y$ in L.

If L is known to be geometric then show also that the following conditions are equivalent to the preceding ones:

(e) every coline is covered by exactly two copoints;

(f) $|\mu(\hat{0}, \hat{1})| = 1$;

(g) $\Delta(\bar{L})$ has the homology of a sphere.

7.48. Let L be a geometric lattice of rank r. For each basis $B \subseteq L^1$ let $\sum^B$ be the subcomplex of $\Delta(\bar{L})$ generated by the maximal chains $\mathbf{c}_{B,\pi}$, for all $\pi \in S_r$; see (7.43). Show the following.

(a) $\sum^B$ is homeomorphic to the $(r-2)$-sphere.

(b) $\Delta(\bar{L}) = \cup \sum^B$, with union over all nbc-bases B.

(c) The fundamental cycle of $\sum^B$ is equal to the elementary cycle ρ_B (up to sign, see (7.44)).

(d) If ρ_B is expressed in the basis $\{\rho_A : A \in \mathbf{nbc}\}$ of Theorem 7.9.3 then all coefficients must equal -1, 0, or $+1$.

7.49. (Björner, 1982) Let $\mathscr{B}$ be a neat base-family in a geometric lattice L of rank r, as defined in Exercise 7.15. Show that the elementary cycles $\{\rho_A : A \in \mathscr{B}\}$ form a basis for $H_{r-2}(L)$.

7.50. (Björner, 1982; Lindström, 1981. See also section 6.5 of White, 1987.) The set of bases $\mathscr{B}_L$ of a geometric lattice L has the structure of a simple matroid induced by linear independence of the elementary cycles ρ_B, $B \in \mathscr{B}_L$, in the free Abelian group $H_{r-2}(\bar{L})$.

Show the following.

(a) For every subset $\mathscr{F} \subseteq \mathscr{B}_L$, if $\{\rho_B : B \in \mathscr{F}\}$ is a basis of $H_{r-2}(\bar{L})$ then $\mathscr{F}$ is a (matroid) basis of $\mathscr{B}_L$.

(b) If L is the lattice of the 3-uniform matroid of size 6, then the matroid $\mathscr{B}_L$ is not regular.

(c) Deduce from (b) that the converse to (a) is in general false.

(d) $\mathscr{B}_L$ is 2-partitionable. (A matroid $M(S)$ is 2-*partitionable* if for every $x \in S$ there is a partition $S - x = S_1 \cup S_2, S_1 \cap S_2 = \varnothing$, such that $x \notin \bar{S}_1$ and $x \notin \bar{S}_2$.)

7.51. Let $M = M(S)$ be a simple matroid and L the corresponding geometric lattice
of flats. The non-spanning subsets of S form a simplicial complex $NS(M)$.
  (a) (Folkman, 1966) Show that $NS(M)$ and $\Delta(\bar{L})$ have isomorphic homology
      groups in all dimensions.
  (b) (Folkman, 1966) Deduce Theorem 7.9.1 from (a).
  (c) (Lakser, 1971) Show that $NS(M)$ and $\Delta(\bar{L})$ are of the same homotopy type.
7.52. (Björner, 1982; 1984; Wachs & Walker, 1986) Let M be an infinite matroid of
rank r and let L be the corresponding geometric lattice.
  (a) Show that $H_i(IN(M)) = 0$ for $i < r - 1$ and $H_i(\Delta(\bar{L})) = 0$ for $i < r - 2$.
  (b) Show that $H_{r-1}(IN(M))$ and $H_{r-2}(\Delta(\bar{L}))$ are free Abelian groups and
      determine their ranks.
  (c) Define shellability for infinite finite-dimensional simplicial complexes in a
      reasonable way. Prove the basic properties of this concept.
  (d) Show that $IN(M)$ and $\Delta(\bar{L})$ are shellable.
  (e) Define $BC_\omega(M)$ and $\overline{BC}_\omega(M)$ for a well-ordering ω of the ground set.
      Develop the basic theory of infinite broken circuit complexes.
  (f) Show that $\overline{BC}_\omega(M)$ is shellable and compute its homology.
7.53. (Baclawski, 1975; Björner, 1982; Orlik & Solomon, 1980) Let L be a geometric
lattice. For $1 \leq k \leq r$ let $D_k(L)$ be the Abelian group freely generated by all
k-chains $x_1 < x_2 < \ldots < x_k$ in $L - \hat{0}$. Put $D_0(L) = \mathbb{Z}$, and $D_k(L) = 0$ for all $k < 0$
and all $k > r$. For $2 \leq k \leq r$ define a group homomorphism $d_k^W : D_k(L) \to D_{k-1}(L)$
on the basis elements by

$$d_k^W(x_1 < x_2 < \ldots < x_k) = \sum_{i=1}^{k-1} (-1)^i (x_1 < \ldots < \hat{x}_i < \ldots < x_k),$$

and extend d_k^W linearly to all of $D_k(L)$. Let $d_k^W = 0$ for all other k. Then
$d_{k-1}^W \circ d_k^W = 0$, for all $k \in \mathbb{Z}$ (check this). The homology of this algebraic
chain complex,

$$H_k^W(L) = \text{Ker } d_k^W / \text{Im } d_{k+1}^W,$$

is the *Whitney homology* of L.
  (a) Show that

$$H_k^W(L) \cong \begin{cases} \bigoplus_{x \in L - \hat{0}} H_{k-2}(\hat{0}, x) & \text{if } k \neq 0 \\ \mathbb{Z} & \text{if } k = 0. \end{cases}$$

  (b) Conclude that the definition (7.45) of Whitney homology is equivalent to
      the one given here.

Section 7.10

7.54. (Orlik, Solomon & Terao, 1984; Jambu & Terao, 1989) Let L be a geometric
lattice and e an atom that is not an isthmus. Let $L - e$ and L/e respectively
denote the geometric lattices of the deletion of e and of the contraction by e.
Show that there exist linear maps giving short exact sequences of algebras:
  (a) $0 \to \mathscr{A}(L - e) \to \mathscr{A}(L) \to \mathscr{A}(L/e) \to 0$;
  (b) $0 \to H^W(L - e) - H^W(L) \to H^W(L/e) \to 0$.
7.55. (Björner & Ziegler, 1991) Let $RC_\pi(L)$ be a rooted complex in a geometric
lattice L, as defined in Exercise 7.16. Show that $\{\bar{e}_A : A \in RC_\pi(L)\}$ is a linear

basis for the Orlik–Solomon algebra $\mathscr{A}(L)$. (Hint: An alternative to a direct argument is to use Theorem 7.10.2(i) together with Exercises 7.16(f) and 7.49.)

References

Abels, H. (1989). The gallery distance of flags, Preprint 89-029, University of Bielefeld.

Abels, H. (1990). The geometry of the chamber system of a semimodular lattice, Preprint 90-001, University of Bielefeld.

Baclawski, K. (1975). Whitney numbers of geometric lattices, *Adv. Math.* **16**, 125–38.

Baclawski, K. (1980). Cohen–Macaulay ordered sets, *J. Algebra* **63**, 226–58.

Baclawski, K. (1982). Cohen–Macaulay connectivity and geometric lattices, *Europ. J. Comb.* **3**, 293–305.

Barcelo, H. (1990). On the action of the symmetric group on the free Lie algebra of the partition lattice, *J. Comb. Theory Ser. A* **55**, 93–129.

Barcelo, H. & Bergeron, N. (1990). The Orlik–Solomon algebra on the partition lattice and the free Lie algebra, *J. Comb. Theory Ser. A* **55**, 80–92.

Beissinger, J. S. (1982). On external activity and inversions in trees, *J. Comb. Theory, Ser. B* **33**, 87–92.

Björner, A. (1979). Homology of matroids, Preprint, Mittag–Leffler Institute. This reference is a preprint version of sections 7.2–7.5 of this chapter.

Björner, A. (1980a). Shellable and Cohen–Macaulay partially ordered sets, *Trans. Amer. Math. Soc.* **260**, 159–183.

Björner, A. (1980b). Some matroid inequalities, *Discrete Math.* **31**, 101–3.

Björner, A. (1982). On the homology of geometric lattices, *Algebra Universalis* **14**, 107–28.

Björner, A. (1984). Some combinatorial and algebraic properties of Coxeter complexes and Tits buildings, *Adv. Math.* **52**, 173–212.

Björner, A. (1991). Topological Methods, in R. Graham, M. Grötschel & L. Lovász (eds), *Handbook of Combinatorics*. North-Holland (to appear).

Björner, A., Frankl, P. & Stanley, R. P. (1987). The number of faces of balanced Cohen–Macaulay complexes and a generalized Macaulay theorem, *Combinatorica* **7**, 23–34.

Björner, A., Garsia, A. M. & Stanley, R. P. (1982). An introduction to Cohen–Macaulay partially ordered sets, in I. Rival (ed.), *Ordered Sets*, pp. 583–615. Reidel, Dordrecht.

Björner, A., Las Vergnas, M., Sturmfels, B., White, N., & Ziegler, G. (1991). *Oriented Matroids*, Cambridge University Press.

Björner, A. & Stanley, R. P. (1985). Unpublished.

Björner, A. & Ziegler, G. M. (1991). Broken circuit complexes: Factorizations and generalizations, *J. Comb. Theory Ser. B* **51**, 96–126..

Bourbaki, N. (1970). *Algèbre, Chapitres 1 à 3*, Hermann, Paris. English translation (1974): *Algebra, Chapters 1–3*, Addison-Wesley, Reading, Mass.

Brylawski, T. (1971). A combinatorial model for series–parallel networks, *Trans. Amer. Math. Soc.* **154**, 1–22.

Brylawski, T. (1977a). The broken-circuit complex, *Trans. Amer. Math. Soc.* **234**, 417–33.

Brylawski, T. (1977b). Connected matroids with the smallest Whitney numbers, *Discrete Math.* **18**, 243–52.

Brylawski, T. & Oxley, J. (1980). Several identities for the characteristic polynomial of a combinatorial geometry, *Discrete Math.* **31**, 161–70.

Brylawski, T. & Oxley, J. (1981). The broken-circuit complex: Its structure and factorizations, *Europ. J. Comb.* **2**, 107–21.

Calderbank, A. R., Hanlon, P. & Robinson, R. W. (1986). Partitions into even and odd block size and some unusual characters of the symmetric groups, *Proc. London Math. Soc.* (3) **53**, 288–320.

Cordovil, R. (1985). On simplicial matroids and Sperner's Lemma, in L. Lovász & A. Recski (eds), *Matroid Theory and its Applications* (Szeged, 1982), Colloq. Math. Soc. János Bolyai 40, pp. 97–105. North-Holland, Amsterdam and Budapest.

Crapo, H. H. (1969). The Tutte polynomial, *Aequationes Math.* **3**, 211–29.

Danaraj, G. & Klee, V. (1974). Shellings of spheres and polytopes, *Duke Math. J.* **41**, 443–51.

Danaraj, G. & Klee, V. (1978). Which spheres are shellable?, in B. Alspach *et al.* (eds), *Algorithmic Aspects of Combinatorics*, Ann. Discrete Math. **2**, pp. 33–52. North-Holland.

Dawson, J. E. (1981). A construction for a family of sets and its application to matroids, in Lecture Notes in Mathematics **884**, pp. 136–47. Springer-Verlag, Berlin.

Dawson, J. E. (1982). Matroid bases, opposite families and some related algorithms, in Lecture Notes in Mathematics **952**, pp. 225–38. Springer-Verlag, Berlin.

Dawson, J. E. (1984). A collection of sets related to the Tutte polynomial of a matroid, in Lecture Notes in Mathematics **1073**, pp. 193–204. Springer-Verlag, Berlin.

Dowling, T. A. (1980). On the independent set numbers of a finite matroid, in Ann. Discrete Math. **8**, pp. 21–8. North-Holland.

Dowling, T. A. & Wilson, R. M. (1974). The slimmest geometric lattices, *Trans. Amer. Math. Soc.* **196**, 203–15.

Folkman, J. (1966). The homology groups of a lattice, *J. Math. Mech.* **15**, 631–6.

Gel'fand, I. M. & Zelevinsky, A. V. (1986). Algebraic and combinatorial aspects of the general theory of hypergeometric functions, *Funct. Anal. and its Applic.* **20**, 183–97.

Gessel, I. & Wang, D.-L. (1979). Depth-first search as a combinatorial correspondence, *J. Comb. Theory Ser. A* **26**, 308–13.

Greene, C. & Zaslavsky, T. (1983). On the interpretation of Whitney numbers through arrangements of hyperplanes, zonotopes, non-Radon partitions, and orientations of graphs, *Trans. Amer. Math. Soc.* **280**, 97–126.

Grünbaum, B. (1967). *Convex Polytopes*, Wiley-Interscience, London.

Halsey, M. D. (1987). Line closed combinatorial geometries, *Discrete Math.* **65**, 245–8.

Hanlon, P. (1984). The characters of the wreath product group acting on the homology groups of the Dowling lattices, *J. Algebra* **91**, 430–63.

Hanlon, P. (1991). The generalized Dowling lattices, *Trans Amer. Math. Soc.* (to appear).

Heron, A. P. (1972). Matroid polynomials, in D. J. A. Welsh & D. R. Woodall (eds), *Combinatorics*, pp. 164–202. Institute of Mathematics and Its Applications, Southend-on-Sea, U.K.

Hibi, T. (1989). What can be said about pure 0-sequences? *J. Comb. Theory Ser. A* **50**, 319–22.

Jambu, M. & Leborgne, D. (1986). Fonction de Möbius et arrangements d'hyperplans, *C.R. Acad. Sci. Paris* **303**, 311–14.

Jambu, M. & Terao, H. (1989). Arrangements of hyperplanes and broken-circuits, in R. Randell (ed.), *Singularities*, (Iowa City, 1986), Contemporary Mathematics **90**, pp. 147–62. American Mathematical Society, Providence, Rhode Island.

Lakser, H. (1971). The homology of a lattice, *Discrete Math.* **1**, 187–92.

Laurent, M. & Deza, M. (1989). Bouquets of geometric lattices: Some algebraic and topological aspects, *Discrete Math.* **75**, 279–313.

Lehrer, G. I. (1987). On the Poincaré series associated with Coxeter group actions on complements of hyperplanes, *J. London Math. Soc.* **36**, 275–94.

Lehrer, G. I. & Solomon, L. (1986). On the action of the symmetric group on the cohomology of the complement of its reflecting hyperplanes, *J. Algebra* **104**, 410–24.

Lindström, B. (1981). Matroids on the bases of simple matroids, *Europ. J. Comb.* **2**, 61–3.

Mahoney, C. (1985). On the unimodality of the independent set numbers of a class of matroids, *J. Comb. Theory Ser. B* **39**, 77–85.

Mason, J. H. (1972). Matroids: Unimodal conjectures and Motzkin's theorem, in D. J. A. Welsh & D. R. Woodall (eds), *Combinatorics*, pp. 207–21. Institute of Mathematics and Its Applications, Southend-on-Sea, U.K.

Maurer, S. B. (1973). Matroid basis graphs I, *J. Comb. Theory Ser. B* **14**, 216–40.

McMullen, P. (1970). The maximum number of faces of a convex polytope, *Mathematika* **17**, 179–84.

Munkres, J. R. (1984). *Elements of Algebraic Topology*, Addison-Wesley, Menlo Park, California.

Orlik, P. (1989). *Introduction to Arrangements*, C.B.M.S. Regional Conference Series in Mathematics **72**, American Mathematical Society, Providence, Rhode Island.

Orlik, P. & Solomon, L. (1980). Combinatorics and topology of complements of hyperplanes, *Invent. Math.* **56**, 167–89.

Orlik, P., Solomon, L. and Terao, H. (1984). Arrangements of hyperplanes and differential forms, in C. Greene (ed.), *Combinatorics and Algebra* (Boulder, 1983), Contemporary Mathematics **34**, American Mathematical Society, Providence, Rhode Island.

Provan, J. S. (1977). Decomposition, shellings and diameters of simplicial complexes and convex polyhedra, Thesis, Cornell University, Ithaca, New York.

Provan, J. S. & Billera, L. J. (1980). Decompositions of simplicial complexes related to diameters of convex polyhedra, *Math. Oper. Res.* **5**, 576–94.

Provan, J. S. & Billera, L. J. (1982). Leontief substitution systems and matroid complexes, *Math. Oper. Res.* **7**, 81–7.

Purdy, G. (1982). The independent sets of rank k of a matroid, *Discrete Math.* **38**, 87–91.

Purtill, M. (1986). Unpublished.

Rota, G.-C. (1964). On the foundations of combinatorial theory: I, Theory of Möbius functions, *Z. Wahrscheinlichkeitstheorie und Verw. Gebiete* **2**, 340–68.

Rotman, J. J. (1985). Homology groups of Steiner systems. *J. Algebra* **92**, 128–49.

Sagan, B. E. (1986). Shellability of exponential structures, *Order* **3**, 47–54.

Stanley, R. P. (1971). Modular elements of geometric lattices, *Algebra Universalis* **1**, 214–17.

Stanley, R. P. (1972). Supersolvable lattices, *Algebra Universalis* **2**, 197–217.

Stanley, R. P. (1974). Finite lattices and Jordan–Hölder sets, *Algebra Universalis* **4**, 361–71.

Stanley, R. P. (1977). Cohen–Macaulay complexes, in M. Aigner (ed.), *Higher Combinatorics*, pp. 51–62. Reidel, Dordrecht and Boston.

Stanley, R. P. (1979). Balanced Cohen–Macaulay complexes, *Trans. Amer. Math. Soc.* **249**, 139–57.

Stanley, R. P. (1982). Some aspects of groups acting on finite posets, *J. Comb. Theory Ser. A* **32**, 132–61.

Stanley, R. P. (1983). *Combinatorics and Commutative Algebra*, Progress in Mathematics **41**, Birkhäuser, Boston.

Stanley, R. P. (1986). *Enumerative Combinatorics*, Vol. I, Wadsworth & Brooks/Cole, Monterey, California.

Stanley, R. P. (1987). Unpublished.

Wachs, M. L. (1986). Quotients of Coxeter complexes and buildings with linear diagram, *Europ. J. Comb.* **7**, 75–92.

Wachs, M. L. & Walker, J. W. (1986). On geometric semilattices, *Order* **2**, 367–85.

Wegner, G. (1984). Kruskal–Katona's theorem in generalized complexes, in *Finite and Infinite Sets*, Vol. 2, Coll. Math. Soc. János Bolyai **37**, pp. 821–7. North-Holland, Amsterdam and Budapest.

Welsh, D. J. A. (1976). *Matroid Theory*, Academic Press, London.

White, N. (ed.) (1986). *Theory of Matroids*, Cambridge University Press.

White, N. (ed.) (1987). *Combinatorial Geometries*, Cambridge University Press.

Whitney, H. (1932). A logical expansion in mathematics, *Bull. Amer. Math. Soc.* **38**, 572–9.

Wilf, H. S. (1976). Which polynomials are chromatic? in *Theorie Combinatorie*, Tomo 1 (Rome, 1973), pp. 247–56. Accad. Naz. Lincei, Rome.

Wilf, H. S. (1977). A note on $P(-\lambda; G)$, *J. Comb. Theory Ser. B* **22**, 296.

Zaslavsky, T. (1981). The slimmest arrangements of hyperplanes: II. Basepointed geometric lattices and Euclidean arrangements, *Mathematika* **28**, 169–90.

Zaslavsky, T. (1983). The slimmest arrangements of hyperplanes. I: Geometric lattices and projective arrangements, *Geom. Dedicata* **14**, 243–59.

Zelevinsky, A. V. (1990). Geometry and combinatorics related to vector partition functions, in *Topics in Algebra*), Banach Center Publications, Vol. 26, Part 2, PWN-Polish Scientific Publishers, Warsaw, pp. 501–10.

8

Introduction to Greedoids

ANDERS BJÖRNER and GÜNTER M. ZIEGLER

8.1. Introduction

Greedoids were invented around 1980 by B. Korte and L. Lovász. Originally, the main motivation for proposing this generalization of the matroid concept came from combinatorial optimization. Korte and Lovász had observed that the optimality of a 'greedy' algorithm could in several instances be traced back to an underlying combinatorial structure that was not a matroid – but (as they named it) a 'greedoid'. In subsequent research greedoids have been shown to be interesting also from various non-algorithmic points of view.

The basic distinction between greedoids and matroids is that greedoids are modeled on the *algorithmic construction* of certain sets, which means that the *ordering of elements* in a set plays an important role. Viewing such ordered sets as words, and the collection of words as a formal language, we arrive at the general definition of a greedoid as a finite language that is closed under the operation of taking initial substrings and satisfies a matroid-type exchange axiom. It is a pleasant feature that greedoids can also be characterized in terms of set systems (the unordered version), but the language formulation (the ordered version) seems more fundamental.

Consider, for instance, the algorithmic construction of a spanning tree in a connected graph. Two simple strategies are: (1) pick one edge at a time, making sure that the current edge does not form a circuit with those already chosen; (2) pick one edge at a time, starting at some given node, so that the current edge connects a visited node with an unvisited node. These well known strategies are used respectively in Kruskal's and in Prim's minimal spanning tree algorithms. In both cases, the collection of feasible sequences of edges, i.e. sequences that are generated by the allowed strategy, forms a greedoid. However, in the first case, but not in the second, any permutation of a feasible sequence of edges is also feasible, so that ordering is irrelevant.

This is so because the first greedoid, but not the second, is a matroid. The optimality of Prim's algorithm, which is not explained by matroid theory, is indeed covered by greedoid theory.

In this chapter we shall give an introduction to greedoids. Our aim is to explain the basic ideas and to give a few glimpses of more specialized topics. In spite of its youth the subject is already large enough to make a complete account impossible in the available space. Due to the space limitation we have frequently chosen to omit detailed proofs, particularly when good proofs exist in the literature. Also, to unburden the main text, all references to original papers and additional comments are gathered in the 'Notes and Comments' at the end of the chapter.

Here is an outline of the contents. Section 8.2 discusses the axiomatics of greedoids and explains the equivalence of the ordered and unordered versions of the concept. Many of the basic definitions in the area are given here. In particular, the important class of interval greedoids is defined.

Many examples of greedoids are described in section 8.3. One of the interesting features of the greedoid concept is that it admits such a variety of combinatorial examples in addition to matroids: branchings in graphs, order ideals in posets, convex hull closures in Euclidean and other spaces, Gaussian elimination sequences, retract sequences, and many more.

In section 8.4 various structural properties of greedoids as combinatorial systems are discussed. Just like matroids, greedoids have cryptomorphic descriptions in terms of a rank function and a closure operator. Deletion, contraction, and some other operations on greedoids are defined, as well as a suitable notion of connectivity.

Connections with combinatorial optimization are presented in section 8.5. For a certain kind of objective function, the greedy algorithm is optimal over a greedoid. In fact, greedoids can be characterized in terms of this algorithmic property. Examples of greedoid optimization include, e.g., Dijkstra's shortest path algorithm. Linear objective functions pose special problems, which are briefly discussed.

Section 8.6 discusses a certain polynomial that is associated with every greedoid. It is a greedoid version of the Tutte polynomial of matroid theory. The polynomial has applications of an algorithmic and of a probabilistic nature. For instance, it is possible to express in terms of this polynomial the probability that rank will not decrease if elements are independently deleted with probability p. Finally, there is a brief discussion of what aspects of matroid duality can be said to exist for general greedoids.

Antimatroids form a special class of interval greedoids with considerable additional structure. They are discussed in section 8.7 as dual objects to convex geometries. Among the interval greedoids, matroids and antimatroids are from several points of view opposite classes. Each is connected with a closure operator, which for matroids abstracts *linear span* and for antimatroids abstracts *convex hull* in Euclidean spaces.

In section 8.8 the connections between greedoids and posets (particularly lattices) are discussed in some detail. Each greedoid has a poset of flats, which in general is not a lattice. For interval greedoids the poset of flats is a semimodular lattice, and every finite semimodular lattice arises in this way.

The following additional topics are briefly discussed in section 8.9: (1) the characterization of certain classes of greedoids by excluded minors; (2) the maximum number of feasible pivots needed to move from one basis to any other basis in a greedoid; (3) examples of greedoid languages that allow repetition of letters within feasible words.

8.2. Definitions and Basic Facts

8.2.A. Ordered and Unordered Versions

There are two equivalent definitions of greedoids, one as set systems and the other as languages. We will start by defining and discussing greedoids as set systems. The equivalence of the two approaches will be heavily used later by freely choosing, depending on context, whatever formulation seems more convenient or natural.

In the following, we will work over a finite ground set E. The set of all subsets of E will be denoted by 2^E, and a *set system* over E is a non-empty family $\mathscr{F} \subseteq 2^E$.

8.2.1. Definition. A *greedoid* is a pair $(E, \mathscr{F})$, where $\mathscr{F} \subseteq 2^E$ is a set system satisfying the following conditions.

(G1) For every non-empty $X \in \mathscr{F}$ there is an $x \in X$ such that $X - x \in \mathscr{F}$.

(G2) For $X, Y \in \mathscr{F}$ such that $|X| > |Y|$, there is an $x \in X - Y$ such that $Y \cup x \in \mathscr{F}$.

The axiom (G2) is the usual matroid exchange axiom. In fact, every matroid is a greedoid, and a greedoid is a matroid exactly if it is *hereditary*, that is, if the axiom

(M1) If $X \in \mathscr{F}$ and $Y \subseteq X$, then $Y \in \mathscr{F}$.

is satisfied. (M1) is a strengthening of (G1); (M1) and (G2) together define a matroid.

Many examples of greedoids that are not matroids will be given in the next section. To i-llustrate the definition, let us now look at one of these.

Let $\Gamma = (V, E, r)$ be a rooted graph, and let $\mathscr{F}$ be the family of subtrees in Γ that contain the root node r. We think of these subtrees as edge sets, so $\mathscr{F} \subseteq 2^E$. Now, if $X \neq \varnothing$ is such a tree, then it must have at least one leaf

Figure 8.1.

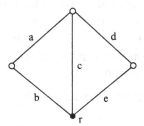

other than r, and if x is the edge adjacent to such a leaf then also $X - x \in \mathscr{F}$. Also, the cardinality $|X|$ of a subtree X equals the number of vertices other than r that are reached by X. Consequently, if $|X| > |Y|$ there must be some node $v \in V - r$ that is reached by X but not by Y. Follow the unique path in X from r to v and let x be the first edge of that path with a vertex not in Y. Then clearly $Y \cup x$ is also a subtree in $\mathscr{F}$. We have verified axioms (G1) and (G2), so $(E, \mathscr{F})$ is a greedoid. The greedoids that arise in this way (called 'undirected branching greedoids') will be further discussed in section 8.3.C. For the particular greedoid given by the rooted graph (V, E, r) in Figure 8.1 we observe e.g. that $\{b\}, \{a, c\}, \{b, c, d\} \in \mathscr{F}$ and $\{a\}, \{a, d\}, \{a, b, c\} \notin \mathscr{F}$.

The axiom (G1) states that $\mathscr{F}$ is an *accessible* set system. It implies – because E is finite and $\mathscr{F}$ non-empty – that $\mathscr{F}$ contains the empty set. In fact, by (G1) every $X \in \mathscr{F}$ can be dismantled by successively removing elements to get a sequence $\varnothing = X_0 \subset X_1 \subset \ldots \subset X_k = X$, where every X_i is a set in $\mathscr{F}$ of cardinality i, $0 \leq i \leq k$. But the same also follows from (G2): if we assume $\varnothing \in \mathscr{F}$, then repeated application of (G2) implies the existence of a sequence $\varnothing = X_0 \subset X_1 \subset \ldots \subset X_k = X$, where $X_i \in \mathscr{F}$ and $|X_i| = i$ for $1 \leq i \leq k$. Hence, in Definition 8.2.1, (G1) could be replaced by the weaker axiom.

(G1') $\varnothing \in \mathscr{F}$.

Just as for matroids, it is again sufficient (using (G1)) to require the exchange property of (G2) only for $|X| = |Y| + 1$:

(G2') For $X, Y \in \mathscr{F}$, $|X| = |Y| + 1$, there is an $x \in X - Y$ such that $Y \cup x \in \mathscr{F}$.

The axioms (G1) and (G2') together define greedoids as well as (G1') and (G2). However, (G1') and (G2') together clearly do not suffice.

The following terminology will be used. For greedoids the sets in $\mathscr{F}$ are called *feasible* (rather than '*independent*'). As usual, the matroid exchange axiom (G2) implies that the (inclusion-wise) maximal feasible sets, the *bases*, have the same size r; $r = r(\mathscr{F})$ is called the *rank* of the greedoid $(E, \mathscr{F})$. For an arbitrary subset A of the ground set E we define its *rank* by $r(A) = \max\{|X|: X \subseteq A, X \in \mathscr{F}\}$. Thus A is feasible if and only if $r(A) = |A|$, and it is a basis if and only if $r(A) = |A| = r(\mathscr{F})$. The characteristic properties of the greedoid rank function will be discussed in section 8.4.A.

A *basis of a subset* $A \subseteq E$ is a maximal feasible subset of A. Equivalently, this is an $X \in \mathscr{F}$ such that $X \subseteq A$ and $r(X) = r(A)$, because the exchange axiom (G2) implies that every maximal feasible subset of A has size $r(A)$. In fact, for any set system $(E, \mathscr{F})$ the property

(B) For any subset $A \subseteq E$ all maximal feasible subsets of A have the same cardinality

is implied by the exchange axiom (G2). On the other hand, (B) together with (G1) does not imply (G2), as shown by $E = \{1, 2, 3\}$ and $\mathscr{F} = 2^E - \{\{1, 3\}, \{2, 3\}\}$. See also Exercise 8.1.

A *coloop* in a greedoid $(E, \mathscr{F})$ is an element $x \in E$ that is contained in every basis, and a *loop* is an element that is contained in no basis. If x is a loop then $r(\{x\}) = 0$, but not conversely. Another difference from the matroid case is that $r(\{x\}) = 0$ is possible for a coloop x. These facts are particularly easy to visualize for branching greedoids, cf. section 8.3.C. We will sometimes write just $\cup \mathscr{F}$ for the union of all feasible sets. Clearly, x is a loop if and only if $x \in E - \cup \mathscr{F}$. (Note that in matroid theory a coloop is often called an 'isthmus').

We will now describe the equivalent 'ordered' version of greedoids, in terms of exchange languages. For the finite ground set E, let E^* denote the free monoid of all *words* over the alphabet E. We use Greek letters α, β, γ, ... for words in E^* and Latin letters x, y, z, ... for 'letters', i.e. elements of E. The concatenation of α and β (the string α followed by the string β) will be denoted by $\alpha\beta$. For any word $\alpha \in E^*$, $|\alpha|$ denotes the *length* of α, i.e. the number of (not necessarily distinct) letters in α. The *support* $\tilde{\alpha}$ of α is the set of letters in α. A word α is called *simple* if it does not contain any letter more than once, i.e. if $|\alpha| = |\tilde{\alpha}|$.

A *language* $\mathscr{L}$ over E is a non-empty set $\mathscr{L} \subseteq E^*$ of words over the alphabet E; it is called *simple* if every word in $\mathscr{L}$ is simple. Every simple language over a finite set E is again finite. Let E_s^* denote the (finite) set of simple words in E^*. By the *support* $\tilde{\mathscr{L}}$ of the language $\mathscr{L}$ we mean the set system $\tilde{\mathscr{L}} = \{\tilde{\alpha}: \alpha \in \mathscr{L}\}$.

8.2.2. Definition. A *greedoid language* over a finite ground set E is a pair $(E, \mathscr{L})$, where $\mathscr{L}$ is a simple language $\mathscr{L} \subseteq E_s^*$ satisfying the following conditions.

(L1) If $\alpha = \beta\gamma$ and $\alpha \in \mathscr{L}$, then $\beta \in \mathscr{L}$, i.e. every beginning section of a word in $\mathscr{L}$ is again in $\mathscr{L}$;

(L2) If α, $\beta \in \mathscr{L}$ and $|\alpha| > |\beta|$, then α contains a letter x such that $\beta x \in \mathscr{L}$.

Here (L1) states that $\mathscr{L}$ is a (*left*) *hereditary* language; (L2) is an exchange axiom. Again, it would be sufficient to require that (L2) holds for $|\alpha| = |\beta| + 1$.

The words in $\mathcal{L}$ are called *feasible*. The maximal words in $\mathcal{L}$ (that is, the words that do not have extensions in $\mathcal{L}$) are called *basic words*. We call a language *pure* if all its maximal words have the same length. In particular, as a consequence of exchange axiom (L2), greedoid languages are pure. The common length of all basic words is called the *rank* of the greedoid $(E, \mathcal{L})$.

Let us illustrate Definition 8.2.2 by again considering a rooted graph $\Gamma = (V, E, r)$, as in the discussion following Definition 8.2.1. A string $x_1 x_2 \dots x_k$ of distinct edges $x_i \in E$ will be considered feasible if the subgraph $\{x_1, x_2, \dots, x_{i-1}\}$ connects the root node r to one endpoint of x_i but not the other, for $1 \leq i \leq k$. It is instructive to check that the language $\mathcal{L}$ of such feasible strings is a greedoid language. For instance, in Figure 8.1 we find that b, ca, $cbd \in \mathcal{L}$, but a, ac, $bdc \notin \mathcal{L}$.

Definitions 8.2.1 and 8.2.2 are tied together by:

8.2.3. Proposition. *Greedoids and greedoid languages are equivalent in the following sense.*

(i) *If $(E, \mathcal{L})$ is a greedoid language, then the support $\tilde{\mathcal{L}}$ is a greedoid.*

(ii) *If $(E, \mathcal{F})$ is a greedoid, then*

$$\mathcal{L}(\mathcal{F}) = \{x_1, x_2 \dots x_k \in E_s^* : \{x_1, x_2, \dots, x_i\} \in \mathcal{F} \text{ for } 1 \leq i \leq k\}$$

is a greedoid language.

(iii) *Furthermore, $\mathcal{L}(\tilde{\mathcal{L}}) = \mathcal{L}$ and $\tilde{\mathcal{L}}(\mathcal{F}) = \mathcal{F}$, so these constructions give a one-to-one correspondence between greedoids and greedoid languages.*

The verification of parts (i) and (ii) is straightforward and very easy. For part (iii), the only point that requires a small argument is the inclusion $\mathcal{L}(\tilde{\mathcal{L}}) \subseteq \mathcal{L}$, which follows by induction on the length of words from the exchange axiom (L2) together with the simplicity of the language $\mathcal{L}$.

In view of this equivalence between greedoids (as set systems) and greedoid languages, the two concepts will from now on be used interchangeably. For a greedoid G we will freely write $G = (E, \mathcal{F}) = (E, \mathcal{L})$, and if the ground set is clear from the context G will often be denoted by just $\mathcal{F}$ or $\mathcal{L}$. (If the ground set is not given by context, one can always take $E = \cup \mathcal{F}$ or $E = \cup \tilde{\mathcal{L}}$ to recover it, except for loops.)

It is often convenient to think of a greedoid $(E, \mathcal{F})$ as a poset $(\mathcal{F}, \subseteq)$, with the partial order given by inclusion. This poset has a least element $\varnothing$, and every unrefinable chain from $\varnothing$ to a maximal element B (i.e. a basis) has the same length $r(\mathcal{F}) = |B|$. More generally, if $A \subseteq C$, where A, $C \in \mathcal{F}$, then every unrefinable chain $A = A_0 \subset\cdot A_1 \subset \dots \subset\cdot A_k = C$ in $\mathcal{F}$ has the same length $k = |C - A|$, since A can be repeatedly augmented from C one element at a time.

For instance, let $E = \{a, b, c, d\}$ and consider the greedoid $\mathcal{F} = 2^E -$

Figure 8.2.

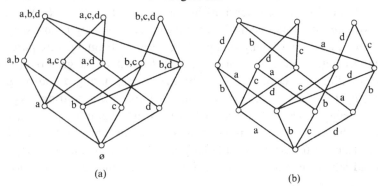

(a)   (b)

$\{\{a, b, c, d\}, \{a, b, c\}, \{c, d\}\}$. (A systematic reason why this *is* a greedoid is given in Exercise 8.4.) The poset $\mathscr{F}$ is depicted in Figure 8.2a where for simplicity set brackets are omitted.

To see clearly the connection with the language version $(E, \mathscr{L})$ of $\mathscr{F}$, reformulate the poset $\mathscr{F}$ of Figure 8.2a into an abstract edge-labeled poset as in Figure 8.2b. Here each covering edge $X \subset\!\!\cdot\, Y$ is labeled by the single element of $Y - X$. Then the words in $\mathscr{L} = \mathscr{L}(\mathscr{F})$ can be read off as the sequences of labels along unrefinable chains starting at the bottom. For instance, the basic words beginning with '*b*' are: *bad*, *bcd*, *bda*, and *bdc*. Also, the feasible set corresponding to an element in this abstract labeled poset can be reconstructed as the set of labels on any unrefinable chain from the bottom to that element.

We remark that non-isomorphic greedoids can have isomorphic unlabeled posets of feasible sets (non-trivial examples are mentioned at the end of section 8.4.E; see also section 8.8.D). Hence, the edge labels as in Figure 8.2(b) are essential for such a poset representation of a greedoid.

8.2.B. Interval Greedoids and Antimatroids

The following 'interval property' characterizes a very large class of greedoids that covers many of the main examples (see section 8.3). Greedoids with this property, usually called *interval greedoids*, behave better than general greedoids in many respects. In some types of study the interval property has to be assumed to obtain meaningful results.

8.2.4. Definition. A greedoid $(E, \mathscr{F})$ has the *interval property* if $A \subseteq B \subseteq C$, $A, B, C \in \mathscr{F}$, $x \in E - C$, $A \cup x \in \mathscr{F}$, and $C \cup x \in \mathscr{F}$ imply that $B \cup x \in \mathscr{F}$. Equivalently, in terms of greedoid language $(E, \mathscr{L})$ this means that αx, $\alpha\beta\gamma x \in \mathscr{L}$ implies $\alpha\beta x \in \mathscr{L}$.

We observe that the greedoid in Figure 8.2 does not have the interval property, since e.g. $\emptyset$, $\{c\}$, and $\{b, c\}$ are feasible, and the first and third can be augmented by d but not the second. Clearly, every greedoid of rank less than three has the interval property.

The following exchange property characterization of interval greedoids is often useful.

8.2.5. Proposition. *A hereditary language* $(E, \mathscr{L})$ *is an interval greedoid if and only if it satisfies the following strong exchange property:*

(L2′) *If* α, $\beta \in \mathscr{L}$ *and* $|\alpha| > |\beta|$, *then* α *contains a subword* α' *of length* $|\alpha'| = |\alpha| - |\beta|$ *such that* $\beta\alpha' \in \mathscr{L}$.

Here a *subword* of $\alpha = x_1 x_2 \ldots x_n$ is a not necessarily consecutive substring of α, i.e. a word of the form $\alpha' = x_{i_1} x_{i_2} \ldots x_{i_k}$ with $1 \leq i_1 < i_2 < \ldots < i_k \leq n$. Obviously, the axiom (L2′) implies the regular exchange property (L2).

Proof. We will here prove that (L2′) implies the interval property. The proof of the converse will be postponed until section 8.8.C.

Suppose that αx, $\alpha\beta\gamma x \in \mathscr{L}$. Strong exchange gives, since $\mathscr{L}$ is simple, that $\alpha x \beta\gamma \in \mathscr{L}$. Since $\mathscr{L}$ is left hereditary, $\alpha\beta$, $\alpha x \beta \in \mathscr{L}$, and a second application of strong exchange yields $\alpha\beta x \in \mathscr{L}$. $\square$

A relatively special but very important class of interval greedoids is the class of antimatroids. Some of their special properties will be discussed in section 8.7.

8.2.6. Definition. A greedoid $(E, \mathscr{F})$ is called an *antimatroid* if it satisfies the following *interval property without upper bounds*: if $A \subseteq B$, $A, B \in \mathscr{F}$, $x \in E - B$, and $A \cup x \in \mathscr{F}$, then $B \cup x \in \mathscr{F}$. Equivalently, in terms of greedoid language, if αx, $\alpha\beta \in \mathscr{L}$ and $x \notin \tilde{\beta}$, then $\alpha\beta x \in \mathscr{L}$.

In many cases, the easiest way to recognize an antimatroid is via the following characterization.

8.2.7. Proposition. *Let* $\mathscr{F} \subseteq 2^E$ *be a set system. Then the following conditions are equivalent.*

(i) $(E, \mathscr{F})$ *is an antimatroid.*
(ii) $\mathscr{F}$ *is accessible and closed under union.*
(iii) $\emptyset \in \mathscr{F}$, *and* $\mathscr{F}$ *satisfies the following exchange axiom.*
(A) *For* X, $Y \in \mathscr{F}$ *such that* $X \nsubseteq Y$, *there is an* $x \in X - Y$ *such that* $Y \cup x \in \mathscr{F}$.

Proof. (ii) $\Rightarrow$ (iii). Suppose that $\mathscr{F}$ is accessible (i.e. satisfies axiom (G1)) and is closed under union (i.e. A, $B \in \mathcal{F}$ implies $A \cup B \in \mathcal{F}$). Let X, $Y \in \mathcal{F}$ such that $X \not\subseteq Y$. Accessibility means that we can find a sequence $\emptyset = X_0 \subset X_1 \subset \ldots \subset X_k = X$ such that $X_i \in \mathcal{F}$ and $|X_i| = i$, for $0 \le i \le k$. Let i be the least integer for which $X_i \not\subseteq Y$. Then $Y \cup X_i = Y \cup x \in \mathcal{F}$, where $x \in X_i - Y \subseteq X - Y$.

(iii)$\Rightarrow$(i). Axiom (A) implies axiom (G2), and axiom (G1') is assumed, so $\mathcal{F}$ is a greedoid. Suppose that A, B, $A \cup x \in \mathcal{F}$, $A \subset B$, and $A \cup x \not\subseteq B$. By axiom (A), the set B can be augmented from the set $(A \cup x) - B = \{x\}$, so $B \cup x \in \mathcal{F}$. This proves the interval property without upper bounds.

(i)$\Rightarrow$(ii). We leave this step, which is similar to the other two, as an exercise for the reader. $\square$

Notice that, since $\mathcal{F}$ is closed under union, every subset in an antimatroid has a unique basis.

The following result expresses some of the ways in which interval greedoids and antimatroids are related. The proof is a simple exercise with the interval property, with and without upper bounds.

8.2.8. Proposition. *Let $(E, \mathcal{F})$ be a greedoid. Then*

(i) *$(E, \mathcal{F})$ is an antimatroid if and only if it is an interval greedoid and has a unique basis;*

(ii) *$(E, \mathcal{F})$ is an interval greedoid if and only if the restriction to each feasible set $X \in \mathcal{F}$, meaning $\{Y \in \mathcal{F} : Y \subseteq X\}$, is an antimatroid.*

A greedoid $(E, \mathcal{F})$ is said to be *full* if $E \in \mathcal{F}$. It follows from the preceding result that if an antimatroid has no loops then it is full. In any case, an antimatroid has one and only one basis, namely $\cup \mathcal{F}$. We remark in this connection that the wealth of examples of antimatroids, all with only one basis, shows that greedoids cannot in general be reconstructed from or axiomatically characterized in terms of their set of bases. On the other hand, a greedoid is of course completely determined by its basic words.

For a general set system $\mathcal{F} \subseteq 2^E$ such that $\emptyset \in \mathcal{F}$, define its *accessible kernel* (with some mild abuse of set notation) by

$$A(\mathcal{F}) = \{\{x_1, \ldots, x_k\} \in \mathcal{F} : \{x_1, \ldots, x_i\} \in \mathcal{F} \text{ for all } 1 \le i \le k\},$$

or – recursively – by: $X \in A(\mathcal{F})$ iff $X \in \mathcal{F}$ and $X = \emptyset$ or there is an $x \in X$ such that $X - x \in A(\mathcal{F})$. The *hereditary closure* of a set system $\mathcal{F}$ is defined as

$$H(\mathcal{F}) = \{Y \subseteq E : Y \subseteq X \text{ for some } X \in \mathcal{F}\}.$$

Thus, for every greedoid $\mathcal{F}$, (G1) states that $A(\mathcal{F}) = \mathcal{F}$, and $\mathcal{F}$ is a matroid precisely when $H(\mathcal{F}) = \mathcal{F}$. But we note that in general for a greedoid $\mathcal{F}$, $H(\mathcal{F})$ need not be the collection of feasible sets of a greedoid.

8.3. Examples

In this section we shall survey some major classes of greedoids. Many other classes are known, and the reader will find new examples constructed in nearly every paper on the subject.

8.3.A. Matroids

As was remarked before, the independent sets of a matroid form the feasible sets of a greedoid. Much of the terminology for greedoids is adapted from matroid theory, so that there is no translation problem. In particular, the rank function and bases of a matroid and its associated greedoid coincide. Matroids are clearly interval greedoids. In fact, they can be characterized as greedoids satisfying the 'interval property without lower bounds': If $B \subseteq C$, $B, C \in \mathscr{F}$ and $x \in E - C$, then $C \cup x \in \mathscr{F}$ implies $B \cup x \in \mathscr{F}$. This is equivalent to the statement that $B \subseteq C$, $C \in \mathscr{F}$ implies $B \in \mathscr{F}$, i.e. that $\mathscr{F}$ is hereditary.

Matroids give rise to greedoids in more than one way. For example, the following construction produces 'twisted matroids'. Let $M = (E, \mathscr{I})$ be a matroid; choose an independent set $A \in \mathscr{I}$. We define a simple language $\mathscr{L}_{M,A}$ by $\mathscr{L}_{M,A} = \{a_1 \ldots a_k \in E_s^*: A \Delta \{a_1, \ldots, a_i\} \in \mathscr{I}$ for all $1 \leq i \leq k\}$. Here, $A \Delta B$ denotes the symmetric difference $(A - B) \cup (B - A)$. This language is clearly left hereditary and the exchange axiom (L2) can be checked. For $A = \emptyset$ we get the standard matroid greedoid $\mathscr{L} = \mathscr{L}_{M,\emptyset}$. However, $\mathscr{L}_{M,A}$ depends heavily on A, and is in general not an interval greedoid. The feasible sets of the twisted matroid $\mathscr{L}_{M,A}$ are the sets whose symmetric difference with A is independent. The basic words describe the ways to move from A to a basis of $E - A$ through a sequence of intermediate independent sets.

Some greedoids are related to matroids in the following way: the greedoid $G = (E, \mathscr{F})$ is a *slimming* of the matroid $M = (E, \mathscr{I})$ if G and M have the same set of bases (which implies that $\mathscr{F} \subseteq \mathscr{I}$). For instance, the twisted matroid $(E, \mathscr{L}_{M,A})$ defined above is a slimming of the direct sum of the free matroid on A and the restriction $M - A$. We will encounter another example in section 8.3.C.

8.3.B. Antimatroids

Antimatroids are in several ways 'opposite' to matroids. For example, while matroids are precisely the greedoids satisfying the interval property without lower bounds, antimatroids are precisely those characterized by the interval property without upper bounds. Also, whereas the matroid closure operator is characterized by the MacLane exchange axiom, the closure operator of antimatroids can be characterized by the opposite 'anti-exchange' axiom (see section 8.7.A).

We shall now describe several classes of antimatroids occurring 'in nature'. They are all easy to identify using the following criterion (Proposition 8.2.7): a set system $\mathscr{F} \subseteq 2^E$ is an antimatroid exactly if it is accessible and closed under union.

(1) Let $P = (E, \leq)$ be a finite partially ordered set and $\mathscr{F}$ the set of ideals of E (a subset $A \subseteq E$ is an *ideal* if $x \leq y \in A$ implies $x \in A$). Then $(E, \mathscr{F})$ is an antimatroid, the *poset greedoid* of P. In this case $\mathscr{F}$ is closed both under union and intersection and hence $(\mathscr{F}, \subseteq)$ forms a distributive lattice. Conversely, by a theorem of G. Birkhoff, every finite distributive lattice occurs this way. The basic words of the poset greedoid are the linear extensions of P.

(2) Let $\Gamma = (V, E, r)$ be a finite rooted graph, and let $V' = V - r$ be the set of vertices distinct from the root r. Then the *vertex search greedoid* of Γ is $(V', \mathscr{F})$, where $\mathscr{F}$ is given by $\mathscr{F} = \{X \subseteq V' : X \cup r$ is the vertex set of a connected subgraph of $G\}$. If Γ is connected, the basic words of this antimatroid correspond to the orderings in which nodes are visited by the standard search procedures starting at r.

(3) In the case of a rooted digraph $\Delta = (V, E, r)$, we again let V' be the set of vertices distinct from the root r, and $\mathscr{F} = \{X \subseteq V' : X \cup r$ is the vertex set of a tree in Δ that is directed away from $r\}$. Then $(E, \mathscr{F})$ is the *vertex search greedoid* of the digraph Δ.

(4) Let E be the vertex set of a tree and $\mathscr{F}$ the collection of complements of subtrees. Again, $(E, \mathscr{F})$ is an antimatroid, the *vertex pruning greedoid* of the tree. The same construction can be repeated for the edge set of a tree, to get the *edge pruning greedoid* of the tree, also an antimatroid.

(5) Both the vertex pruning and the edge pruning greedoids of trees are special cases of the *simplicial vertex pruning greedoids* of graphs. A vertex of a graph (V, E) is *simplicial* if all its neighbors are pairwise adjacent. Successive removal of simplicial vertices gives a hereditary language $(V, \mathscr{L})$ that is easily seen to be an antimatroid.

For $\mathscr{L}$ to be non-trivial a sufficient supply of simplicial vertices in (V, E) and its subgraphs is needed. This is guaranteed if the graph is *chordal*, meaning that no induced subgraph on k vertices is a k-cycle, for $k \geq 4$. Chordal graphs are characterized by the property that every induced subgraph has a simplicial vertex. It follows that the simplicial vertex pruning greedoid of a graph is full if and only if the graph is chordal.

(6) Our final example is crucial for the geometric interpretation of antimatroids. Let E be a finite subset of $\mathbb{R}^n$, and for $A \subseteq E$ define $\bar{A}$ to be the convex hull of A intersected with E. We call $A \subseteq E$ *convex* if $A = \bar{A}$, and define $\mathscr{F}$ to be the family $\mathscr{F} = \{X \subseteq E \mid E - X$ is convex$\}$. Then $(E, \mathscr{F})$ is an antimatroid, the *convex pruning greedoid* on E. This

example in fact generalizes in a straightforward way to oriented matroids with an appropriate notion of convexity.

Antimatroids have a lot of additional structure, which makes them quite special among greedoids. We will study antimatroids in greater detail in section 8.7, and proceed here to describe more general classes of greedoids.

8.3.C. Branching Greedoids

Let $\Delta = (V, E, r)$ be a finite rooted directed graph. Let $\mathscr{F}$ be the collection of edge sets of trees in Δ that contain the root and are directed away from it (such trees are called *branchings* or *arborescences*). Then $(E, \mathscr{F})$ is a greedoid, the *directed branching greedoid* (or *line search greedoid*) on Δ. Every non-empty tree in $\mathscr{F}$ has a leaf, which can be removed to get another tree in $\mathscr{F}$. This verifies axiom (G1). To check (G2), one observes that for two trees X and Y in Δ, $|X| > |Y|$ implies that X reaches a vertex v that Y does not reach ($|X|$ is the number of vertices of $V - r$ reached by X). Now the first arc along the path in X from r to v that reaches a vertex not reached by Y can be added to Y.

Figure 8.3 illustrates a particular rooted digraph Δ and the associated branching greedoid $\mathscr{F}$. This greedoid is of rank 2 with bases $\{a, d\}$, $\{a, c\}$, $\{b, c\}$, and $\{b, d\}$. The language is $\mathscr{L} = \{\varnothing, a, b, c, ac, ad, bc, bd, ca, cb\}$.

The rooted digraph in Figure 8.8a on p. 315 gives a branching greedoid of rank 6. That greedoid has two loops and two coloops, which shows that these greedoid concepts do not in this case have their standard graph-theoretic meaning.

It is clear, as was also observed with the example of Figure 8.2, that an analogous construction works for every finite rooted *undirected* graph $\Gamma = (V, E, r)$. The construction then yields the *undirected branching greedoid* $(E, \mathscr{F})$, where $\mathscr{F}$ is the set of trees in Γ that contain the root. We note that – ignoring the root – the graph Γ also gives rise to the graphic matroid. $(E, \mathscr{I})$. If Γ is connected, then the bases of the branching greedoid and of the graphic matroid are the same, namely the spanning trees of Γ. Hence,

Figure 8.3.

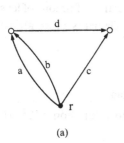

(a)   (b)

$(E, \mathscr{F})$ is a slimming of $(E, \mathscr{I})$, or equivalently, the hereditary closure of $\mathscr{F}$ is the graphic matroid.

The common algorithmic search procedures on a rooted graph (directed or not) visit the nodes one at a time so that the currently visited node (originally just r) is at each stage reached along some edge from a previously visited node. It is clear that the basic words of the associated branching greedoid record the sequences of *edges* generated by such search procedures. Similarly, the associated vertex search greedoid (defined in section 8.3.B) records the possible orders in which the *nodes* are reached.

Both directed and undirected branching greedoids are interval greedoids. If an edge x is a legal continuation at any given stage, that means that x leads from a visited node v to an unvisited node u. But then clearly x will remain a legal choice at a later stage if and only if u remains unvisited. This verifies the interval property.

For many purposes branching greedoids can serve as 'canonical examples' of greedoids. Being easy to represent graphically they play a role similar to the role graphic matroids play in matroid theory. However, branching greedoids are relatively well behaved and do not exhibit all the pathologies that can occur. For example, the intervals in the poset $(\mathscr{F}, \subseteq)$ are distributive lattices – given any two branchings $X \subseteq Y$ in Δ, the interval $[X, Y]$ of $\mathscr{F}$ corresponds to the order ideals of $Y \backslash X$, ordered by 'precedence along the paths in Y emanating from the root'. Greedoids with the property that all the intervals in $\mathscr{F}$ are distributive are called *local poset greedoids*. This name comes from the fact that $\mathscr{F}$ is a local poset greedoid if and only if the restriction of $\mathscr{F}$ to any feasible set is a poset greedoid ('restriction' will be defined in section 8.4.D as a straightforward generalization of the matroid operation). From this it is easy to see that all local poset greedoids are interval greedoids.

The following class of greedoids is closely related to the branching greedoids. Let (V, E, r) be a rooted undirected graph, and let $\mathscr{A}$ be the collection of edge sets of all connected subgraphs covering r. Then $(E, \mathscr{A})$ is a greedoid, in fact an antimatroid. An analogous construction associates an antimatroid with every rooted directed graph.

The construction of these greedoids is part of a much more general procedure: if $(E, \mathscr{F})$ is a greedoid and $\mathscr{A}$ is the collection of unions of feasible sets from $\mathscr{F}$, then $(E, \mathscr{A})$ is an antimatroid. The elements of $\mathscr{A}$ are called the *partial alphabets* of $(E, \mathscr{F})$. See Exercise 8.5.

8.3.D. Polymatroid Greedoids

A pair (E, f), consisting of a finite ground set and a function $f: 2^E \to \mathbb{N}$, is called a *polymatroid* if for all $X, Y \subseteq E$:

(PM1) $f(\emptyset) = 0$;

(PM2) $X \subseteq Y$ implies $f(X) \leq f(Y)$;

(PM3) $f(X \cap Y) + f(X \cup Y) \leq f(X) + f(Y)$.

Polymatroids are generalizations of matroids: f is the rank function of a matroid if in addition $f(X) \leq |X|$ for all $X \subseteq E$.

Polymatroids give rise to greedoids in the following way. Suppose that (E, f) is a polymatroid, and let

$$\mathscr{L} = \{x_1 x_2 \ldots x_k : f(\{x_1, \ldots, x_i\}) = i \text{ for } 1 \leq i \leq k\}.$$

Then $(E, \mathscr{L})$ is a greedoid, called a *polymatroid greedoid*. Such greedoids are local poset greedoids (as defined in section 8.3.C), and hence they have the interval property.

We give three examples of polymatroid greedoids that have been discussed before:

(1) If (E, f) is a matroid, then the polymatroid greedoid is the greedoid usually associated with this matroid (cf. section 8.3.A). In fact, the above construction will reconstruct any greedoid $\mathscr{L}$ from its rank function f.

(2) Let $\Gamma = (V, E, r)$ be a rooted undirected graph. For $X \subseteq E$, let $f(X)$ be the number of vertices in $V \backslash r$ covered by X. Then it is easy to check that (E, f) is a polymatroid. The associated polymatroid greedoid is the undirected branching greedoid of Γ, since if $\{x_1, \ldots, x_{i-1}\}$ is a tree in Γ containing the root r, then the same is true for $\{x_1, \ldots, x_i\}$ if and only if x_i covers exactly one additional vertex.

In contrast, directed branching greedoids are not in general polymatroid greedoids.

(3) Poset greedoids are polymatroid greedoids. The corresponding rank function measures the size $f(X)$ of the ideal in P generated by a subset X of $P = (E, \leq)$.

8.3.E. Faigle Geometries

For this class of greedoids the ground set is assumed to be partially ordered in a way that is suitably compatible with the greedoid structure. Both matroids and poset greedoids belong to this class.

A *Faigle geometry* is a triple $(E, \mathscr{L}, \leq)$ where $(E, \mathscr{L})$ is a greedoid language and $(E, \leq)$ a poset such that:

(F1) for $x_1 x_2 \ldots x_k \in \mathscr{L}$, $x_i \leq x_j$ implies $i \leqslant j$ (that is, the ordering of every word in $\mathscr{L}$ is compatible with the partial order on E);

(F2) if A, B are ideals in $(E, \leq)$ with $A \subseteq B$, then every $p \in A$ that occurs in every maximum length word in $B^* \cap \mathscr{L}$ also occurs in every maximum length word in $A^* \cap \mathscr{L}$.

We note that when the poset $(E, \leq)$ is an antichain (meaning that $x \leq y$ implies $x = y$), the axiom (F1) is vacuously fulfilled, and (F2) implies that $(E, \mathcal{L})$ is a matroid. In general, if $(E, \mathcal{L}, \leq)$ is a Faigle geometry then $(E, \mathcal{L})$ is an interval greedoid. However, not every interval greedoid admits the structure of a Faigle geometry (e.g. some branching greedoids do not).

There is a second rank function f on a Faigle geometry $(E, \mathcal{L}, \leq)$, which is in general different from the greedoid rank function r of $(E, \mathcal{L})$. This *ideal rank* function $f: 2^E \rightarrow \mathbb{N}$ is defined by

$$f(X) = r(I(X)), \text{ for } X \subseteq E,$$

where $I(X)$ denotes the ideal generated by X, i.e. $I(X) = \{y \in E: y \leq x$ for some $x \in X\}$.

The ideal rank f of a Faigle geometry is a polymatroid rank function, i.e. it satisfies axioms (PM1)–(PM3) of section 8.3.D. The corresponding polymatroid greedoid is also a Faigle geometry $(E, \mathcal{L}', \leq)$ over the same poset and with the same ideal rank function, and $\mathcal{L} \subseteq \mathcal{L}'$. However, in general $\mathcal{L} \neq \mathcal{L}'$, which shows that a Faigle geometry is not uniquely determined by its ideal rank function.

8.3.F. Retract Greedoids

A *retract* of a poset $(E, \leq)$ is a subposet $Q \subseteq E$ such that there is an order-preserving map $r: E \rightarrow Q$ with $r(x) = x$ for all $x \in Q$. In this case r is called a *retraction* of E to Q. We are going to consider retracts such that $|Q| = |E| - 1$, that is, Q corresponds to the deletion of a single element x from E. There are two cases; either x and $r(x)$ are not comparable, or one of x and $r(x)$ covers the other. In the second case Q is the poset obtained from E by deleting a meet or join irreducible element x (i.e. an element with a unique cover or a unique cocover). We will call this a *monotone* retract.

This situation gives rise to the following two greedoids: the *retract greedoid* $(E, \mathcal{L})$ given by

$$\mathcal{L} = \{x_1 x_2 \ldots x_k: \text{for } 1 \leq i \leq k, E - \{x_1, \ldots, x_i\} \text{ is a retract of}$$
$$E - \{x_1, \ldots, x_{i-1}\}\}$$

and the *dismantling greedoid* $(E, \mathcal{L}')$ defined by

$$\mathcal{L}' = \{x_1 x_2 \ldots x_k: \text{for } 1 \leq i \leq k, E - \{x_1, \ldots, x_i\} \text{ is a monotone retract of}$$
$$E - \{x_1, \ldots, x_{i-1}\}\}.$$

For instance, consider the poset shown in Figure 8.4. Here, $b \in \mathcal{L}$, $ab \notin \mathcal{L}$, and $acb \in \mathcal{L}$. Also, $cd \in \mathcal{L}'$, $ced \notin \mathcal{L}'$, and $cead \in \mathcal{L}'$.

This example shows that the greedoids $(E, \mathcal{L})$ and $(E, \mathcal{L}')$ in general fail to have the interval property. Also, they may have many loops. Observe that by definition $\mathcal{L}' \subseteq \mathcal{L}$.

Figure 8.4.

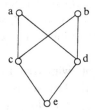

There is a straightforward generalization of retract sequences and dismantling sequences from posets (i.e. comparability graphs) to finite digraphs, which leads to a generalization of the corresponding greedoids. In this case, one works with digraphs with a loop at every vertex and defines retracts as above, using graph maps instead of order preserving maps. For monotone retracts, the additional requirement is that every vertex is mapped to an adjacent vertex.

The following construction gives a very general framework for retract greedoids. Let Φ be a set of mappings of a finite set E into itself, which is closed under composition and contains the identity (in other words, Φ is a submonoid of E^E). Call a subset $X \subseteq E$ a *retract* if $X = \phi(E)$ for some idempotent element $\phi \in \Phi$. Now, define a left hereditary language by

$$\mathscr{L} = \{x_1\, x_2\, ...\, x_k\colon \text{for } 1 \leq i \leq k,\ E - \{x_1, ..., x_k\} \text{ is a retract}\}.$$

Then $(E, \mathscr{L})$ is greedoid. Taking Φ to be the monoid of order preserving self-maps of a poset $(E, \leq)$ we get the special retract greedoids that were originally defined.

A general formulation of dismantling greedoids along similar lines is also possible.

8.3.G. Transposition Greedoids

A greedoid $(E, \mathscr{F})$ is said to have the *transposition property* if it satisfies the axiom

(TP) If A, $A \cup x$, $A \cup y \in \mathscr{F}$ and $A \cup x \cup y \notin \mathscr{F}$, then $A \cup x \cup B \in \mathscr{F}$ implies $A \cup y \cup B \in \mathscr{F}$, for all $B \subseteq E - (A \cup x \cup y)$.

If an accessible set system has the transposition property then it is a greedoid, but the converse does not hold. Hence, the axioms (G1) and (TP) form an axiom system for a proper subclass that we call *transposition greedoids*.

The best known way to prove that a retract greedoid actually *is* a greedoid is to verify (G1) and (TP). In fact, both retract greedoids and dismantling greedoids are transposition greedoids. Examples of greedoids that lack the transposition property can be found among the twisted matroids.

Let us now verify (TP) for an arbitrary interval greedoid $(E, \mathcal{F})$. In the given situation $A \cup y$ can be augmented from the larger set $A \cup x \cup B$ to a set $A \cup y \cup B' \in \mathcal{F}$, where $B' \subseteq B \cup x$ and $|B'| = |B|$. Because of the interval property, $x \notin B'$, i.e. $B' = B$. So, all interval greedoids have the transposition property.

For later use we record the following trivial strengthening of the preceding paragraph: property (TP) with the last words replaced by 'for all $B \subseteq E - A'$, is satisfied by every interval greedoid. Note that this stronger formulation is not possible for transposition greedoids in general.

8.3.H. Gaussian Greedoids

The Gaussian algorithm for solving systems of equations gives rise to greedoids in the following way. Let $M = (m_{ij})$ be an $m \times n$ matrix over an arbitrary field. Perform Gaussian elimination working downward row by row from the top and keep track of the column indices of the pivot elements. Each possible such procedure gives rise to a sequence of column indices (for row 1, row 2, and so on), and these sequences are the basic words of the *Gaussian elimination greedoid* $(E, \mathcal{F})$ of M, $E = \{1, 2, ..., n\}$. Equivalently, this greedoid can be defined directly by

$$\mathcal{F} = \{A \subseteq E: \text{the submatrix } M_{\{1, 2, ..., |A|\}, A} \text{ is non-singular}\}.$$

Gaussian elimination greedoids are not in general transposition greedoids, as may be checked for the matrix

$$N = \begin{pmatrix} 1 & 1 & 0 & 1 \\ 0 & 0 & 1 & 0 \\ 1 & 0 & 0 & 0 \end{pmatrix}$$

whose associated greedoid $(\{1, 2, 3, 4\}, \mathcal{F})$ has $\{1\}, \{2\}, \{1, 3, 4\} \in \mathcal{F}$, but $\{1, 2\}, \{2, 3, 4\} \notin \mathcal{F}$.

We shall now discuss a special case of Gaussian elimination greedoids and then generalize their construction. Suppose that $\Gamma = (V, U, E)$ is a bipartite graph, $E \subseteq V \times U$, and fix an ordering $u_1, u_2, ..., u_n$ of the elements of the color class U. Now, let

$$\mathcal{F} = \{A \subseteq V: A \text{ can be matched to } \{u_1, u_2, ..., u_{|A|}\} \text{ in } \Gamma\}.$$

Then $(V, \mathcal{F})$ is a greedoid, called a *medieval marriage greedoid*.

Every medieval marriage greedoid is a Gaussian elimination greedoid over a suitable field. To see this, take the $U \times V$ incidence matrix and replace the elements that are unity, if necessary, by algebraically independent field elements. The matrix N given above is the incidence matrix of a bipartite graph, which shows that in general medieval marriage greedoids also lack the transposition property.

A more general class of greedoids is obtained as follows. Suppose that $M_i = (E, \mathscr{I}_i)$, $i = 0, 1, \ldots, m$, is a sequence of matroids on the same ground set E such that (1) if $A \subseteq E$ is closed in M_{i-1} then it is closed in M_i, for $1 \leqslant i \leqslant m$, and (2) rank $M_i = i$, for $0 \leqslant i \leqslant m$. A greedoid $(E, \mathscr{F})$ of rank m is then defined by $\mathscr{F} = \{A \subseteq E: A$ is a basis of $M_{|A|}\}$. Greedoids of this kind are called *Gaussian*. Clearly Gaussian elimination greedoids are special cases of Gaussian greedoids (take for M_i the column matroid determined by the first i rows of the given matrix). Notice that in a Gaussian greedoid the matroids M_i can be uniquely recovered from $\mathscr{F}$, namely, M_i is the hereditary closure of the feasible sets of cardinality i. Also, every matroid M is a Gaussian greedoid (for M_i take the rank i truncation of M).

8.3.I. Relations Among Classes of Greedoids

We have discussed several cases where one class of greedoids is seen to be a generalization or a specialization of another class. For an overview, these containment relations between classes of greedoids are gathered in the form of a poset diagram in Figure 8.5. (The containment of the class of matroids

Figure 8.5.

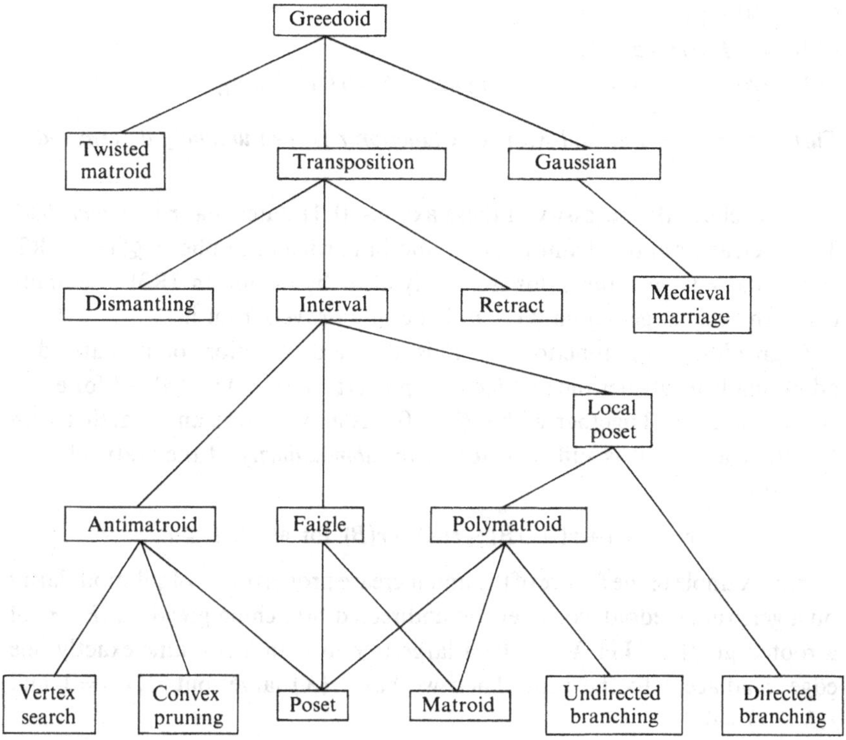

in the classes of twisted matroids and Gaussian greedoids is not indicated in the diagram.)

8.4. Structural Properties

In this section we will discuss the greedoid rank function and the closure operator to which it gives rise. Just like matroids, greedoids have cryptomorphic definitions in terms of rank and closure. Also, various elementary constructions on greedoids will be defined.

8.4.A. Rank Function

Recall that the rank function of a greedoid $(E, \mathscr{F})$ is defined by $r(A) = \max\{|X| : X \subseteq A, X \in \mathscr{F}\}$, for $A \subseteq E$. Clearly, $\mathscr{F} = \{A \subseteq E: r(A) = |A|\}$, which means that the greedoid $\mathscr{F}$ is completely determined by its rank function. This proves the last sentence of the following result.

8.4.1. Theorem. *A function $r: 2^E \to \mathbb{N}$ is the rank function of a greedoid if and only if for all $A, B \subseteq E$ and $x, y \in E$:*

(R1) $r(A) \leq |A|$;
(R2) $A \subseteq B$ implies $r(A) \leq r(B)$;
(R3) $r(A) = r(A \cup x) = r(A \cup y)$ implies $r(A) = r(A \cup \{x, y\})$.

Furthermore, the greedoid with rank function r is then uniquely determined.

Let us check the necessity of these axioms. (R1) states that r is *subcardinal*. This is clear from the definition of r, and in particular implies $r(\varnothing) = 0$. (R2) states that r is *monotone*; this is equally clear by definition. (R3) essentially codes the exchange axiom (G2) and is easily proved from it.

A greedoid rank function on E is the rank function of a matroid if additionally it satisfies the *unit increase property*: $r(A \cup x) \leq r(A) + 1$ for every $A \subseteq E$ and $x \in E$. Together with $r(\varnothing) = 0$ this implies (R1), and together with the other axioms it is sufficient to prove *submodularity* of the matroid rank function:

$$r(A \cap B) + r(A \cup B) \leq r(A) + r(B), \text{ for all } A, B \subseteq E.$$

For an example of the failure of the unit increase property and of submodularity on a general greedoid, consider the undirected branching greedoid $(E, \mathscr{F})$ of a rooted graph Γ. Let $X \in \mathscr{F}$ be a large tree in Γ that contains exactly one edge x adjacent to the root. Then we have $r(X)$ large, but $r(\{x\}) = 1$ and $r(X - x) = 0$.

8.4.B. Closure Operator

Using the rank function, we define the *(rank) closure operator* $\sigma : 2^E \to 2^E$ of a greedoid $(E, \mathscr{F})$ by

$$\sigma(A) = \{x \in E: r(A \cup x) = r(A)\}.$$

Clearly, σ is *increasing*: $A \subseteq \sigma(A)$. Furthermore, for all $A \subseteq E$,

$$r(A) = r(\sigma(A)). \tag{8.1}$$

To see this, suppose $r(A) < r(\sigma(A))$ and let X and Y be bases of A and $\sigma(A)$ respectively. Then X can be augmented by some $y \in Y$ so that $X \cup y$ is feasible and of cardinality $r(A) + 1$. But then $r(A \cup y) \geq r(X \cup y) = r(A) + 1$, which means that $y \notin \sigma(A)$, contradicting $y \in Y \subseteq \sigma(A)$.

As a consequence of the preceding we find that σ is *idempotent*:

$$\sigma\sigma(A) = \sigma(A), \tag{8.2}$$

for all $A \subseteq E$. Namely, if $x \in \sigma\sigma(A)$, then $r(\sigma(A)) = r(A) \leq r(A \cup x) \leq r(\sigma(A) \cup x) = r(\sigma(A))$, where the first equality is by (8.1) and the last by the definition of the closure of $\sigma(A)$. So, $x \in \sigma\sigma(A)$ implies $x \in \sigma(A)$.

A serious shortcoming of greedoid closure is that it is *not necessarily monotone*, i.e. $A \subseteq B$ does not in general imply $\sigma(A) \subseteq \sigma(B)$. For an easy counterexample, take the full greedoid with exactly one basic word xy. In this greedoid, which is both a poset greedoid and a branching greedoid, $\sigma(\varnothing) = \{y\}$, $\sigma(\{x\}) = \{x\}$. The *closed sets* (i.e. $A = \sigma(A)$) of this greedoid are $\{x\}$, $\{y\}$, and $\{x, y\}$.

The failure of monotonicity means that greedoid closure is not a closure operator in the usual sense (i.e. as defined in section 8.7.A). Many of the characteristic properties of ordinary closure are absent; for instance the intersection of two closed sets is not always closed, and the closed sets ordered by inclusion do not form a lattice. It can be shown that greedoid closure σ is monotone only in the matroid case.

In spite of what has just been said, it turns out that greedoid closures can be axiomatically characterized in a way that is reminiscent of matroid closure.

8.4.2. Theorem. *A function* $\sigma : 2^E \to 2^E$ *is the rank closure operator of a greedoid if and only if for all A, $B \subseteq E$ and x, $y \in E$ the following conditions hold.*

(RC1) $A \subseteq \sigma(A)$.

(RC2) $A \subseteq B \subseteq \sigma(A)$ *implies* $\sigma(B) = \sigma(A)$.

(RC3) *Suppose* $x \notin A$, $z \notin \sigma(A \cup x - z)$ *for all* $z \in A \cup x$. *Then* $x \in \sigma(A \cup y)$ *implies* $y \in \sigma(A \cup x)$.

Furthermore, the greedoid with closure σ is uniquely determined.

Suppose that $(E, \mathscr{F})$ is a greedoid with closure operator σ. Now, $\mathscr{F} = \{A \subseteq E: x \notin \sigma(A - x)$ for all $x \in A\}$, so the greedoid can be uniquely reconstructed from σ. Axiom (RC2), a weak form of monotonicity, is easy to verify for σ using property (8.1). Axiom (RC3) states that if $x \notin A$ and $A \cup x \in \mathcal{F}$, then $x \in \sigma(A \cup y)$ implies $y \in \sigma(A \cup x)$. This generalizes the MacLane exchange axiom for matroid closures, where the requirement $A \cup x \in \mathcal{F}$ is replaced by $x \notin \sigma(A)$. The MacLane exchange axiom is in general not satisfied by greedoid closure.

It turns out to be convenient to consider also the *monotone closure operator* $\mu: 2^E \rightarrow 2^E$ obtained from σ by the following construction:

$$\mu(A) = \cap \{\sigma(X): A \subseteq \sigma(X), X \subseteq E\}.$$

It is easy to check that $A \subseteq \mu(A) \subseteq \sigma(A)$ and $\mu\mu(A) = \mu(A)$ for all $A \subseteq E$, and that $A \subseteq B$ implies $\mu(A) \subseteq \mu(B)$. So, μ is a closure operator in the usual sense. However, μ does not determine $\mathcal{F}$, since $\mu = \mathrm{id}$ for all full greedoids.

8.4.C. Rank and Closure Feasibility

We have seen that greedoid rank and closure lack some of the good behaviour of their matroid counterparts. However, loosely speaking, they behave better on certain subsets of the greedoid than on others. This motivates the introduction of two special feasibility concepts, which are trivial for matroids, but give useful structural information about greedoids.

Let $G = (E, \mathcal{F})$ be a greedoid, and define the *basis rank* of $A \subseteq E$ by

$$\beta(A) = \max \{|A \cap X| : X \in \mathcal{F}\}.$$

Equivalently, $\beta(A)$ is the maximal size of the intersection of A with a basis. It is clear that $\beta(A) \geq r(A)$, for all $A \subseteq E$. For matroids the equality always holds, but this is not true for greedoids. For suppose that $A \subseteq B$, $A \notin \mathcal{F}$, $B \in \mathcal{F}$; then $r(A) < |A| = \beta(A)$, since $A = A \cap B$. Hence, $\beta(A) = r(A)$ for all subsets A if and only if G is a matroid.

A set $A \subseteq E$ is called *rank feasible* if $\beta(A) = r(A)$, that is, if $|A \cap X| \leq r(A)$ for all $X \in \mathcal{F}$. The collection of all rank feasible subsets is denoted by $\mathcal{R}$ or $\mathcal{R}_G$. It is clear that $\mathcal{F} \subseteq \mathcal{R}$, and that for a full greedoid $\mathcal{F} = \mathcal{R}$ (because then $\beta(A) = |A|$). Also, we have seen that $\mathcal{R}_G = 2^E$ if and only if G is a matroid. The collection $\mathcal{R}_G$ is an interesting set system associated with the greedoid G. Many properties valid for $\mathcal{F}$ generalize to $\mathcal{R}$.

Here is a characterization of rank feasible sets.

8.4.3. Proposition. *Let* $(E, \mathcal{F})$ *be a greedoid,* $A \subseteq E$. *Then the following conditions are equivalent:*

  (i) *A is rank feasible;*
  (ii) *$r(A \cup X) \leq r(A) + |X|$, for all $X \subseteq E - A$;*
  (iii) *$X \subseteq A \subseteq \mu(X)$, for some $X \in \mathcal{F}$.*

Property (ii) shows that r has the unit-increase property on rank feasible sets. Also, it allows us to conclude that for $A, B \subseteq E$:

$$\beta(A \cup B) + r(A \cap B) \leq \beta(A) + \beta(B),$$

which implies that on rank feasible sets, $r = \beta$ is semimodular. Property (iii) easily implies that $\mathscr{R}$ is an accessible set system. However, in general $(E, \mathscr{R})$ is not itself a greedoid.

In some situations it is useful to consider a stronger property than rank feasibility: $A \subseteq E$ is said to be *closure feasible* if $A \subseteq \sigma(X)$ implies $A \subseteq \mu(X)$ for all subsets $X \subseteq E$. Equivalently, A is closure feasible if $A \subseteq \sigma(X)$ and $X \subseteq Y \subseteq E$ imply $A \subseteq \sigma(Y)$. We denote the collection of closure feasible sets by $\mathscr{C}$ or $\mathscr{C}_G$. Here are some basic properties.

8.4.4. Proposition. *Every closure feasible set is rank feasible:* $\mathscr{C} \subseteq \mathscr{R}$. *Furthermore, the set system $\mathscr{C}$ is closed under union.*

8.4.5. Proposition. *The following conditions are equivalent:*
   (i) *$(E, \mathscr{F})$ is an interval greedoid;*
   (ii) *$\mathscr{F} \subseteq \mathscr{C}$;*
   (iii) *$\mathscr{C} = \mathscr{R}$.*

Since $\mathscr{R}$ is an accessible set system, it follows from Propositions 8.2.7, 8.4.4, and 8.4.5 that $(E, \mathscr{C}) = (E, \mathscr{R})$ is an antimatroid if $(E, \mathscr{F})$ has the interval property.

8.4.D. Constructions

The basic matroid construction of deletion, contraction, truncation, and direct sum generalize to greedoids in the following way.

Let $G = (E, \mathscr{F})$ be a greedoid and $A \subseteq E$. Define

$$\mathscr{F} \setminus A = \{X \subseteq E - A : X \in \mathscr{F}\}, \tag{8.3}$$

and, if A is feasible,

$$\mathscr{F}/A = \{X \subseteq E - A : X \cup A \in \mathscr{F}\}. \tag{8.4}$$

It is not hard to check that the set systems obtained are in both cases greedoids on the ground set $E' = E - A$. $G \setminus A = (E', \mathscr{F} \setminus A)$ is said to be obtained from G by *deletion* of A, or by *restriction* to $E - A$, and $G/A = (E', \mathscr{F}/A)$ by *contraction* of A. Also, by a *minor* of $(E, \mathscr{F})$ we shall mean any restriction of a contraction, i.e. any greedoid of the form $(E - (A \cup A'), (\mathscr{F}/A) \setminus A')$, where $A \in \mathscr{F}$ and $A' \subseteq E - A$.

Observe in this connection that restriction and contraction commute:

$$(\mathscr{F}/A) \setminus A' = (\mathscr{F} \setminus A')/A = \{X \subseteq E - (A \cup A') : X \cup A \in \mathscr{F}\},$$
for $A \cap A' = \varnothing$, $A \in \mathscr{F}$ and $A' \subseteq E$.

Thus minors can equivalently be defined as contractions of restrictions. Minors of minors of a greedoid are again minors of the greedoid, and hence 'being a minor of' defines a partial order on the set of isomorphism types of greedoids. For more about this, see section 8.9.A.

Let r, r' and r'' denote the rank functions of G, $G \setminus A$, and G/A, respectively. Then for all $X \subseteq E - A$:

$$r'(X) = r(X) \quad \text{and} \quad r''(X) = r(X \cup A) - r(A). \tag{8.5}$$

Definition (8.4) produces a greedoid only if A is feasible, since otherwise $\varnothing \notin \mathscr{F}/A$. It is possible to extend the definition of contraction $\mathscr{F}/A$ in the following cases: (i) $\mathscr{F}$ is an arbitrary greedoid and A is rank feasible, or (ii) $\mathscr{F}$ is an interval greedoid and A is an arbitrary subset. (However, the contraction $\mathscr{F}/A$ is *not* a meaningful concept for general $\mathscr{F}$ and A.) In case (i) one extends the rank formula (8.5), while in case (ii) one picks a basis B of A, contracts by B, and then deletes $A - B$. It is instructive to try to visualize the second case in terms of branching greedoids.

It is easy to formulate the ordered versions of deletion and contraction. For instance, the *contraction* of a greedoid language $(E, \mathscr{L})$ by the feasible word α is defined by

$$\mathscr{L}/\alpha = \{\beta \in E_s^* : \alpha\beta \in \mathscr{L}\}, \tag{8.6}$$

which is clearly again a greedoid language.

If $(E, \mathscr{F})$ is a greedoid of rank r and $0 \leq k \leq r$, then the *k-truncation*

$$\mathscr{F}^{(k)} = \{X \in \mathscr{F} : |X| \leq k\} \tag{8.7}$$

is a greedoid as well.

There are two ways to define the sum of greedoids. Let $G_1 = (E_1, \mathscr{F}_1)$ and $G_2 = (E_2, \mathscr{F}_2)$ be two greedoids on disjoint ground sets. Then their *direct sum* is the greedoid $G_1 \oplus G_2 = (E_1 \cup E_2, \mathscr{F}_1 \oplus \mathscr{F}_2)$, where

$$\mathscr{F}_1 \oplus \mathscr{F}_2 = \{X_1 \cup X_2 : X_1 \in \mathscr{F}_1 \text{ and } X_2 \in \mathscr{F}_2\}. \tag{8.8}$$

Their *ordered sum* is the greedoid $G_1 \otimes G_2 = (E_1 \cup E_2, \mathscr{F}_1 \otimes \mathscr{F}_2)$, where

$$\mathscr{F}_1 \otimes \mathscr{F}_2 = \mathscr{F}_1 \cup \{B \cup X : B \text{ is a basis of } \mathscr{F}_1, X \in \mathscr{F}_2\}. \tag{8.9}$$

Clearly, $G_1 \oplus G_2$ and $G_1 \otimes G_2$ have the same family of bases. Also, the language of feasible words of $G_1 \oplus G_2$ is the shuffle product of the two component languages.

All constructions discussed in this section take interval greedoids into interval greedoids, and similarly for local poset greedoids. The same is true for antimatroids, except that the k-truncation of an antimatroid is in general only an interval greedoid. Conversely, the restriction of an interval greedoid to any feasible set is an antimatroid (cf. Proposition 8.2.8). A special operation for antimatroids, called *trace*, will be defined in section 8.7.C.

Unfortunately, the duality operation of matroid theory has no counterpart for greedoids. Only a weak notion of duality operation exists for general greedoids; see section 8.6.C.

8.4.E. Connectivity

The concept of the connectivity of greedoids is modeled to generalize the graph-theoretic connectivity of rooted graphs in the case of branching greedoids. In this context a rooted digraph $\Delta = (V, E, r)$ is called *connected* (or 1-*connected*) if there is a directed path from the root to every vertex. More generally, Δ is k-*connected* if every vertex $v \in V$ can be reached from the root by a directed path after removal of at most $k - 1$ vertices in $V - \{r, v\}$. Equivalently, by Menger's theorem, Δ is k-connected if there are k vertex-disjoint directed paths from the root to every vertex $v \in V - r$ such that (r, v) is not an arc in E. Similar definitions apply to rooted undirected graphs.

The digraph Δ is connected if and only if the associated vertex search greedoid is full. In contrast, the connectedness of Δ is not encoded in the directed branching greedoid of Δ.

The case of higher connectivity $(k > 1)$ suggests the following definition.

8.4.6. Definition. Let $(E, \mathscr{F})$ be a greedoid of rank r, and let $X \in \mathscr{F}$. A set $A \subseteq E - X$ is called *free over* X if for every $B \subseteq A$, $X \cup B$ is feasible. The greedoid $(E, \mathscr{F})$ is called k-*connected* $(1 \leq k \leq r)$ if for every $X \in \mathscr{F}$ there is a free set A over X of size $\min\{k, r - r(X)\}$. Equivalently, $(E, \mathscr{F})$ is k-connected if for every $X \in \mathscr{F}$ there is a $Y \in \mathscr{F}$ such that $X \subseteq Y, |Y - X| = \min\{k, r - r(X)\}$ and the interval $[X, Y]$ of the poset $(\mathscr{F}, \subseteq)$ is Boolean.

Obviously, every greedoid is 1-connected, and every k-connected greedoid is also $(k - 1)$-connected for $k \geq 2$. Matroids are r-connected (i.e. maximally connected), but this does not fully characterize matroids.

If A is free over $X \in \mathscr{F}$, then it is contained in the set

$$\Gamma(X) = E \backslash \sigma(X) = \{a \in E - X : X \cup a \in \mathscr{F}\}$$

of *continuations* of X. In antimatroids, we know (e.g. from Lemma 8.7.9) that the free sets over X are exactly the subsets of $\Gamma(X)$. Thus an antimatroid is k-connected if and only if for all $X \in \mathscr{F}$,

$$|\Gamma(X)| \geq \min\{k, r - r(X)\}.$$

The following result shows that (for $k > 1$) Definition 8.4.6 describes a reasonable generalization of graph connectivity.

8.4.7. Proposition. *Let* $\Delta = (V, E, r)$ *be a connected rooted digraph. Then the following are equivalent for $k > 1$:*

(i) Δ *is k-connected;*
(ii) *the branching greedoid on* Δ *is k-connected;*
(iii) *the vertex search greedoid on* Δ *is k-connected.*

Proof. Since (ii)⇔(iii) is a special case of Proposition 8.8.9, we will only demonstrate (i)⇔(iii).

If the digraph Δ is not k-connected, then there is a cut set $A \subseteq V' = V - r$ of size $|A| < k$ that separates a vertex $v \in V' - A$ from the root. Consider the feasible set $X = \cup \{Y \in \mathscr{F} : Y \subseteq V' - A\}$ of the vertex search greedoid $(V', \mathscr{F})$. We have $|A| + r(X) < r = |V'|$, since $v \notin X \cup A$. But the free sets over X are contained in $\Gamma(X) \subseteq A$, and $|A| < \min\{k, r - r(X)\}$. Thus $(V', \mathscr{F})$ is not k-connected.

Conversely, if the vertex search greedoid is not k-connected, then there is a set $X \in \mathscr{F}$ such that $|\Gamma(X)| < \min\{k, r - r(X)\}$. But then $|X \cup \Gamma(X)| = r(X) + |\Gamma(X)| < r$, so $\Gamma(X)$ is a cut set of size less than k that separates all vertices of $V' - (X \cup \Gamma(X))$ from the root. $\square$

In concluding this section, we observe that the case of *undirected* rooted graphs and their associated greedoids can be reduced to the previous *directed* case by a standard graph-theoretic construction: for an undirected rooted graph $\Gamma = (V, E, r)$ let $\Delta = (V, E', r)$ be the rooted digraph on the same vertex set that has a pair of antiparallel arcs for every edge of Γ. Then Γ is k-connected if and only if Δ is k-connected, and the vertex search greedoids of Γ and Δ coincide. The branching greedoids of Γ and Δ differ already in the size of their ground sets, but the associated posets $(\mathscr{F}, \subseteq)$ are isomorphic. This proves the analog of Proposition 8.4.7 for undirected graphs via the observation that k-connectedness of a greedoid can be determined from the unlabeled poset $(\mathscr{F}, \subseteq)$ alone.

8.5. Optimization on Greedoids

As mentioned in the introduction, greedoids were originally developed to give a unified approach to the optimality of various *greedy algorithms* known in combinatorial optimization. Such algorithms can be loosely characterized as having locally optimal strategy and no backtracking.

In this section we will formulate a greedy algorithm for hereditary languages, define *compatible* objective functions on such languages, and then characterize greedoids as those languages on which the greedy algorithm is optimal for all compatible objective functions. The well known algorithmic characterization of matroids in terms of *linear* objective functions is here viewed in a broader context.

To illustrate the results, we will discuss Kruskal's and Prim's algorithms for minimal spanning trees and Dijkstra's shortest path algorithm as instances of greedoid optimization.

8.5.A. The Greedy Algorithm

In the following, let $(E, \mathscr{L})$ be a simple hereditary language over a finite ground set E. We do not assume *a priori* that $\mathscr{L}$ is pure. As usual, maximal words in $\mathscr{L}$ are called *basic*. We will be interested in the following optimization problem.

Given an objective function $\omega : \mathscr{L} \to \mathbb{R}$, *find a basic word* α *that maximizes* $\omega(\alpha)$.

The greedy approach to this problem is expressed by the following algorithm.

GREEDY: (1) Put $\alpha_0 := \varnothing$ and $i := 0$.
        (2) Given α_i, choose $x_{i+1} \in E$ such that
           (i) $\alpha_i x_{i+1} \in \mathscr{L}$
           (ii) $\omega(\alpha_i x_{i+1}) \geqq \omega(\alpha_i y)$, for all $y \in E$ such that $\alpha_i y \in \mathscr{L}$.
        (3) Put $\alpha_{i+1} := \alpha_i x_{i+1}$.
        (4) If the word α_{i+1} is not basic, put $i := i + 1$ and go to (2).
        (5) If α_{i+1} is basic, print $\alpha := \alpha_{i+1}$ and stop.

Whether GREEDY works (that is, whether the *greedy solution* α produced by GREEDY actually maximizes $\omega(\alpha)$) must obviously depend on both $\mathscr{L}$ and ω – they have to be 'compatible'.

8.5.1. Definition. An objective function $\omega : \mathscr{L} \to \mathbb{R}$ is *compatible* with $\mathscr{L}$ if it satisfies the following conditions: for $\alpha x \in \mathscr{L}$ such that $\omega(\alpha x) \geqq \omega(\alpha y)$ for every $\alpha y \in \mathscr{L}$ ('x is a best choice after α'),

(C1) $\alpha \beta x \gamma \in \mathscr{L}$ and $\alpha \beta z \gamma \in \mathscr{L}$ imply that $\omega(\alpha \beta x \gamma) \geqq \omega(\alpha \beta z \gamma)$ ('x is a best choice at every later stage'), and

(C2) $\alpha x \beta z \gamma \in \mathscr{L}$ and $\alpha z \beta x \gamma \in \mathscr{L}$ imply that $\omega(\alpha x \beta z \gamma) \geqq \omega(\alpha z \beta x \gamma)$ ('it is always better to choose x first and z later than the other way round').

Of course, if ω is *stable*, in the sense that $\omega(\alpha)$ only depends on the underlying set $\tilde{\alpha}$, then (C2) is vacuously satisfied.

The following main theorem characterizes greedoids algorithmically.

8.5.2. Theorem. *Suppose* $(E, \mathscr{L})$ *is a simple hereditary language. Then* $(E, \mathscr{L})$ *is a greedoid if and only if GREEDY gives an optimal solution for every compatible objective function on* $\mathscr{L}$.

Proof. (1) We will show that GREEDY works on interval greedoids – the proof for general greedoids is similar but more complicated. All examples discussed below involve interval greedoids and are therefore covered by this proof.

Let $(E, \mathscr{L})$ be an interval greedoid, ω a compatible objective function and γ a greedy solution. Choose an optimal solution δ so that the common prefix with γ is of maximal length, i.e. $|\alpha|$ is maximal with $\gamma = \alpha\gamma'$ and $\delta = \alpha\delta'$. We claim that $\gamma = \alpha = \delta$.

If this is not the case, then $\gamma' = x\gamma''$, $\delta' = y_1 y_2 \ldots y_n$ where $x \neq y_1$. Now augment αx from $\beta = \alpha y_1 \ldots y_n$, using the strong exchange property (L2') for interval greedoids: $\alpha xy_1 \ldots \hat{y}_k \ldots y_n \in \mathscr{L}$, for some $1 \leq k \leq n$. Here, '$\hat{y}_k$' denotes that y_k is deleted.

For $1 \leq i \leq k-1$ define

$$\beta_i = \alpha y_1 \ldots y_{i-1} xy_i \ldots \hat{y}_k \ldots y_n,$$

and let

$$\beta_k = \alpha y_1 \ldots y_{k-1} xy_{k+1} \ldots y_n.$$

We know that $\beta_1 = \alpha xy_1 \ldots \hat{y}_k \ldots y_n \in \mathscr{L}$, and augmenting $\alpha y_1 \ldots y_i$ from $\beta_i \in \mathscr{L}$, using strong exchange (L2'), we get $\beta_{i+1} \in \mathscr{L}$. Hence, $\beta_1, \beta_2, \ldots, \beta_k \in \mathscr{L}$.

Now, x is a 'best choice' after α (since γ is greedy), so conditions (C1) and (C2) give

$$\omega(\beta_1) \geq \omega(\beta_2) \geq \ldots \geq \omega(\beta_k),$$

and (C1) implies

$$\omega(\beta_k) = \omega(\alpha y_1 \ldots y_{k-1} xy_{k+1} \ldots y_n)$$
$$\geq \omega(\alpha y_1 \ldots y_{k-1} y_k y_{k+1} \ldots y_n) = \omega(\delta).$$

Hence, $\omega(\beta_1) \geq \omega(\delta)$. But then β_1 is an optimal basic word having a longer common prefix αx with γ than does δ. This contradicts the choice of δ.

(2) For the converse, we define *generalized bottleneck functions* on $\mathscr{L}$: they are the objective functions of the form $\omega(x_1 x_2 \ldots x_n) = \min\{f_1(x_1), \ldots, f_n(x_n)\}$, where the $f_i: E \rightarrow \mathbb{R}$ $(1 \leq i \leq r)$ are functions satisfying $f_i(x) \leq f_{i+1}(x)$ for every $x \in E, 1 \leq i < r$. Here r denotes the maximal length of a word in $\mathscr{L}$. Generalized bottleneck functions are compatible with all hereditary languages, as is easily checked.

Now, suppose that $\alpha, \beta \in \mathscr{L}$ and $|\alpha| = k > m = |\beta|$. We want to show that there is some $x \in \tilde{\alpha}$ such that $\beta x \in \mathscr{L}$. For this, let $A = \tilde{\alpha} \cup \tilde{\beta}$ and define a generalized bottleneck function ω by

$$f_1(x) = \ldots = f_k(x) = \begin{cases} 0 & \text{if } x \notin A, \\ 1 & \text{if } x \in A, \end{cases}$$

$$f_{k+1}(x) = \ldots = f_r(x) = \begin{cases} 1 & \text{if } x \notin A, \\ 2 & \text{if } x \in A. \end{cases}$$

Let $\delta = \alpha\delta'$ be a basic word extending α. Then $\omega(\delta) = 1$. Next, let $\gamma = \beta x_1 x_2 \ldots x_p$ be a greedy solution extending β. Such solutions clearly exist. Since GREEDY is assumed to be optimal, we have

$$1 = \omega(\delta) \leq \omega(\gamma) \leq f_{m+1}(x_1),$$

which, since $m + 1 \leq k$, implies that $x_1 \in A$. Now, $\beta x_1 \in \mathscr{L}$, since $\mathscr{L}$ is hereditary, and therefore $x_1 \in A - \tilde{\beta} \subseteq \tilde{\alpha}$, since $\mathscr{L}$ is simple. This completes the proof. □

8.5.B. Examples

8.5.3. Example. (Matroid optimization) An objective function $\omega \colon \mathscr{L} \to \mathbb{R}$ is called *linear* if it is of the form

$$\omega(x_1 x_2 \ldots x_n) = \sum_{i=1}^{n} u(x_i)$$

for some given weight function $u \colon E \to \mathbb{R}$.

If $(E, \mathscr{L})$ is a matroid then all linear objective functions are compatible and hence can be greedily optimized. One easily checks condition (C1), and (C2) is clear since linear objective functions are stable.

For example, if $\Gamma = (V, E)$ is an undirected connected graph with weight function $u \colon E \to \mathbb{R}$, then a *minimal spanning tree* (i.e. a spanning tree $T \subseteq E$ minimizing $\sum_{e \in T} u(e)$) will be obtained by applying Kruskal's algorithm. From a greedoid point of view, we apply GREEDY to the linear objective function $\omega(T) = -\sum_{e \in T} u(e)$ on the graphic matroid $(E, \mathscr{L})$ associated with Γ.

8.5.4. Example. (Breadth-first-search) Let $(E, \mathscr{L})$ be the branching greedoid of a connected rooted digraph $\Delta = (V, E, r)$, and assume that $d \colon E \to \mathbb{R}^+$ is a length function on the arcs. Define an objective function $\omega \colon \mathscr{L} \to \mathbb{R}$ by

$$\omega(x_1 x_2 \ldots x_n) = -\sum_{i=1}^{n} d(r, v_i),$$

where the $v_i = \mathrm{head}(x_i)$ are the nodes reached by the branching $x_1 \ldots x_n$, and $d(r, v_i)$ is the sum of the lengths of the arcs on the unique path from r to v_i in $x_1 x_2 \ldots x_n$.

We have to check that ω is compatible with $\mathscr{L}$: (C2) is again clear because ω is stable; the easy argument for (C1) is an instructive exercise.

Hence, the theorem implies that GREEDY, which executes Breadth-first-search on Δ, finds a spanning arborescence that minimizes the sum of the distances from the root. Such an arborescence must, in fact, also minimize each individual distance, since, as can easily be seen, there exist spanning arborescences that simultaneously minimize all the distances from the root to the other vertices. This particular instance of GREEDY gives the shortest path algorithm of Dijkstra.

It is noteworthy that the objective function $-\omega$ (corresponding to *Depth-first-search*) is *not* compatible with the branching greedoid language. For instance, GREEDY fails to optimize $-\omega$ for the greedoid shown in Figure 8.6.

Figure 8.6.

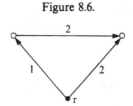

8.5.C. Linear Objective Functions

Linear objective functions (as defined in Example 8.5.3) cannot in general be greedily optimized over greedoids. However, for some special linear functions and for some special greedoids the situation is better.

Let $(E, \mathcal{F})$ be a greedoid, $\mathcal{R} \subseteq 2^E$ its collection of rank feasible sets, and $u: E \to \mathbb{R}$ a function. The linear objective function $\omega(x_1 x_2 \ldots x_n) = \sum_{i=1}^{n} u(x_i)$ is called $\mathcal{R}$-compatible if $\{x \in E: u(x) \geq c\} \in \mathcal{R}$, for all $c \in \mathbb{R}$, that is, if all the *level sets* of u are rank feasible.

In the situation of the preceding paragraph, suppose that $c_1 > c_2 > \ldots > c_k$ are the values assumed by u, and let $C_i = \{x \in E: u(x) \geq c_i\}$. Clearly, GREEDY will first pick a basis of C_1, then augment it to a basis of C_2, and so on. Hence, if B is a greedy basis then $|B \cap C_i| = \text{rank } C_i$, for $1 \leq i \leq k$. Since $C_i \in \mathcal{R}$, an arbitrary basis B' must satisfy $|B' \cap C_i| \leq \text{rank } C_i$, $1 \leq i \leq k$. It easily follows that $\omega(B') \leq \omega(B)$, i.e. we have proven the following.

8.5.5. Proposition. *Let $(E, \mathcal{F})$ be a greedoid. Then GREEDY is optimal for every $\mathcal{R}$-compatible linear objective function.*

As observed in section 8.4.C, $(E, \mathcal{F})$ is a matroid if and only if $\mathcal{R} = 2^E$, that is, if and only if every linear objective function is $\mathcal{R}$-compatible. So Proposition 8.5.5 again generalizes, but in a different way from Theorem 8.5.2, the fact that matroids have the property that all linear objective functions can be greedily optimized.

However, not every greedoid with that property is a matroid, as we shall now see.

8.5.6. Proposition. *Let $(E, \mathcal{F})$ be a greedoid. Then GREEDY is optimal for every linear objective function if and only if the hereditary closure $(E, \mathcal{H}(\mathcal{F}))$ is a matroid and every set that is closed in $(E, \mathcal{F})$ is also closed in $(E, \mathcal{H}(\mathcal{F}))$.*

An example of a greedoid that satisfies these conditions is the undirected branching greedoid of a connected rooted graph $\Gamma = (V, E, r)$, for which the hereditary closure is the corresponding graphic matroid. Greedy optimization of linear objective functions over this branching greedoid is equivalent to Prim's algorithm for finding a minimal spanning tree in Γ.

8.6. The Greedoid Polynomial

Every greedoid has an associated polynomial that reflects some of its combinatorial structure. In this section we will present the basic properties of this polynomial and also discuss the related notions of greedoid invariants and dual complexes.

8.6.A. A Greedoid 'Tutte' Polynomial

Let $G = (E, \mathcal{F})$ be a greedoid with $n = |E|$ and $r = \text{rank } G$. Give the underlying set E a total ordering Ω. This induces a total ordering of the set $\mathcal{B}_G$ of bases of G as follows: $B <_\Omega B'$ if the lexicographically first feasible permutation of B is lexicographically smaller than the lexicographically first feasible permutation of B'.

For instance, consider the branching greedoid of the directed graph shown in Figure 8.7. There are two bases: $B_1 = \{a, b\}$ and $B_2 = \{a, c\}$. If Ω is $a < b < c$ then $B_1 < B_2$, but if Ω is $b < c < a$ then $B_2 < B_1$.

Now, for a basis $B \in \mathcal{B}_G$ we shall say that $x \in E - B$ is *externally active in* B if $B < (B \cup x) - y$, for all $y \in B$ such that $(B \cup x) - y$ is a basis. Let $\text{ext}_\Omega(B)$ denote the set of externally active elements, and define

$$\lambda_{G,\Omega}(t) = \sum_{B \in \mathcal{B}_G} t^{|\text{ext}_\Omega(B)|}. \qquad (8.10)$$

Let us again exemplify with the small branching greedoid above:

Ω	$\text{ext}_\Omega(B_1)$	$\text{ext}_\Omega(B_2)$	$\lambda_{G,\Omega}(t)$
$a < b < c$	$\{c\}$	$\varnothing$	$1 + t$
$b < c < a$	$\varnothing$	$\{b\}$	$1 + t$

The crucial combinatorial fact about this notion of external activity in bases is stated in the following lemma. Recall that a subset $A \subseteq E$ is said to be *spanning* if it contains a basis.

Figure 8.7.

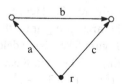

8.6.1. Lemma. *For each spanning set A there exists a unique basis B such that $B \subseteq A \subseteq B \cup \text{ext}_\Omega(B)$.*

The basis B with this property is the first one (in the ordering induced by Ω) that is contained in A. The partitioning of the set $\mathcal{S}$ of spanning sets into Boolean intervals implies part (iii) of the following theorem, which in turn implies part (i).

8.6.2. Theorem. *Let $G = (E, \mathcal{F})$ be a greedoid of rank r and cardinality n.*
(i) *$\lambda_G(t) := \lambda_{G,\Omega}(t)$ is independent of the ordering Ω of E.*
(ii) *$\lambda_G(t)$ is a monic polynomial of degree $n - r$ with non-negative integer coefficients.*
(iii) *If G has s_j spanning sets of size j, for $r \le j \le n$, then*

$$\lambda_G(1 + t) = \sum_{i=0}^{n-r} s_{r+i} t^i.$$

(iv)
$$\lambda_G(t) = \begin{cases} \lambda_{G/e}(t) & \text{if } \{e\} \in \mathcal{F} \text{ and } e \text{ is a coloop,} \\ \lambda_{G/e}(t) + \lambda_{G\backslash e}(t) & \text{if } \{e\} \in \mathcal{F} \text{ and } e \text{ is not a coloop,} \\ \lambda_{G_1}(t)\lambda_{G_2}(t) & \text{if } G \text{ is the direct or ordered sum of} \\ & G_1 \text{ and } G_2. \end{cases}$$

The polynomial $\lambda_G(t)$ is a greedoid counterpart to the Tutte polynomial. If G is a matroid with Tutte polynomial $T_G(x, y)$ and dual matroid G^*, then $T_G(1, t) = \lambda_G(t)$ and $T_G(t, 1) = \lambda_{G^*}(t)$.

Every greedoid of positive rank must have some feasible singleton. Therefore part (iv) of the theorem gives a recursive algorithm for computing the polynomial λ_G for any greedoid G. The algorithm will stop at the trivial greedoids of cardinality k and rank zero, whose polynomial is t^k. In general, if G has k loops and G' is obtained by deleting these loops, then $\lambda_G(t) = t^k \lambda_{G'}(t)$, since G is then the direct sum of G' and the loops. For instance, if G is of rank one with k feasible singletons, then $\lambda_G = t^{n-k}(1 + t + \dots + t^{k-1})$. Notice (e.g. from (8.10)) that if G is full (i.e. $E \in \mathcal{F}$) then $\lambda_G(t) = 1$.

Let us as a small example compute λ_G for the branching greedoid G in Figure 8.8a.

One sees that the arc e is a feasible coloop (i.e. it emanates from the root and lies in every spanning arborescence). Hence, it may be contracted away without affecting λ_G. Now, $G/e \cong G_1 \oplus G_2$, where G_1 is the branching greedoid in Figure 8.7 and G_2 that in Figure 8.8b. Hence, $\lambda_G = (1 + t)\lambda_{G_2}$, and deleting the two loops in G_2 (i.e. the arcs going into the root) we get $\lambda_{G_2} = t^2 \lambda_{G_3}$, where G_3 is the greedoid in Figure 8.8c. Using deletion–contraction or simple counting, part (iv) or part (iii) of Theorem 8.6.2 quickly gives $\lambda_{G_3} = 2t + t^2$. Hence, $\lambda_G(t) = (1 + t)t^2(2t + t^2) = 2t^3 + 3t^4 + t^5$.

Figure 8.8.

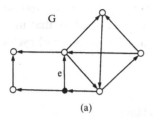

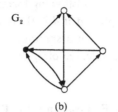

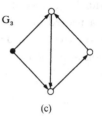

(a)　　　　　　　　(b)　　　　　　　　(c)

We have seen that the degree of $\lambda_G(t)$ as well as the coefficients of $\lambda_G(1 + t)$ have direct combinatorial meaning. Also the subdegree of $\lambda_G(t)$ has an interesting interpretation in terms of a certain algorithmic property. By the *subdegree* of a non-zero polynomial $c_0 + c_1 t + \dots + c_k t^k$ we mean the least integer d such that $c_d \neq 0$.

Suppose that we want to design an algorithm that for arbitrary subsets $A \subseteq E$ decides whether A is a spanning set in the greedoid $G = (E, \mathscr{F})$. Think of A as being represented by its incidence vector $\mathscr{X}_A$ (a 0–1-vector of length n with elements equal to unity in the positions corresponding to A), and suppose that the algorithm can read $\mathscr{X}_A$ by inspecting only one position at a time. If the best such algorithm can decide whether A is spanning or not after k inspections for all $A \subseteq E$, and if $k - 1$ inspections will not suffice, then we say that k is the *argument complexity* of the spanning property in G.

For instance, it is easy to check that the argument complexity is 4 for the branching greedoid in Figure 8.8c. In other words, it would be redundant to inspect all five arcs in order to decide algorithmically whether a subset of arcs contains a directed path from the root to every other node.

8.6.3. Proposition. *The argument complexity of the spanning property in G is $n - d$, where d is the subdegree of $\lambda_G(t)$.*

The result can be made more precise, since the method of proof implies an explicit optimal algorithm that will decide whether an arbitrary subset A is spanning after at most $n - d$ inspections of $\mathscr{X}_A$. Briefly, here is what to do.

(1) Pick a feasible singleton $\{e\}$ in G and read the corresponding position $\mathscr{X}_e$ of $\mathscr{X}_A$.

(2) If $\mathscr{X}_e = 1$ and rank $G > 1$, put $G := G/e$ and go to (1).

(3) If $\mathscr{X}_e = 1$ and rank $G = 1$, print 'A spans' and stop.

(4) If $\mathscr{X}_e = 0$ and e is not a coloop, put $G := G \backslash e$ and go to (1).

(5) If $\mathscr{X}_e = 0$ and e is a coloop, print 'A does not span' and stop.

There is one more characterization of the subdegree of $\lambda_G(t)$, which for branching greedoids takes the following form: If G is the branching greedoid of a rooted directed graph and if d is the least number of edges that must be removed in order to obtain a spanning and acyclic (no directed cycles) subgraph, then d is the subdegree of $\lambda_G(t)$.

8.6.B. Invariants and Reliability

Suppose that ϕ is a function that associates some complex number with each greedoid $G = (E, \mathscr{F})$. For instance, $\phi(G)$ could be the number of feasible sets, the number of bases, or the cardinality of the ground set. Such a function ϕ is called the *invariant* if the following axioms are satisfied.

(I1) $\phi(G) = \phi(G/e)$ if $\{e\}$ is a feasible coloop.

(I2) $\phi(G) = \phi(G/e) + \phi(G\backslash e)$ if $\{e\}$ is feasible and not a coloop.

(I3) $\phi(G) = \phi(G_1)\phi(G_2)$ if G is the direct or ordered sum of G_1 and G_2.

(I4) $\phi(G_1) = \phi(G_2)$ if G_1 and G_2 are isomorphic.

(I5) $\phi(G) \neq 0$ for at least one greedoid G.

Let G_0^n denote the (up to isomorphism) unique greedoid of rank zero and cardinality n. If $\phi(G_0^1) = z$, then $\phi(G_0^n) = z^n$. For $n \geq 1$ this is a direct consequence of axiom (I3). For $n = 0$, $\phi(G_0^0) = 1$ follows from (I3) and (I5) taken together. Notice that this together with (I1) implies that $\phi(G) = 1$ for every full greedoid G.

There is a close connection between invariants and the polynomial $\lambda_G(t)$.

8.6.4. Proposition. *Every invariant ϕ is an evaluation of the greedoid polynomial. More precisely, if $\phi(G_0^1) = z \in \mathbb{C}$, then $\phi(G) = \lambda_G(z)$ for all greedoids G.*

Proof. By definition the invariant ϕ enjoys the same recursive properties as the polynomial evaluation $\lambda_G(z)$. Hence, the two will coincide for all greedoids if they coincide for greedoids of rank zero. But we have already seen that this is the case. $\square$

Simple examples of greedoid invariants are the number of bases $(= \lambda_G(1))$ and the number of spanning sets $(= \lambda_G(2))$. The following probabilistic example is, however, more interesting.

Let $G = (E, \mathscr{F})$ be a greedoid of rank r and cardinality n. Suppose that each element of E is colored red with probability p and blue with probability $1 - p$, for some real number $0 < p < 1$. The coloring of each element is assumed to be independent of the coloring of the others. Let $\pi_G(p)$ denote the probability of the event that the set of blue elements is spanning in G: $\pi_G(p) = \text{Prob(blue spans)}$.

For instance, if G is the circuit matroid of a connected graph, then $\pi_G(p)$ is the probability that the blue edges will connect all vertices; if G is the branching greedoid of a directed rooted graph then $\pi_G(p)$ is the probability that each node can be reached along a path of blue edges from the root; if G is the k-truncation of a poset greedoid then $\pi_G(p)$ is the probability that the order filter generated by the red elements has size at most $n - k$; if G is the k-truncation of a Euclidean convex pruning greedoid then $\pi_G(p)$ is the probability that the convex hull of the red points has size at most $n - k$. In examples such as these one may think of the greedoid as some abstract stochastic system in which components may 'fail' independently of each other with probability p, and the question is to assess the probability that the system will still 'operate' in an appropriate sense (the 'damage' caused by failed components is sufficiently limited).

To analyze the function $\pi_G(p)$, define $\phi_p(G) = p^{r-n}(1-p)^{-r}\pi_G(p)$. We claim that $\phi_p(G)$ *is an invariant.* Only axiom (I2) will be verified here; verification of the other axioms is either similar or trivial.

Suppose that $\{e\}$ is feasible and not a coloop. Then Prob(blue spans) = Prob(e is blue and blue spans) + Prob(e is red and blue spans), or equivalently, $\pi_G(p) = (1-p)\pi_{G/e}(p) + p\pi_{G\backslash e}(p)$. Multiplication of this equation by $p^{r-n}(1-p)^{-r}$ gives $\phi_p(G) = \phi_p(G/e) + \phi_p(G\backslash e)$, so (I2) is satisfied.

Since ϕ_p is an invariant, and clearly $\phi_p(G_0^1) = p^{-1}$ (all subsets of G_0^1 are spanning), we conclude that $\phi_p(G) = \lambda_G(p^{-1})$ for all greedoids G. Hence, we have proven the following.

8.6.5. Proposition. *The probability that the set of blue elements is spanning in G is* $p^{n-r}(1-p)^r\lambda_G(p^{-1})$.

For instance, using our previous calculation we conclude that the blue arcs will reach every node of the graph in Figure 8.8a with probability $(1-p)^6(1+3p+2p^2)$.

8.6.C. Duality

What parts, if any, of the matroid duality operation remain valid for greedoids? The answer to this question will depend on what we mean by a duality operation $G \rightarrow G^*$ that sends a greedoid $G = (E, \mathscr{F})$ to some structured set system $G^* = (E, \mathscr{F}^*)$ on the same ground set. To get reasonably general positive answers we must unfortunately give up the requirement that the dual $(E, \mathscr{F}^*)$ itself is a greedoid. Then there are two weak notions of greedoid duality.

The first is the complementation construction $\mathscr{F}^c = \{E - X : X \in \mathscr{F}\}$. For antimatroids (Proposition 8.7.3), this leads to a dual object that is a convex geometry (it is a greedoid if and only if the original antimatroid is a poset greedoid). A pleasant property of this duality is that the original antimatroid can be uniquely recovered from its dual. A definite disadvantage is that the construction gives nothing of apparent interest for general greedoids, with the following exception.

8.6.6. Theorem. *If* $(E, \mathscr{F})$ *is a full Gaussian greedoid, then so is also* $(E, \mathscr{F}^c)$.

The second notion of duality, and the one that we will briefly discuss here, associates with an arbitrary greedoid a dual object that is a shellable simplicial complex. One price paid for the generality is that the original greedoid cannot be uniquely reconstructed from its dual. On the other hand, this duality operation commutes with deletion and contraction in the desired way and also generalizes some other matroid properties to arbitrary greedoids. Last but not least, it throws additional light on the greedoid polynomial.

In the following discussion hereditary set systems will be called *simplicial complexes*, and we will assume familiarity with the concepts of *shellability* and *shelling polynomial* of simplicial complexes. These and other related notions are defined and discussed in Chapter 7 of this book.

For a greedoid $G = (E, \mathscr{F})$, define its *dual complex* $G^* = (E, \mathscr{F}^*)$ by $\mathscr{F}^* = \{A: A \subseteq E - B \text{ for some basis } B \in \mathscr{B}_G\}$. So, $\mathscr{F}^*$ is the hereditary closure of the family of complements of the bases of G. It is therefore a pure simplicial complex. Clearly, if G is a matroid then G^* is the dual matroid in the usual sense.

As a simplicial complex, a matroid can be characterized either by the exchange property (by definition) or else by being sufficiently shellable (by Theorem 7.3.4). These two properties, exchange and shellability, go different ways in the more general picture: the former is found in all greedoids and the latter in their duals.

8.6.7. Theorem. *The dual complex* G^* *of a greedoid* $G = (E, \mathscr{F})$ *is shellable, and its shelling polynomial is* $\lambda_G(t)$. *Furthermore, if* $e \in E$, *then*

(i) $G^* \backslash e = (G/e)^*$, *if* $\{e\}$ *is feasible;*
(ii) $G^*/e = (G \backslash e)^*$, *if* e *is not a coloop.*

The deletion and contraction operation on simplicial complexes should be understood in the natural way: $G^* \backslash e = \{A \subseteq E - e: A \in \mathscr{F}^*\}$ and $G^*/e = \{A \subseteq E - e: A \cup e \in \mathscr{F}^*\}$ if $\{e\} \in \mathscr{F}^*$. The special requirements on the element e in (i) and (ii) ensure that in each case contraction is well defined.

Let us sketch the proof of Theorem 8.6.7 to the extent that its relevance for the greedoid polynomial becomes clear. Start by assigning a total ordering Ω to the ground set E. As explained in the first paragraph of section 8.6.A, this induces a total ordering of the set $\mathscr{B}$ of bases of G. The first main fact is that the corresponding ordering of the basis complements is a shelling order. In particular, G^* is shellable. The second main fact is that the restriction of the facet $E - B$ induced by this shelling order (as defined in section 7.2) is $E - (B \cup \text{ext}_\Omega(B))$. Consequently, $\lambda_G(t) = \lambda_{G,\Omega}(t)$, as defined in (8.10), is the shelling polynomial. Also, it follows that Lemma 8.6.1 is a special case of Proposition 7.2.2.

8.7. Antimatroids

Antimatroids were defined in section 8.2 as greedoids that have the interval property without upper bounds. They were characterized in Proposition 8.2.7 as accessible set systems closed under union and by a special exchange property. Several examples were given in section 8.3.B.

In this section we shall discuss some of the special structure of antimatroids, which makes them an exceptional class of greedoids. It turns out that antimatroids model some combinatorial properties of the *convex hull* operator in Euclidean spaces much like matroids model the combinatorial properties of the *linear span* operator.

8.7.A. The Duality with Convex Geometries

A *closure operator* on a finite set E is an increasing, monotone, and idempotent function $\tau: 2^E \to 2^E$. This means that for all $A, B \subseteq E$:

(CO1) $A \subseteq \tau(A)$;
(CO2) $A \subseteq B$ implies $\tau(A) \subseteq \tau(B)$;
(CO3) $\tau\tau(A) = \tau(A)$.

Fixed sets $A = \tau(A)$ are called *closed*, and it follows from the axioms that the family $\mathscr{C}$ of closed sets is preserved under intersection (i.e. $A, B \in \mathscr{C} \Rightarrow A \cap B \in \mathscr{C}$). Conversely, if $\mathscr{C}' \subseteq 2^E$ is a set system preserved under intersection then $\tau(A) = \cap \{C \in \mathscr{C}' : A \subseteq C\}$ is a closure operator, and this gives a one-to-one correspondence between closure operators and intersection-invariant set systems containing E. In particular, a closure operator can be specified by giving the family $\mathscr{C}$ of closed sets.

The closure operator $\tau(A) = \{x \in E : r(A \cup x) = r(A)\}$ of a matroid is characterized by the additional *MacLane exchange axiom*:

(E) If $x, y \notin \tau(A)$ and $y \in \tau(A \cup x)$, then $x \in \tau(A \cup y)$.

Now, let E be a finite subset of $\mathbb{R}^n$ and for subsets $A \subseteq E$ let $\tau(A) = E \cap \text{conv}(A)$, where $\text{conv}(A)$ denotes the convex hull of A in the usual sense of Euclidean geometry, i.e. $\tau(A) = \{x \in E : x = \Sigma \lambda_i a_i, \ a_i \in A, \ 0 \leq \lambda_i \leq 1, \ \Sigma \lambda_i = 1\}$. It is a very

Figure 8.9.

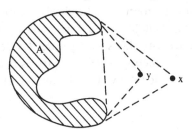

interesting fact that this convex hull closure τ satisfies a property opposite to (E), which we call the *anti-exchange axiom*:

(AE)   If $x, y \notin \tau(A)$, $x \neq y$, and $y \in \tau(A \cup x)$, then $x \notin \tau(A \cup y)$.

An intuitive illustration of this axiom is given in Figure 8.9.

This leads to the following general definition.

8.7.1. Definition. A *convex geometry* is a pair (E, τ) where E is a finite set and τ is a closure operator on E satisfying the anti-exchange condition (AE).

To stay close to the geometric intuition it could have seemed natural to demand that $\tau(\varnothing) = \varnothing$, and even that $\tau(\{x\}) = \{x\}$ for all $x \in E$, in a convex geometry. However, it will soon appear that from a greedoid point of view such restrictions are unwise.

As the following characterization shows, convex geometries have several of the well known properties of Euclidean convexity, for instance with respect to the role of extreme points. For a general closure operator $\tau: 2^E \rightarrow 2^E$, a point $x \in A$ is called an *extreme point* of $A \subseteq E$ if $x \notin \tau(A - x)$. The set of extreme points of A is denoted by ex(A). Observe that for general closures (e.g. for matroid closures), ex(A) = $\varnothing$ is possible for sets $A \nsubseteq \tau(\varnothing)$.

8.7.2. Proposition. *Let $\tau: 2^E \rightarrow 2^E$ be a closure operator on a finite set E. Then the following conditions are equivalent.*

   (i) *(E, τ) is a convex geometry.*
  (ii) *For all closed sets $A \subset B$ there exists $x \in B - A$ such that $A \cup x$ is closed.*
 (iii) *For every closed set $A \subset E$ there exists $x \in E - A$ such that $A \cup x$ is closed.*
 (iv) *All maximal chains of closed sets, $\tau(\varnothing) = A_0 \subset A_1 \subset \ldots \subset A_k = E$, have the same length $k = |E - \tau(\varnothing)|$.*
  (v) *$A = \tau(\text{ex}(A))$, for every closed set A.*
 (vi) *Every $A \subseteq E$ has a unique minimal spanning subset (i.e. the family $\{S \subseteq A: \tau(S) = \tau(A)\}$ has an inclusion-wise least member).*

Proof. (i)⇒(ii). Suppose C is a minimal closed set such that $A \subset C \subseteq B$, and let $x \in C - A$. Then $A \cup x$ is closed. For, if $y \in \tau(A \cup x) - (A \cup x)$, then by the anti-exchange condition (AE) $x \notin \tau(A \cup y)$; hence $A \subset \tau(A \cup y) \subset C$, which contradicts the minimality of C.

(ii)⇔(iii). Condition (iii) is a specialization of (ii). Suppose now that (iii) holds, and let $A \subset B$ be closed sets. By repeated use of (iii) we can find a chain $A = A_0 \subset A_1 \subset \ldots \subset A_s = E$ of closed sets A_i with $|A_i| = |A| + i$, $0 \le i \le s = |E - A|$. Since $B \cap A_0 = A$ and $B \cap A_s = B$, we can find some i such that $|B \cap A_i| = |A| + 1$. Since $B \cap A_i = A \cup x$ is closed (being the intersection of two closed sets), (ii) follows.

(ii)⇒(iv). A reformulation of (ii) is that an inclusion $A \subset B$ is a covering in the lattice of closed sets if and only if $|B - A| = 1$, from which (iv) immediately follows.

(iv)⇒(v). Let A be a closed set. One easily sees that $x \in ex(A)$ if and only if $A - x$ is closed. Hence, assuming (iv), we have that $ex(A) = \cup\{A - B: A$ covers B in the lattice of closed sets$\}$. Now, if $\tau(ex(A)) \subset A$, then $\tau(ex(A)) \subseteq B = A - x$, for some $x \in ex(A)$, which contradicts $x \in \tau(ex(A))$. Hence, $\tau(ex(A)) = A$.

(v)⇒(vi). Let $D = ex(\tau(A))$. Then, clearly, $D \subseteq S$ for every spanning subset $S \subseteq A$. Also, $\tau(D) = \tau(A)$, by (v).

(vi)⇒(i). Suppose that axiom (AE) fails, i.e. we have $x, y \notin \tau(A)$, $x \ne y$, $y \in \tau(A \cup x)$, and $x \in \tau(A \cup y)$. Then $\tau(A \cup x) = \tau(A \cup y)$. Let D be the unique minimal spanning subset of $\tau(A \cup x)$. Then, since $A \cup x$ and $A \cup y$ are spanning, we get $D \subseteq (A \cup x) \cap (A \cup y) = A$, and hence $\tau(D) \subseteq \tau(A) \subset \tau(A \cup x)$, a contradiction. $\square$

Here are a few examples of convex geometries (E, τ).

(1) Let $P = (E, \le)$ be a finite poset, and for $A \subseteq E$ define $\tau(A) = \{x \in E: x \ge y$ for some $y \in A\}$. The closed sets in this geometry are the *order filters* (or, *dual ideals*) of P.

(2) Let $P = (E, \le)$ again be a finite poset and take the *interval closure* $\tau(A) = \{x \in E: y_1 \le x \le y_2$ for some $y_1, y_2 \in A\}$.

(3) Let E be the edge set (or, vertex set) of a tree T and for $A \subseteq E$ let $\tau(A)$ be the smallest subtree of T that contains A. The closed sets of this geometry are the subtrees of T, and the extreme points of a subtree are its leaves.

(4) Let E be the arc set of an acyclic digraph (i.e. a directed graph with no directed cycles), and let $\tau(A)$ be the transitive closure of $A \subseteq E$. In particular, if E is the set of comparability relations of a poset, then the closed sets of this geometry can be identified with the subposets, and the extreme points of a subposet are its covering relations.

(5) Let E be a finite subset of $\mathbb{R}^n$, and for $A \subseteq E$ let $\tau(A) = E \cap \text{conv}(A)$ be the Euclidean convex hull closure. This example, which we already used to motivate the anti-exchange axiom (AE), is particularly important for providing geometric intuition.

The preceding list of examples of convex geometries shows considerable overlap with the examples of antimatroids given in section 8.3.B. In fact, the two concepts are completely equivalent, in the sense of the following duality.

8.7.3. Proposition. *Let E be a finite set and $\mathscr{F} \subseteq 2^E$. Then $(E, \mathscr{F})$ is an antimatroid if and only if $\mathscr{F}^c = \{E - X : X \in \mathscr{F}\}$ is the family of closed sets of a convex geometry. Hence, there is a one-to-one correspondence $\mathscr{F} \leftrightarrow \mathscr{F}^c$ between antimatroids and convex geometries on E.*

Proof. Condition (iii) of Proposition 8.7.2 shows that a set system $\mathscr{C} \subseteq 2^E$ is the family of closed sets of a convex geometry if and only if $\mathscr{C}$ is closed under intersection and for every $A \in \mathscr{C}$ there exists $B \in \mathscr{C}$ such that $A \subset B$ and $|B| = |A| + 1$. This means precisely that the family of set complements is closed under union and accessible, i.e. it is an antimatroid. $\square$

The duality with convex geometries is very useful and illuminating for the study of antimatroids. Examples are often easily recognized by their closure operator, and the geometric intuition provided by the dual point of view is most valuable. From now on we will say that a subset A of an antimatroid is *convex* if A is closed in the dual convex geometry, i.e. if $E - A$ is feasible.

8.7.B. Some Characterizations of Antimatroids

In this section we shall give some additional characterizations of antimatroids both as set systems and as languages.

Let E be a finite set and let H be a mapping that associates with each element $x \in E$ a set system $H(x) \subseteq 2^{E-x}$. This defines a left hereditary language:

$$\mathscr{L}_H = \{x_1 x_2 \ldots x_k \in E_s^* : \text{for all } 1 \leq i \leq k \text{ there is a set } A \in H(x_i)$$
$$\text{such that } A \subseteq \{x_1, x_2, \ldots, x_{i-1}\}\}. \tag{8.11}$$

The system $H = (H(x))_{x \in E}$ will be called an *alternative precedence system*, and the language $\mathscr{L}_H$ that it generates an *alternative precedence language*. In the context of scheduling and searching procedures it is often natural to obtain feasible sequences this way: an item x becomes legal once at least one 'alternative precendence set' has already been processed. Examples will be discussed after this result.

8.7.4. Proposition. *Let $\mathscr{L} \subseteq E_s^*$ be a finite simple language. Then the following conditions are equivalent.*

(i) *$(E, \mathscr{L})$ is an antimatroid.*

(ii) *$\mathscr{L}$ is an alternative precedence language.*

(iii) *$\varnothing \in \mathscr{L}$ and $\mathscr{L}$ satisfies the following exchange axiom:*

(A') *for $\alpha, \beta \in \mathscr{L}$ such that $\tilde{\alpha} \nsubseteq \tilde{\beta}$, there is some $x \in \tilde{\alpha} - \tilde{\beta}$ such that $\beta x \in \mathscr{L}$.*

(Note that (A') is the ordered version of axiom (A) in Proposition 8.2.7.)

Proof. (i) $\Rightarrow$ (ii). Define an alternative precedence system by $H(x) = \{\tilde{\alpha}: \alpha x \in \mathscr{L}\}$. Then clearly $\mathscr{L} \subseteq \mathscr{L}_H$. Conversely, suppose $x_1 x_2 \ldots x_k \in \mathscr{L}_H$. By induction on k we may assume that $x_1 x_2 \ldots x_{k-1} \in \mathscr{L}$. By definition of $\mathscr{L}_H$ there exists some $\alpha \in \mathscr{L}$ such that $\alpha x_k \in \mathscr{L}$ and $\tilde{\alpha} \subseteq \{x_1, \ldots, x_{k-1}\}$. Since $\mathscr{L}$ is closed under union, we get that $\{x_1, \ldots, x_k\} = \{x_1, \ldots, x_{k-1}\} \cup \widetilde{\alpha x_k} \in \mathscr{L}$. Hence, by Proposition 8.2.3, $x_1 x_2 \ldots x_k \in \mathscr{L}$. So, $\mathscr{L}_H \subseteq \mathscr{L}$.

(ii) $\Rightarrow$ (iii). Suppose $\alpha, \beta \in \mathscr{L} = \mathscr{L}_H$, and $\tilde{\alpha} \nsubseteq \tilde{\beta}$, $\alpha = x_1 x_2 \ldots x_k$. Let j be minimal such that $x_j \notin \tilde{\beta}$. Then for some $A \in H(x_j)$, $A \subseteq \{x_1, \ldots, x_{j-1}\} \subseteq \tilde{\beta}$; hence $\beta x_j \in \mathscr{L}_H$.

(iii) $\Rightarrow$ (i). This was proven in the unordered version in Proposition 8.2.7. $\square$

Let us find alternative precedence systems giving rise to some of the familiar antimatroids.

(1) Let $P = (E, \leq)$ be a finite poset, and for $x \in E$ let $H(x) = \{\{y \in E: y < x\}\}$. Then $(E, \mathscr{L}_H)$ is the poset greedoid.

(2) Let $(V', \mathscr{L})$ be the vertex search greedoid of a rooted graph (V, E, r), $V' = V - r$. If $x \in V'$ is adjacent to the root let $H(x) = \varnothing$; otherwise let $H(x)$ consist of singletons, one for each neighbor of x. Then $\mathscr{L} = \mathscr{L}_H$.

(3) Let E be a finite subset of $\mathbb{R}^n$, and let $(E, \mathscr{L})$ be the convex pruning greedoid. For each $x \in E$ let $H(x)$ consist of the intersections of $E - x$ with closed halfspaces having x on the boundary. Then $\mathscr{L} = \mathscr{L}_H$.

The feasible sets of an antimatroid $(E, \mathscr{F})$ ordered by inclusion form a lattice, with lattice operations: $X \vee Y = X \cup Y$, and $X \wedge Y$ is the unique basis of $X \cap Y$. Lattices of this kind can be characterized in purely lattice-theoretical terms.

A finite lattice L is said to be *join-distributive* (or, *locally free*) if for every $x \in L - \{\hat{1}\}$ the interval $[x, j(x)]$ is Boolean, where $j(x)$ is the join of all elements in L that cover x. Clearly, every distributive lattice is join-distributive, and every join-distributive lattice is semimodular. In particular, every join-distributive lattice is *graded*, i.e. there exists a rank function $r: L \to \mathbb{N}$ satisfying $r(\hat{0}) = 0$ and $r(x) = r(y) + 1$ whenever x covers y.

8.7.5. Proposition. *Let $\mathcal{F} \subseteq 2^E$ be an accessible set system. Then the following conditions are equivalent.*

(i) $(E, \mathcal{F})$ *is an antimatroid.*

(ii) $(\mathcal{F}, \subseteq)$ *is a join-distributive lattice.*

(iii) $(\mathcal{F}, \subseteq)$ *is a semimodular lattice.*

Proof. (i)⇒(ii). The sets that cover $X \in \mathcal{F}$ in an antimatroid lattice $(\mathcal{F}, \subseteq)$ are of the form $X \cup x_i$, for some $x_i \in E - X$, $1 \le i \le t$. Since $\mathcal{F}$ is closed under unions, $X \cup \{x_{i_1}, x_{i_2}, ..., x_{i_v}\} \in \mathcal{F}$, for all $1 \le i_1 < i_2 < ... < i_v \le t$. Hence, $(\mathcal{F}, \subseteq)$ is join-distributive.

(ii)⇒(iii). Every join-distributive lattice is semimodular.

(iii)⇒(i). The assumption implies that cardinality is a semimodular rank function on the lattice $(\mathcal{F}, \subseteq)$, i.e. $X \subset Y$ is a covering only if $|X| + 1 = |Y|$ and for all $X, Y \in \mathcal{F}$: $|X \wedge Y| + |X \vee Y| \le |X| + |Y|$.

Suppose that $X, Y \in \mathcal{F}$ and $X \nsubseteq Y$. Take a saturated chain $X \wedge Y = A_0 \subset A_1 \subset ... \subset A_k = X$ of sets $A_i \in \mathcal{F}$, $|A_i| = |X \wedge Y| + i$. Let j be maximal such that $A_j \subseteq Y$. Then, clearly $A_j = Y \wedge A_{j+1}$, and by semimodularity $1 \le |Y \vee A_{j+1}| - |Y| \le |A_{j+1}| - |Y \wedge A_{j+1}| = 1$. Hence, for $x \in A_{j+1} - A_j \subseteq X - Y$ we have $Y \cup x = Y \cup A_{j+1} = Y \vee A_{j+1} \in \mathcal{F}$. Thus, axiom (A) has been verified and, by Proposition 8.2.7, $(E, \mathcal{F})$ is an antimatroid. $\square$

Antimatroids are related to join-distributive lattices in a stronger sense than that expressed by the previous result. The two concepts are essentially equivalent.

8.7.6. Theorem. *A finite lattice L is join-distributive if and only if L is isomorphic to the lattice $(\mathcal{F}, \subseteq)$ of feasible sets of some antimatroid $(E, \mathcal{F})$.*

Proof. We shall merely sketch the construction, leaving the verification of a crucial lemma aside.

Let L be a finite graded lattice and let $M(L)$ denote the set of meet-irreducible elements in L. The lemma we need is that L is join-distributive if and only if the natural map $T: x \mapsto \{y \in M(L): y \ngeq x\}$ embeds L into the Boolean lattice $2^{M(L)}$ preserving both rank and joins.

Now, if L is a join-distributive lattice, let $\mathcal{F} := T(L) \subseteq 2^{M(L)}$. From the properties of T one concludes that $\mathcal{F}$ is accessible and closed under union, i.e. an antimatroid, and that T gives an isomorphism $L \cong (\mathcal{F}, \subseteq)$. $\square$

This representation theorem is a natural extension of G. Birkhoff's theorem, which says that L is distributive if and only if L is isomorphic to the lattice of ideals of some poset. As a direct consequence of Birkhoff's theorem we derive the following.

8.7.7. Proposition. *A greedoid* $(E, \mathscr{F})$ *is a poset greedoid if and only if* $\mathscr{F}$ *is closed under both union and intersection.*

Although quite special, poset greedoids generate all other antimatroids as homomorphic images.

8.7.8. Proposition. *Let* $f\colon P \rightarrow E$ *be a function from a finite poset* P *to a finite set* E, *and let* $\mathscr{F} = \{f(A) \subseteq E \colon A$ *is an ideal in* $P\}$. *Then* $(E, \mathscr{F})$ *is an antimatroid. Furthermore, every antimatroid is induced in this way by a map from some poset.*

8.7.C. Circuits

Let $(E, \mathscr{F})$ be an antimatroid, and let τ denote the convex closure operator of the dual convex geometry. Recall the notions of X-free sets and extreme points defined in sections 8.4.E and 8.7.A respectively. Also, for $X \in \mathscr{F}$ let $\Gamma(X) = \{a \in E - X \colon X \cup a \in \mathscr{F}\}$.

For a subset $A \subseteq E$, we define the *trace*, $\mathscr{F} \colon A = \{X \cap A \colon X \in \mathscr{F}\}$. Since $\mathscr{F} \colon A$ is accessible and closed under union, $(E, \mathscr{F} \colon A)$ is again an antimatroid.

8.7.9. Lemma. *For* $A \subseteq E$, *the following conditions are equivalent.*
  (i) $\mathscr{F} \colon A = 2^A$.
  (ii) $A = \Gamma(X)$ *for some* $X \in \mathscr{F}$.
  (iii) A *is free over* X *for some* $X \in \mathscr{F}$.
  (iv) $A = \mathrm{ex}(C)$ *for some convex set* C.
  (v) $a \notin \tau(A - a)$ *for all* $a \in A$.
  (vi) $E - A$ *is closed (i.e.* $\sigma(E - A) = E - A$).

We leave the easy verification as an exercise.

A subset $A \subseteq E$ is called *free* if it satisfies the conditions in Lemma 8.7.9. Subsets of free sets are again free. Minimal non-free sets are called *circuits*. A 1-element circuit is the same thing as a loop. (Notice that this terminology is consistent with the matroid case. If $(E, \mathscr{F})$ is a matroid then the trace $\mathscr{F} \colon A$ equals the restriction to A; hence free (in the sense of (i)) means independent. Notice also that for general interval greedoids the trace operation does not necessarily produce a greedoid.)

Let C be a circuit in the antimatroid $(E, \mathscr{F})$. Then $a \in \tau(C - a)$ for some $a \in C$, by condition (v). Let $x \in C - a$, and put $B = C - \{a, x\}$. Then $a, x \notin \tau(B)$, since $B \cup a$ and $B \cup x$ are free, and $a \in \tau(B \cup x)$. Hence, by anti-exchange, $x \notin \tau(B \cup a) = \tau(C - x)$.

We have shown that each circuit C has a unique element a such that $a \in \tau(C - a)$. This is called the *root* of C.

8.7.10. Lemma. *For $a \in C \subseteq E$, the following conditions are equivalent.*

(i) C *is a circuit with root* a.

(ii) *If* $B \subseteq C$ *and* $x \in C - B$, *then* $x \in \tau(B) \Leftrightarrow B = C - a$.

(iii) $\mathscr{F} : C = 2^C - \{\{a\}\}$.

Proof. (i)$\Rightarrow$(ii). For $B \cup x = C$ this has already been shown. If $B \cup x \neq C$, then $B \cup x$ is free, hence $x \notin \tau(B)$.

(ii)$\Rightarrow$(iii). Reformulate (ii) as follows: if $B \subseteq C$ then $B = \tau(B) \cap C \Leftrightarrow B \neq C - a$. Since $E - \tau(B) \in \mathscr{F}$, this implies (iii).

(iii)$\Rightarrow$(i). Every proper subset of C is clearly free. $\qquad \Box$

Let us exemplify these definitions, using the familiar antimatroids.

(1) For a poset greedoid, the free sets are the antichains, and the circuits are the pairs $\{a, b\} \subseteq E$ such that $a < b$. The root of such a circuit is b, the larger point.

(2) For the interval closure greedoid of a poset, the free sets are the unions of two antichains, and the circuits are the triples $a < b < c$, with the root in the middle.

(3) For the vertex pruning greedoid of a tree, the free sets are the sets of endpoints (leaves) of subtrees, and the circuits are the triples of vertices that lie on some path, the middle vertex being the root.

(4) For a Euclidean convex pruning greedoid on $E \subseteq \mathbb{R}^n$, a subset $A \subseteq E$ is free if every point of A is an extreme point of the convex hull of A. A circuit consists of the vertices of a simplex together with a point in the relative interior of the simplex. The interior point is the root of the circuit. So the size of a circuit is at least 3 and at most $n + 2$.

It is clear from these examples that the circuits of an antimatroid do not determine the greedoid. For instance, a poset $P = (E, \leq)$ and the dual poset $P^* = (E, \geq)$ in general have different poset greedoids, but these greedoids have the same circuits. However, an antimatroid *is* determined by its *rooted circuits* (C, a), i.e. pairs such that C is a circuit with root a.

8.7.11. Proposition. *Let* $(E, \mathscr{F})$ *be an antimatroid and* $A \subseteq E$. *Then* $A \in \mathscr{F}$ *if and only if* $C \cap A \neq \{a\}$, *for every rooted circuit* (C, a).

Proof. The condition is necessary for a feasible set, by condition (iii) of Lemma 8.7.10. To prove sufficiency, suppose that $A \notin \mathscr{F}$, or, equivalently, that $E - A$ is not convex. Let $x \in A \cap \tau(E - A)$, and let $D = \text{ex}(\tau(E - A))$. That is, D is the unique minimal spanning subset of $E - A$ (cf. Proposition 8.7.2). Since $x \notin D$ and $x \in \tau(D)$, the set $D \cup x$ is not free, and hence contains some circuit C. Since D is free (being the set of extreme points of a convex set), we conclude that (C, x) is a rooted circuit. But $C \cap A = \{x\}$, since by construction $C - x \subseteq D \subseteq E - A$. ☐

The previous result suggests that a characterization of antimatroids in terms of rooted circuits might be possible. This is indeed so, as shown by the following axiomatization, which has a curious resemblance to the circuit axioms for matroids.

8.7.12. Theorem. Let $\mathscr{C} \subseteq \{(C, a): a \in C \subseteq E\}$ be a family of rooted subsets of a finite set E. Then $\mathscr{C}$ is the family of rooted circuits of an antimatroid if and only if the following two conditions hold.

(CI1) If (C_1, a), $(C_2, a) \in \mathscr{C}$, then $C_1 \not\subseteq C_2$.
(CI2) If (C_1, a_1), $(C_2, a_2) \in \mathscr{C}$, $a_1 \neq a_2$, and $a_1 \in C_1 \cap C_2$, then there exists $(C, a_2) \in \mathscr{C}$ such that $C \subseteq C_1 \cup C_2 - a_1$.

8.8. Poset of Flats

The geometric lattice of a matroid has two different generalizations in greedoid theory, which coincide exactly for the class of interval greedoids. Neither of them determines the associated greedoid completely. Nevertheless, a substantial part of the structure theory of greedoids is captured by its order and lattice theoretic aspects.

8.8.A. Poset Representations and Flats

We will now construct the poset of flats of a greedoid as its 'most efficient' poset representation, define the poset of closed sets of a greedoid (which requires a more complicated ordering than inclusion), and then study the canonical map from the poset of flats to the poset of closed sets.

In section 8.2 we described how a greedoid $(E, \mathscr{F})$ can be described by the Hasse diagram of the poset $(\mathscr{F}, \subseteq)$, in which every edge (covering relation) $X <\cdot Y$ is labeled by the 1-element set $Y - X$. From this labeled poset, the language $\mathscr{L}$ can explicitly be read off: the words of $\mathscr{L}$ are uniquely given by the label sequences along the unrefinable chains in the poset $(\mathscr{F}, \subseteq)$ that start at the least element $\varnothing$.

However, it is more efficient to allow larger sets of labels; often the greedoid can be given by the Hasse diagram of a smaller poset P with least element $\hat{0}$, whose edges $s <\cdot t$ are labeled by sets $\lambda(s <\cdot t) \subseteq E$ of alternative labels. A *poset representation* of a greedoid $(E, \mathscr{L})$ is such a set-labeled poset from which $\mathscr{L}$ arises (without repetitions) as the collection of words along unrefinable chains starting at $\hat{0}$ that pick exactly one letter from each label set. That is,

$$\mathscr{L} = \{x_1 x_2 \ldots x_k : x_i \in \lambda(s_{i-1} <\!\cdot\, s_i) \text{ for } 1 \leq i \leq k, \text{ where } k \geq 0 \text{ and}$$
$$\hat{0} = s_0 <\!\cdot\, s_1 <\!\cdot\, \ldots <\!\cdot\, s_k \text{ is a chain in } P\}. \tag{8.12}$$

Here repetitions do not occur if and only if $\lambda(s <\!\cdot\, t_1) \cap \lambda(s <\!\cdot\, t_2) = \varnothing$ whenever $t_1 \neq t_2$ both cover $s \in P$.

For example, Figure 8.10 gives two poset representations of the greedoid that we have previously described in Figure 8.2. Set brackets for the label sets are again omitted.

Figure 8.10.

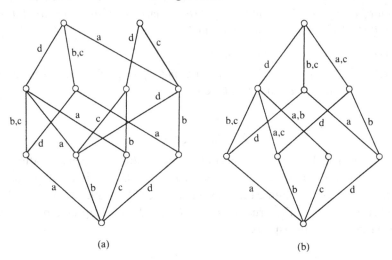

(a)   (b)

How do such poset representations arise? Whenever the same set of words can be read off above two different poset elements, these elements can be identified, resulting in a smaller poset and a more efficient poset representation. Now every poset element s corresponds to a set of words – the words in $\mathscr{L}$ that are coded along maximal chains from $\hat{0}$ to s. Thus two poset elements s_1 and s_2 can be identified if the words corresponding to s_1 and s_2 have the same continuations. This suggests the following construction for the most efficient (or universal) poset representation, using the contraction of greedoid languages as defined by (8.6).

8.8.1. Definition. Let $(E, \mathscr{L})$ be a greedoid. We define an equivalence relation on $\mathscr{L}$ by

$$\alpha \sim \beta \Leftrightarrow \mathscr{L}/\alpha = \mathscr{L}/\beta, \tag{8.13}$$

that is, α and β are equivalent if they have the same set of continuations.

The equivalence classes $[\alpha] \in \mathscr{L}/\!\sim$ with respect to this relation are the *flats* of the greedoid $\mathscr{L}$. The *poset of flats* of the greedoid $(E, \mathscr{L})$ is

$$\Phi = (\mathscr{L}/\!\sim, \leq),$$

where the flats are ordered by

$$[\alpha] \leq [\beta] \Leftrightarrow \alpha\gamma \sim \beta, \text{ for some } \gamma \in \mathscr{L}/\alpha. \tag{8.14}$$

The *labeled poset of flats* $\hat{\Phi}$ is the poset Φ together with the edge labeling λ of the Hasse diagram of Φ that associates to every covering relation $[\alpha] <\cdot [\beta]$ in Φ the set

$$\lambda([\alpha] < [\beta]) = \{x \in E : \alpha x \sim \beta\}. \tag{8.15}$$

The verification that $\leq$ and λ are well defined by (8.14) and (8.15) is straightforward.

Since $\alpha \sim \beta$ implies $|\alpha| = |\beta|$, the rank function on $\mathscr{L}$ carries over to $\mathscr{L}/\sim$, with $r([\alpha]) = |\alpha|$, for all $\alpha \in \mathscr{L}$. This makes Φ into a graded poset of rank r; its unique minimal element is $[\varnothing]$ and its unique maximal element is the equivalence class of all basic words.

Now since words with the same support are always related ($\tilde{\alpha} = \tilde{\beta}$ implies $\alpha \sim \beta$), one can use the equivalence between greedoids and greedoid languages, as described in section 8.2, to give an equivalent definition of the poset of flats in set-theoretic terms. For this, let $(E, \mathscr{F})$ be a greedoid, and for $X, Y \in \mathscr{F}$ define $X \sim Y$ if $\mathscr{F}/X = \mathscr{F}/Y$. From this we get a poset of flats $\tilde{\Phi} = (\mathscr{F}/\sim, \leq)$, where '$\leq$' is the partial order induced by inclusion, that is, $[X] \leq [Y]$ if and only if $X \cup Z \sim Y$ for some $Z \in \mathscr{F}/X$. Note that this in particular implies $[X] \leq [Y]$ whenever $X \subseteq Y$.

8.8.2. Proposition. *The map given by* $[\alpha] \mapsto [\tilde{\alpha}]$ *is an isomorphism of posets* $\Phi \cong \tilde{\Phi}$.

This proposition (which, using Proposition 8.2.3, is easy to verify) shows that there is an essentially unique concept of the poset of flats $\Phi \cong \tilde{\Phi}$ and of the labeled poset of flats $\hat{\Phi}$.

The labeled poset $\hat{\Phi} = (\Phi, \lambda)$ is *universal* as a poset representation of $\mathscr{L}$, in the sense that for every poset representation (P, λ') of $\mathscr{L}$, there is a unique, order preserving, rank preserving surjective map

$$f: P \to \Phi$$

such that for $s, t \in P$, $s <\cdot t$ implies that

$$\lambda'(s < t) \subseteq \lambda(f(s) < f(t)).$$

For example, consider again the greedoid depicted in Figure 8.2. Its labeled poset of flats is given by Figure 8.10b. The canonical maps from the poset representations in Figures 8.2b and 8.10a to the universal representation are easy to see.

Here are descriptions of the poset of flats Φ for some important classes of greedoids.

(1) In the case of matroids, Φ is (isomorphic to) the geometric lattice of flats, since for two independent sets X and Y, $X \sim Y$ holds exactly if X and Y have the same closure. The edge labeling of $\hat{\Phi}$ is then given by

$$\lambda([X] <\cdot [Y]) = \sigma(Y) - \sigma(X).$$

(2) If $(E, \mathcal{F})$ is a greedoid with only one basis, then $\Phi \cong (\mathcal{F}, \subseteq)$, and λ is the labeling by 1-element sets discussed in section 8.2. This includes the case of all antimatroids, so by Theorem 8.7.6 we see that the poset of flats of an antimatroid is a join-distributive lattice.

More generally, for every greedoid $(E, \mathcal{F})$ the canonical surjective poset map $f: (\mathcal{F}, \subseteq) \rightarrow \Phi$, defined by $f(X) = [X]$, is injective on each interval $[X, Y]$ in $\mathcal{F}$.

(3) Let $(E, \mathcal{F})$ be a branching greedoid on a rooted undirected or directed graph, as in section 8.3.C. Clearly, two branchings X and Y are related, $X \sim Y$, if and only if they reach the same set of vertices. Thus there is a bijection between $\mathcal{F}/\sim$ and feasible vertex sets. One sees from this that the poset of flats of the branching greedoid is isomorphic to the poset of feasible sets of the associated vertex search greedoid, and is hence a join-distributive lattice.

8.8.B. Poset of Closed Sets

From the example in Figure 8.10b we can see that for greedoids without the interval property the flats cannot be identified with closed sets – the greedoid has four flats, but only three closed sets of rank 1. What is the structure on the collection of closed sets? How does it relate to the poset of flats?

It is not natural to order the closed sets by inclusion – the resulting posets have little structure and do not seem to encode relevant information. An instructive example is the full greedoid with exactly one basic word: xy, whose poset of flats is a 3-element chain but whose closed sets are $\{x\}$, $\{y\}$ and $\{x, y\}$. Instead, examples such as this suggest ordering the closed sets by

$$A \leq B \text{ if } B \text{ contains a basis of } A.$$

Equivalently, we could put

$$A \leq B \text{ if } r(A \cap B) = r(A). \tag{8.16}$$

Clearly, this generalizes the matroid case. However, it turns out that for non-interval greedoids, the relation '$\leq$' defined by (8.16) is not in general transitive. (For example, if $E = \{a, b, c, d, e\}$ and $(E, \mathcal{F})$ is the greedoid defined by $\mathcal{F} = 2^E - \{\{a, b\}, \{b, c, d\}\}$, then $\{a, b\}$, $\{b, c, d\}$, and $\{c, d, e\}$ are closed sets of ranks 1, 2, and 3, respectively. Here $\{a, b\} \leq \{b, c, d\}$ and $\{b, c, d\} \leq \{c, d, e\}$, but $r(\{a, b\} \cap \{c, d, e\}) = r(\varnothing) = 0 < 1 = r(\{a, b\})$.) We are therefore led to consider the transitive closure of the relation defined by (8.16).

8.8.3. Definition. The *poset of closed sets* of a greedoid is the set $\mathscr{Cl} = \{\sigma(A) : A \subseteq E\}$ together with the partial order for which $A \leq B$ holds if and only if there are closed sets $A_0 = A$, A_1, ..., $A_k = B$ such that for all i ($1 \leq i \leq k$), A_i contains a basis of A_{i-1}, that is, $r(A_i \cap A_{i-1}) = r(A_{i-1})$.

The poset of closed sets has reasonable combinatorial properties. It is graded, and the poset and greedoid rank functions coincide for it. It is clear that $A \subseteq B$ implies $A \leq B$, but not conversely.

For the greedoid of Figure 8.2, whose poset of flats is given by Figure 8.10b, the poset of closed sets is drawn in Figure 8.11.

Figure 8.11.

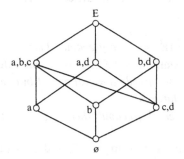

For matroids, antimatroids, and branching greedoids, the poset of flats Φ (as constructed before) and the poset of closed sets $\mathscr{Cl}$ are canonically isomorphic. This is explained by the following result.

Let $(E, \mathscr{L})$ be a greedoid. If $\alpha \sim \beta$ for α, $\beta \in \mathscr{L}$, then $\tilde{\alpha} \sim \tilde{\beta}$ by Proposition 8.8.2, from which $\sigma(\tilde{\alpha}) = \sigma(\tilde{\beta})$ follows. Hence, we have a well defined map

$$\phi : [\alpha] \mapsto \sigma(\tilde{\alpha})$$

from flats to closed sets.

8.8.4. Theorem. *The map $\phi : \Phi \to \mathscr{Cl}$ is order preserving, rank preserving, and surjective. Furthermore, the following conditions are equivalent:*

(i) *ϕ is an isomorphism of posets;*
(ii) *ϕ is injective;*
(iii) *$(E, \mathscr{L})$ is an interval greedoid.*

The greedoid of Figure 8.2 again illustrates this: the map ϕ from Φ (Figure 8.10b) to $\mathscr{Cl}$ (Figure 8.11) is obvious. The interval property fails (e.g. c, $dac \in \mathscr{L}$, but $dc \notin \mathscr{L}$) and ϕ is not injective ($[c] \neq [d]$, although $\phi([c]) = \phi([d]) = \{c, d\}$).

Observe that in general the composite map

$$\mathscr{F} \to \mathscr{F}/\sim \,\cong\, \Phi \to \mathscr{Cl}$$

is the closure operator σ on $\mathscr{F}$. This illustrates the divergence of the concepts of flats and closed sets in non-interval greedoids: the closure operator factorizes over the flats.

8.8.C. Interval Greedoids and Semimodular Lattices

Interval greedoids are intimately related to semimodular lattices both via the poset of feasible sets $(\mathscr{F}, \subseteq)$ and via the poset of flats Φ. In fact, it is reasonable to view interval greedoids as the combinatorial models for semimodular lattices much as matroids and antimatroids are the combinatorial models for geometic and join-distributive lattices respectively.

Here are some key facts about the lattice property and semimodularity in posets of feasible sets.

8.8.5. Proposition. *Let $(E, \mathscr{F})$ be a greedoid. Then*
  (i) *$(E, \mathscr{F})$ is an antimatroid $\Leftrightarrow$ $(\mathscr{F}, \subseteq)$ is a semimodular lattice;*
 (ii) *$(E, \mathscr{F})$ is an interval greedoid $\Leftrightarrow$ all closed intervals $[\varnothing, X]$ in $(\mathscr{F}, \subseteq)$ are semimodular lattices;*
(iii) *if $(E, \mathscr{F})$ is an interval greedoid, then $(\mathscr{F}, \subseteq)$ is a meet-semilattice.*

Proof. Part (i) is from Proposition 8.7.5, and part (ii) follows from it via Proposition 8.2.8.

For part (iii), let $\hat{\mathscr{F}}$ denote the poset $(\mathscr{F}, \subseteq)$ with a maximal element $\hat{1}$ adjoined. To see that meets (greatest lower bounds) exist in $(\mathscr{F}, \subseteq)$, it suffices (by a standard lattice-theoretical argument) to show that any pair $X, Y \in \hat{\mathscr{F}}$ has a join $X \vee Y$ (least upper bound) in $\hat{\mathscr{F}}$. Now, if X and Y have an upper bound Z in $(\mathscr{F}, \subseteq)$, so that $X \cup Y \subseteq Z \in \mathscr{F}$, then from Propositions 8.2.7(ii) and 8.2.8(ii) we conclude that $X \cup Y \in \mathscr{F}$ and therefore $X \vee Y = X \cup Y$. If X and Y do not have an upper bound in $(\mathscr{F}, \subseteq)$, then $X \vee Y = \hat{1}$. $\square$

We remark in connection with the preceding result that for non-interval greedoids the meet operation on $(\mathscr{F}, \subseteq)$ may or may not exist (both cases occur).

Some of the special structure of the poset $(\mathscr{F}, \subseteq)$ for interval greedoids carries over to the poset of flats Φ, and to its labeled version $\hat{\Phi}$.

8.8.6. Lemma. *Let $(E, \mathscr{L})$ be an interval greedoid, and $\alpha x, \alpha y \in \mathscr{L}$ with $[\alpha x] \neq [\alpha y]$. Then $\alpha x y, \alpha y x \in \mathscr{L}$ and $[\alpha x y] = [\alpha y x]$.*

Proof. This is a reformulation of the transposition property for interval greedoids, as observed in the last paragraph of section 8.3.G. $\square$

The lemma shows that for interval greedoids, Φ is a *semimodular poset*: if $s \in \Phi$ is covered by two different elements $t_1, t_2 \in \Phi$, then there is an element $u \in \Phi$ covering both t_1 and t_2. More precisely, the following is true.

8.8.7. Theorem. *If $(E, \mathscr{F})$ is an interval greedoid, then the poset of flats Φ is a semimodular lattice. Conversely, every finite semimodular lattice arises as the poset of flats of some interval greedoid.*

Proof. (1) To prove the lattice property, we use the following simple lemma: a finite poset P having a least element is a lattice if, for $s_1, s_2, t \in P$, the join $s_1 \vee s_2$ exists whenever s_1 and s_2 both cover t.

Hence, here we only have to show that in the situation of Lemma 8.8.6, $[\alpha xy]$ is the only minimal upper bound for $[\alpha x]$ and $[\alpha y]$. Assume that on the contrary $[\alpha x\gamma] = [\alpha y\delta]$ is a different minimal upper bound (so that $[\alpha x\gamma] \not\geqslant [\alpha xy]$), with $\gamma = c_1 c_2 \ldots c_k$. We will prove by induction that $\alpha xc_1 \ldots c_i y \in \mathscr{L}$ and $\alpha xyc_1 \ldots c_i \in \mathscr{L}$, for $0 \leqslant i \leqslant k$, which will lead to a contradiction.

The case $i = 0$ is clear. For $i > 0$ we know by induction that $\alpha xc_1 \ldots c_{i-1} y \in \mathscr{L}$ and $\alpha xyc_1 \ldots c_{i-1} \in \mathscr{L}$, and hence $[\alpha xc_1 \ldots c_{i-1} y] = [\alpha xyc_1 \ldots c_{i-1}] \geqslant [\alpha xy]$. On the other hand, for $\alpha xc_1 \ldots c_i \in \mathscr{L}$ we have $[\alpha xc_1 \ldots c_i] \not\geqslant [\alpha xy]$ by assumption. Hence, $[\alpha xc_1 \ldots c_{i-1} y] \neq [\alpha xc_1 \ldots c_i]$, and Lemma 8.8.6 implies that $\alpha xc_1 \ldots c_i y \in \mathscr{L}$. Augmenting $\alpha xyc_1 \ldots c_{i-1}$ from this yields $\alpha xyc_1 \ldots c_i \in \mathscr{L}$, and the induction is complete.

We have in particular shown that $\alpha xc_1 \ldots c_k y = \alpha x\gamma y \in \mathscr{L}$. Since $[\alpha x\gamma] = [\alpha y\delta]$ if follows that $\alpha y\delta y \in \mathscr{L}$, which is impossible since $\mathscr{L}$ is a simple language.

(2) For the converse, let L be a finite semimodular lattice, and let $E = J(L)$ be the set of join-irreducible elements of L, that is, the set of those lattice elements that cover exactly one element in L. We label the edges in the Hasse diagram of L by

$$\lambda(s <\cdot t) = \{p \in E : s \vee p = t\}.$$

Let $\mathscr{L}$ be the left hereditary language defined by the poset representation (L, λ) as in (8.12). Then $(E, \mathscr{L})$ is an interval greedoid, and L is isomorphic to its poset of flats. We leave the straightforward verification to the reader. $\square$

Contrary to what one might expect, semimodularity of Φ does *not* characterize the interval greedoids. This is shown e.g. by the non-interval greedoid with 6 basic words, xab, xba, yab, yba, xay, and yax, whose poset of flats is a semimodular lattice (the unique non-modular semimodular lattice of rank 3 and order 7).

8.8.8. Lemma. *Let $(E, \mathscr{F})$ be an interval greedoid and $\hat{\Phi} = (\Phi, \lambda)$ the labeled poset of flats. If $t_1, t_2 \in \Phi$ and $t_1 \wedge t_2 <\cdot t_1$, then $t_2 <\cdot t_1 \vee t_2$ and*

$$\lambda(t_1 \wedge t_2 <\cdot t_1) \subseteq \lambda(t_2 <\cdot t_1 \vee t_2). \tag{8.17}$$

Proof. The first claim is true in any semimodular lattice. If $t_1 \wedge t_2 <\cdot t_2$, then
(8.17) follows directly from Lemma 8.8.6. In general, (8.17) has to be proven
by induction on $\text{rank}(t_2) - \text{rank}(t_1 \wedge t_2)$, using Lemma 8.8.6 repeatedly. $\square$

We conclude with the following application of the results of this section.

Proof of Proposition 8.2.5 (necessity). Let $(E, \mathscr{L})$ be an interval greedoid and
let $\alpha = a_1 a_2 \dots a_k$, $\beta = b_1 b_2 \dots b_l \in \mathscr{L}$, $k > l$. Now, α determines an unrefinable
chain $\hat{0} = s_0 <\cdot s_1 <\cdot \dots <\cdot s_k$ in Φ by $s_i = [a_1 a_2 \dots a_i]$ for $0 \le i \le k$. Let $t = [\beta]$.
Since, by Theorem 8.8.7, Φ is a semimodular lattice, each step in the following
chain is an equality or a covering:

$$t = t \vee s_0 \le t \vee s_1 \le \dots \le t \vee s_k.$$

Now let $1 \le i_1 < i_2 < \dots < i_m \le k$ be the sequence of those indices i_j for which

$$t \vee s_{i_j - 1} <\cdot t \vee s_{i_j}.$$

Thus we have an unrefinable chain

$$t <\cdot t \vee s_{i_1} <\cdot t \vee s_{i_2} <\cdot \dots <\cdot t \vee s_{i_m} = t \vee s_k.$$

Clearly, $m = \text{rank}(t \vee s_k) - \text{rank}(t) \ge k - l$.

Now, $a_i \in \lambda(s_{i-1} <\cdot s_i)$ implies by Lemma 8.8.8 that $a_{i_j} \in \lambda(t \vee s_{i_j - 1} <\cdot t \vee s_{i_j})$
for $1 \le j \le m$. Hence, the definition of Φ allows us to read off $\beta a_{i_1} a_{i_2} \dots a_{i_m} \in \mathscr{L}$,
where $\alpha' = a_{i_1} a_{i_2} \dots a_{i_m}$ is a subword of α, of length $|\alpha'| = m \ge k - l$. $\square$

8.8.D. Poset Properties

The unlabeled poset $(\mathscr{F}, \subseteq)$ carries important but incomplete information
about a greedoid $(E, \mathscr{F})$. We will here discuss which greedoid properties and
invariants are *poset properties*, that is, completely determined by the abstract
poset $(\mathscr{F}, \subseteq)$ and not requiring explicit knowledge of the set system $\mathscr{F}$.

We have seen that $(\mathscr{F}, \subseteq)$ is a finite graded poset of rank r and size $|\mathscr{F}|$,
with minimum element $\hat{0} = \varnothing$ and $|\mathscr{B}|$ maximal elements. The number of
unrefinable chains from $\hat{0}$ to some $X \in \mathscr{F}$ is $|\mathscr{L}|$. From this it is clear that r,
$|\mathscr{F}|$, $|\mathscr{B}|$, $|\mathscr{L}|$, and the number of basic words are poset properties.

As a contrast, $|E|$ and $|\cup \mathscr{F}|$ are *not* poset properties. This follows e.g. from
the observation, made in section 8.4.E, that the branching greedoids of a
rooted graph and its associated digraph have isomorphic posets $(\mathscr{F}, \subseteq)$.
Thus, any greedoid property that distinguishes the two branching greedoids
cannot be a poset property. For example, although $\lambda(1)$ is the number of
bases, the greedoid polynomial $\lambda(t)$ is not a poset invariant, and neither are
the evaluations $\lambda(2)$ and $\lambda(0)$ (i.e. the number of spanning sets and the Euler
characteristic of the dual complex).

The interval property and that of being a matroid, antimatroid, poset, or
local poset greedoid are all poset properties. This follows from the following

information, which was gathered in earlier sections:

$(E, \mathscr{F})$ $(\mathscr{F}, \subseteq)$

interval greedoid $\Leftrightarrow$ all intervals are semimodular lattices

local poset greedoid $\Leftrightarrow$ all intervals are distributive lattices

matroid $\Leftrightarrow$ all intervals are Boolean lattices

antimatroid $\Leftrightarrow$ semimodular lattice

poset greedoid $\Leftrightarrow$ distributive lattice

Next we note that k-connectivity is a poset property by definition. In fact, we can define a ranked poset P of rank r with minimal element $\hat{0}$ to be *k-connected* if for every $X \in P$ there is an element $Y \geq X$ in P such that the interval $[X, Y]$ of P is a Boolean lattice of rank $\min\{k, r - r(X)\}$. With this, a greedoid $(E, \mathscr{F})$ is k-connected if and only if the poset $(\mathscr{F}, \subseteq)$ of feasible sets is k-connected.

8.8.9. Proposition. *For a k-connected greedoid $(E, \mathscr{F})$, the poset Φ of flats is also k-connected. If $(E, \mathscr{F})$ is an interval greedoid, then the converse is also true.*

Proof. The first part follows from the remark, made near the end of section 8.8.A, that every restriction of the natural map $(\mathscr{F}, \subseteq) \to \Phi$ to an interval is injective. The second part follows from Lemma 8.8.8. $\square$

Observe that this proposition implies in particular the equivalence (ii)$\Leftrightarrow$(iii) of Proposition 8.4.7, since the poset of flats of a branching greedoid equals the poset of feasible sets of the associated vertex search greedoid.

8.9. Further Topics

8.9.A. Excluded Minor Characterizations

For each of the main classes of greedoids arising among the examples, there is a representation problem: how do we recognize whether an abstractly given greedoid is isomorphic to some greedoid in that class? This problem is in most cases unsolved. For instance, no effective way is known for telling whether a given antimatroid can be represented as the convex pruning greedoid of a point set in $\mathbb{R}^n$.

There are two main ways of characterizing a class of greedoids, either by structural conditions or by excluded minors. The following are examples of structural characterizations: Let $G = (E, \mathscr{F})$ be a greedoid. Then

(1) G is an interval greedoid if and only if $X, Y \subseteq Z$ implies $X \cup Y \in \mathscr{F}$, for all $X, Y, Z \in \mathscr{F}$;

(2) G is a local poset greedoid if and only if $X, Y \subseteq Z$ implies $X \cup Y$, $X \cap Y \in \mathscr{F}$, for all $X, Y, Z \in \mathscr{F}$.

These statements are merely reformulations of information from Propositions 8.2.7, 8.2.8, and 8.7.7. The following result is more substantial.

8.9.1. Theorem. *G is a directed branching greedoid if and only if G is a local poset greedoid and* $\sigma(X) \cap \sigma(Y) \subseteq \sigma(X \cup Y) \subseteq \sigma(X) \cup \sigma(Y)$, *for all* X, $Y \in \mathscr{F}$.

Hereditary classes of greedoids, i.e. classes closed under taking minors, can be specified by listing the minimal non-members. (Here 'minimal' is to be understood as referring to the partial ordering of isomorphism classes of greedoids induced by the relation 'is a minor of', as discussed in section 8.4.D.) These minimal non-members are the *excluded minors* of the hereditary class, and clearly a greedoid is a member of the class precisely when none of its minors is among the excluded ones. Examples of hereditary classes are interval greedoids, local poset greedoids, directed and undirected branching greedoids, polymatroid greedoids, antimatroids, and matroids.

There is no *a priori* reason to expect the list of excluded minors for a hereditary class to be finite or effectively describable. For the classes that were mentioned the following can be said.

Let $E = \{a, b\}$ and $\mathscr{F} = 2^E - \{\{b\}\}$, $\mathscr{F}' = 2^E - \{\{a, b\}\}$. Clearly, $G = (E, \mathscr{F})$ is not a matroid and $G' = (E, \mathscr{F}')$ is not an antimatroid, so they are among the excluded minors for these classes. In fact, they are the only excluded minors.

8.9.2. Proposition.
   (i) *A greedoid is a matroid if and only if it has no minor isomorphic to G.*
   (ii) *A greedoid is an antimatroid if and only if it has no minor isomorphic to G'.*
   (iii) *The classes of interval greedoids, directed branching greedoids, undirected branching greedoids, local poset greedoids, and polymatroid greedoids cannot be characterized by a finite set of excluded minors.*

Proof. (i) In a greedoid that is not a matroid there are feasible sets having non-feasible subsets. Pick one such feasible set X of minimal cardinality. Then for some $a \in X$ the subset $X - a$ is non-feasible, and accessibility gives that $X - b$ is feasible for some $b \in X$. The choice of X implies that $Y = X - \{a, b\}$ is feasible. Now, restriction to X and contraction by Y produces a minor isomorphic to G.

(ii) The lack of minors of type G' is equivalent to the following property: if X, Y and $X \cap Y$ are feasible and $|X| = |Y| = |X \cap Y| + 1$, then $X \cup Y$ is feasible. This implies that the poset of feasible sets is a semimodular lattice, which by Proposition 8.7.5 implies that the greedoid is an antimatroid.

(iii) For $k = 1, 2, \ldots$, let α be a simple word of length k not containing the letters x and y. Let G_k denote the full greedoid with exactly two basic words: $x \alpha y$ and $y \alpha x$. The following facts are easy to verify: (1) G_k lacks the interval property; (2) every proper minor of G_k is a branching greedoid (both directed and undirected). It follows that G_k is a minimal non-member for each of the five hereditary classes, and hence that $(G_k)_{k \geq 1}$ is an infinite list of excluded minors.

$\square$

In spite of what has just been shown in turns out that undirected branching greedoids and local poset greedoids *can* be characterized by finite sets of excluded minors. The requirement for this is that attention must be restricted to the class of interval greedoids only.

Let $E = \{x, y, z\}$ and define greedoids $G_i = (E, \mathscr{F}_i)$, $i = 1, 2, 3, 4$, by (with some obvious simplifications of set notation)

$$\mathscr{F}_1 = 2^E - \{z\},$$
$$\mathscr{F}_2 = 2^E - \{z, xz, yz\},$$
$$\mathscr{F}_3 = 2^E - \{z, yz, xyz\},$$
$$\mathscr{F}_4 = 2^E - \{xyz\}.$$

In Figure 8.12 these greedoids G_1–G_4 are represented as a vertex pruning of a tree, a poset, a directed branching greedoid, and a graphic matroid, respectively.

Figure 8.12.

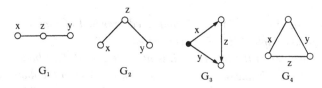

8.9.3. Theorem. *Let G be an interval greedoid. Then*

(i) *G is a local poset greedoid if and only if G has no minor isomorphic to G_1.*

(ii) *G is an undirected branching greedoid if and only if G has no minor isomorphic to G_1, G_2, G_3, or G_4.*

8.9.B. Diameter of the Basis Graph

Let $(E, \mathscr{F})$ be a greedoid of rank r, and $\mathscr{B}$ its set of bases. Two bases X, Y are *adjacent* if they differ in exactly one element and their intersection is feasible, that is, $|X \cap Y| = |X| - 1$ and $X \cap Y \in \mathscr{F}$. By definition, the *basis graph* of $(E, \mathscr{F})$ has vertex set $\mathscr{B}$, and two bases are joined by an edge whenever they are adjacent.

We ask under which conditions the basis graph of a greedoid is connected and what can be said about the diameter of basis graphs.

For matroids, the feasibility condition $X \cap Y \in \mathscr{F}$ is always true. Then the exchange axiom (G2) produces a sequence of adjacent bases between X and Y, which shows that the basis graph of a matroid is always connected and

Figure 8.13.

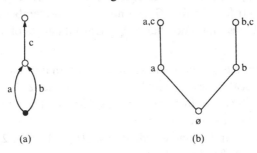

(a)   (b)

has diameter at most r. On the other hand, the distance between any two disjoint bases of a matroid is exactly r.

Questions of the connectedness and the diameter for basis graphs of general greedoids are less trivial. For example the digraph of Figure 8.13a has a branching greedoid (Figure 8.13b) with disconnected basis graph; the intersection of its two bases is not feasible. The branching greedoid of the digraph in Figure 8.3a has rank $r = 2$, whereas its basis graph is a path of length 3; here the diameter of the basis graph is larger than the rank.

In general, the following can be said.

8.9.4. Theorem. Let $(E, \mathscr{F})$ be a 2-connected greedoid of rank r.
  (i) The basis graph of $(E, \mathscr{F})$ is connected.
 (ii) The diameter of the basis graph is at most $2^r - 1$. This bound is sharp.
(iii) If $(E, \mathscr{F})$ is a branching greedoid of rank $r > 0$, then the diameter of the basis graph is at most $r^2 - r + 1$. The bound is best-possible.

Note that the branching greedoid of Figure 8.3 is 2-connected and serves as an extremal example for the case $r = 2$. In fact, this case can be used to prove connectedness of the basis graph, and the bound in (ii) on its diameter, by induction on r. Sharpness of the bound is then established by an explicit construction.

To prove (iii), one has to exploit the surjective map of section 8.8.A from the branching greedoid $\mathscr{F}$ to its poset of flats Φ, corresponding to the vertex search greedoid of the graph. The poset Φ – a semimodular, coatomic lattice – has special properties that can be lifted back to $\mathscr{F}$ and used there to construct paths in the basis graph of $\mathscr{F}$.

In general, higher connectivity of a greedoid decreases the possible diameter of its basis graph, although most arguments for k-connected greedoids with $k \geq 3$ require the interval property.

8.9.5. Proposition. Let $(E, \mathscr{F})$ be a k-connected interval greedoid of rank r, where $2 \leq k \leq r$. Then the diameter of its basis graph is at most $2^{r-k+1} \cdot k - 1$.

This bound is not sharp in general. However, for $k = 2$ it reduces to the sharp bound of Theorem 8.9.4(ii). For $k = r$ it states that the diameter of the basis graph of an r-connected interval greedoid is at most $2r - 1$. This bound is sharp, even for branching greedoids. It is, however, still higher than the bound r for matroids.

8.9.C. Non-simple Greedoids: Chip Firing Games and Coxeter Groups

The way in which a greedoid is defined in section 8.2 requires that all feasible words are *simple* (i.e. have no letter occurring more than once). If this requirement is dropped, one gets a more general notion of greedoids, and such *non-simple* greedoids arise naturally in some examples.

Let E be a finite alphabet and $\mathscr{L} \subseteq E^*$ a finite language. Consider these axioms (the notation is explained in section 8.2):

(L1) if $\alpha = \beta\gamma$ and $\alpha \in \mathscr{L}$, then $\beta \in \mathscr{L}$;

(L2) if $\alpha, \beta \in \mathscr{L}$ with $|\alpha| > |\beta|$, then α contains a letter x such that $\beta x \in \mathscr{L}$;

(L2') if $\alpha, \beta \in \mathscr{L}$ with $|\alpha| > |\beta|$, then α contains a subword α' of length $|\alpha'| = |\alpha| - |\beta|$ such that $\beta\alpha' \in \mathscr{L}$.

In this section only (and in Exercises 8.36–8.38) we shall use the following definitions: $(E, \mathscr{L})$ is a *greedoid* if it satisfies (L1) and (L2), and a *strong greedoid* if it satisfies (L1) and (L2'). Thus, what was called a 'greedoid' and an 'interval greedoid' in section 8.2 would be called respectively a 'simple greedoid' and a 'simple strong greedoid' here.

Much of the theory of simple greedoids, as developed in previous sections, breaks down for non-simple greedoids. This is to a large extent due to the lack of an unordered, or set-theoretic, version. However, some parts of the theory that rely only on the ordered, or language-theoretic, version, survive the generalization. In particular, each greedoid $(E, \mathscr{L})$ has a *poset of flats* Φ (and a labelled poset of flats $\hat{\Phi}$), defined exactly as in Definition 8.8.1.

A proper subclass of greedoids, for which there is an unordered version, is given by the following definition: a finite language $(E, \mathscr{L})$ is a *polygreedoid* if it satisfies (L1) and

(L2'') If $\alpha, \beta \in \mathscr{L}$, $|\alpha| > |\beta|$, then there is some letter x, occurring more times in α than in β, such that $\beta x \in \mathscr{L}$.

Clearly, all simple greedoids are polygreedoids. Also, the polygreedoids for which all permutations of a feasible word are feasible are equivalent to the 'integral polymatroids' of J. Edmonds.

The two exchange axioms (L2') and (L2'') are logically independent. Any simple greedoid without the interval property satisfies (L2'') but not (L2').

Conversely, an example will be given after Proposition 8.9.7 of a greedoid that satisfies (L2') but not (L2'').

Define the *support* $\tilde{\alpha}$ of a word α as the *multiset* of letters in α. For example, if $\alpha = $ loophole then $\tilde{\alpha} = \{e, h, l^2, o^3, p\}$. The support of a language $\mathscr{L}$ is the multiset system $\tilde{\mathscr{L}} = \{\tilde{\alpha}: \alpha \in \mathscr{L}\}$.

In general, a greedoid $\mathscr{L}$ cannot be uniquely recovered from its support $\tilde{\mathscr{L}}$, but this *is* the case if $\mathscr{L}$ is a polygreedoid. The multiset systems that are the supports of polygreedoids can be characterized by accessibility and a suitable exchange axiom, and this permits an equivalent unordered formulation; see Exercise 8.37 for the precise statement (which extends Proposition 8.2.3). Because of this special property, polygreedoids occupy a middle ground between simple greedoids and general greedoids, and several facts from simple greedoid theory have straightforward extensions to polygreedoids.

Let us look at two situations where non-simple greedoids arise.

Suppose that we have a finite connected rooted graph (V, E, r) and an integer $k > 0$. This gives rise to the following *chip firing game*.

Think of the graph as drawn on a desk top. We have k chips that during the game are placed on and moved around among the vertices. By a *chip configuration* we mean a multiset $A: V \to \mathbb{N}$, $|A| = k$, which denotes that $A(v)$ chips are lying on vertex v.

When the game starts all k chips lie in a pile on the root vertex r (other initial positions work equally well). At this time or any later time a *legal move* consists in firing a legal vertex. By this is meant the following. For a given chip configuration A, a vertex v is *legal* if $A(v) \geq \deg(v)$, i.e. if there are at least as many chips on v as there are neighbors. To *fire* v then means to remove $\deg(v)$ chips from v and distribute them along the adjacent edges to v's neighbors, one to each. The game will *terminate* when a chip configuration is reached that permits no further legal move.

For instance, consider the graph in Figure 8.14, and let $k = 4$. If b is the root, the game will terminate after 2 moves. If a is the root the game will terminate after 6 moves. Finally, if c or d is the root the game will go on for ever, no matter how it is played.

The chip firing game determines a language $\mathscr{L} \subseteq V^*$ of legal firing sequences: $x_1 x_2 \ldots x_k \in \mathscr{L}$ if x_i is a legal vertex in the chip configuration

Figure 8.14.

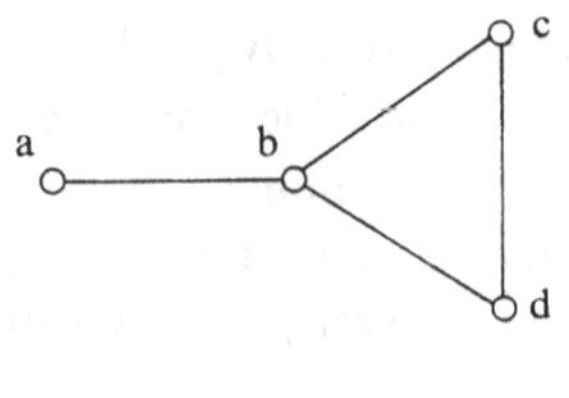

obtained after firing x_1, x_2, ..., x_{i-1}, for $1 \leq i \leq k$. This language is hereditary (i.e. satisfies (L1)), and in general not simple. For instance, if b is the root of the graph in Figure 8.14 and $k = 6$, then $babcdbaa \in \mathscr{L}$ but $babcb \notin \mathscr{L}$.

Assume from now on that the chip firing game terminates for some sequence of legal moves.

8.9.6. Proposition. *The chip firing language* $(V, \mathscr{L})$ *is a strong polygreedoid. Its poset of flats is isomorphic to the poset of chip configurations.*

It is easy to verify axiom (L2″), and an approach to axiom (L2′) is suggested in Exercise 8.38. The poset of chip configurations consists of the legal configurations (those which can occur in a game) ordered as follows: $A \leq B$ if and only if some sequence of legal moves transforms A into B.

The preceding result contains some combinatorial information about the chip firing game that is not *a priori* evident. For instance, one sees that every maximal sequence of legal moves has the same length, and also that there is a unique final chip configuration to which all such sequences lead.

The other example where non-simple greedoids arise comes from group theory. Let W be a group, S a generating subset, and assume that all elements in S are of order 2 (i.e. $s^{-1} = s$ for all $s \in S$). Then every group element $w \in W$ can be expressed as a product $w = s_1 s_2 \ldots s_k$, $s_i \in S$, and we call such an expression *reduced* if k is minimal, i.e. if w cannot be obtained as the product of a shorter sequence of generators. The reduced expressions can be thought of as words in the alphabet S, and the collection of all reduced expressions for all group elements forms a language $\mathscr{L} \subseteq S^*$. The language of reduced expressions $(S, \mathscr{L})$ has the following strong heredity property: if $\alpha = \beta\gamma, \alpha \in \mathscr{L}$, then, $\beta, \gamma \in \mathscr{L}$. In particular, it satisfies axiom (L1).

The pair (W, S) is called a *Coxeter group* if all relations among the generators are implied by pairwise relations of the form $(st)^{m(s,t)} = e$, s, $t \in S$. Examples of finite Coxeter groups are the symmetry groups of regular convex polytopes and the Weyl groups of simple complex Lie groups. In fact, every finite Coxeter group is of either of these types or a direct product of such.

The set W of group elements in a Coxeter group has a well known partial ordering, called *weak order* (or *weak Bruhat order*), which can be defined as follows: for u, $w \in W$, $u \leq w$ if some reduced expression for u ($u = s_1 \ldots s_i$) can be extended to a reduced expression for w ($w = s_1 \ldots s_i \ldots s_k$, $i \leq k$). The weak ordering of W is a graded lattice, if W is finite.

The symmetric group Σ_n of all permutations of $\{1, 2, \ldots, n\}$, together with the generating set of all adjacent transpositions $S = \{(i, i+1) : 1 \leq i \leq n-1\}$, is a Coxeter group. For example, if $n = 3$ and $S = \{a, b\}$ one sees that the language of reduced expressions is $\mathscr{L} = \{\varnothing, a, b, ab, ba, aba, bab\}$, and the weak order is the hexagon lattice.

8.9.7. Proposition. *The language of reduced expressions* $(S, \mathscr{L})$ *of a finite Coxeter group* (W, S) *is a strong greedoid. Its poset of flats is isomorphic to the weak ordering of* W.

Coxeter group languages are not in general polygreedoids. For example, in the language of Σ_3 given above, 'ba' can be augmented from 'aba' only by 'b', which occurs exactly once in both words.

The Coxeter group greedoids have an interesting geometric interpretation. In fact, they could be defined geometrically with no reference to group theory. We will sketch this geometric picture in one tangible special case only.

Let (W, S) be the symmetry group of the three-dimensional cube C. This is a Coxeter group of order 48, $|S| = 3$, and its language $(S, \mathscr{L})$ of reduced expressions is of rank 9 with 42 basic words. Let Δ be the barycentric subdivision of C's boundary. Then Δ consists of 48 triangles, and we pick one of these as a *root*; see Figure 8.15 where the root triangle is shaded.

Figure 8.15.

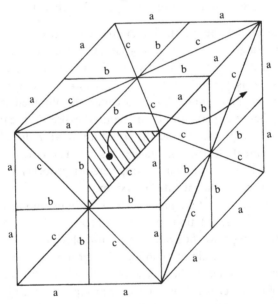

Consider now all *walks* from the root triangle T_0, by which we mean sequences of triangles $(T_0, T_1, ..., T_k)$ such that T_{i-1} and T_i are adjacent (share an edge) for $1 \leqq i \leqq k$. If the edges of Δ are labeled by a, b, and c as in Figure 8.15, then there is an obvious one-to-one correspondence between the walks from T_0 and the set S^* of all words in the alphabet $S = \{a, b, c\}$. For instance, the walk indicated in the figure corresponds to the word 'acabc'.

Call a walk $(T_0, T_1, ..., T_k)$ *geodesic* if no shorter walk from T_0 to T_k exists. The *geodesic language* $\mathscr{L}' \subseteq S^*$ consists of all words that correspond to geodesic walks from T_0. It is clear by symmetry that $\mathscr{L}'$ does not depend on the choice of T_0.

The basic fact now is that $\mathscr{L}'$ is isomorphic to $\mathscr{L}$, i.e. the Coxeter group greedoid of the cube group is the same thing as the geodesic language of the subdivided cube. Similarly, the greedoid of any finite Coxeter group can be obtained as the geodesic language of a simplicial sphere, which in the case of polytopal groups is the subdivision of the boundary of the corresponding regular polytope.

The examples we have discussed in this section show that even when the alphabet E is finite it makes sense to consider *infinite* greedoids $\mathscr{L} \subseteq E^*$. A non-terminating chip firing game played on a finite graph, and also an infinite Coxeter group (W, S) with finite S (e.g. an affine Weyl group or the symmetry group of a sufficiently regular tesselation of $\mathbb{R}^d$), gives rise to such an infinite non-simple greedoid.

8.10. Notes and Comments

Section 8.1

Greedoids were introduced by Korte & Lovász (1981) and their basic properties were developed in Korte & Lovász (1983, 1984a). The theory has since then been extensively developed by its creators and others. A book-length exposition will appear in Korte, Lovász & Schrader (1991).

There had been some earlier attempts to develop order dependent versions of matroids, by Dunstan, Ingleton & Welsh (1972) and by Faigle (1979, 1980), but the more comprehensive work of Korte & Lovász seems originally to have been independent of these antecedents. In Korte & Lovász (1985a) Faigle's structures are shown to correspond to a certain class of interval greedoids (cf. section 8.3.E).

This chapter was written in 1986–7, and it covers most of the basic properties of greedoids known at that time. The forthcoming monograph by Korte, Lovász & Schrader (1991) will presumably be more comprehensive.

Section 8.2

Most of the material here is from the early papers of Korte & Lovász (1981, 1984a). However, Proposition 8.2.5 (due to Björner and Lovász) is from Björner (1985), and Proposition 8.2.7 is from Björner (1985) and Korte & Lovász (1984b). Interval greedoids without loops were studied under the name *selectors* in Crapo (1984); see also Korte & Lovász (1985c).

Antimatroids have been written about under several names: *alternative precedence structure (APS) greedoids, upper interval greedoids, anti-exchange greedoids, shelling structures,* and *locally free selectors.* Of these, the name *shelling structure* is very unfortunate, since *shelling* has a precise and well established meaning in combinatorics that finds use also in greedoid theory (cf. section 8.6.C). To add to the confusion, in some papers convex geometries (the dual objects to antimatroids) are called 'antimatroids'.

While on the subject of names, it could be mentioned that the suitability of the word *greedoid* itself has been heatedly debated. If already the name *matroid* is 'ineffably cacophonic', as was claimed by Crapo & Rota (1970), then *greedoid* is undoubtedly much worse. Alternative names which have been proposed for related structures include *exchange language* (Björner, 1985), *selector* (Crapo, 1984), and *exchange system* (Brylawski & Dieter, 1988). However, the name *greedoid* is distinctive and catchy, albeit slightly frivolous, and there is no doubt that it is here to stay.

Section 8.3

Section 8.3.A: Twisted matroids were defined in Björner (1985), and slimmed matroids were defined in Korte & Lovász (1984c) where also several procedures for 'slimming' a matroid were discussed. A notion of 'trimmed' matroids, equivalent to the notion of 'meet' defined in Exercise 8.11, appears in Korte & Lovász (1985b, 1989b).

Section 8.3.B: These standard examples of antimatroids and many others are described in e.g. Björner (1985), Edelman & Jamison (1985), and Korte & Lovász (1984a, 1984b). See also the comments for section 8.7.

Sections 8.3.C and 8.3.D: Branching greedoids originate in Korte & Lovász (1981, 1984a); polymatroid greedoids and local poset greedoids in Korte & Lovász (1985b).

Section 8.3.E: Faigle geometries were defined by Faigle (1979, 1980). The connection with greedoids was studied in Korte & Lovász (1985a).

Sections 8.3.F and 8.3.G: For retract greedoids, see Crapo (1984) and Korte & Lovász (1985c, 1986a). Transposition greedoids and dismantling greedoids were defined in Korte & Lovász (1986a). The original example of dismantling sequences is due to Duffus & Rival (1978).

Section 8.3.H: Gaussian greedoids are due to Goecke (1986, 1988). The concept was rediscovered by Serganova, Bagotskaya, Levit & Losev (1988). Axiomatic as well as algorithmic characterizations of this class of greedoids are known; see Goecke (1986, 1988) and Exercise 8.34. Medieval marriage greedoids were defined in Korte & Lovász (1986a). The name was coined by J. Edmonds in reference to some generic 'medieval' king, in whose opinion a sequence of suitors is feasible if and only if they will marry his daughters in decreasing order of age.

Section 8.3.I: Figure 8.5 is adapted from information in Korte & Lovász (1985e, 1986a).

Several examples of greedoids are known which have not been discussed here. For instance, Korte & Lovász (1984a, 1986a) have described several other classes of greedoids arising in graph theory: ear decomposition greedoids, blossom greedoids (Edmonds' matching algorithm), perfect elimination greedoids, series–parallel reduction greedoids, etc. Goecke, Korte & Lovász (1989) provide an extensive survey of examples.

Section 8.4

All material in sections 8.4.A–8.4.D comes from Korte & Lovász (1983), except that Theorem 8.4.1 was proven in Korte & Lovász (1984b). An alternative closure operation, called *kernel closure*, which is idempotent, monotone for interval greedoids, but not in general increasing, is defined and studied in Schmidt (1985a); see Exercise 8.9.

Two-connectivity in greedoids was defined and studied in Korte & Lovász (1985d), *k*-connectivity in Björner, Korte & Lovász (1985). The connectivity properties of branching greedoids are studied in more detail in Ziegler (1988).

Section 8.5

The material in sections 8.5.A and 8.5.B is from Korte & Lovász (1981, 1984a); see also Goecke, Korte & Lovász (1989). It should be said that there exist optimal discrete algorithms that are of a greedy nature, but which do not come from an underlying greedoid structure. For instance, no greedoid can be discerned behind the greedy algorithm for knapsack problems of Magazine, Nemhauser & Trotter (1975), also treated in Hu & Lenard (1976).

The algorithms of Dijkstra, Kruskal, and Prim are discussed in every book on combinatorial optimization; see Tarjan (1983) and the interesting historical discussion in Graham & Hell (1985). Korte & Lovász (1981, 1984a) show that some machine scheduling algorithms of Lawler also fit into the greedoid framework – the question is of optimizing some generalized bottleneck function over a poset greedoid.

The fact that Depth-first-search is not compatible with branching greedoids was pointed out by Korte & Lovász (1981, 1984a). They remark that the problem is NP-hard, since it includes the problem of finding a Hamiltonian path.

The results about linear objective functions in section 8.5.C are from Korte & Lovász (1984c). The optimization of linear objective functions over greedoids is also discussed in Brylawski (1991), Faigle (1985), Goecke (1986, 1988), Goecke, Korte & Lovász (1989), Goetschel (1986), and Serganova, Bagotskaya, Levit & Losev (1988). In connection with linear objective functions Gaussian greedoids have special properties; see the cited papers by Brylawski and by Goecke, and also Exercise 8.34.

Bagotskaya, Levit & Losev (1988, 1990a) define structures called 'fibroids', designed to incorporate some optimization features of dynamic programming. Fibroids contain Gaussian greedoids as a special case. Optimization over (W, P)-matroids (see the remarks below for section 8.9) is discussed in Zelevinsky & Serganova (1989). See also the work of Bouchet (1987).

Section 8.6
The material in this section is from sections 5–6 of Björner, Korte & Lovász (1985). From a matroid-theoretic point of view sections 8.6.A and 8.6.B extend parts of the theory of Tutte polynomials and Tutte–Grothendieck invariants to all greedoids. For these topics see Chapter 6 and section 7.3 of this book.

A general discussion of the concept of argument complexity can be found in Chapter 8 of Bollobás (1978). The $d = 0$ case of Proposition 8.6.3 appears in Björner, Korte & Lovász (1985).

There is a large literature on the reliability analysis of stochastic networks and other systems. See Colbourn (1987) for more information and references in this area.

The lack of a fully-fledged duality operation on the class of all greedoids has been noticed by several authors. It appears that it is only when the demand for total symmetry between primal and dual is abandoned that some interesting remnants of duality in greedoids can be discerned. Interesting axiomatic discussions of duality (for matroids and some other set systems) appear in Kung (1983) and in Bland & Dietrich (1987, 1988).

A 2-variable greedoid 'Tutte' polynomial, defined by the corank-nullity formula

$$f_G(t, z) = \sum_{A \subseteq E} t^{r(E) - r(A)} z^{|A| - r(A)},$$

has been studied by Gordon & McMahon (1989); see also Gordon & Traldi (1989), Chaudhuri & Gordon (1991), and Gordon (1990). Its relationship to the polynomial studied here is $\lambda_G(t) = f_G(0, t - 1)$, as can be seen from Theorem 8.6.2(iii).

Section 8.7
Convex geometries were independently discovered by Edelman (1980) and Jamison (1982), and later studied by them jointly. Edelman & Jamison (1985) gives a good overview of their work on convex geometries, and all section 8.7A is from that paper except the duality with antimatroids (Proposition 8.7.3), which was first observed by Björner (1985). All examples of antimatroids that have been mentioned in the text were originally known to Edelman and Jamison as examples of convex geometries. The example of convex hull closure in $\mathbb{R}^n$ (and hence, dually, convex pruning antimatroids) was generalized to oriented matroids in Edelman (1982).

We have included only a limited number of references for antimatroids

90(mainly those taking a greedoid point of view). See Edelman & Jamison (1985) and Edelman (1986) for more extensive bibliographies (stressing the convex geometry or lattice-theoretic point of view).

Theorem 8.7.6 showing the equivalence of antimatroids and join-distributivity is due to Edelman (1980), in a dual version for convex geometries. The result was rediscovered by Crapo (1984) using another terminology. The characterization of antimatroids as alternative precedence languages is due to Korte & Lovász (1984a, b). The characterization by exchange axiom (A) as well as by semimodularity of $\mathscr{F}$ is from Björner (1985). Proposition 8.7.7 is from Korte & Lovász (1985b), and Proposition 8.7.8 from Korte & Lovász (1986b); see also Edelman & Saks (1986). For Birkhoff's theorem, see Birkhoff (1967).

All of section 8.7.C is from Korte & Lovász (1984b), except for Theorem 8.7.12 which is due to Dietrich (1987).

Section 8.8

The concept of poset representations of greedoids, the definition of the poset of flats, and basic properties of Φ and $\hat{\Phi}$ are from Björner (1985). The poset of closed sets was first defined for interval greedoids (as in Exercise 8.26) by Korte & Lovász (1983), and then in general (as in Definition 8.8.3) by Björner, Korte & Lovász (1985). The relationship between flats and closed sets (Theorem 8.8.4), and also Proposition 8.8.5, is from Björner, Korte & Lovász (1985).

A notion of a 'lattice of flats' for selectors was defined by Crapo (1984); this can be shown to be equivalent to our poset of flats $\hat{\Phi}$ for the case of interval greedoids. Theorem 8.8.7 (in terms of selectors) is due to Crapo (1984). The lattice-theoretic lemma used in our proof is from Björner, Edelman & Ziegler (1990). A different proof for the lattice property (but not semimodularity), using the poset of closed sets, was given in Korte & Lovász (1983).

Poset properties were first studied in Ziegler (1988), from where Proposition 8.8.9 is taken.

Section 8.9

The excluded minor characterizations of local poset greedoids and undirected branching greedoids (Theorem 8.9.3) are due to Korte & Lovász (1985b) and Schmidt (1985a, 1988), respectively. See also Goecke & Schrader (1990) for a shorter proof. The characterization of directed branching greedoids in Theorem 8.9.1 appears in Schmidt (1985b).

The minor poset of isomorphism classes has long been studied for matroids and particularly for graphs. From work of N. Robertson and P. Seymour it is known that infinite antichains do not exist in the minor poset of all graphs. However, infinite antichains do exist in the minor poset of all matroids; see White (1986), p. 155. Also, infinite antichains of branching greedoids exist (cf. Exercise 8.27).

The basis graph of a greedoid was introduced in Korte & Lovász (1985d), where it was shown that a 2-connected greedoid has a connected basis graph. Basis graphs of matroids had earlier been studied by Maurer (1973). The bounds on the diameter for greedoids (Theorems 8.9.4 and 8.9.5) are due to Ziegler (1988). A higher-dimensional analog of the basis graph, the *basis polyhedron*, was investigated by Björner, Korte & Lovász (1985), and it was shown that a k-connected greedoid has a (topologically) $(k-2)$-connected basis polyhedron.

Non-simple greedoids were first studied in Björner (1985), and this was mainly motivated by the example of Coxeter group greedoids. See Björner (1984, 1985) for more information about Coxeter groups and their weak partial ordering, and for references to the extensive literature about these topics. The greedoids of graph chip firing games were discovered by Björner, Lovász & Shor (1988). The greedoid rank (length of the game) in terms of the size of the graph has been studied by Tardos (1988) and Eriksson (1989). Polygreedoids appeared in Björner (1985); for integral polymatroids see e.g. White (1987), p. 181. Faigle (1985) also discusses non-simple greedoids.

A different connection between greedoids and Coxeter groups was discovered by Gel'fand & Serganova (1987a, b). For each finite Coxeter group (W, S) and each subset P of the generating set S, they define a class of subsets of the family of left cosets $W^P = W/\langle P \rangle$, the members of which they call (W, P)-*matroids*. The definition involves a certain minimality condition in terms of Bruhat order on W^P. For the symmetric group $W = \Sigma_n$ and $P = \{(i, i+1): 1 \leqslant i \leqslant n-1, i \neq k\}$, the (W, P)-matroids are precisely the ordinary matroids of rank k given by their bases (as sets). The characterization of matroids obtained this way is equivalent to that of Gale (1968): for every ordering of the ground set there is a point-wise minimal basis. For $P = \{(i, i+1): k+1 \leqslant i \leqslant n-1\}$ the (W, P)-matroids are the Gaussian greedoids of rank k given by their basic words. Taking W to be the symmetry group of a cube and for a certain choice of P, the (W, P)-matroids coincide with the symmetric matroids of Bouchet (1987). (For those who undertake to read the papers of Gel'fand and Serganova, let us point out that part (b) of Theorem 2 in (1987a) and the definition of Bruhat order given there are incorrectly stated.) Many details about (W, P)-matroids can also be found in Zelevinsky & Serganova (1989).

Exercises

The following propositions, theorems, and lemmas which were stated without proof or with incomplete proof in the text, make suitable exercises: 8.2.3, 8.2.7, 8.2.8, 8.4.1–8.4.5, 8.5.6, 8.6.1–8.6.3, 8.7.8, 8.7.9, 8.8.2, 8.8.4, and 8.9.6.

8.1. Show that a set system $(E, \mathcal{F})$ is a greedoid if and only if it satisfies the two axioms:
(G1'') $\varnothing \in \mathcal{F}$ and for all X, $Y \in \mathcal{F}$ such that $Y \subset X$ there is an $x \in X - Y$ such that $X - x \in \mathcal{F}$.
   (B) For any subset $A \subseteq E$ all maximal feasible subsets of A have the same cardinality.

8.2. (Korte & Lovász, 1986a) Let $\mathcal{F} \subseteq 2^E$, and consider the following axiom.
(G2'') If $A \subseteq E$, $x, y, z \in E - A$ such that $A \cup x$, $A \cup y$, $A \cup x \cup z \in \mathcal{F}$, and $A \cup x \cup y \notin \mathcal{F}$, then $A \cup y \cup z \in \mathcal{F}$.
Show that the axioms (G1) and (G2'') together define greedoids.

8.3. Prove the following sharpening of the strong exchange property (L2') of Proposition 8.2.5 for interval greedoids $(E, \mathcal{L})$:
let α, $\beta \in \mathcal{L}$, $\alpha = x_1 x_2 \ldots x_n$, and $|\alpha| - |\beta| = k > 0$; among all strings $(i_1, i_2, \ldots, i_k)$ such that $\beta x_{i_1} x_{i_2} \ldots x_{i_k} \in \mathcal{L}$, the lexicographically first one satisfies $i_1 < i_2 < \ldots < i_k$.

8.4. (Korte & Lovász, 1984a) Let E be finite and $\mathcal{P} \subseteq 2^E$ a system of non-empty sets such that $|A| = |B|$ and $|A - B| = 1$ imply $A \cap B \in \mathcal{P}$, for all A, $B \in \mathcal{P}$.
   (a) Show that $(E, 2^E - \mathcal{P})$ is a greedoid (called a *paving greedoid*).
   (b) Show that $(E, 2^E - \mathcal{P})$ in general lacks the transposition property.

8.5. (Crapo, 1984; Korte & Lovász, 1985c) Let $(E, \mathcal{F})$ be a greedoid and $\mathcal{A} = \{\cup \mathcal{F}' : \mathcal{F}' \subseteq \mathcal{F}\}$. Show that
   (a) $(E, \mathcal{A})$ is an antimatroid,
   (b) $\mathcal{F} \subseteq \mathcal{A} \subseteq \mathcal{R}$, if $(E, \mathcal{F})$ is an interval greedoid,
   (c) both inclusions in part (ii) can be strict.

8.6. Let $\Gamma = (V, E, r)$ be a finite, connected undirected graph with root $r \in V$. Let $(E, \mathcal{F}_b)$ be the branching greedoid on Γ, and $(V - r, \mathcal{F}_s)$ the vertex search greedoid.
   (a) The join irreducibles (elements covering exactly one element) in the poset $(\mathcal{F}_b, \subseteq)$ are the paths in Γ starting at r. Hence, the join irreducibles form an order ideal in $\mathcal{F}_b$.
   (b) The join irreducibles in $(\mathcal{F}_s, \subseteq)$ are the induced paths (i.e. without chords) in Γ, starting at r.
   (c) The meet irreducibles (elements covered by exactly one element) in $(\mathcal{F}_b, \subseteq)$ correspond to the bridges in Γ.
   (d) The meet irreducibles in $(\mathcal{F}_s, \subseteq)$ of corank at least 2 correspond to cut vertices in Γ; those of corank 1 correspond to non-cut-vertices.

8.7. (Korte & Lovász, 1983) Show that the rank closure operator σ of a greedoid G is monotone only if G is a matroid.

8.8. (Schmidt, 1985a, b)
   (a) Prove that directed and undirected branching greedoids satisfy $\sigma(X) \cap (Y) \subseteq \sigma(X \cup Y)$, for X, $Y \in \mathcal{F}$.
   (b) Show also that $\sigma(X \cup Y) \subseteq \sigma(X) \cup \sigma(Y)$ holds for directed branching greedoids, but not in general for undirected branching greedoids.

8.9. (Schmidt, 1985a, b) The closure operator for greedoids, which is given by

$$\sigma(A) = \cup\{X \subseteq E : r(A \cup X) = r(A)\},$$

has some shortcomings (cf. section 8.4.B). As an alternative, the *kernel closure operator* $\lambda : 2^E \to 2^E$, defined by

$$\lambda(A) = \cup\{X \in \mathcal{F}: r(A \cup X) = r(A)\},$$

has been proposed. Define the *kernel* of a subset $A \subseteq E$ by

$$\ker(A) = \cup\{X \in \mathcal{F}: X \subseteq A\}.$$

(a) Give a graph-theoretic description of $\sigma(A)$, $\lambda(A)$, and $\ker(A)$ for a set A in a branching greedoid. Exemplify with a branching greedoid that $A \subseteq \lambda(A)$ may fail.

(b) Show that the operators σ, λ, ker: $2^E \to 2^E$ satisfy the relations:

$$\lambda\sigma = \ker \sigma = \lambda,$$

$$\sigma\lambda = \sigma \ker = \sigma.$$

(c) Deduce that

$$\lambda^2 = \lambda \text{ (i.e. } \lambda \text{ is idempotent)},$$

$$\lambda\sigma\lambda = \lambda,$$

$$\sigma\lambda\sigma = \sigma.$$

(d) There is a canonical bijection between σ-closed sets and λ-closed sets, and

$$r(\lambda(A)) = r(\sigma(A)) = r(A), \quad \text{for all } A \subseteq E.$$

(e) In an interval greedoid

$$\beta(\lambda(A)) = r(A), \quad \text{for all } A \subseteq E.$$

(f) The kernel closure operator λ is monotone if and only if $(E, \mathcal{F})$ is an interval greedoid.

(g) For σ-closed sets $A, B \in \mathscr{C}\ell$, $A \leq B$ implies $\lambda(A) \subseteq \lambda(B)$. The converse holds for interval greedoids, but fails in general.

8.10. (Korte & Lovász, 1983) Show that the monotone closure operator μ of full greedoid $(E, \mathcal{F})$ satisfies $\mu(A) = A$, for all $A \subseteq E$.

8.11. (Korte & Lovász, 1989b) Let $(E, \mathcal{M})$ and $(E, \mathcal{A})$ be respectively a matroid and an antimatroid on the same ground set E, with closure operators $\sigma_{\mathcal{M}}$ and $\sigma_{\mathcal{A}}$. Define a language $(E, \mathcal{L})$ by

$$\mathcal{L} = \{x_1 x_2 \ldots x_k \in E^*: x_i \notin \sigma_{\mathcal{A}}(\sigma_{\mathcal{M}}(\{x_1, \ldots, x_{i-1}\})) \text{ for all } 1 \leq i \leq k\}.$$

(a) Show that $(E, \mathcal{L})$ is an interval greedoid. The corresponding set system, denoted by $(E, \mathcal{M} \wedge \mathcal{A})$, is called the *meet* of $(E, \mathcal{M})$ and $(E, \mathcal{A})$.

(b) Verify that $\mathcal{M} \cap \mathcal{A} \subseteq \mathcal{M} \wedge \mathcal{A} \subseteq \mathcal{M}$.

(c) Show that $(E, \mathcal{M} \cap \mathcal{A})$ is not a greedoid in general, but that if it is a greedoid, then $(E, \mathcal{M} \cap \mathcal{A}) = (E, \mathcal{M} \cap \mathcal{A}^*) = (E, \mathcal{M} \wedge \mathcal{A}^*)$ for the antimatroid $\mathcal{A}^* = \{\cup \mathcal{F}': \mathcal{F}' \subseteq \mathcal{M} \cap \mathcal{A}\}$.

(d) Show that if $(E, \mathcal{M})$ is any matroid, and $(E, \mathcal{A})$ is a poset greedoid, then $(E, \mathcal{M} \wedge \mathcal{A})$ is a Faigle geometry.

(e) Show that every directed branching greedoid arises as a meet.

(f) Show that every polymatroid greedoid arises as a meet.

8.12. (Korte & Lovász, 1983) For a greedoid $(E, \mathcal{F})$, let $A_1, \ldots, A_n \subseteq E$. A *feasible system of representatives* for $\{A_1, \ldots, A_n\}$ in $(E, \mathcal{F})$ is a set $X \in \mathcal{F}$ for which there is a bijection $\phi: X \to \{A_1, \ldots, A_n\}$ with $x \in \phi(x)$ for all $x \in X$.

Suppose that $(E, \mathcal{F})$ is an interval greedoid and $A_1, \ldots, A_n$ are rank feasible.

Show that $\{A_1, \ldots, A_n\}$ has a feasible system of representatives if and only if

$$r(A_{i_1} \cup \ldots \cup A_{i_k}) \geqq k,$$

for all $1 \leqq i_1 < \ldots < i_k \leqq n$.

8.13. Let $K \subseteq \mathbb{R}^n$ be a convex body, and $E \subseteq \mathbb{R}^n - K$ a finite set. Let $\mathscr{F} \subseteq 2^E$ consist of those subsets A that are disjoint from the convex hull of $K \cup (E - A)$.
   (a) Prove that $(E, \mathscr{F})$ is an antimatroid.
   (b) Prove that the class of such antimatroids is hereditary (i.e. closed under taking minors).

8.14. (Björner, Korte & Lovász, 1985) Say that a greedoid $(E, \mathscr{F})$ is *weakly k-connected* if $r(E - A) = r(E)$ for all $A \subseteq E$ with $|A| < k$.
   (a) Show that an undirected branching greedoid (or graphic matroid) is weakly k-connected if and only if the underlying graph is k-edge-connected.
   (b) Show that the number of bases in a weakly k-connected greedoid of rank r is at least $\binom{k+r-1}{r}$.

8.15. (Korte & Lovász, 1989b) Let $(E, \mathscr{F})$ be a greedoid and A a closure feasible subset of E. Show that $(E, \mathscr{F}')$ is a full greedoid, but not in general an antimatroid, for

$$\mathscr{F}' = \mathscr{F} \cup \{B \subseteq E : \sigma(B) \supseteq A\}.$$

As a special case, conclude that every greedoid is a truncation of a full greedoid.

8.16. Let $(E, \mathscr{F})$ be a greedoid of rank r. Show that $(E, \mathscr{F})$ is the r-truncation of an antimatroid if and only if for all $X, Y \in \mathscr{F}$, $|X \cup Y| \leqq r$ implies $X \cup Y \in \mathscr{F}$. In this case, construct the smallest and the largest antimatroid on E whose r-truncation is $(E, \mathscr{F})$.

8.17. (Björner, Korte & Lovász, 1985) Let $(E, \mathscr{F})$ be an interval greedoid. Show that
   (a) $(E, \mathscr{F})$ is a matroid if and only if $\{x\} \in \mathscr{F}$ for all $x \in \cup \mathscr{F}$;
   (b) if $A \in \mathscr{F}$, then the free sets over A are the independent sets of a matroid.

8.18. (Korte & Lovász, 1986a) Show that an accessible set system with the transposition property (TP) of section 8.3.G is a greedoid.

8.19. (Korte & Lovász, 1983; Goecke, 1986) Prove that a greedoid $(E, \mathscr{F})$ has the interval property if and only if $\mathscr{F}/X_1 = \mathscr{F}/X_2$ for all subsets $A \subseteq E$ and all bases X_1 and X_2 of A.

8.20. One might have hoped for the following weak form of greedoid duality: if $\mathscr{B} \subseteq 2^E$ is the set of bases of a greedoid then $\{E - B : B \in \mathscr{B}\}$ is the set of bases of some other greedoid, not necessarily unique. Show that this is false.

8.21. Let $\tau : 2^E \to 2^E$ be a closure operator on a finite set E. Show that τ satisfies the anti-exchange condition if and only if every closed set other than $\tau(\varnothing)$ has at least one extreme point.

8.22. (Korte & Lovász, 1984b) Let E be a finite set, and for each $x \in E$ let $H(x) \subseteq 2^{E-x}$ be some set system. Define a left hereditary language

$$\mathscr{L}^H = \{x_1 x_2 \ldots x_k \in E_s^* : \text{for all } 1 \leqq i \leqq k \text{ and all } A \in H(x_i), A \nsubseteq \{x_{i+1}, \ldots, x_k\}\}.$$

   (a) Show that $(E, \mathscr{L}^H)$ is an antimatroid.
   (b) Show that every antimatroid arises in this way.
   (c) Suppose that $(E, \mathscr{L}_K)$ is an alternative precedence language determined by some system K. For each $K(x)$ describe $H(x)$ so that $\mathscr{L}^H = \mathscr{L}_K$.

8.23. For a full antimatroid $(E, \mathcal{F})$, let Δ consist of those subsets of E that are free and convex. Show that

(a) Δ is a simplicial complex,

(b) $\sum(-1)^i f_i = 0$, where f_i is the number of sets in Δ of cardinality i,

(c) Δ is contractible (in the topological sense).

(d) Let h be the maximum cardinality of a set in Δ. Show that h is the *Helly number* of $(E, \mathcal{F})$, meaning that h is the least integer such that, for any family of convex sets, if each subfamily of size h has non-empty intersection then the whole family has non-empty intersection.

(Part (b) is an unpublished theorem of J. Lawrence; see Edelman & Jamison (1985). The proof of Theorem 7.4 in Björner, Korte & Lovász (1985) can be adapted to prove the contractibility of Δ. Part (d) is due to A. Hoffman and R. Jamison; see Edelman & Jamison (1985).)

8.24. (a) Show that the greedoid polynomial of the branching greedoid of a rooted connected graph is independent of the root.

(b) Show that the analogous statement for the branching greedoid of a strongly connected digraph is false.

8.25. (Björner, 1985) Show that the poset representations of a greedoid form a lattice when they are ordered by $(P_1, \lambda_1) \leq (P_2, \lambda_2)$ if and only if there is a rank-preserving poset map $f: P_1 \rightarrow P_2$ satisfying $\lambda_1(x <\cdot y) \subseteq \lambda_2(f(x) <\cdot f(y))$ for $x, y \in P_1$ and $x <\cdot y$. Show that every such map is necessarily surjective. Identify the universal representation (poset of flats) in terms of this lattice.

8.26. (Korte & Lovász, 1983) According to Theorems 8.8.4 and 8.8.7, the poset of closed sets $(\mathcal{C}\ell, \leq)$ of an interval greedoid is a semimodular lattice. Show that its meet operation is given by

$$A \wedge B = \sigma(A \cap B), \quad A, B \in \mathcal{C}\ell.$$

8.27. Construct an infinite sequence of branching greedoids G_i, $i = 1, 2, \ldots$, such that G_i is not a minor of G_j for all $i \neq j$.

8.28. Define a graph over the set $\mathcal{B}$ of bases of a greedoid G by letting (B_1, B_2) be an edge when $|B_1 - B_2| = 1$, $B_1, B_2 \in \mathcal{B}$. (This graph contains the basis graph as a subgraph.) Prove the bounds

$$r - |A \cap B| \leq d(A, B) \leq r - r(A \cap B),$$

for the graph distance $d(A, B)$ between two bases A and B, where $r = $ rank G. In particular, the diameter of the graph is at most r.

8.29. Define the *basic word graph* of a greedoid $(E, \mathcal{L})$ as follows. The vertices are the basic words, and two basic words are adjacent if the corresponding maximal chains in the poset $(\mathcal{F}, \subseteq)$ differ in exactly one element (equivalently, if one arises from the other by exchanging two consecutive letters or by exchanging the last letter).

(a) (Korte & Lovász, 1985d) Show that the basic word graph is connected if $(E, \mathcal{L})$ is 2-connected.

(b) If $(E, \mathcal{L})$ is an antimatroid of rank r, show that the basic word graph is connected and has diameter at most $\binom{r}{2}$. This bound is best-possible.

8.30. (Korte & Lovász, 1984b) Show that convex pruning greedoids have the following property, not shared by general antimatroids: if $(A \cup x, x)$ and $(A \cup y, y)$ are rooted circuits, then there exists a unique subset $A' \subseteq A$ such that $(A' \cup x \cup y, y)$ is a rooted circuit.

8.31. For an antimatroid $(E, \mathcal{F})$, let c be the minimum and C the maximum size of a circuit.

(a) (Björner, Korte & Lovász, 1985) Show that $(E, \mathcal{F})$ is k-connected if and only if $C \geq k + 1$.

(b) (Björner & Lovász, 1987) Show that $C - 1$ is the *Carathéodory number* of $(E, \mathcal{F})$, i.e. the least integer such that if x lies in the convex hull of $A \subseteq E$, then there is some subset $A' \subseteq A$ of size at most $C - 1$ such that x lies in the convex hull of A'.

8.32. (Korte & Lovász, 1984c) Let $(E, \mathcal{F})$ be a greedoid. Given a linear objective function, the *worst-out greedy algorithm* starts with the complete ground set E and at each step eliminates the worst possible element so that the remaining set is still spanning (contains a basis). Show that every linear objective function can be optimized over $(E, \mathcal{F})$ by the worst-out greedy algorithm if and only if the hereditary closure $(E, \mathcal{H}(\mathcal{F}))$ is a matroid.

8.33. (a) For an interval greedoid, show that every $\mathcal{R}$-compatible linear objective function is compatible in the sense of Definition 8.5.1.

(b) Give an example of a non-interval greedoid and a linear function that is $\mathcal{R}$-compatible but not compatible.

8.34. (Serganova, Bagotskaya, Levit & Losev, 1988) Let $\mathcal{F} \subseteq 2^E$ be an accessible set system. Show that the following are equivalent.

(a) $(E, \mathcal{F})$ is a Gaussian greedoid.

(b) For any linear objective function the greedy algorithm constructs a sequence of sets $A_i \in \mathcal{F}$, $i = 1, ..., r$, such that A_i is optimal in the class $\mathcal{F}_i = \{X \in \mathcal{F} : |X| = i\}$ for all $i = 1, ..., r = \max\{|X| : X \in \mathcal{F}\}$.

(c) For $X, Y \in \mathcal{F}$, $|X| = |Y| + 1$, there is an $x \in X - Y$ such that $Y \cup x \in \mathcal{F}$ and $X - x \in \mathcal{F}$.

(d) For $X, Y \in \mathcal{F}$, $|X| > |Y|$, there is a subset $A \subset X - Y$, $|A| = |X| - |Y|$, such that $Y \cup A \in \mathcal{F}$ and $X - A \in \mathcal{F}$.

Furthermore, show that the condition $|X| = |Y| + 1$ in (c) cannot be relaxed to $|X| > |Y|$.

(The equivalence of (a), (b), and (c) is proved in the cited source. The same authors have subsequently proved (personal communication) the equivalence of (c) and (d), two versions of what they call the 'fork axiom'. The equivalence of (a) and (b) was also observed by Brylawski (1991).)

8.35. Does there exist a non-Gaussian greedoid $(E, \mathcal{F})$ for which $\mathcal{H}(\mathcal{F}_i)$, the hereditary closure of the feasible i-sets, is a matroid, for $i = 0, 1, ..., r = \text{rank} (\mathcal{F})$?

The remaining three exercises concern non-simple greedoids (section 8.9.C).

8.36. (Björner, 1985, extending Korte and Lovász, 1984a)

(a) Show that the greedy algorithm will optimize any compatible objective function $w: \mathcal{L} \to \mathbb{R}$ over a polygreedoid $(E, \mathcal{L})$.

(b) Show that the greedy algorithm will optimize any generalized bottleneck function (defined in the proof of Theorem 8.5.2) over a greedoid, whether simple or not.

8.37. (Björner, 1985; extending Korte & Lovász, 1984a) For multisets A, B: $E \rightarrow \mathbb{N}$ define *inclusion* $A \subseteq B$ by $A(e) \leq B(e)$ for all $e \in E$, and *cardinality* $|A| = \sum_{e \in E} A(e)$. Identify elements $e \in E$ with their characteristic functions χ_e: $E \rightarrow \{0, 1\}$. For a finite non-empty multiset system $\mathscr{F} \subseteq \mathbb{N}^E$, E finite, consider the following axioms.

(P1) For all $A \in \mathscr{F}$, $A \neq \varnothing$, there exists $B \subseteq A$ such that $|B| = |A| - 1$ and $B \in \mathscr{F}$.

(P2) If $A, B \in \mathscr{F}$ and $|A| > |B|$, then there exists an element $e \in E$ such that $A(e) > B(e)$ and $B + e \in \mathscr{F}$.

Prove the following.

(a) If $(E, \mathscr{L})$ is a polygreedoid, then the support $\tilde{\mathscr{L}}$ satisfies axioms (P1) and (P2).

(b) If $(E, \mathscr{F})$ is a multiset system satisfying (P1) and (P2) then the language

$$\mathscr{L}(\mathscr{F}) = \{x_1 x_2 \ldots x_k \in E^*: \widetilde{x_1 \ldots x_i} \in \mathscr{F} \text{ for } 1 \leq i \leq k\}$$

is a polygreedoid.

(c) These operations are mutually inverse: $\mathscr{L}(\tilde{\mathscr{L}}) = \mathscr{L}$ and $\widetilde{\mathscr{L}(\mathscr{F})} = \mathscr{F}$.

8.38. A finite language $\mathscr{L} \subseteq E^*$, not necessarily simple, is called an A-*language* if it is hereditary (axiom (L1)) and satisfies the following axiom.

(L2''') If α, αx, $\alpha \beta \in \mathscr{L}$ and the letter x does not occur in β, then $\alpha x \beta$, $\alpha \beta x \in \mathscr{L}$ and $\alpha x \beta \gamma \in \mathscr{L}$ if and only if $\alpha \beta x \gamma \in \mathscr{L}$, for all $\gamma \in E^*$.

Prove the following (cf. section 8.9.C).

(a) Every A-language is a strong greedoid.

(b) Every graph chip firing language is an A-language.

(c) A simple A-language is the same thing as an antimatroid.

References

Bagotskaya, N. V., Levit, V. E. & Losev, I. S. (1988). On a generalization of matroids that preserves the applicability of the method of dynamic programming (in Russian), in *Sistemy Peredachi i Obrabotki Informatsii, Chast* 2, pp. 33–6, Institut Problem Peredachi Informatsii, Academy of Sciences, Moscow.

Bagotskaya, N. V., Levit, V. E. & Losev, I. S. (1990a). Fibroids, *Automatics and Telemechanics* (translation of Russian journal *Automatika i Telemechanika*) (to appear).

Bagotskaya, N. V., Levit, V. E. & Losev, I. S. (1990b). Branching greedy algorithm and its combinatorial structure, preprint.

Birkhoff, G. (1967). *Lattice Theory*, Amer. Math. Soc. Colloq. Publ. XXV, 3rd edition. American Mathematical Society, Providence, Rhode Island.

Björner, A. (1984). Orderings of Coxeter groups, in C. Greene (ed.), *Proceedings of the A.M.S.-N.S.F. Conference on Combinatorics and Algebra* (Boulder, 1983), Contemporary Mathematics 34, pp. 175–95. American Mathematical Society, Providence, Rhode Island.

Björner, A. (1985). On matroids, groups and exchange languages, in L. Lovász & A. Recski (eds), *Matroid Theory and its Applications* (Szeged, 1982), Colloq. Math. Soc. János Bolyai 40, pp. 25–60. North Holland, Amsterdam and Budapest.

Björner, A. & Lovász, L. (1987). Pseudomodular lattices and continuous matroids, *Acta Sci. Math. Szeged* 51, 295–308.

Björner, A., Edelman, P. & Ziegler, G. M. (1990). Hyperplane arrangements with a lattice of regions, *Discrete and Comp. Geom.* 5, 263–88.

Björner, A., Korte, B. & Lovász, L. (1985). Homotopy properties of greedoids, *Adv. Appl. Math.* 6, 447–94.

Björner, A., Lovász, L. & Shor, P. W. (1988). Chip-firing games on graphs, *Europ. J. Comb.* (to appear).

Bland, R. G. & Dietrich, B. L. (1987). A unified interpretation of several combinatorial dualities, preprint, Cornell University.

Bland, R. G. & Dietrich, B. L. (1988). An abstract duality, *Discrete Math.* **70**, 203–8.

Bollobás, B. (1978). *Extremal Graph Theory*, Academic Press, New York and London.

Bouchet, A. (1987). Greedy algorithm and symmetric matroids, *Math. Progr.* **38**, 147–59.

Boyd, E. A. (1987a). Optimization problems on greedoids, Ph.D. Thesis, M.I.T.

Boyd, E. A. (1987b). An algorithmic characterization of antimatroids, preprint, Rice University.

Brylawski, T. H. (1991). Greedy families for linear objective functions, *Studies Appl. Math.* **84**, 221–9.

Brylawski, T. H. & Dieter, E. (1988). Exchange systems, *Discrete Math.* **69**, 123–51.

Chang, G. J. (1986). MPP-greedoids, Preprint 86436-OR, Institut für Operations Research, Bonn.

Chaudhary, S. & Gordon, G. (1991). Tutte polynomials for trees, *J. Graph Theory* (to appear).

Colbourn, C. J. (1987). *The Combinatorics of Network Reliability*, Oxford University Press.

Crapo, H. (1984). Selectors. A theory of formal languages, semimodular lattices, branchings and shelling processes, *Adv. Math.* **54**, 233–77.

Crapo, H. & Rota, G.-C. (1970). *Combinatorial Geometries*, M.I.T. Press, Cambridge, Mass.

Dietrich, B. L. (1986). A unifying interpretation of several combinatorial dualities, Ph.D. Thesis, Cornell University.

Dietrich, B. L. (1987). A circuit set characterization of antimatroids, *J. Comb. Theory Ser. B* **43**, 314–21.

Dietrich, B. L. (1989). Matroids and antimatroids – A survey, *Discrete Math.* **78**, 223–37.

Ding, L.-Y. & Yue, M.-Y. (1987). On a generalization of the Hall–Rado theorem to greedoids, *Asia–Pacific J. Oper. Res.* **4**, 28–38.

Dress, A. W. M. & Wenzel, W. (1990). Valuated matroids – a new look at the greedy algorithm, *Appl. Math. Letters* **3**, 33–5.

Duffus, D. & Rival, I. (1978). Crowns in dismantlable partially ordered sets, in *Combinatorics* (Keszthely, 1976), Colloq. Math. Soc. János Bolyai **18**, Vol. I, pp. 271–92. North-Holland, Amsterdam.

Dunstan, F. D. J., Ingleton, A. W. & Welsh, D. J. A. (1972). Supermatroids, in *Combinatorics, Proceedings of the Conference on Combinatorial Mathematics* (Oxford 1972), pp. 72–122. The Institute of Mathematics and Its Applications, Southend-on-Sea, U.K.

Edelman, P. (1980). Meet-distributive lattices and the anti-exchange closure. *Algebra Universalis* **10**, 290–9.

Edelman, P. (1982). The lattice of convex sets of an oriented matroid. *J. Comb. Theory Ser. B* **33**, 239–44.

Edelman, P. (1986). Abstract convexity and meet-distributive lattices, *Contemporary Math.* **57**, 127–50.

Edelman, P. & Jamison, R. (1985). The theory of convex geometries, *Geom. Dedicata* **19**, 247–70.

Edelman, P. & Saks, M. (1986). Combinatorial representation and convex dimension of convex geometries, *Order* **5**, 23–32.

Edmonds, J. (1971). Matroids and the greedy algorithm, *Math. Progr.* **1**, 127–36.

Eriksson, K. (1989). No polynomial bound for the chip firing game on directed graphs, *Proc. Amer. Math. Soc.* (to appear).

Faigle, U. (1979). The greedy algorithm for partially ordered sets, *Discrete Math.* **28**, 153–9.

Faigle, U. (1980). Geometries on partially ordered sets, *J. Comb. Theory Ser. B* **28**, 26–51.

Faigle, U. (1985). On ordered languages and the optimization of linear functions by greedy algorithms, *J. Assoc. Comp. Mach.* **32**, 861–70.

Faigle, U. (1987). Exchange properties of combinatorial closure spaces, *Discrete Appl. Math.* **15**, 240–60.

Faigle, U., Goecke, O. & Schrader, R. (1986). Church–Rosser decomposition in combinatorial structures, Preprint 86426-OR, Institut für Operations Research, Bonn.

Gale, D. (1968). Optimal assignments in an ordered set: an application of matroid theory,

J. Comb. Theory **4**, 176–80.

Gel'fand, I. M. & Serganova, V. V. (1987a). On the general definition of a matroid and a greedoid, *Soviet Mathematics Doklady* **35**, 6–10.

Gel'fand, I. M. & Serganova, V. V. (1987b). Combinatorial geometries and torus strata on homogeneous compact manifolds, *Russian Math. Surveys* **42**, 133–68.

Goecke, O. (1986). Eliminationsprozesse in der kombinatorischen Optimierung – ein Beitrag zur Greedoidtheorie, Dissertation, Universität Bonn.

Goecke, O. (1988). A greedy algorithm for hereditary set systems and a generalization of the Rado–Edmonds characterization of matroids, *Discrete Appl. Math.* **20**, 39–49.

Goecke, O., Korte, B. & Lovász, L. (1989). Examples and algorithmic properties of greedoids, in B. Simione (ed.), *Combinatorial Optimization*, Lecture Notes in Mathematics 1403, pp. 113–61. Springer-Verlag, New York.

Goecke, O. & Schrader, R. (1990). Minor characterization of undirected branching greedoids – a short proof, *Discrete Math.* **82**, 93–9.

Goetschel, R. H. Jr. (1986). Linear objective functions on certain classes of greedoids, *Discrete Appl. Math.* **14**, 11–16.

Gordon, G. (1990). A Tutte polynomial for partially ordered sets, preprint, Lafayette College.

Gordon, G. & McMahon, E. (1989). A greedoid polynomial which distinguishes rooted arborescences, *Proc. Amer. Math. Soc.* **107**, 287–98.

Gordon, G. & Traldi, L. (1989). Polynomials for directed graphs, preprint, Lafayette College.

Graham, R. L. & Hell, P. (1985). On the history of the minimum spanning tree problem, *Ann. History of Computing* **7**, 43–57.

Hu, T. C. & Lenard, M. L. (1976). Optimality of a heuristic algorithm for a class of knapsack problems, *Oper. Res.* **24**, 193–6.

Iwamura, K. (1988a). Contraction greedoids and a Rado–Hall type theorem, Preprint 88502-OR, Institut für Operations Research, Bonn.

Iwamura, K. (1988b). Primal–dual algorithms for the lexicographically optimal base of a submodular polyhedron and its relation to a poset greedoid, Preprint 88507-OR, Institut für Operations Research, Bonn.

Iwamura, K. & Goecke, O. (1985). An application of greedoids to matroid theory, preprint, Josai University, Sakado, Saitama, Japan.

Jamison, R. E. (1982). A perspective on abstract convexity: Classifying alignments by varieties, in D. C. Kay and M. Breem (eds), *Convexity and Related Combinatorial Geometry*, pp. 113–50. Dekker, New York.

Korte, B. & Lovász, L. (1981). Mathematical structures underlying greedy algorithms, in *Fundamentals of Computation Theory* (Szeged, 1981), Lecture Notes in Computer Science **117**, pp. 205–9. Springer-Verlag, New York.

Korte, B. & Lovász, L. (1983). Structural properties of greedoids, *Combinatorica* **3**, 359–74.

Korte, B. & Lovász, L. (1984a). Greedoids, a structural framework for the greedy algorithm, in W. R. Pulleyblank (ed.), *Progress in Combinatorial Optimization, Proceedings of the Silver Jubilee Conference on Combinatorics* (Waterloo, 1982), pp. 221–43. Academic Press, London, New York, and San Francisco.

Korte, B. & Lovász, L. (1984b). Shelling structures, convexity and a happy end, in B. Bollobás (ed.), *Graph Theory and Combinatorics, Proceedings of the Cambridge Combinatorial Conference in Honour of Paul Erdös* (1983), pp. 219–32. Academic Press, New York and London.

Korte, B. & Lovász, L. (1984c). Greedoids and linear objective functions, *SIAM J. Algebraic and Discrete Math.* **5**, 229–38.

Korte, B. & Lovász, L. (1985a). Posets, matroids and greedoids, in L. Lovász & A. Recski (eds), *Matroid Theory and its Applications* (Szeged, 1982), Colloq. Math. Soc. János Bolyai **40**, pp. 239–65. North Holland, Amsterdam and Budapest.

Korte, B. & Lovász, L. (1985b). Polymatroid greedoids, *J. Comb. Theory Ser. B* **38**, 41–72.

Korte, B. & Lovász, L. (1985c). A note on selectors and greedoids, *Europ. J. Comb.* **6**, 59–67.

Korte, B. & Lovász, L. (1985d). Basis graphs of greedoids and two-connectivity, *Math. Progr.*

Study **24**, 158–65.

Korte, B. & Lovász, L. (1985e). Relations between subclasses of greedoids, *Z. Oper. Res. Ser. A* **29**, 249–67.

Korte, B. & Lovász, L. (1986a). Non-interval greedoids and the transposition property, *Discrete Math.* **59**, 297–314.

Korte, B. & Lovász, L. (1986b). Homomorphisms and Ramsey properties of antimatroids, *Discrete Appl. Math.* **15**, 283–90.

Korte, B. & Lovász, L. (1986c). On submodularity in greedoids and a counterexample, *Sezione di Matematica Applicata*, Dipartimento di Matematica, Universitá di Pisa **127**, 1–13.

Korte, B. & Lovász, L. (1989a). The intersection of matroids and antimatroids, *Discrete Math.* **73**, 143–57.

Korte, B. & Lovász, L. (1989b). Polyhedral results for antimatroids, in *Combinatorial Mathematics: Proceedings of the Third International Conference* (1985), Annals of the New York Academy of Sciences **555**, pp. 283–95.

Korte, B., Lovász, L. & Schrader, R. (1991). *Greedoid Theory*, Algorithms and Combinatorics **4**, Springer-Verlag (to appear).

Kung, J. P. S. (1983). A characterization of orthogonal duality in matroid theory, *Geom. Dedicata* **15**, 69–72.

Magazine, M., Nemhauser, G. L. & Trotter, L. E. (1975). When the greedy solution solves a class of knapsack problems, *Oper. Res.* **23**, 207–17.

Maurer, S. B. (1973). Matroid basis graphs I, II, *J. Comb. Theory Ser. B* **14**, 216–40; **15**, 121–45.

Schmidt, W. (1985a). Strukturelle Aspekte in der kombinatorischen Optimierung: Greedoide auf Graphen, Dissertation, Universität Bonn.

Schmidt, W. (1985b). Greedoids and searches in directed graphs, *Discrete Math.* (to appear).

Schmidt, W. (1985c). A min-max theorem for greedoids, Preprint 85396-OR, Institut für Operations Research, Bonn.

Schmidt, W. (1988). A characterization of undirected branching greedoids, *J. Comb. Theory Ser. B* **45**, 160–84.

Schrader, R. (1986). Structural theory of discrete greedy procedures, Habilitationsschrift, Universität Bonn.

Serganova, V. V., Bagotskaya, N. V., Levit, V. E. & Losev, I. S. (1988). Greedoids and the greedy algorithm (in Russian), in *Sistemy Peredachi i Obrabotki Informatsii, Chast 2*, pp. 49–52. Institut Problem Peredachi Informatsii, Academy of Sciences, Moscow.

Tardos, G. (1988). Polynomial bound for a chip firing game on graphs, *SIAM J. Discrete Math.* **1**, 397–8.

Tarjan, R. E. (1983). *Data structures and network algorithms*, C.B.M.S–N.S.F. Regional Conference Series in Applied Mathematics **44**, Society for Industrial and Applied Mathematics, Philadelphia.

Welsh, D. J. A. (1976). *Matroid Theory*, Academic Press, London.

White, N. (ed.) (1986). *Theory of Matroids*, Cambridge University Press.

White, N. (ed.) (1987). *Combinatorial Geometries*, Cambridge University Press.

Zelevinsky, A. V. & Serganova, V. V. (1989). Combinatorial optimization on Weyl groups, greedy algorithms and generalized matroids (in Russian), preprint, Nauchnyi Soviet po Kompleksnoi Probleme 'Kibernetika', Academy of Sciences, Moscow.

Ziegler, G. M. (1988). Branchings in rooted graphs and the diameter of greedoids, *Combinatorica* **8**, 214–37.

Index